24th Ethylene Producers' Conference 2012

Topical Conference at the 2012 AIChE Spring Meeting and 8th Global Congress on Process Safety

Houston, Texas, USA
1-5 April 2012

ISBN: 978-1-62276-564-5

Printed from e-media with permission by:

Curran Associates, Inc.
57 Morehouse Lane
Red Hook, NY 12571

Some format issues inherent in the e-media version may also appear in this print version.

Printed by Curran Associates, Inc. (2012)

AIChE
3 Park Avenue
New York, NY 10016-5991

Phone: (203) 702-7660
Fax: (203) 775-5177

www.aiche.org

Additional copies of this publication are available from:

Curran Associates, Inc.
57 Morehouse Lane
Red Hook, NY 12571 USA
Phone: 845-758-0400
Fax: 845-758-2634
Email: curran@proceedings.com
Web: www.proceedings.com

TABLE OF CONTENTS

AIChE Paper Number - 247009

RERATING ETHYLENE PLANT EQUIPMENT FOR OPTIMUM RESULTS

Daniel J. Renard
Manager of Application Engineering and Marketing, Engineered Solutions
Elliott Group

Prepared for Presentation at the 2012 Spring National Meeting
Houston, Tx (February 12, 2012)

RERATING ETHYLENE PLANT EQUIPMENT FOR OPTIMUM RESULTS

Daniel J. Renard
Manager of Application Engineering and Marketing, Engineered Solutions
Elliott Group

Abstract: Rerating of turbomachinery equipment installed in ethylene plants is a well established method of improving plant operation, whether it is capacity changes, feedstock flexibility, energy savings or a combination of all three. By utilizing modern compressor and steam turbine component technology, rerating already installed equipment can achieve the desired plant operating objectives.

Through careful planning beginning with review of the process parameters to execution of the turnaround, rerating of existing equipment can provide optimal process performance and efficiency improvements accomplished in a normal maintenance shutdown.

Recent case study will provide example of Elliott's experience in achieving the desired process operation and improved turbomachinery performance using modern component technology while meeting normal shutdown schedules.

Paper will provide users and process designer's criteria helpful in evaluating the rerating of centrifugal compressors and steam turbines for future operating conditions using recent case study and considerations for project execution. With this information equipment engineers will be able to work with process engineers to optimize performance to meet the operating parameters and execute the project within the required timeline.

Ethylene plants have continued to evolve over time as dictated by market conditions including capacity changes, changes to feedstock and environmental and energy related programs. This is particularly true today as users are looking for added feedstock flexibility in their process to take advantage of current market conditions.

Figure 1 shows the trend in ethylene plant sizes over the last several years.

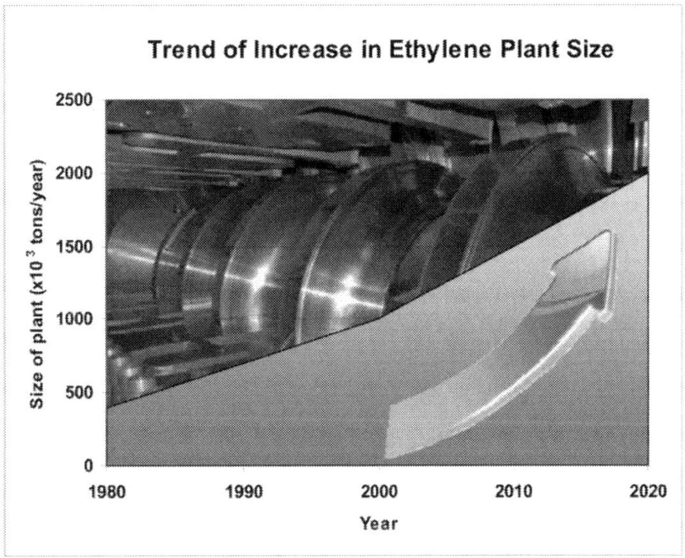

Figure 1 – Trend in Ethylene Plant Size

As existing ethylene plants have been required to change to meet market conditions so has compressor and steam turbine technology continued to evolve allowing many older plants to maintain a viable and competitive market position through the rerating and upgrading of the plant installed turbomachinery. Rerating of installed equipment when carefully planned can help to minimize the turnaround cycle due to the reuse of existing casings and not requiring changes to existing piping and foundation.

Elliott Company has in excess of 1100 units installed in ethylene plants worldwide with many of these units being rerated over the years and in some cases multiple times. Over the last 15 years, Elliott has rerated more than 120 units installed in 19 plants. The majority of the units rerated were due to capacity increases but many of the rerated units included Elliott's latest technology with improved efficiency, which helped to reduce the power consumption requirements.

Planning a plant expansion requires a clear understanding of the operating objectives to be attained and requires a close working relationship among all the involved parties to achieve a successful installation. Depending on the nature of the process changes projects of this type can extend between 18 and 24 months from the initial involvement of the equipment supplier to shipment of equipment hardware. Elliott's experience has shown the earlier the turbomachinery supplier is involved in the process operation feasibility review the better the success rate in achieving the stated goals. Depending on the changes to the process operation usually dictates a very interactive relationship between the end user process, engineering and operations personnel and the equipment application engineers. Initially the involvement of this group evolves around several feasibility iterations of the desired process parameters until final agreement is reached.

Crack Gas Service

Crack gas turbomachinery configuration varies from installation to installation. Crack gas (CG) trains usually consist of large volume capacity compressors driven by large power steam turbines. Usually a minimum of two compressor cases with a speed increasing gear between the cases up to four compressor bodies have been utilized with the most common being a three casing arrangement. Three compressor casings most likely have a double flow compressor for the LP section. Three and four compressor casing arrangements provide the user with more flexibility when considering process changes then a two compressor casing arrangement. In most cases, interstage pressures can be modified to accommodate the change in casing flow and head to optimize the performance for the train.

As noted above the majority of users today are looking to take advantage of changing market conditions particularly feedstock flexibility which dictates changes to the gas composition and plant process parameters. Hence these process changes require review of the turbomachinery performance to achieve optimum plant operation.

Figure 2 – 3 Compressor Casing Crack Gas Train

Figure 3 – 4 Compressor Casing Crack Gas Train

Refrigeration Service

Unlike the CG compressors, refrigeration compressors are usually single casing configurations with multiple side-streams or sideloads and/or extraction streams. Propylene units are generally larger volume compressor casings than the corresponding ethylene compressor. However, both applications require unique analysis when evaluating side-stream mixing performance to ensure correct interstage pressures and flows can be met. As noted above refrigeration compressors have multiple nozzle connections that make the ability to reuse casings a project objective.
See Figures 4 and 5 for example of refrigeration compressors with multiple side-streams.

Refrigeration service can be demanding

- **Low Temperature**
- **High Mole Weight**
- **Significant side-stream capacity**
- **Accurate fluid dynamic design**

Figure 4 – Compressors for Refrigeration Service

Figure 5 – Large Casing Compressors Needed For Refrigeration Service

Centrifugal Compressor Technology

Let's review in more detail some of the reason compressors in ethylene plant service are able to be continuously rerated.

Centrifugal compressors by design can cover a wide volume capacity range and through the continuous development of compressor technology have expanded the flow capacity range to allow the rerating of the same units several times. Thus is the advantage of centrifugal compressors. Over the years of manufacturing centrifugal compressors, if the compressor supplier has designed the product line with an eye towards continuous improvement, the advances in technology both engineering and manufacturing have created opportunities to continuously improve and expand the operating range for not only newly manufactured compressors but also for installed units.

Table 1 - Elliott Large Compressor Frame Nominal Operating Parameters

	Nom. Imp. O.D.	Nom. ICFM	Nom. RPM
46M/MB	27.6"	11900 – 38100	7500
56M/MB	31.7"	15700 – 50400	6500
60M/MB	36.5"	20800 – 67000	5600
70M/MB	42.0"	27500 – 88000	4900
78M/MB	48.3"	36400 – 117000	4300
88M/MB	55.5"	48100 – 154000	3700
103M	63.8"	63600 – 203000	3200
110M	73.4"	84100 – 270000	2800

Part of Elliott's compressor development program over the years as been the "evolution" or optimization in design of the entire aerodynamic flow path. The aerodynamic flow path encompasses not only the rotating impeller(s) but includes all the stationary components. The stationary components consist of the diffuser section and the return channel which together make up the stationary diaphragm. The stationary diaphragm and rotating impeller together define an entire "stage". See Figures 6.

Figure 6 – Modern Compressor Technology Matches Rotating And Stationary Components

Although the rotating impeller is obviously critical in obtaining compressor performance, the designer cannot ignore the stationary components and their impact on "stage" overall performance. Modern compressor design technology has been developed and enhanced through the use of such tools as computational fluid dynamics (CFD), finite element analysis (FEA), solids modeling and rotor dynamic analysis tools. Development of these analytical tools provides the engineer a 3-dimensional view of the entire aerodynamic flow path and the effects the design will have on overall performance.

Impeller Advancements – Evolutionary Not Revolutionary

Part of Elliott's compressor development program over the years as been the "evolution" of impeller designs and the creation of impeller "families" within specific impeller designs that allow for wider flow range within a given compressor model. These impeller technology improvements come from a gradual evolution of earlier design developments combined with advances in aerodynamic analytical analysis and testing capabilities.

This has led to significant increases in flow capability coupled with the development of compressor stage head and efficiency to dramatically expand a given compressor's operating range flexibility and overall performance.

Figure 7 is an example of the improvement in performance and extended flow range.

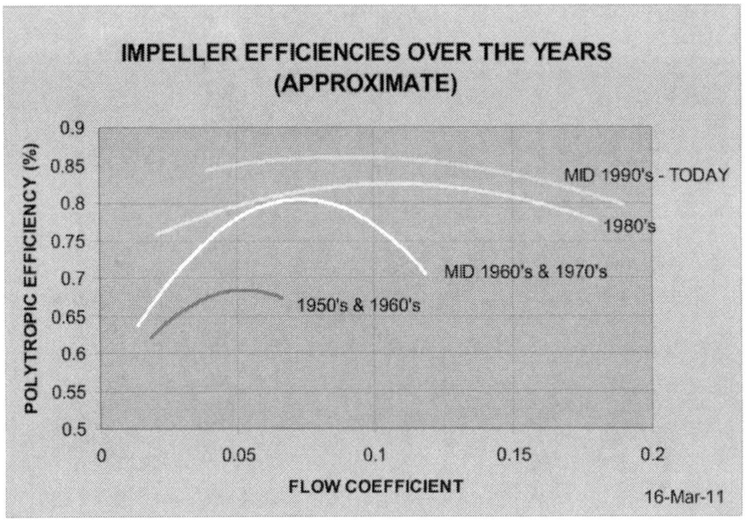

Figure 7 – Impeller Efficiency Improvements Over the Years

Aero components are scaled from frame size to frame size so that impellers have the same geometry regardless of size. Elliott has been successfully scaling components for nearly 40 years used for stage designs in the past. Once the base design is established a scaling of components for various compressor frame sizes can be created.

Table 2 – Scaling Aero Components Used For Over 40 Years

Scaling from One Compressor Frame To The Next Compressor Frame • Maintain Similar Geometric Shape • Impeller Inlet & Discharge Velocity Triangles Are Similar • Impeller Tip Mach Number Is Constant

This same procedure is used to create "families" and hence impeller stage ratings from proven base design. One stage is tested via the single stage test facility in different configurations such as high and low mach numbers. Higher and lower flow stage ratings are derived from the tested components to form a "family" of stages.

Within each family impeller geometry is fixed. Blade heights are varied for higher and lower flows. Both impeller and stationary components are scaled up or down for different compressor models. Stage analytical results are continuously checked and verified against actually aerodynamic performance and field tests.

Figure 8 shows an example of three impeller designations each having a "family" of 5, 5 and 2 impellers respectively with wide variation in flow coefficients for a given compressor model. Impeller flow coefficient is dependent on the impeller inlet volume flow, rotational speed and the impeller diameter. Therefore it is significant that each compressor frame have several impeller designations and families to allow the application engineer a wide variety of impeller options when selecting the rerated compressor configuration.

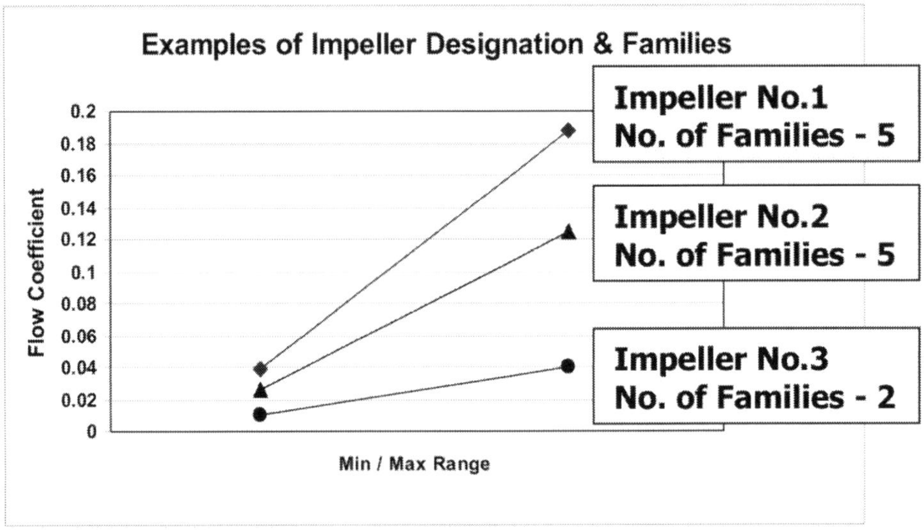

Figure 8 – Every Impeller Design Has Impeller Families

An impeller family will retain the same impeller geometry with basic impeller shroud contour changes to fixed hub contour as shown in Figure 9.

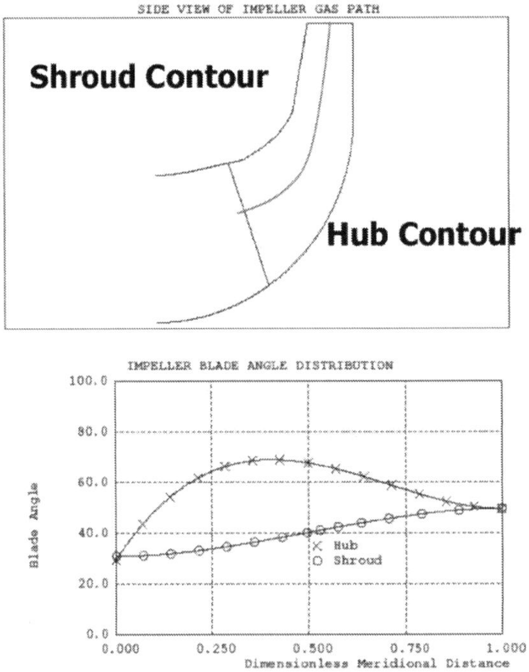

Figure 9 – Impeller Family Design Contour Changes

Not only advancements in design techniques but improvements in manufacturing technology have been implemented to achieve the design performance.

Figure 10 shows an impeller blade 5-axis milling techniques that allow high performance impeller blades to be utilized for higher flow and/or head to create expanded compressor operating range.

Figure 10 - 5 Axis Milled Impeller

Stationary Component Design

Emphasis on improving the performance of the entire flow path, requires the design engineer to consider all the stationary components as well. The diffuser and return channel sections which immediately follow the gas discharge from the impeller are critical to overall stage performance. Both components which make up the stationary diaphragm have been shown to have a significant performance benefit when matched together with the impeller design to remove parasitic losses to improve overall stage performance.

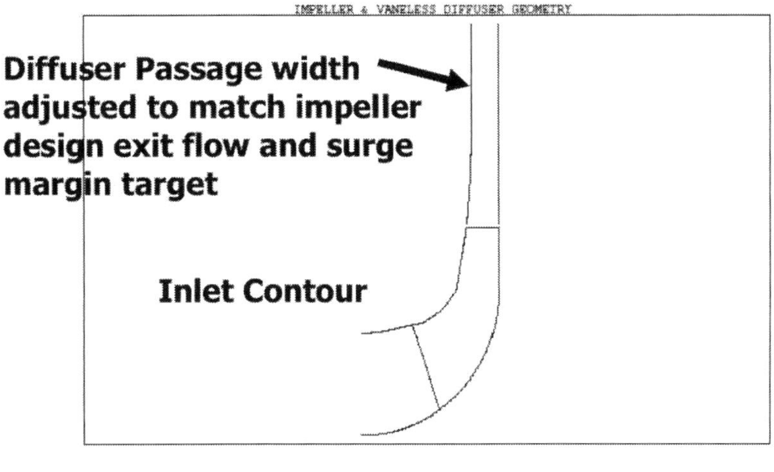

Figure 11 – Diffuser Section Design Critical to Overall Stage Performance

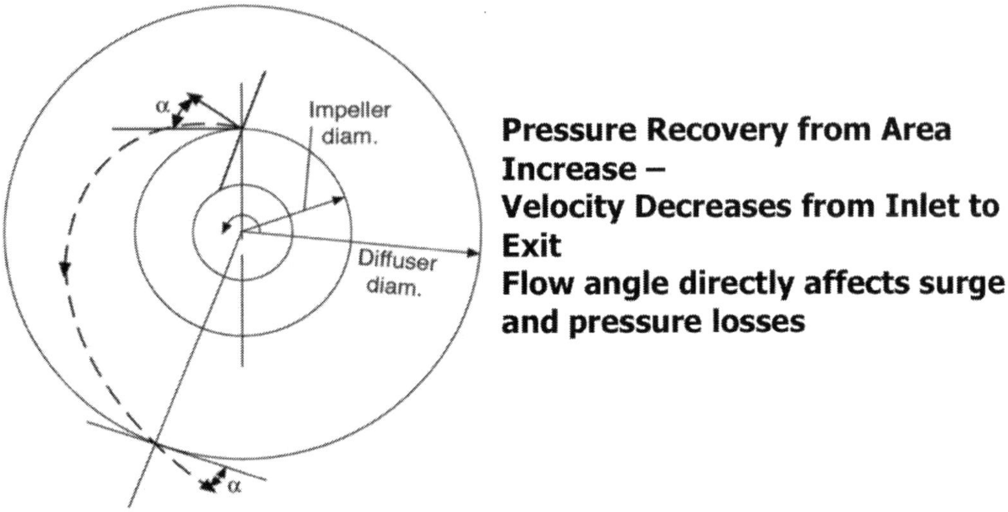

Figure 12 – Impeller Discharge Velocity Triangle and Diffuser Effect On Performance

The stage return channel immediately follows the diffuser section. The return channel is designed to remove the swirl component of velocity which improves performance.

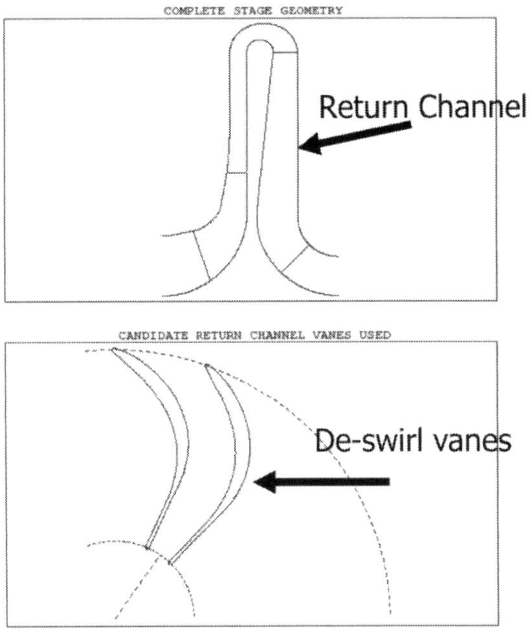

Figure 13 – Return Channel Removes Swirl Component of Velocity

The stationary diaphragms are milled and bolted (see Figure 14) to eliminate rough surface finishes inherent with older cast technology providing stage efficiency benefits.

Figure 14 – Stationary Diaphragms Are Milled

In summary, by continuously creating and extending impeller "families", the application engineer can now select from several impeller/diaphragm (stage) options within a given compressor frame to optimize stage-to-stage performance throughout the compressor aerodynamic flow path.

Advancements in modern technology with higher stage performance and efficiency have provided the means to reuse existing compressor casings while minimizing the power impact on the train driver.

Computational Fluid Dynamic (CFD) Analysis

Computational Fluid Dynamics (CFD) has provided designers with one of the key analytical tools in advancing compressor performance. CFD analysis creates 3-dimensional models allowing the design engineer to model the aerodynamic flow path to optimize performance. These developmental tools along with improvement in manufacturing techniques not only has allowed for the extension of compressor operation but has provided confidence in performance prediction. Designers now have the tools to focus on areas of the flow path that previously could only be verified except via single stage testing. By creating models of the flow path, the designer can see the impact of secondary flow losses in components of the design. Once identified, the designer can make the necessary component changes to the flow path model to minimize or even eliminate the negative impact of these areas of performance loss. CFD analysis allows the engineer to match the impeller performance and stationary diaphragm performance to achieve the optimum overall stage performance.

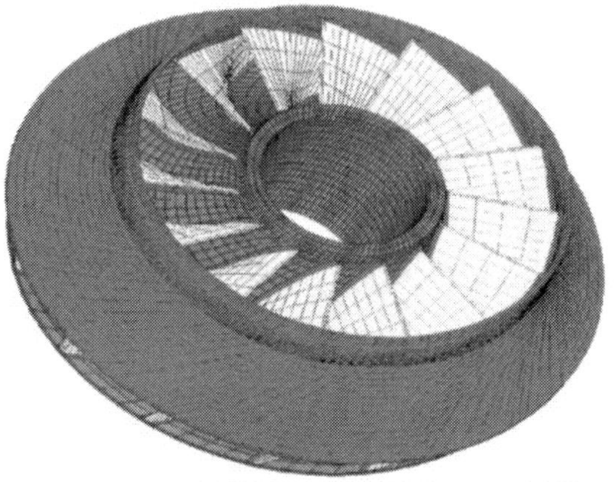

Figure 15 – CFD Analysis of Impeller Design

Benefits of Computational Fluid Dynamics (CFD)

- **Qualitative analysis of flow field**
- **Identify regions of increased loss and separated flow**
- **Reduce regions of secondary flow**
- **Map the stage performance to check flowrate at peak head and choke**
- **Compare alternate designs for best efficiency and overall performance**

Side-stream Mixing Technology

Refrigeration compressors usually have multiple side-streams that require accurate prediction methods. Side-stream mixing calculation techniques have improved greatly through the use of CFD analysis. CFD analysis has provided the design engineer with increased ability to analysis the side-stream mixing that occurs internal to the compressor.

Advanced analytical calculation methods and CFD analysis combined with years of operating experience provides users with the needed reliability in prediction of the compressor performance critical to side-stream performance.

Refrigeration compressors represent a demanding challenge especially for rerating in terms of side-load mixing performance and rotor dynamics particularly when multiple side-streams are required. CFD analysis has enhanced the designer's ability to accurately optimize the mixing of the two flows while minimizing pressure drops. CFD analysis and years of experience in designing refrigeration compressors has provided reliable performance prediction.

Accuracy in side-stream mixing calculations involves not only the impeller performance prediction but the ability to accurately calculate the flow into the compressor through the nozzle and volute configuration. Figures 16 and 17 show how CFD can aid the designer in minimizing side-stream losses through the nozzle and volute.

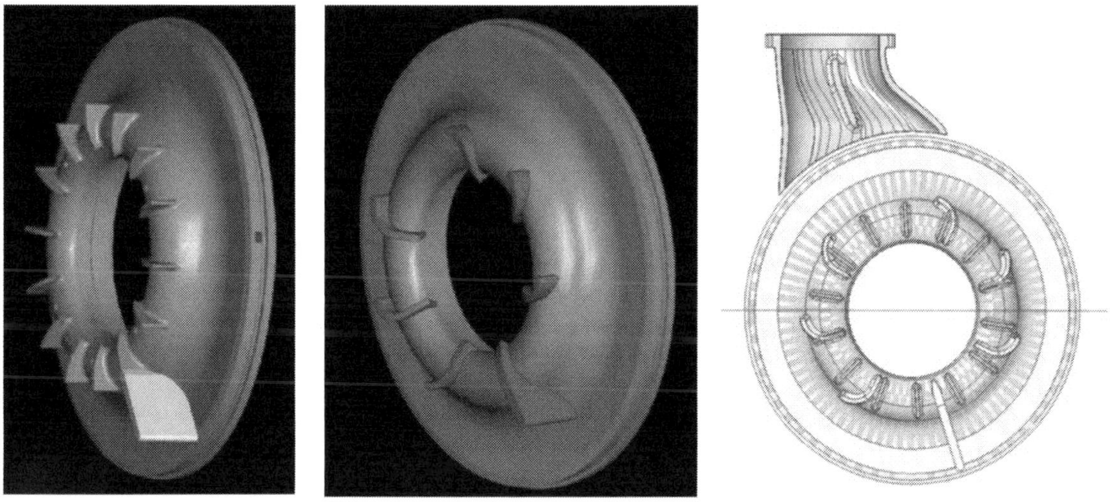

Figure 16 – Design of the Inlet Volute Critical for Accurate Side-Stream Mixing

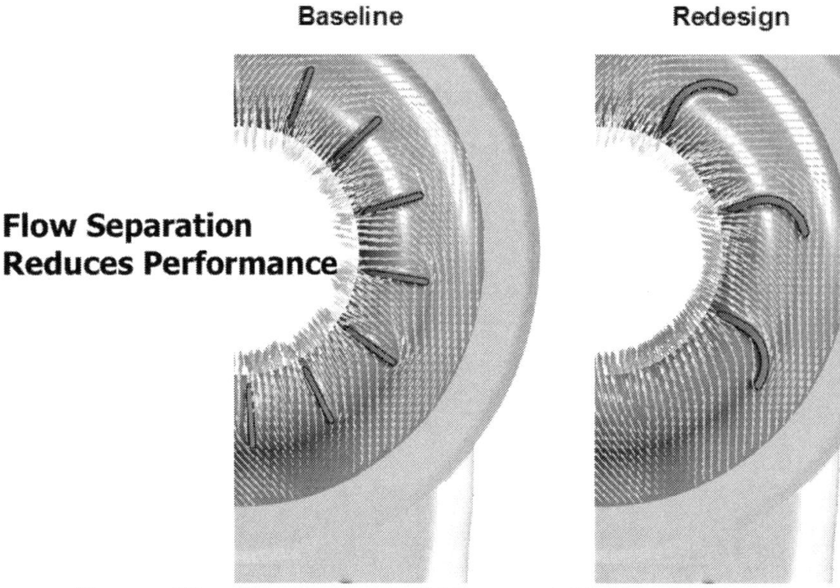

Figure 17 – Comparison of New and Old Design

Once the side-stream nozzle and volute configurations have been reviewed the mixing of the side-stream gas with the flow path can be evaluated. Figure 18 shows how side-stream velocity can impact stage performance. Being able to accurately predict side-stream mixing performance can have a dramatic impact on overall compressor performance and ability to maintain critical side-stream pressures especially when multiple side-streams must be considered as Figure 19 demonstrates.

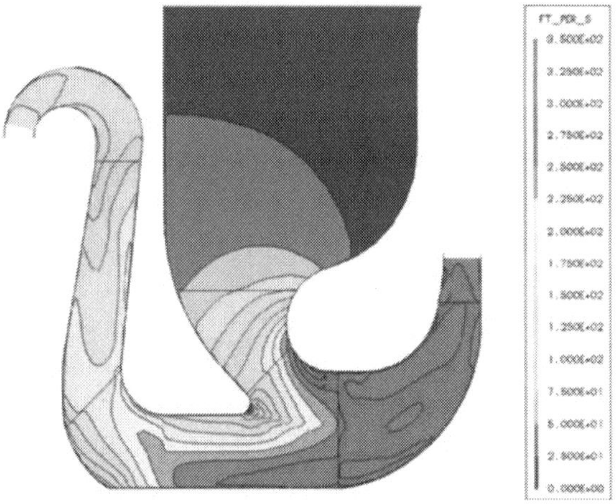

- **Passages are sized to match velocities in mixing region**
- **Minimizes momentum disruption to impeller**
- **Minimizes mixing losses**

Figure 18 – CFD Analysis Side-Stream Mixing Stream Velocity

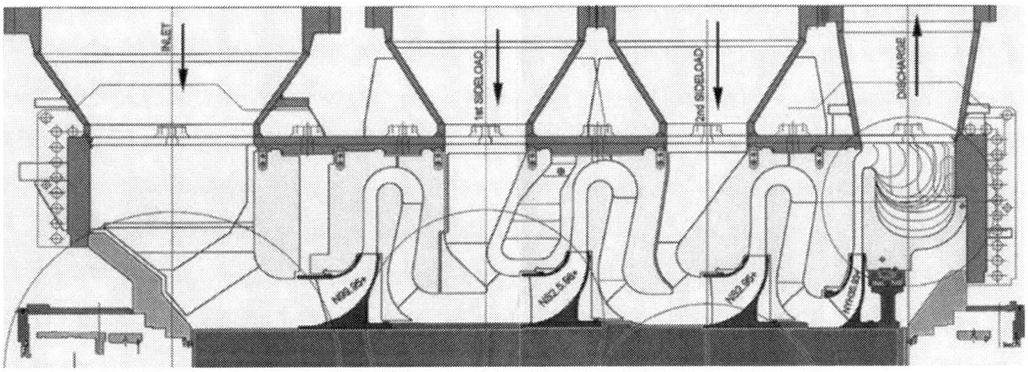

Figure 19 – Refrigeration Compressor With Multiple Side-Streams

Propylene compressors, which are usually larger in size then ethylene compressors, represent a particularly challenge when rerating an existing unit. In many cases, with multiple side-streams, there may only be a single impeller in a specific section. When this is the case, there is usually a pressure tolerance provided by the user for the compressor supplier to consider. Again having various families of impeller designs available improves the likelihood that a solution can be achieved.

Design Verification Process

Whether single stage testing in the lab or multistage compressor shop performance testing or actual field testing, Elliott is continuously evaluating impeller stage performance. Once evaluated, compressor stage ratings are updated and the data is curved fit into the latest revision of the compressor selection program. This ensures the latest stage ratings are used to select and predict future compressor performance.

Design Verification Process

- Single Stage Test
  - Performance maps at range of impeller tip Mach No.
  - Define surge and choke flow limits
- Develop family of Stages from Test
  - Variable impeller blade heights with common blade shape at discrete flow increments
  - Matching diffuser and return channel passage widths to impeller
  - Performance Rating Curves for Computer
- Multistage Factory Test

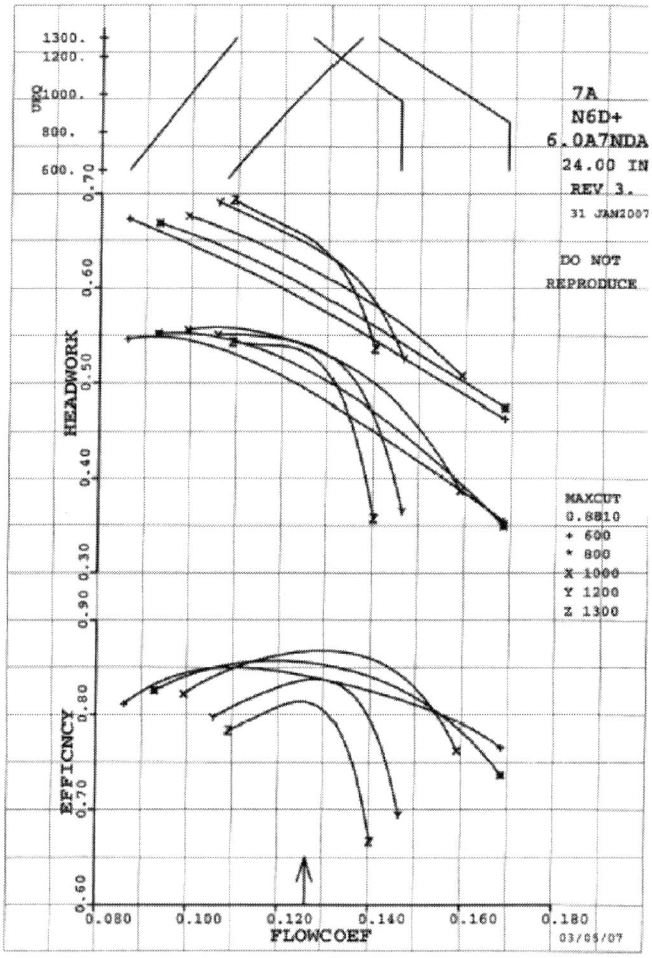

Figure 20 – As Tested Data is Curve Fit Installed Into Selection Program

Steam Turbine Rerate Considerations

Steam turbine technology, like compressors, has also utilized the latest analytical tools and methods to advance performance and reliability in component design specifically rotating blades and stationary diaphragms. Crack gas and propylene drivers are usually high power units that may require significant changes especially when the driven compressor power has increased.

Figure 21 – Steam Turbine Technology Has Improved With CFD Analysis

Even with improvements in stage efficiencies over the years, increase in power usually requires more steam flow area inside the turbine, which may or may not be possible within the physical limits of the existing casing. As noted above significant plant capacity increases resulting in higher compressor flow and therefore power, in most cases, cannot be achieved by the steam turbine without significant increase in steam flow. Casing size limitations provide unique challenges for the application engineer to reconfigure the existing steam flow path to achieve the desired power and speed.

Obviously steam turbine drivers have to match the compressor train speed and with the same steam operating conditions additional steam flow area will be required. And although modern staging can be used to improve the turbine's overall efficiency additional modifications to the existing aero flow path are usual required such as removing stages to increase the steam flow area.

New steam turbine rotating and stationary components can be used to replace existing staging to increase reliability. CFD analysis on high pressure and low pressure staging is used to optimize both the rotating and stationary components.

Figure 22 – High Pressure Stage Design Developed Using CF

Figure 23 – CFD Analysis of Low Pressure Staging

Additional improvements available for existing installed turbines are by replacing labyrinth seals by integrating brush type seals and including tip seals.

Rerate Equipment Review – Step-By-Step Process

Over the years of accumulating experience in rerating turbomachinery, a process for rerating equipment has been developed that can be applied to most applications.

It begins with a feasibility review of new process parameters and simulation of current operating conditions. Depending on the date of the original equipment installation, most likely current plant process operating conditions have changed from the original design.

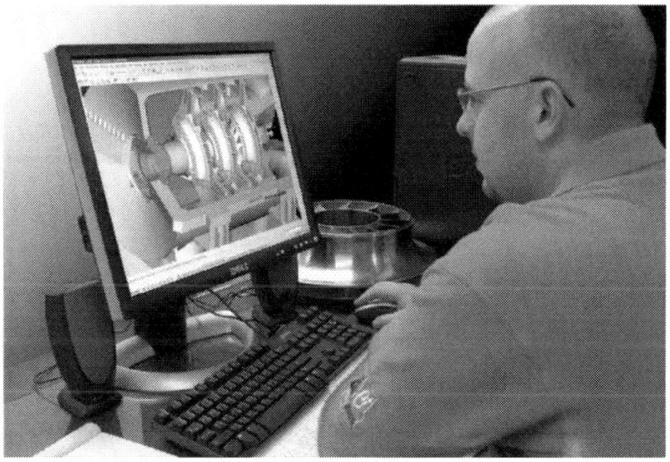

Figure 24 – Rerate Process: Step By Step Review

This initial review may take several iterations with the user, engineering contractor and process personnel. These early discussions usually focus on the process conditions but can and should include discussions of possible hardware changes and plant turnaround schedule, expectations and execution.

Budgetary equipment costs can be developed at this stage. Preliminary review of equipment mechanical items are conducted as well such as preliminary stage spacing, nozzle velocities, preliminary rotor critical speed, shaft end size, etc.

The feasibility/budget phase is usually followed by a detailed technical study with the objective to finalize process operating parameters and equipment selections. In some cases, the user and/or engineering contractor issues a request for quotation (RFQ) with clear definition of the process requirements.

At this stage a more thorough equipment selection is developed by the application engineer who may require modifications to the preliminary selection based the RFQ requirements (i.e. process operating changes, equipment specifications, etc.)

Detailed equipment selections may entail equipment layouts to ensure stage spacing is met and confirm the new and/or reuse of components within the casing. Preliminary mechanical checks are confirmed and additional mechanical checks completed.

Case Study

Original plant was built in the late 60's and had a two body compressor crack gas train as shown below. Over the years, the plant operation had been changed such that the LP compressor had already been rerated three times and the HP compressor and steam turbine driver had been rerated twice previously.

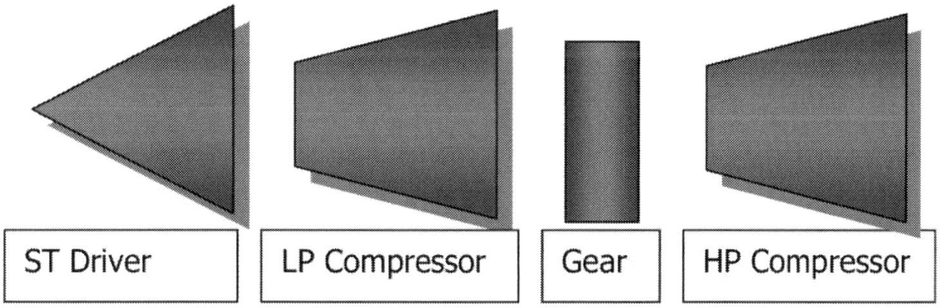

Figure 25 – Two Casing Compressor Crack Gas Train

Project Background & Requirements

- Original Equipment Installed in Late 60's
- LP Casing Previously Rerated Three Times
- HP Casing & ST Previously Rerated Twice
- Flow & Power Increased Approximately 55%
- User wants Additional 58% Flow

The user decided to expand the plant yet again and include feedstock changes. The original plant operating flow and power had already been increased by approximately 55% during the previous equipment rerate when the LP compressor was rerated for the third time and the HP compressor and ST driver had been rerated twice before.

The latest plant operating flow is to be increased by an additional 58%.

Working closely with the user, engineering contractor and process licensor a solution was developed after several iterations of process conditions.

It became clear early in the analysis that such another large flow increase would necessitate the addition of a new crack gas booster compressor. The addition of a new booster compressor in front of the LP and HP compressors reduced the volume flow into the existing units. Reducing the volume flow to these compressors allowed these units to be rerated yet again for a fourth and third time respectively to meet the required plant increase in flow and accommodate the feedstock revision.

By using the latest impeller technology with the highest performance available, the overall train power was limited to an increase of approximately 35%.

Limiting the overall train power increase to only 35%, allowed the steam turbine driver to be rerated for a third time.

The LP & HP compressors required all new rotors and stationary components but with reduced number of stages in each section. The steam turbine required a new rotor with new diaphragms but also reduced number of stages.

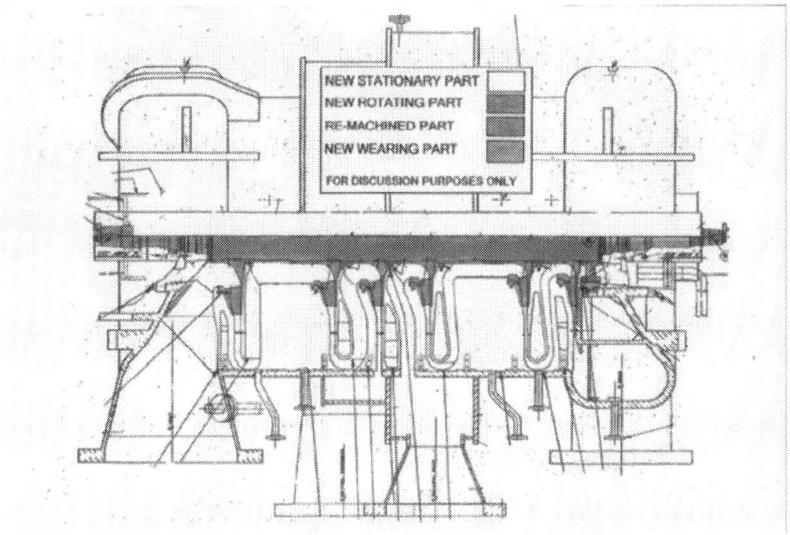

Figure 26 – LP Compressor Casing

The LP compressor is an iso-cooled unit originally manufactured with 8 impellers; now only requires 6 impellers. Special diaphragms were designed to replace the stage spacing with two less stages.

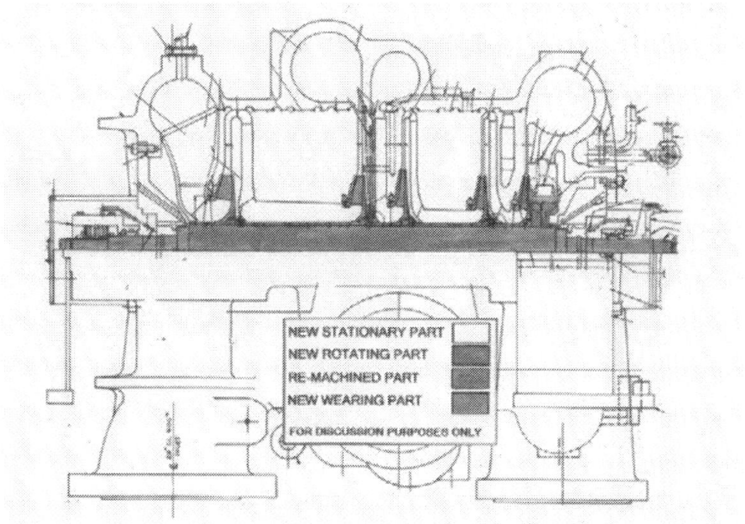

Figure 27 – HP Compressor Casing

The HP compressor was manufactured with 8 stages but for the latest rerate needs only 5 stages.

Thus the user was able to achieve the overall plant operating objectives for capacity increase, feedstock flexibility as well as minimize plant site work by keeping the existing casings with minimal investment in new turbomachinery hardware.

Table 3 – Comparison of Original Conditions to Current Rerate

	LP COMPRESSOR	HP COMPRESSOR	TURBINE
ORIGINAL			
FLOW	**3,227 LB/MIN**	**3,163 LB/MIN**	
POWER	**10,272 HP**	**9,568 HP**	**19,840 HP**
PREVIOUS RERATE			
FLOW	**4,963 LB/MIN**	**4,888 LB/MIN**	
POWER	**15,561 HP**	**14,830 HP**	**30,391 HP**
CURRENT RERATE			
FLOW	**7,834 LB/MIN**	**7,736 LB/MIN**	
POWER	**21,606 HP**	**19,356 HP**	**40,962 HP**
TOTAL TIMES RERATED	**4**	**3**	**3**

It is interesting to note from the original plant installation in the late 60's, the overall train power has increased by nearly two times the original design. By applying the latest in compressor and steam turbine technology, the existing compressors and steam turbine have remained in operation for over 50 years.

References

1. Blahovec, James F., Eads, Kevin S. and Matthews, Terryl (1998),"Guidelines for Specifying and Evaluating New and Rerated Multistage Centrifugal Compressors," Proceedings of the 27th Turbomachinery Symposium
2. Boyce, Douglas W., Mansfield, Donald H. and Pacelli, William G., (2006), "Guidelines for Specifying and Evaluating the Rerating and Reapplication of Steam Turbines," Proceedings of the 35th Turbomachinery Symposium
3. Haight, Brent (2000) "Delivering the EDGE," CompressorTech Two, July-August, Diesel & Gas Turbine Publications
4. Kane, Joe (2003) "Revisiting the EDGE," CompressorTech Two, January-February, Diesel & Gas Turbine Publications

AIChE Paper # 246805

Ethylene Plant Optimization: Footprint Replacements & Parallel Compressor Trains

Prepared for Presentation at the 2012 Spring National Meeting
Houston, Texas, April 02, 2012

MITSUBISHI HEAVY INDUSTRIES
COMPRESSOR CORPORATION

Contents

1. Rerate Options: Revamp vs. Replace vs. Parallel Train

MITSUBISHI HEAVY INDUSTRIES COMPRESSOR CORPORATION

I. Base Plan
by End-user, Process Licensor & Supplier

1. Process Requirement
2. Mechanical Restrictions
3. Plant Turnaround Schedule

II. Option
for Compressor Capacity Expansion

Step 1. Process Adjustment
Step 2. Re-Rotoring (Revamp)
Step 3. Replacement
Step 4. Additional Compressor

III. Scope of Revamp
MCO's Supply Parts

■Step-2: Re-Rotoring
<Compressor & Turbine Rotors & Internal Parts>

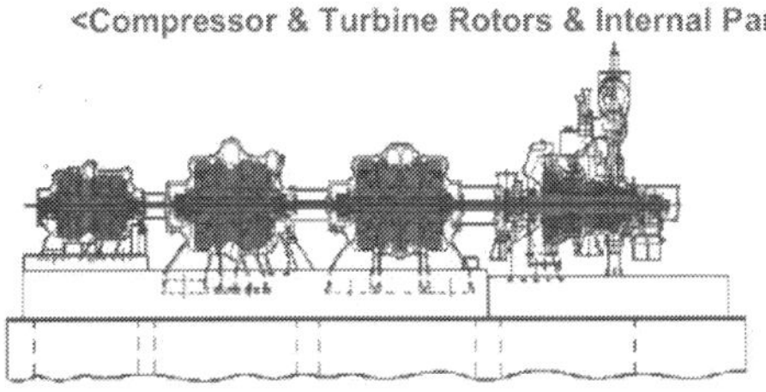

■Step-3: Replacement
<Compressor & Turbine Assemblies>

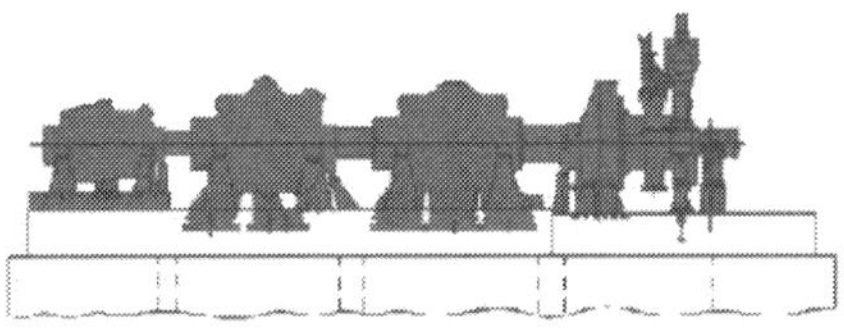

3

2. Requirements and Flexibility vs Limitation

MITSUBISHI HEAVY INDUSTRIES COMPRESSOR CORPORATION

Feature of Approach for Plant Capacity Upgrading

Approach	Modification	Limitation
(1) Changing process parameters	Suction Pressure, Rotating Speed	-Design Pressure, -MCS, -Max. Output Power of Driver
(2) Revamping with internal modifications	Compressor Internal Parts (Impeller, Diaphragm, etc.)	Internal Diameter and Bearing Span of Existing Compressor
(3) Equipment Replacement	Compressor	Size of Existing Foundation, etc.
(4) Parallel Train	N/A New Construction	ex. Site Space for New Construction

4

MHI BUSINESS CONFIDENTIAL

2. Requirements and Flexibility vs Limitation

MITSUBISHI HEAVY INDUSTRIES COMPRESSOR CORPORATION

Example of Initial Investment Cost (300 KTA Ethylene Plant)

Item (Increasing Ratio)		Revamping (+20%)		Replacement (+40%)	
Planning / Engineering		18 %		18 %	
Equipment		70 %		86 %	
	Compressor & Driver		(20 %)		(28 %)
	Furnace, Others		(50 %)		(58 %)
Field Installation / Start-up Work		12 %		14 %	
Total Initial Investment Cost		**100 %**		**118 %**	

Production Increase by Replacement
= 300 KTA x 0.4 = +120 KTA

3. Design Concepts for Compressors and Steam Turbine Replacement

MITSUBISHI HEAVY INDUSTRIES COMPRESSOR CORPORATION

I. Design concept for compressor replacement

1.High Efficiency

⇒Application of High efficiency Aero Parts and Direct Lubrication Bearings

2.Smallest Additional Work

⇒ Compressor replacement with same footprint / same nozzle arrangement using Flexible Frame Design (FFD) and custom-designed nozzle arrangement

3. Design Concepts for Compressors and Steam Turbine Replacement

MITSUBISHI HEAVY INDUSTRIES COMPRESSOR CORPORATION

II. Design concept for steam turbine replacement

1.High Efficiency

⇒Improvement of flow path design

2. Long Term Operation

⇒Coating technology for corrosion-resistance

4.1 Compressors
a) High Efficiency Improvements

MITSUBISHI HEAVY INDUSTRIES COMPRESSOR CORPORATION

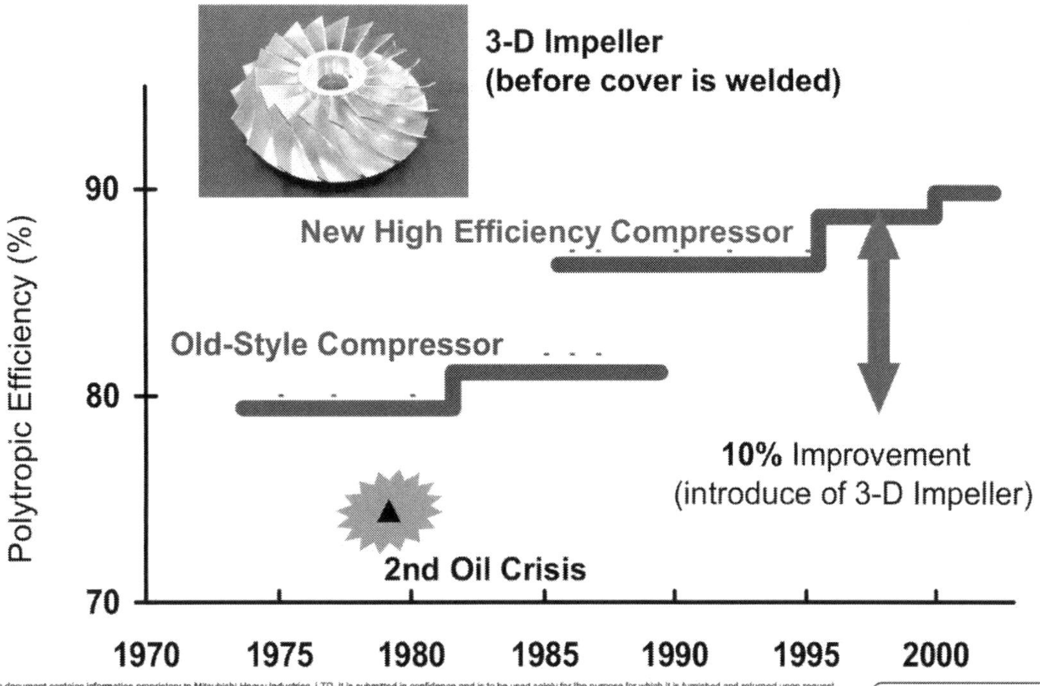

This document contains information proprietary to Mitsubishi Heavy Industries, LTD. It is submitted in confidence and is to be used solely for the purpose for which it is furnished and returned upon request.
This document and such information is not be reproduced, transmitted, disclosed or used otherwise in whole or in part without the written authorization of Mitsubishi Heavy Industries, LTD.
MITSUBISHI HEAVY INDUSTRIES, COMPRESSOR CORPORATION.
Hiroshima Machinery Works
MHI BUSINESS CONFIDENTIAL

4.1 Compressors
a) High Efficiency Improvements

MITSUBISHI HEAVY INDUSTRIES COMPRESSOR CORPORATION

Direct Lubrication Type Bearings

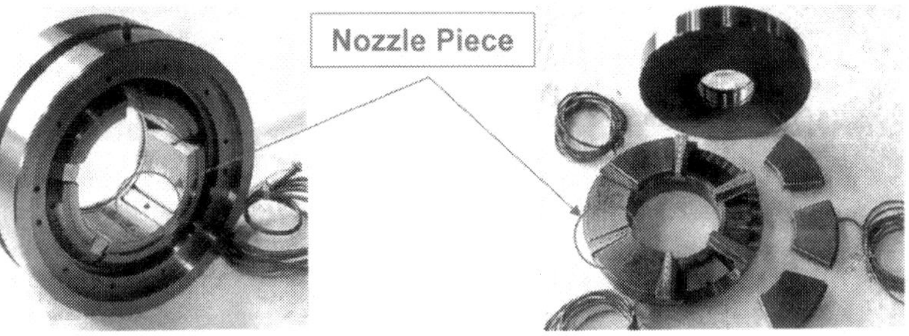

Journal Bearing

Thrust Bearing

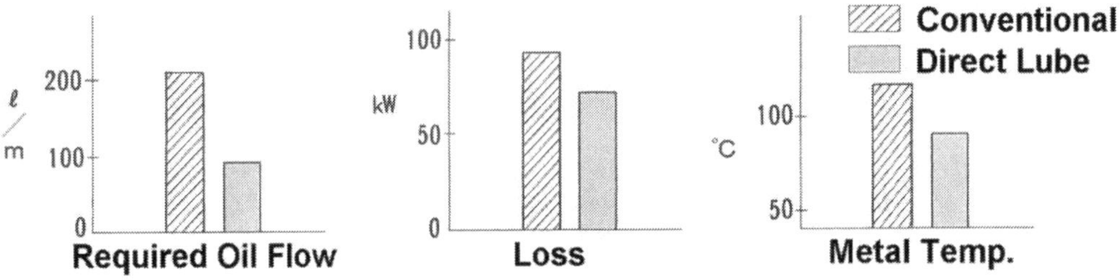

4.1 Compressors
a) High Efficiency Improvements

MITSUBISHI HEAVY INDUSTRIES COMPRESSOR CORPORATION

Actual Result of Past Power Improvement Project

	Existing	After Replacement	Improvement
Plant Capacity	300 KTA	300 KTA	-
Pressure	1.4 → 37.2 bara	1.4 → 37.2 bara	-
Measured Horsepower	25,516 kW @ 5,765 rpm	21,908 kW @ 5,634 rpm	3,608 kW Reduced (14% Reduced)
Steam Consumption (100 barg)	93 ton/hr	79 ton/hr	14 ton/hr Reduced (15% Reduced)

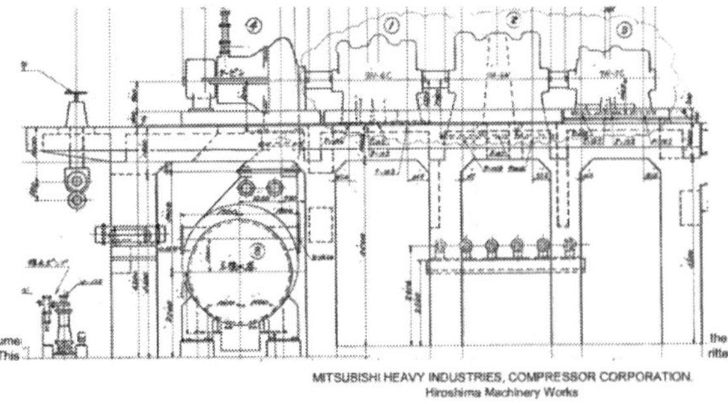

4.1 Compressors

a) High Efficiency Improvements

MITSUBISHI HEAVY INDUSTRIES COMPRESSOR CORPORATION

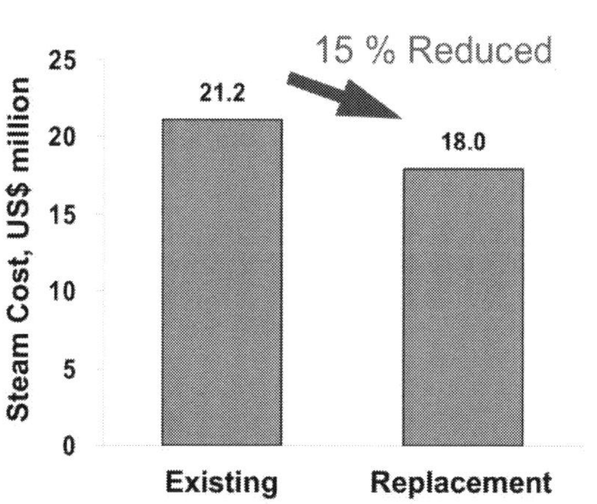

Note : Typical steam cost 26 US$/ton (100 barg)

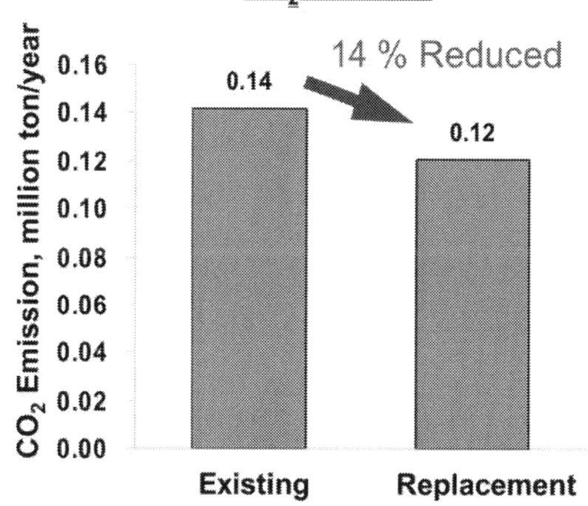

Benefits:

Reduction in utility costs and CO_2 emissions contribute to the company's profitability and brand image with carbon credits.

11 MHI BUSINESS CONFIDENTIAL

b) Footprint Replacement based on Flexible Design for High Efficiency and Rotor Stability Improvement

MITSUBISHI HEAVY INDUSTRIES COMPRESSOR CORPORATION

Replacement with larger equipment

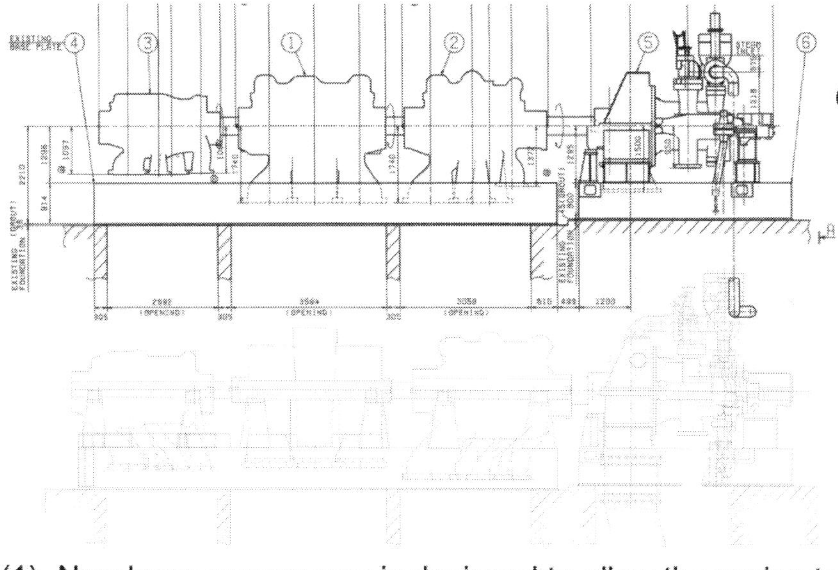

Reference Charge Gas Compressors

600 KTA

800 KTA

(1) New large compressor is designed to allow the casing to put on

(2) Nozzles to set in the narrow opening

12 MHI BUSINESS CONFIDENTIAL

b) Footprint Replacement based on Flexible Design for High Efficiency and Rotor Stability Improvement

MITSUBISHI HEAVY INDUSTRIES COMPRESSOR CORPORATION

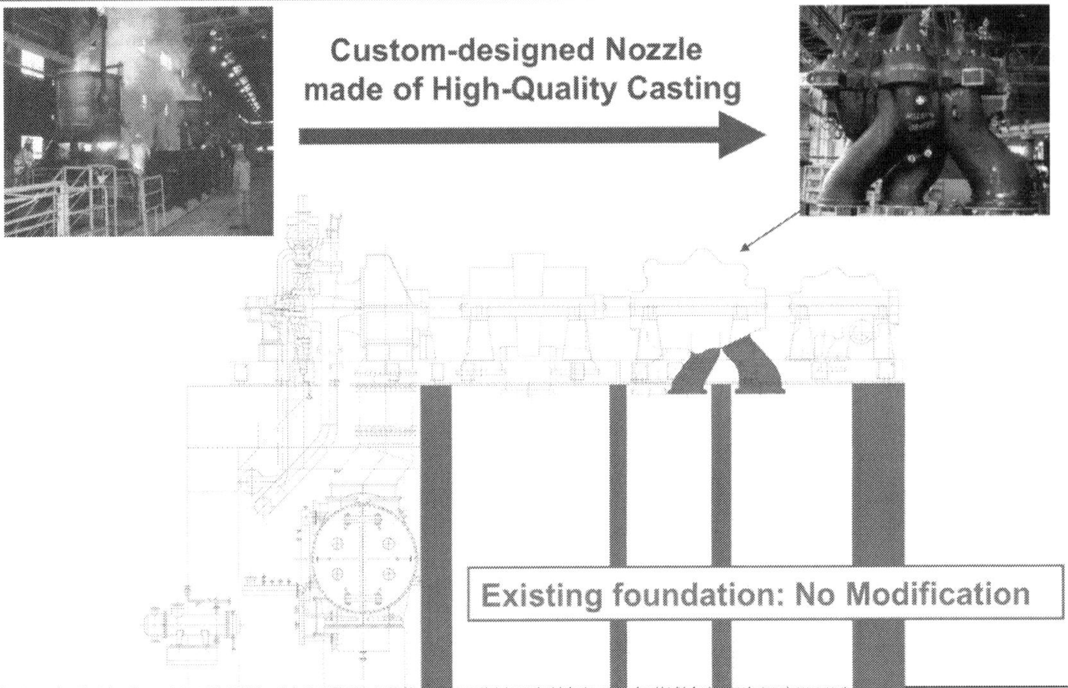

13 MHI BUSINESS CONFIDENTIAL

b) Footprint Replacement based on Flexible Design for High Efficiency and Rotor Stability Improvement

MITSUBISHI HEAVY INDUSTRIES COMPRESSOR CORPORATION

Picture of compressor applied with off-set nozzle to suite for existing piping arrangement

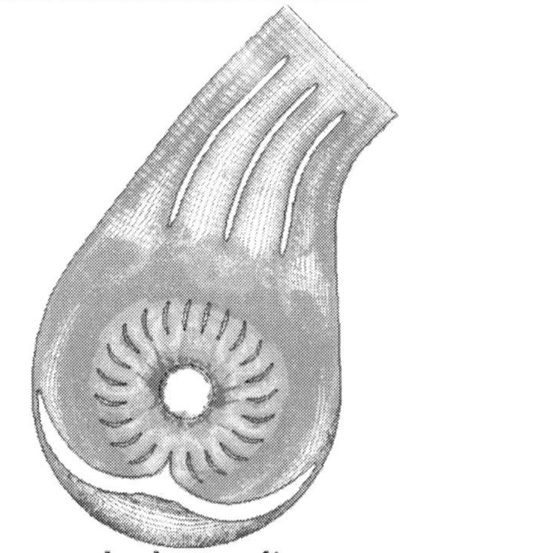

Flow analysis result for offset inlet nozzle →Verified uniform inlet flow to 1st stage impeller

MCO can propose the same foot print replacement with minimum piping & foundation modifications.

14 MHI BUSINESS CONFIDENTIAL

b) Footprint Replacement based on Flexible Design for High Efficiency and Rotor Stability Improvement

MITSUBISHI HEAVY INDUSTRIES COMPRESSOR CORPORATION

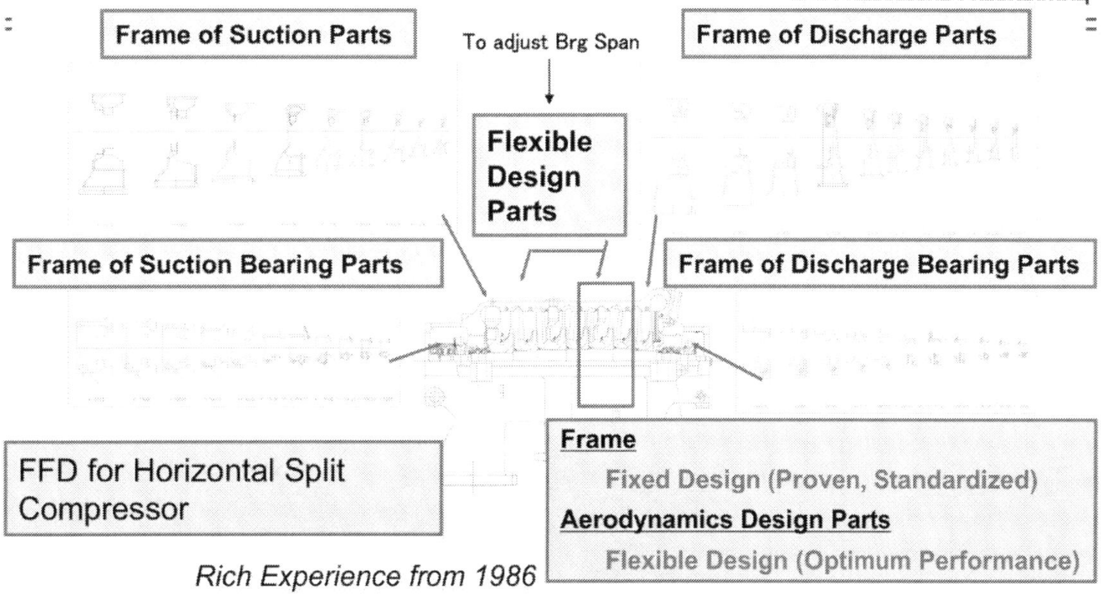

Rich Experience from 1986

Flexible Frame Design (FFD)

MCO can propose optimized aero performance compressors for flexible arrangement based on FFD design.

15 MHI BUSINESS CONFIDENTIAL

b) Footprint Replacement based on Flexible Design for High Efficiency and Rotor Stability Improvement

MITSUBISHI HEAVY INDUSTRIES COMPRESSOR CORPORATION

Re-use of Existing Base-plate

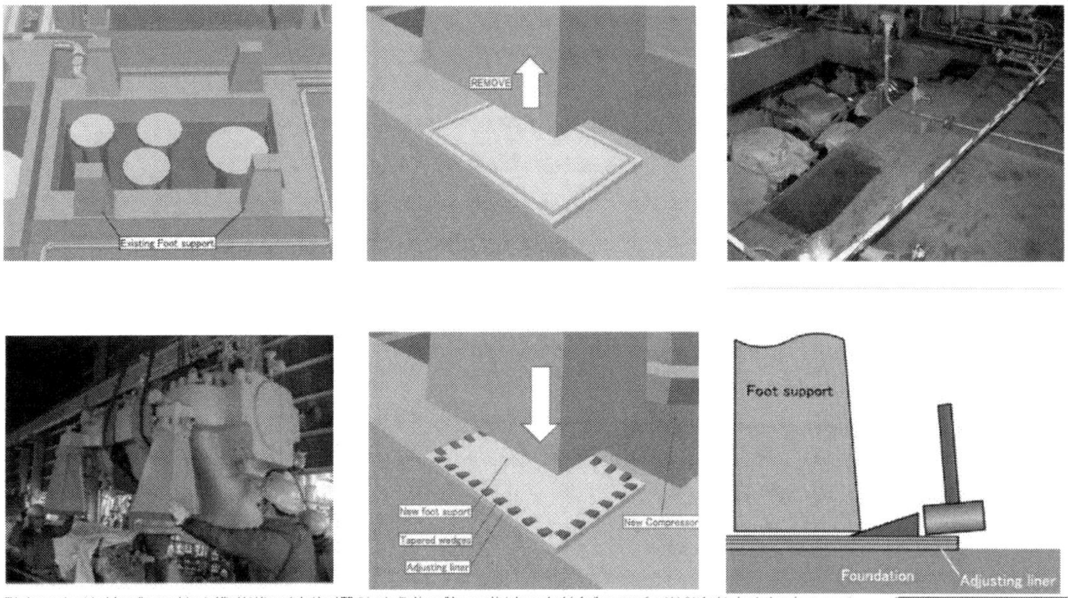

16 MHI BUSINESS CONFIDENTIAL

b) Footprint Replacement based on Flexible Design for High Efficiency and Rotor Stability Improvement

MITSUBISHI HEAVY INDUSTRIES COMPRESSOR CORPORATION

Re-use of Existing Base-plate

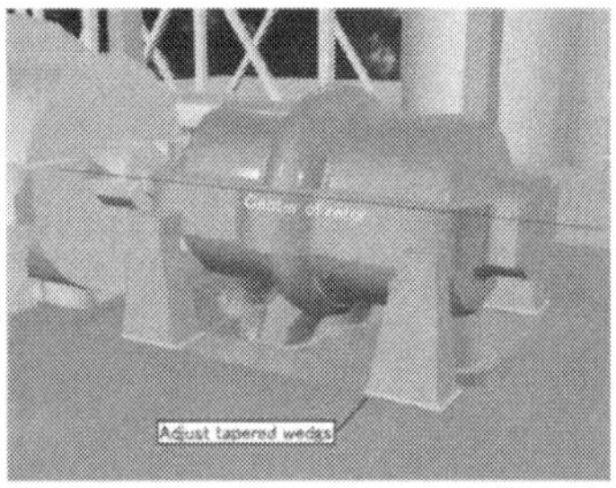

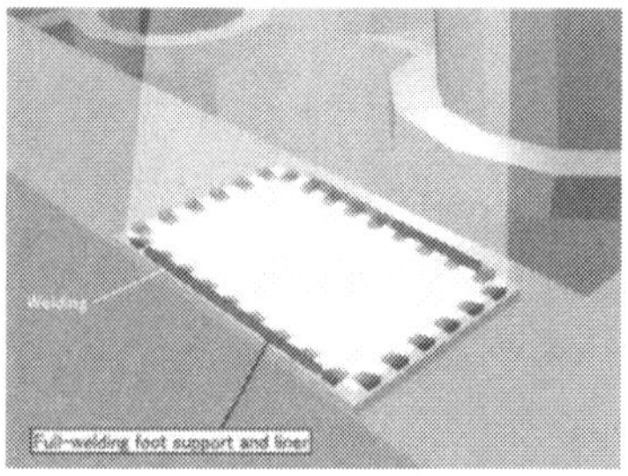

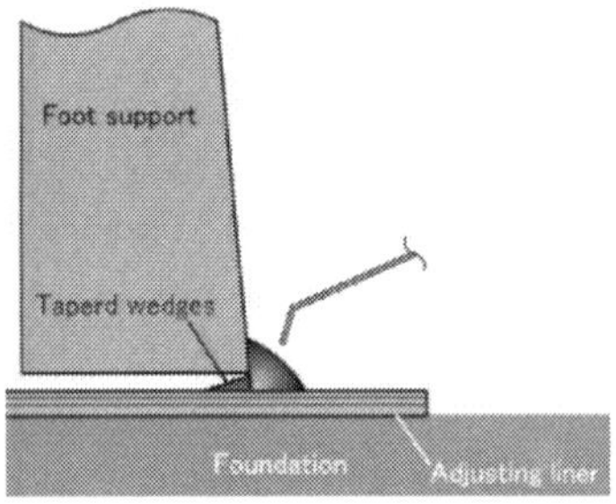

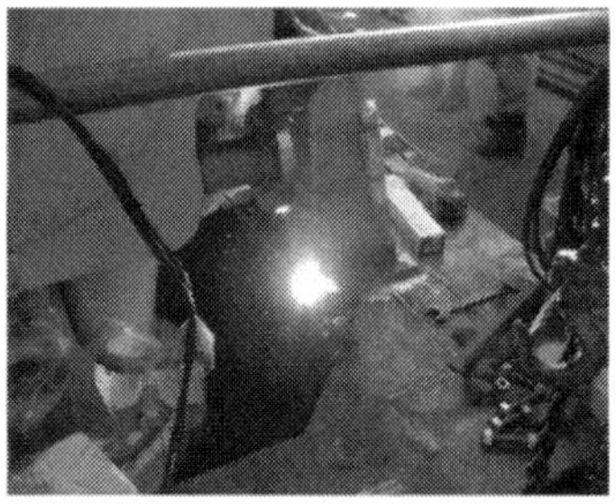

17

MHI BUSINESS CONFIDENTIAL

b) Footprint Replacement based on Flexible Design for High Efficiency and Rotor Stability Improvement

MITSUBISHI HEAVY INDUSTRIES COMPRESSOR CORPORATION

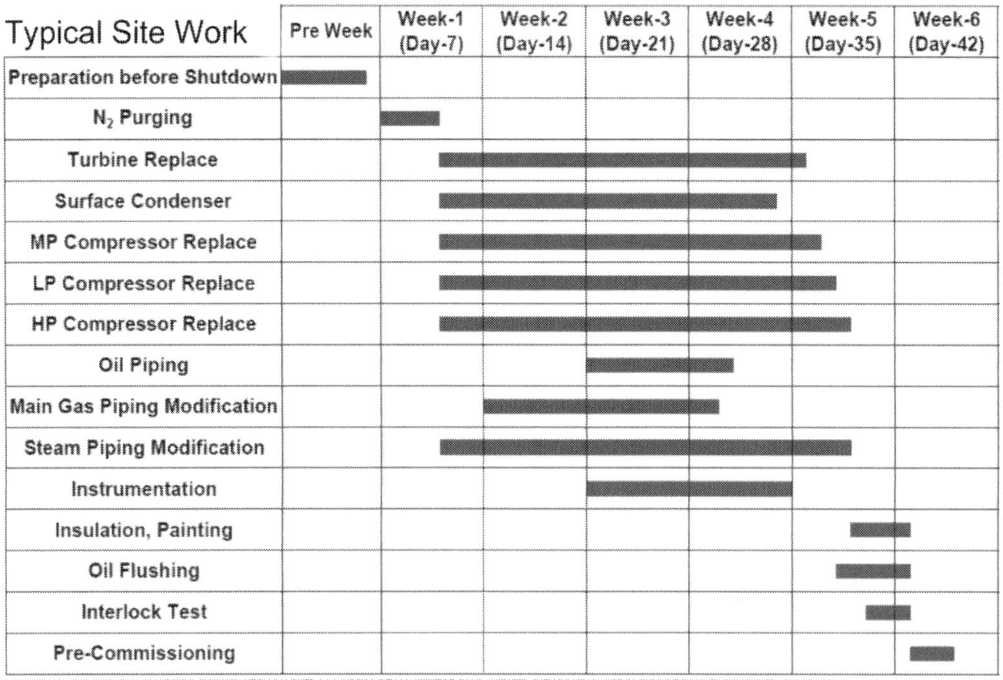

18

MHI BUSINESS CONFIDENTIAL

c) Parallel Train configurations

MITSUBISHI HEAVY INDUSTRIES COMPRESSOR CORPORATION

Case Study: Upgrading History of Ethylene Plant in Korea

Year	Capacity	CGC	PRC	ERC
1991	350 KTA	ST – MP – LP – HP	ST – Comp.	ST – Comp.
1996	600 KTA	ST – MP – LP – HP	ST was revamped – Comp.	ST was revamped – Comp.
2006	800 KTA	ST was re-used – MP – LP – HP	Tertiary compressors (Parallel): M – FC – LP – HP	
2009	1,000 KTA	ST – Compressors were re-used		ST – Comp.

Revamp/Replacement/Addition of Comp/Driver are applied according to customer's request

(shaded box):Replacement

a) Internal Replacement for High Efficiency Improvement

MITSUBISHI HEAVY INDUSTRIES COMPRESSOR CORPORATION

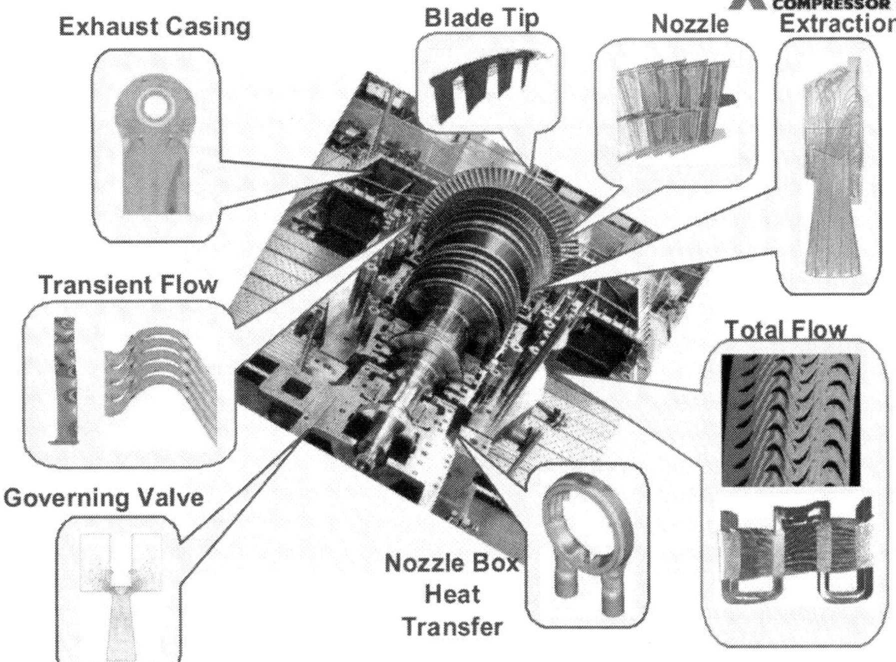

CFD: Computational Fluid Dynamics

a) Internal Replacement for High Efficiency Improvement

MITSUBISHI HEAVY INDUSTRIES COMPRESSOR CORPORATION

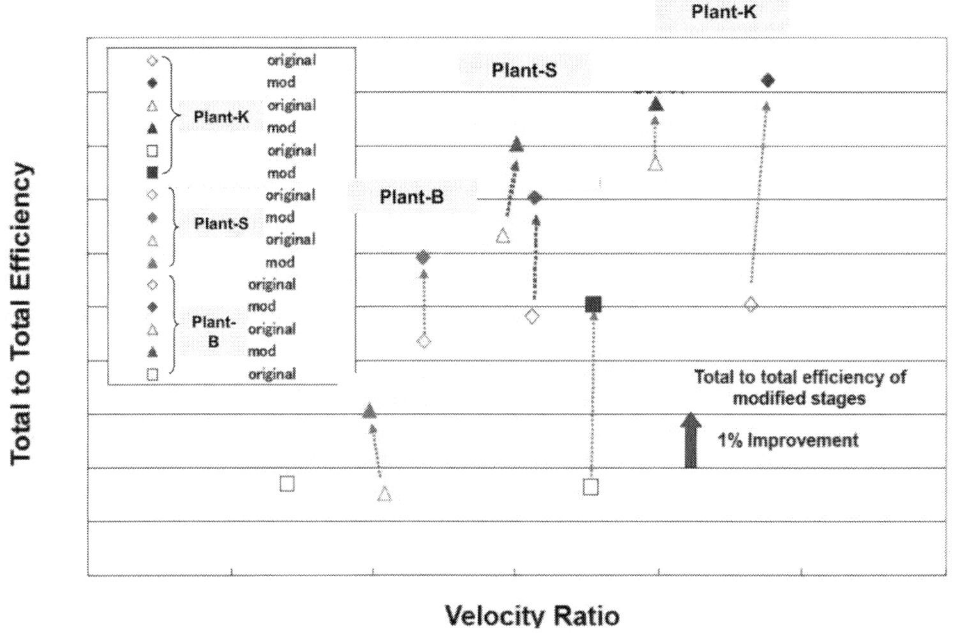

21 MHI BUSINESS CONFIDENTIAL

b) Long-term Operation by New Technologies

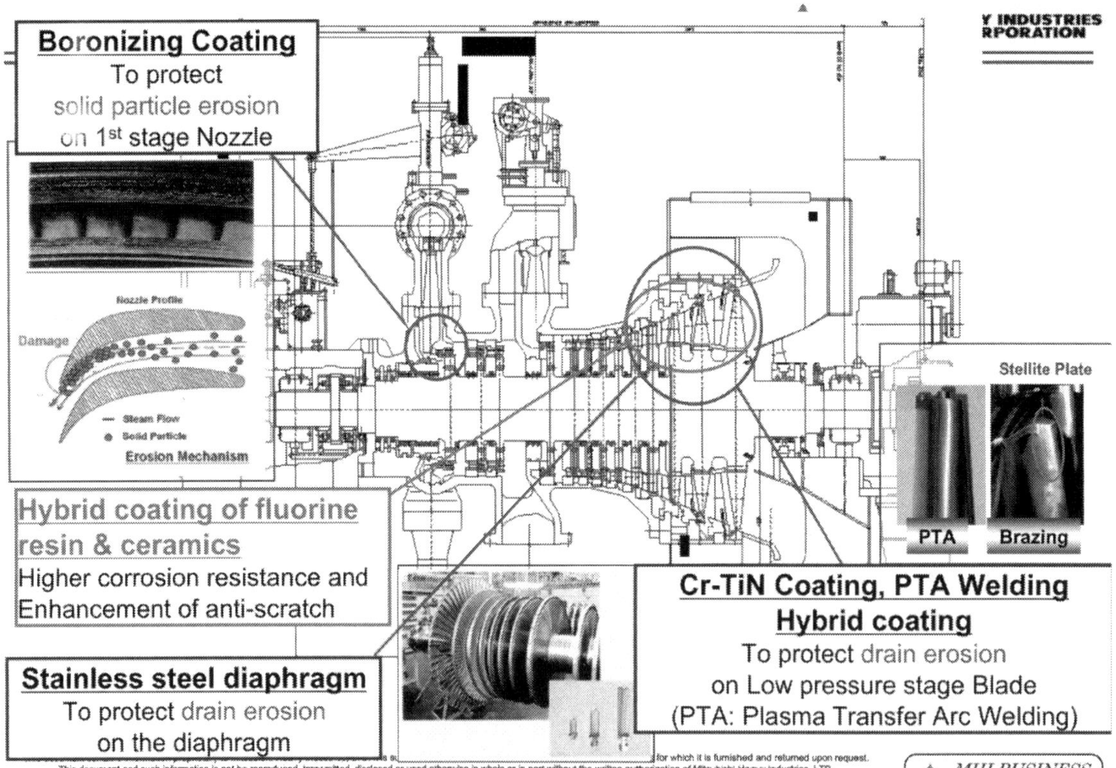

22 MHI BUSINESS CONFIDENTIAL

b) Long-term Operation by New Technologies

Good Experience for Long Term Operation

MITSUBISHI HEAVY INDUSTRIES COMPRESSOR CORPORATION

MCO has the experience of key technologies and long continuous operation without Major maintenance.

Longer maintenance interval

↓

Higher availability of the plant

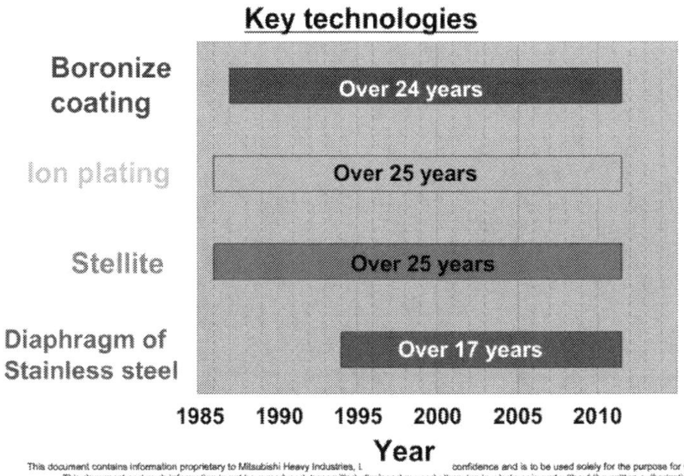

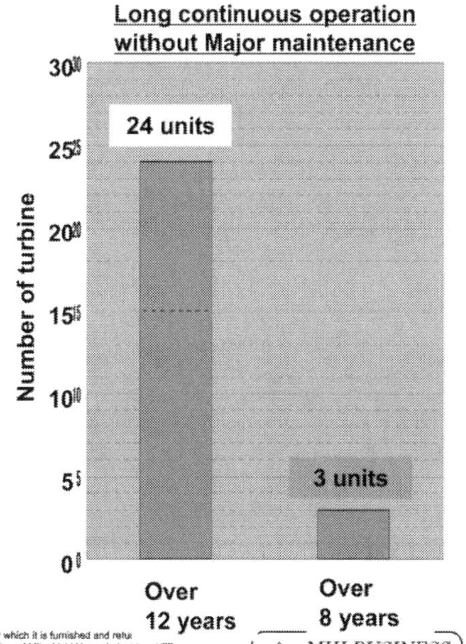

23 MHI BUSINESS CONFIDENTIAL

Conclusion

MITSUBISHI HEAVY INDUSTRIES COMPRESSOR CORPORATION

- Advantage of Replacement
  - ➔ Larger increasing ratio in production (approx. +100%)
  - ➔ Not so higher investment cost
  - ➔ Improve efficiency by application of latest technology
- Footprint Replacement
  - ➔ Custom-designed compressor made by high quality casting
  - ➔ Minimum installation work period as scheduled
- Parallel Train configurations
  - ➔ Increase Plant Capacity if additional area is available

MCO is able to propose whichever revamp/replacement/addition of compressor train according to customer's request

24 MHI BUSINESS CONFIDENTIAL

INSTALLING A CHARGE GAS COMPRESSOR BOOSTER

PAPER # 246768

Giorgio Franceschetti
Olefin Plant Head

Reinaldo Antonio Cardoso
Process Department Head

Prepared for Presentation at the 2012 Spring National Meeting
Houston, Texas, March, 1-5, 2012

INSTALLING A CHARGE GAS COMPRESSOR BOOSTER

Giorgio Franceschetti

Olefin Plant Head

Reinaldo Antonio Cardoso

Process Department Head

Abstract

The Braskem Ethylene Southeast Plant expansion study from 500 MTA to 700 MTA recommended that the charge gas compressor turbine and the 1st and 2nd stages should be replaced. Due to a series of technical and logistical constraints, it was decided to install a booster compressor upstream the first stage of the existing charge gas compressor. Minor modifications to the 4th & 5th stage impellers of the existing charge gas compressor were required to satisfy the final stage discharge pressure requirement

Summary

Braskem UNIB-3 is located in Santo André City, Brazil and produces polymer grade ethylene from full range naphtha cracking and refinery off gas (ROG) recovery.

During the last revamp project the plant capacity was increased from 500 KTA to 700 KTA (570 KTA from naphtha processing and 130 KTA from ROG recovery).

In order to achieve this capacity the cracked gas compressor should be revamped but several technical and physical restrictions were found and the best solution was installing a booster at main compressor suction and modifying the 4th and 5th stages rotors.

After booster startup, the stable operation at higher capacities was achieved, the cracking pressure was significantly reduced and furnaces selectivity was increased as planned in the revamp project.

This work describes the main steps of booster compressor design, including its conceptual, definition of constrains and the development of a dynamic model using MATLAB to simulate transients and define the best control strategy to be used in operation of both machines.

Main compressor evaluation

During the early phases of basic engineering project the need of revamp the cracked gas compressor has become clear for the project team due to the increase in the naphtha and ethane throughput.

A conventional revamp scenario or compressor replacement was considered but some major restrictions were immediately identified:

- The time available for modifications was short once all revamp project should be implemented during a scheduled maintenance shutdown;
- The compressor driver (a steam turbine with medium pressure steam extraction and condensation) was already revamped in previous projects. All asked turbine suppliers were not able to increase the turbine capacity once more or offer a new equipment in the same footprint;
- The gas velocity at first stage nozzle was very high and further capacity increase would lead to higher suction pressures and lower furnace selectivity;
- No room was available for steam condenser revamp or medium pressure steam production increase.

Considering these restrictions it was very unlike the conventional revamp approach or even the equipment replacement. The solution was engineered by the Licensor doing the Basic Engineering Design, Lummus Technology, a CB&I company, based on its extensive design and operation experience.

Booster compressor solution

The proposed alternative by Lummus Technology and Braskem was installing a new booster compressor upstream the main compressor first stage suction to reduce the power requirement for the main compressor (Figure 1).

The booster compressor should have its own suction drum, aftercooler, driver and anti-surge control and the plant should be able to work only with main compressor at low capacity or during start up.

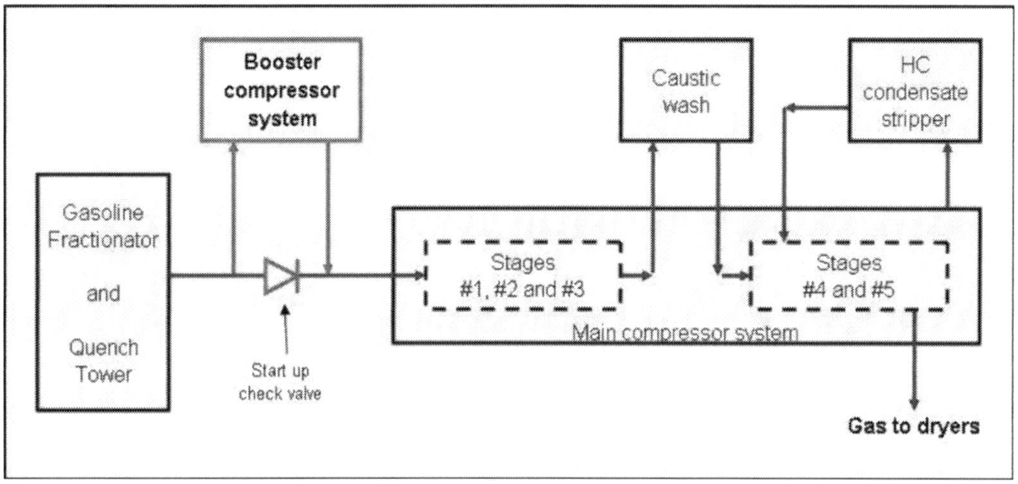

Figure 1: Booster compresssor alternative

The following guidelines were used in the new compressor design:

- The booster compressor should operate at low suction pressure to allow good pyrolysis selectivity and to increase the gas pressure high enough to avoid the power and gas velocity limitations of the existing main compressor.
- A loading share control system should be considered to balance the required power between the two compressors and to keep the suction and intermediate pressure under control.
- The booster compressor should start only after the main compressor reaches stable operation.
- Recycles from minimum flow valves should be avoided in all operating scenarios.

Pressure profile

The definition of booster compressor discharge pressure was a trade-off among a number of restrictions:

- It should be high enough to reduce power requirements of main compressor and avoid the replacement or modifications in its first three stages and in its condenser and steam turbine;
- It should not be so high that maximum work pressure of caustic wash tower located between 3rd and 4th stages of main compressor was exceeded;
- Replacement or modifications in existing aftercoolers and suctions drums should be minimized to reduce project cost and complexity;
- Main compressor discharge pressure should not be reduced to avoid operating problems in condensate stripping and cryogenic section downstream.

Figure 2 compares original and final pressure profiles with maximum allowable working pressures (MAWP).

- The final booster discharge pressure was enough to satisfy all restrictions above but only the 2nd stage aftercooler and the 3rd suction drum were replaced due to MAWP restrictions;
- To keep the 3rd stage pressure bellow the caustic tower MAWP, we reduced the existing charge gas compressor speed and modified the rotors of 4th and 5th stages to achieve required discharge pressure.

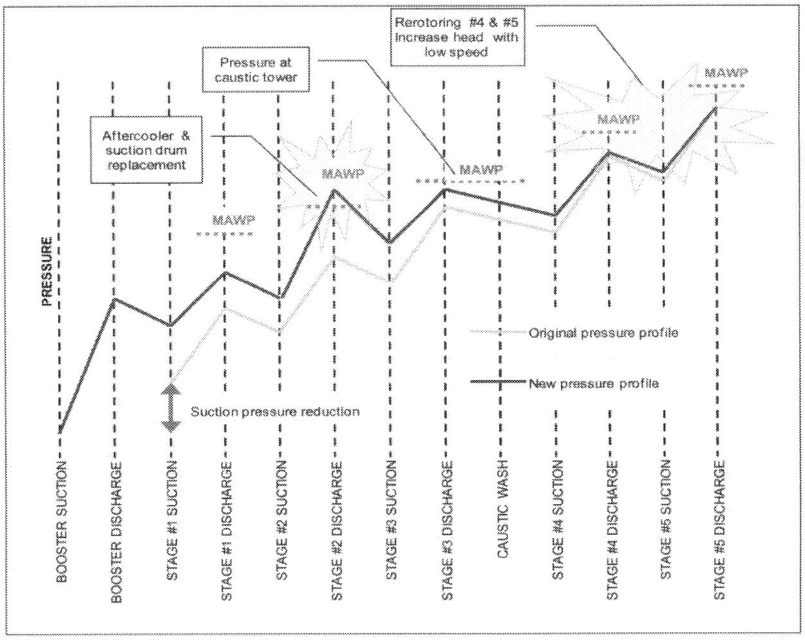

Figure 2: Pressure profiles

Startup, shutdown and normal operation control strategies

After conceptual design, the major concern of project team was define the control strategy to be used during normal operation and major disturbances such as startup, planned and unplanned shutdown.

For normal operation a number of different approaches were analyzed and the project team has decided to control the entire system (booster + main compressor) like a single six stages compressor and use the booster suction pressure control to vary both speeds (booster and main compressor).

A "bias" between the booster speed and main compressor speed was defined and the control system could vary both speeds simultaneously to keep the suction pressure under control.

During startup the main compressor should run at constant speed (suction pressure control in manual mode) and the booster speed should be increased with minimum flow bypass open until the start up check valve closes. Once this valve is closed the booster compressor could be accelerated to the nominal speed and the suction pressure control could be enabled, reducing the speed of the existing charge gas compressor.

The bias between booster speed and main compressor speed should be manually adjusted to balance the compressors and optimize the system performance.

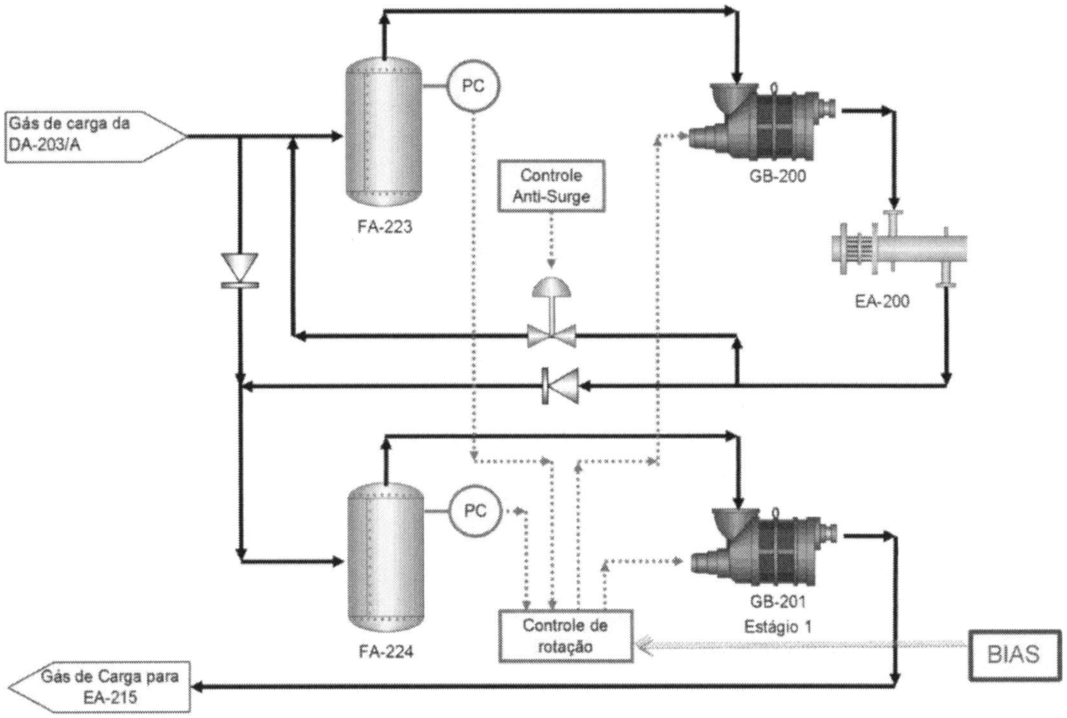

The major concern related with startup scenario was the possible sudden reduction in discharge pressure as soon as the check valve was close due to the fast flowrate increase through the booster. If this reduction was long enough the valve could start an "open and close" cycle and be damaged.

In order to check this scenario a dynamic model was developed in MATLAB by project team and the simulation showed that the system was robust and fast enough to avoid the cycles.

Other transient scenarios were tested in the model such as bias changing, booster startup and shutdown and sudden gas flow reduction with good results.

The main modification in the control strategy identified during these simulations was related with the booster trip scenario when the suction pressure control should be disabled and the main compressor should remain in fixed speed to avoid overspeed shutdown and the gas surplus should be diverted to flare to avoid major disturbances in the plant.

These control strategies and operating procedures were included in the project and the booster compressor is operating since 2009. During this two years the system is operating very well during normal and abnormal conditions as plant throughput reduction, shutdown and start up after power failure, steam boiler upsets and etc. The original control strategy is still in use unchanged.

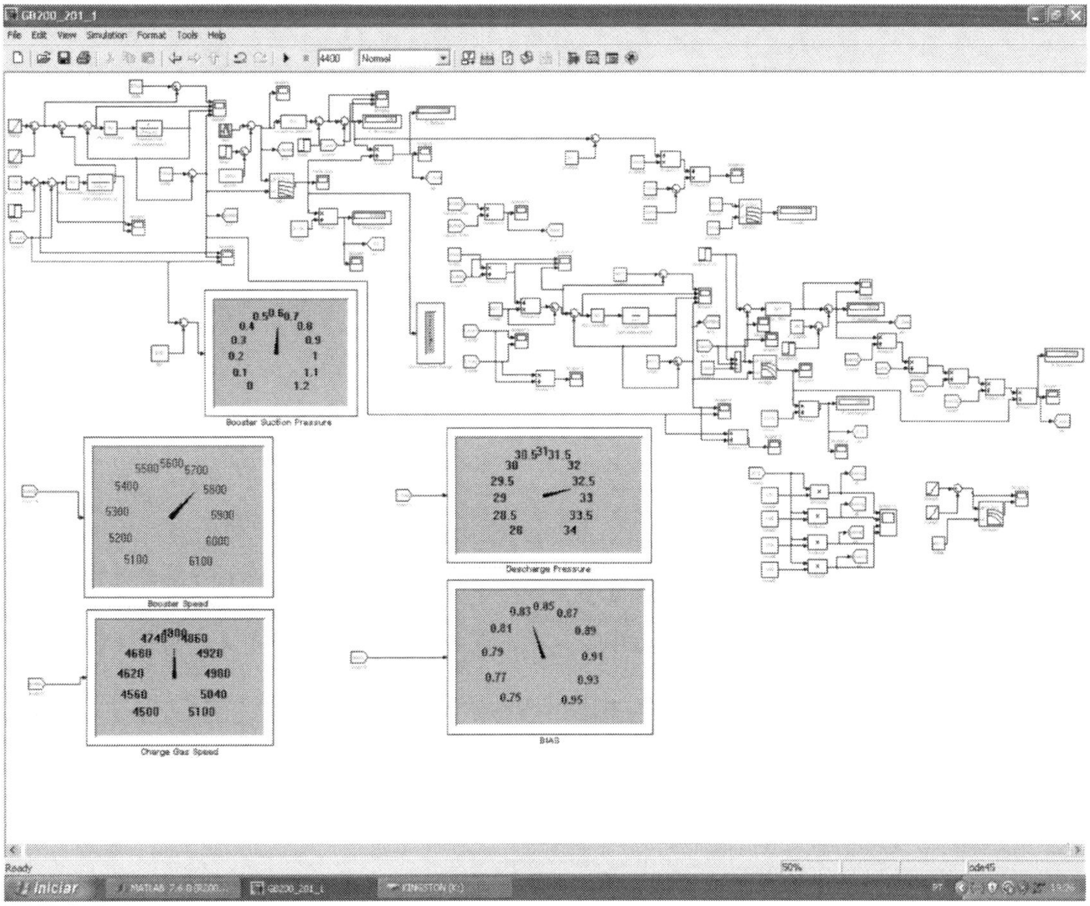

Diagram block (Matlab)

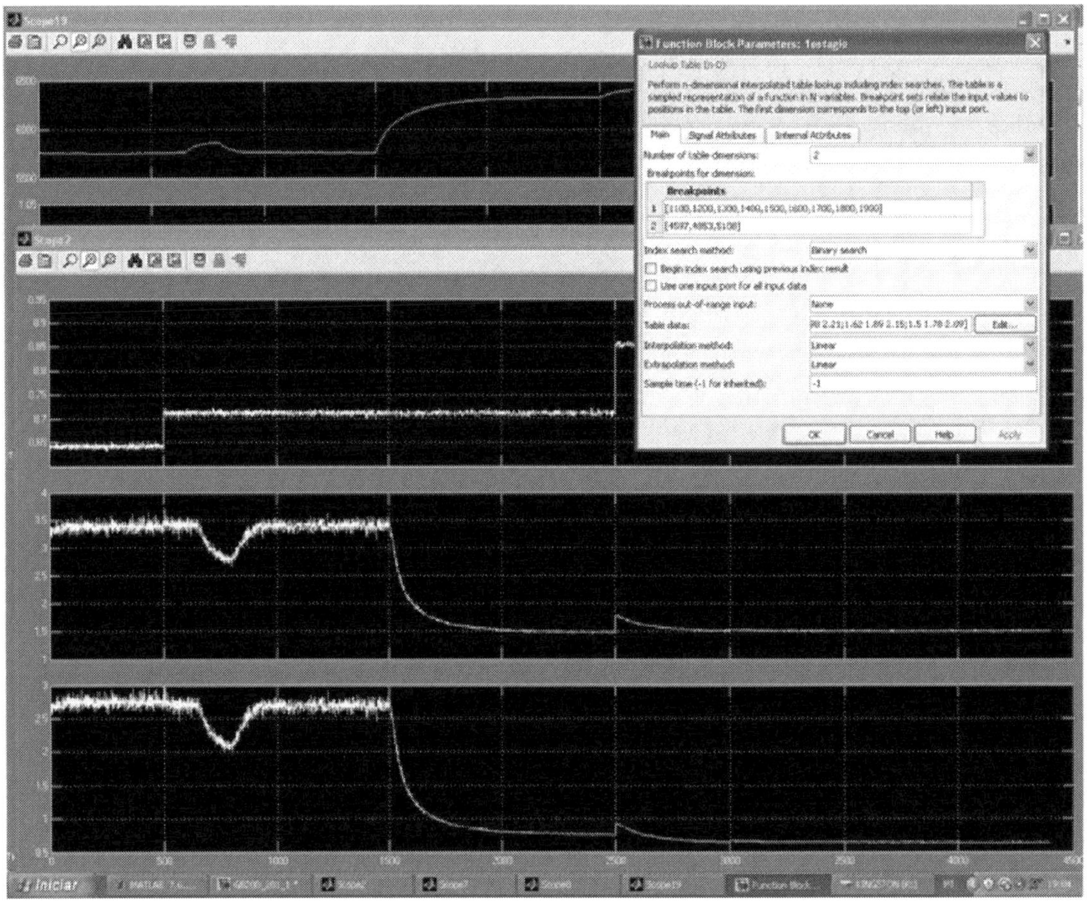

Outputs from dynamic simulation

Conclusion

The booster compressor is an effective alternative to steam crackers revamps when conventional main compressor rotor retrofit or replacement is not possible or uneconomical.

Even a simple sharing loading approach such as booster suction pressure control with fixed bias between compressors speeds is able to control the system in a wide range of operating scenarios.

A special care was taken to avoid conflict between the anti-surge control that is not cited in this paper.

Some tools like dynamic simulation models are very useful in the system analysis and allow early problems identification and correction in an effective, fast a safe way.

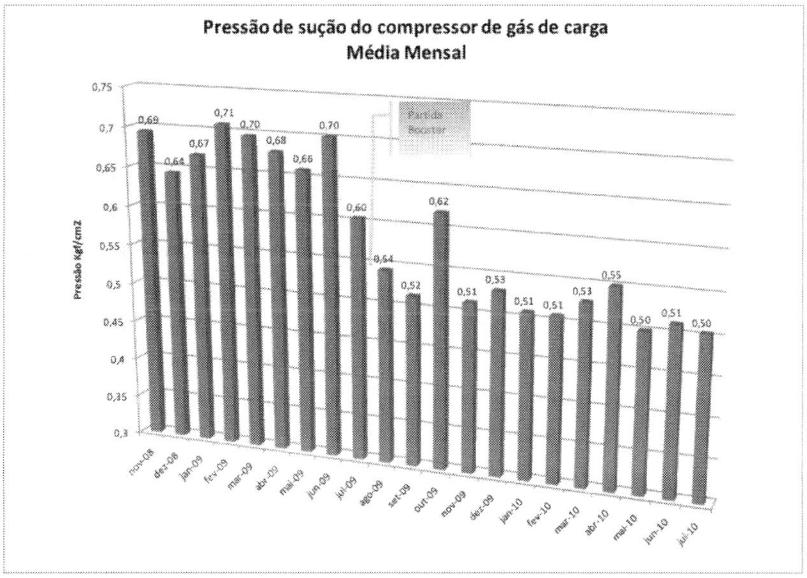

AICHE PAPER # 247047

THREE-DIMENSIONAL IMPELLER MANUFACTURING: A HISTORICAL PERSPECTIVE

PAOLO MOLA

Engineering Manufacturing Technology

GE OIL & GAS

Prepared for Presentation at the 2012 Spring National Meeting

Houston, Texas, April 02, 2012

THREE-DIMENSIONAL IMPELLER MANUFACTURING: A HISTORICAL PERSPECTIVE

PAOLO MOLA

Engineering Manufacturing Technology

GE OIL & GAS

ABSTRACT

Injecting high technology into the products supplied to the market is a competitive advantage of a modern Company compared to its traditional competitors. This article provides a historical perspective on how GE Oil & Gas has played a leading role in the petrochemical market segment as an innovator from very early, "primitive" technology of the past up to the most advanced technology of today.

INTRODUCTION

An impeller is a key rotating component of centrifugal compressors used in Oil & Gas applications that moves the gas through the compression phase, transforming kinetic energy into pressure.

The term "three-dimensional" describes the gas path geometry that employs a twisted design to convert the flow of the gas from axial at the inlet to radial at the exhaust. Generally, this type of impeller is suitable for high flow coefficient applications (phi => 0.03).

Three-dimensional impeller machining has been considered a difficult manufacturing discipline since the very early prototypes were produced for installation in centrifugal compressors. The market demand for special, high flow and high efficiency performance forced manufacturing engineers to find leading edge technology solutions to advanced gas path designs.

Much of GE Oil & Gas' expertise in this highly technical field came from Nuovo Pignone, a compressor manufacturer based in Florence, Italy that GE acquired in the '90s. As a leader in the development of manufacturing technology, evolving from molds and welded fabrication, GE Oil & Gas is today known worldwide as one of the foremost single-piece milled impeller providers to the oil & gas industry. Their unique, patented technique allows producing impellers with three-dimensional blade designs starting from a single forging without any welding. This generates a robust, repeatable and reliable high tech component that is the current state-of-the-art in machining technology.

BACKGROUND

The production of the first three-dimensional impellers (Figure 1) started in the early '60. These impellers were necessarily built using a manufacturing process consisting of turning the finished profile of the disc (hub) and counter disc (shroud), and then assembling the blades by welding them to the sides of the hub and shroud.

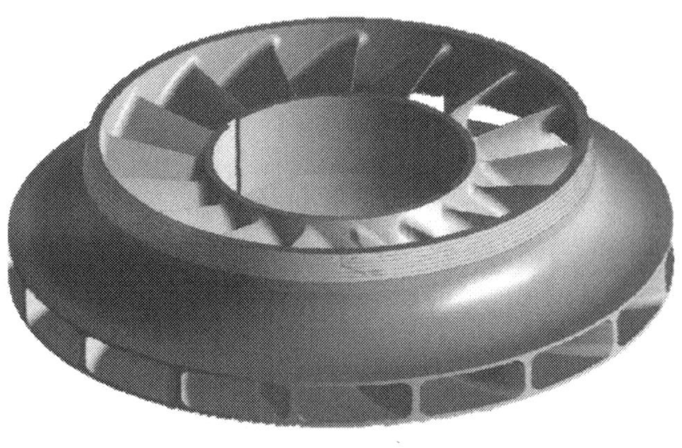

Figure 1 – Typical three-dimensional impeller design

This was the only possible method because the machining systems available at that time did not allow continuous 5-axis machining on continuously variable profiles. Therefore, the welded assembly method proved to be the most suitable approach. With the arrival of numerical control (NC) and CNC, the latest evolution of this technology, it was possible, more efficient and definitely safer and faster, to produce complex shapes using systems that allowed CAD and CAM to work together.

Even though the moulding-plus-welding method allowed the production of a complex impeller, the quality achieved with regard to dimensions and compliance with the design data was undoubtedly very far from what is available today. With the development of linear motors and digital numeric controls, today it is possible to produce the same impeller from a single piece of material and with a higher intrinsic quality level.

In order to allow this new method of machining, it was necessary to process the profiles to define them as points. The output of the calculations carried out by product engineering was a data file; initially a hard copy, easy-to-read listing, which subsequently became available as an electronic file. Through this data file stored on microfilm and its related electronic format (a formatted text file), it was possible to rebuild the surfaces of a blade and gas vane.

The "array" of points made up the "splines" of the blades, with an orientation that went from the leading edge to the trailing edge.

The two splines of the pressure side therefore made up the "ruled" surface of that side, while the two splines of the suction side made up the ruled surface of the other side of the part. (Figure 2)

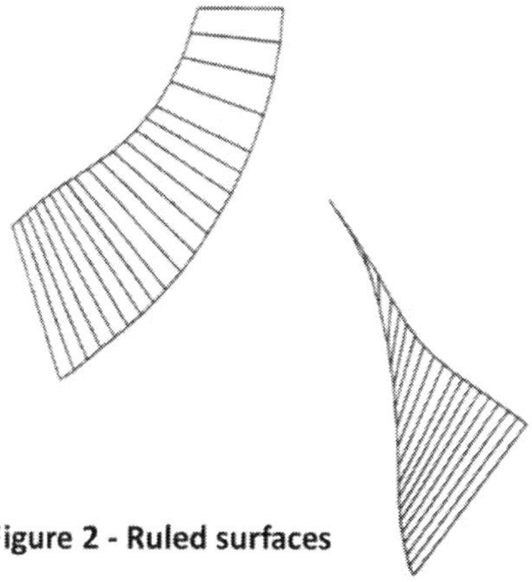

Figure 2 - Ruled surfaces

Similar splines were created on the meridian profile of the gas passage to generate the shroud and hub surfaces.

Just these few elements, which today still constitute the basic elements of the stage design, together with the external profile of straight lines and arcs, were enough to define the impeller.

THE WAY WE WERE:
THE '60S (BI-WELDING PROCESS)

The manufacturing process for the production of MCL compressors started back in the early '60. The method included turning of the finished profile of the disc (hub) and counter disc (shroud). While Hub and Shroud were produced through forging, the blades were cut from a sheet of material with thickness equal to the finished dimension of the blade. Then it was forged with die to create the three-dimensional profile. The leading edge, defined using points, was machined to achieve the final shape. (Figure 3)

At this stage, the assembling of blades was done by spot welding and adjusting all the blades on the counter disc (shroud), followed by welding to the shroud on both sides. Next, a two-element resin was applied to each gas passage, to prevent deformation of the blades and to allow easier management of the coupling profile at the disc (hub) surface on the lathe. This is consequent to the interrupted cut, on a blade with a very low thickness/height ratio, that could have introduced forces capable of folding the blade that had already been welded.

When this critical phase was over, the resin had to be removed, the welding chamfer had to be made and finally, it was possible to weld the blades to the hub. At this point, the impeller went through deburring, i.e., the phase in which the weld beads in the leading edge and trailing edge areas were smoothed, repairing of porosity and defects that might have been spotted by checking with dye penetrant, and then to heat treatment that imparts the final design mechanical characteristics.

The manufacturing process was completed with a dimensional check, balancing and overspeed testing, and a dye penetrant inspection to ensure that cracks were not introduced during the overspeed test.

The handling between the different welding and deburring operations introduces unbalances, which requires a balancing operation in which material is removed by grinding on both the hub and shroud sides.

The lead time necessary to produce an impeller was very long, in fact, about three times longer than today.

All the steps indicated above led to a limitation in use of this technology due to the high cost and dimensional quality issues stemming mostly from deformations related to the manual positioning process and the subsequent welding.

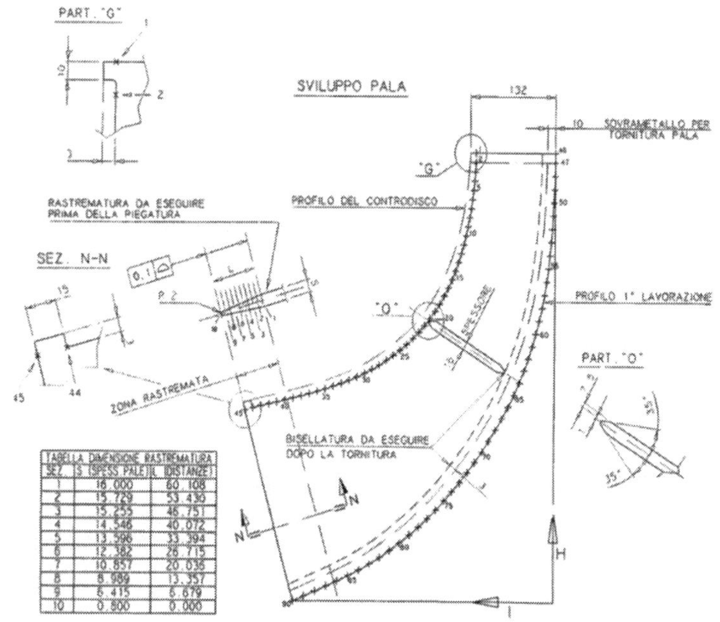

Figure 3 - Two-dimensional blade profiles

Figure 3 - Blade milling on the forging

THE '70S (SINGLE SIDE WELDING – INTERNAL E/O EXTERNAL)

In the mid-70s, the availability of more advanced numerical control systems made it possible to develop methods for economically producing impellers by machining. After the blades were formed directly on the forged ring (Figure 3), it was possible to weld the counter disc to the blade heads from the internal or external sides.

Although this technology was a great improvement, it is necessary to point out that, the production of a 3D impeller with welding inevitably involves deformations from the shape designed in the planning phase.

This can be appreciated if the deformations intrinsic to the process are kept in mind.
This welding process requires pre-heating of the piece, which unavoidably induces shocks in the treated material, creating deformation that is difficult to detect and correct.

Tables that allow the deviation to be corrected in advance were prepared to keep the problem under control.

METHOD AND TOOLS

In the beginning, the tools used were HSS cylindrical mills for roughing and finishing. The mills were of small diameter to access the narrow passage at the leading edge and longer than the maximum height of the blade. For this reason, they had limited stiffness and therefore a low chip removal feed rate. It must be mentioned that the speed of the machine and the control was limited by the calculation time needed by the CNC for checking the trajectory.
Therefore, the material was intentionally removed slowly but with deep milling cuts.
A channel was excavated near the blade side coming down in steps and the slot was widened in the center with "radial" cuts using the entire mill side. The blade side was completed with a finishing mill with a spherical head and a cutting edge length greater than the maximum height of the blade. Substantially, the mill side was sliding on ruled surfaces that defined the blade surfaces.

WHAT'S GOING ON?
LATE '90S (FULL MILLING DEVELOPMENT)

GE Oil & Gas developed technology (US Patent 7305762 B2 - IT MI20021876 A) to eliminate, at least in part and where applicable, the extensive use of welding.
This development started at the end of the '90s, when "affordable" CAD/CAM and manufacturing systems ushered in a new generation of high tech manufacturing systems.

This approach allowed the procurement time to be reduced and eliminated the purchasing of one of the two forged pieces needed for the welded version. As a consequence, a product cost and lead time reduction was achieved.
Obviously, the intrinsic quality of an item produced as a single "piece" compared to an item manufactured using a fabrication process is higher and therefore,

preferable. The market immediately demonstrated interest in this technology, especially for systems that require high reliability due to severe rotordynamic and performance characteristics. The mechanical strength is markedly increased because the fillet radius is uniform and smooth. This keeps the stress concentration factor in the region where the blades are connected to the hub and shroud under control. In fact, with respect to the design CTQ (Critical to Quality factors), they are not only easier to achieve but – and especially – more repeatable and therefore, easier to keep under control in a complex manufacturing system encountered in the production of a customized product.

Practically what becomes a differentiating element, compared to previous technologies, is the gas flow passage that is achieved completely by milling. The order of the remaining operations and the related operating sequence, although greatly simplified and reduced, remains unchanged. The operation starts with a preparation turning that pre-shapes the impeller external surfaces, including the removal of the ring for mechanical testing. In the next step, the gas passage is created on continuous 5-axis machines. In this case, in consideration of the morphologic characteristics of the gas passage and the impeller dimensions, as well as the material that must be machined, a roughing out strategy is chosen that can include 1) the total removal of material with a "traditional" milling method or 2) roughing out with ECDM technology (Electro-Chemical Discharge Machining), which will be illustrated later.

In the first case, the gas passage is opened starting from two holes following a vector along the "bisector" of the gas passage.

The openings made from the inlet to the outlet (Exducer area) and from the outlet to the inlet (Inducer area) meet at a point corresponding to the middle of the gas passage length (Figure 4).

The purpose is to free the tool working area of the chips generated during the cutting operation. The fact that the openings intersect each other makes it possible to remove the chips through the washing process created by the coolant that is flowing under pressure along the flow passage (Figure 5). The result is simpler and safer machining.

Figure 4 – Pre-boring

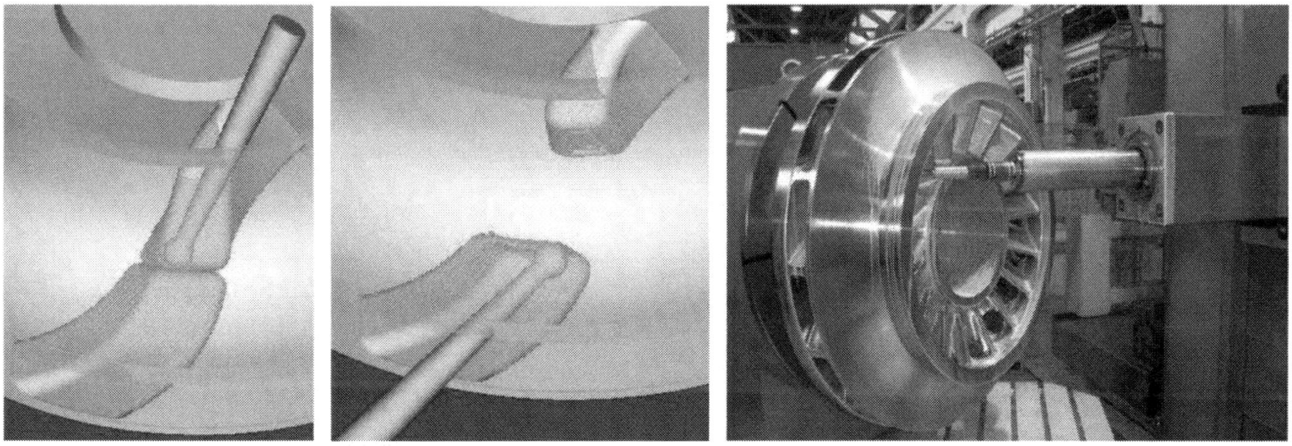

Figure 5 - Vane machining

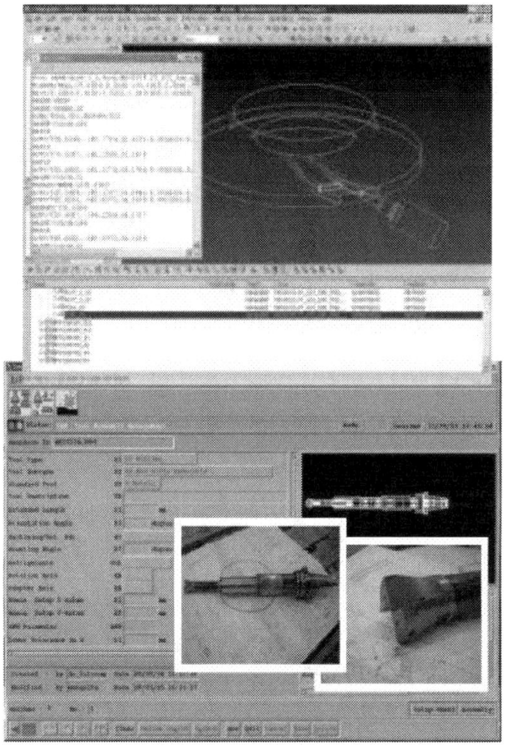

EXTENSIVE USE of CAD/CAM

The CAD/CAM machining preparation starts while the electronic three-dimensional model of the impeller is uploaded in the Engineering exchange area where the raw and finished models are picked up. These elements are necessary to generate the tool-path with the CAM module and for the subsequent simulation with the virtual machining module.

This manufacturing technique allows complex geometries to be produced, but occasionally, tool constrains occur that limit the use of this technology. Whenever there is such a doubt, a preliminary "feasibility study" must be conducted to verify the ability to produce the impeller using the "full milling" method.

The "feasibility study" is performed using a CAM module and an *ad hoc* tool (master model and associated excel sheets). In this way, the machining strategy and technology that offer the highest degree of automation possible for the impeller in question can quickly be made. This method requires the CAD generated 3D model, which provides the CAM engineer with a parametric model that is robust and reliable. The model is generated in accordance with the manufacturing requirements and contains all the specifications needed for the CAM Automation (surface tags, features, etc.) The CAM Automation routine is interfaced with the manufacturing tool database, which catalogs the tooling that has been proven through successful testing. The result of this process is a percentage that indicates the "coverage" factor, i.e., how well the selected tool is able reach "half way" between the Inducer and the Exducer, determining whether or not the technology is applicable.

The basic interface between CAD/CAM and the Machine Tool/CNC is the Post-Processor, a customized software adapted to both the machine and the Numerical Control, and converts the CLDATA (Cutter Location DATA file) output of the CAD/CAM into an ISO language Input file, in addition to supplying a detailed listing of information (machining times, axis stroke range, various diagnostics, etc.).

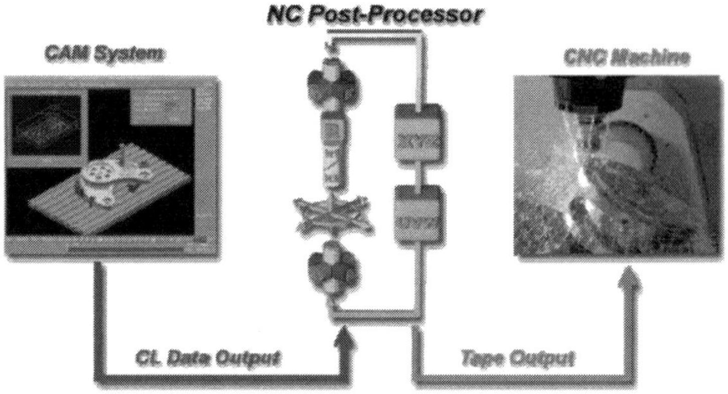

2007 - FULL MILLING PATENT (EXCERPT FROM DISCLOSURE LETTER)

Due to the high density of the gas, rotor stability is a very critical issue in high pressure compressors. Cross-coupled forces, proportional to the gas density, induced by the gas labyrinth can cause unacceptable sub synchronous vibration. The sensitivity of the rotor to these forces increases with rotor flexibility.

Moreover, all impeller dynamic behavior is particularly critical for these applications and the easiest way to improve its stiffness at the same operating conditions (same bearings, temperature, gas pressure, etc.) is to have shafts with large diameters.

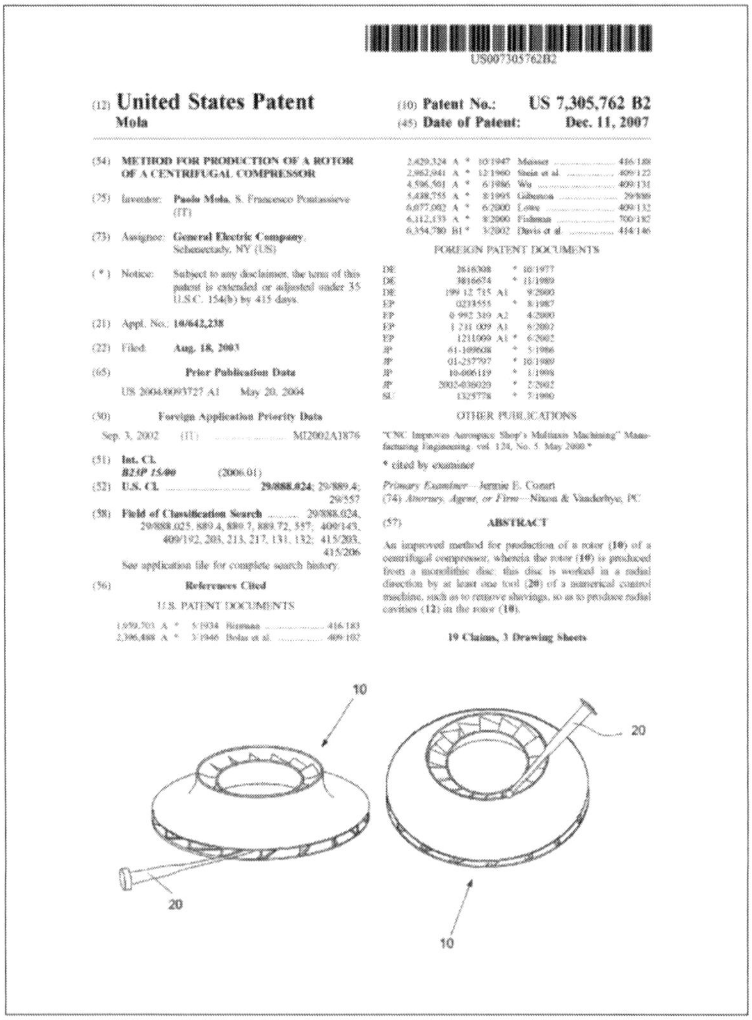

US007305762B2

(12) **United States Patent**
Mola

(10) **Patent No.: US 7,305,762 B2**
(45) **Date of Patent: Dec. 11, 2007**

(54) METHOD FOR PRODUCTION OF A ROTOR OF A CENTRIFUGAL COMPRESSOR

(75) Inventor: **Paolo Mola**, S. Francesco Pontassieve (IT)

(73) Assignee: **General Electric Company**, Schenectady, NY (US)

(*) Notice: Subject to any disclaimer, the term of this patent is extended or adjusted under 35 U.S.C. 154(b) by 415 days.

(21) Appl. No.: **10/642,238**

(22) Filed: **Aug. 18, 2003**

(65) **Prior Publication Data**

US 2004/0093727 A1 May 20, 2004

(30) **Foreign Application Priority Data**

Sep. 3, 2002 (IT) MI2002A1876

(51) **Int. Cl.**
B23P 15/00 (2006.01)

(52) **U.S. Cl.** **29/888.024**; 29/889.4; 29/557

(58) **Field of Classification Search** 29/888.024, 29/888.025, 889.4, 889.7, 889.72, 557; 409/143, 409/192, 203, 213, 217, 131, 132; 415/203, 415/206
See application file for complete search history.

(56) **References Cited**

U.S. PATENT DOCUMENTS

1,959,703 A * 5/1934 Birmann 416/183
2,396,488 A * 3/1946 Bolas et al. 409/102
2,429,324 A * 10/1947 Meisser 416/188
2,962,941 A * 12/1960 Stein et al. 409/122
4,596,501 A * 6/1986 Wu 409/131
5,438,755 A * 8/1995 Giberson 29/889
6,077,002 A * 6/2000 Lowe 409/132
6,112,133 A * 8/2000 Fishman 700/182
6,354,780 B1 * 3/2002 Davis et al. 414/146

FOREIGN PATENT DOCUMENTS

DE 2616308 * 10/1977
DE 3816674 * 11/1989
DE 199 12 715 A1 9/2000
EP 0233555 * 8/1987
EP 0 992 319 A2 4/2000
EP 1 211 009 A1 6/2002
EP 1211009 A1 * 6/2002
JP 61-109608 * 5/1986
JP 01-257707 * 10/1989
JP 10-006119 * 1/1998
JP 2002-036020 * 2/2002
SU 1325778 * 7/1990

OTHER PUBLICATIONS

"CNC Improves Aerospace Shop's Multiaxis Machining" Manufacturing Engineering, vol. 124, No. 5, May 2000.*

* cited by examiner

Primary Examiner—Jermie E. Cozart
(74) *Attorney, Agent, or Firm*—Nixon & Vanderhye, PC

(57) **ABSTRACT**

An improved method for production of a rotor (**10**) of a centrifugal compressor, wherein the rotor (**10**) is produced from a monolithic disc; this disc is worked in a radial direction by at least one tool (**20**) of a numerical control machine, such as to remove shavings, so as to produce radial cavities (**12**) in the rotor (**10**).

19 Claims, 3 Drawing Sheets

State of the art:

The blades are traditionally obtained by milling on the hub and then welding from the inside of the vane to unite the parts. Another technique is used for low flow coefficient impellers in which the vanes are too narrow to perform the welding from inside. In this case, the two parts are slot-welded from the shroud.

After the welding, the assembly is heat

treated to reduce the residual stresses.

Milling machining:
The machining method specified at present in the method, provides for access from inside and from outside until the division plane of both the Exducer and Inducer is reached, by means of terracing on subsequent planes.

Description of fabrication method:
Starting from a monolithic pre-turned center-drilled steel disk, compressor radial impellers are manufactured by milling. Numerical control machines are used.

Results:
The impeller is machined with traditional tools. The main advantages of this technology are:

- No structural discontinuities
- High structural strength
- High dimensional accuracy
- Fine surface quality

All these points are particularly important for high pressure re-injection machines. Due to the high pressure and high density involved, pressure pulsations originating from asymmetries in the aerodynamic field, particularly in the discharge scroll, can cause significant oscillatory forces.

The absence of any metallurgical discontinuity within the impeller is clearly a step forward.
The distortion associated with welding can cause modification of the blade width. With milling, the accuracy increases further allowing more accurate matching between the expected and the tested performance.

CONCLUSION

In order to fulfill the increasing market demand for full milling technology, the GE Oil & Gas Florence workshop, since the start of this improvement, has expanded its fleet of machines able to work with five and more axes simultaneously. This has required extensive investment not only in machining capability, but also in personnel skills and competencies.
Through this milestone project, GE Oil & Gas is now well positioned in its market for single-piece three-dimensional impellers with external diameters ranging from 340 up to 2000 mm.

Always being ready to supply advanced technology solutions to customers repeatably, reliably and with proven technology is a key factor for winning in today's competitive market, especially in the oil & gas industry.
Relying on the competencies and skills developed over the years is the best baseline for future improvements to insure this capability. Continuously investing in Research and Development is not only part of our new product development strategy, but gives GE the ability to keep moving forward. As demonstrated in this short excursion into technology evolution applied to

the machining process, taking the initiative to change the game is more rewarding than looking through the window and observing what's going on. Investing in developing the latest generation of technology and moving beyond the "comfort zone" has allowed GE Oil & Gas to achieve a leadership position in high quality, reliable and repeatable impeller manufacturing. However, the journey does not end here – rather, this is just another milestone in our continuous innovation to deliver value to the oil & gas industry.

Paper 8a

Ethylene Industry Review
"Frack to the Future"

Mark Wegenka

Director

IHS Chemical Consulting

Prepared for Presentation at the 2012 Spring National Meeting

Houston, Texas, April 2, 2012

Ethylene Industry Review

"Frack to the Future"

Mark Wegenka

Director

IHS Chemical Consulting

Abstract:

- Global review of the markets and market outlooks for olefins
    - Ethylene / Propylene / Butadiene
- Overview of demand drivers
- Discussion of investment plans
- Review of trade patterns
- Analysis of relative cost positions
- Discussion of other significant industry dynamics focused on the impact of Shale Gas on the North American olefins business
    - Overview of impact on feedstock (ethane) availability
    - Discussion of resurgence of petrochemical investment plans
    - Review of competitiveness

The development of multiple shale resources has had a profound impact on the North American petrochemical industry. Given the disparity between domestic oil and natural gas prices, the increased production of natural gas liquids has not only yielded a renewed ethylene cost advantage, but has also changed the production dynamics in other petrochemicals. As ethane will continue to be leveraged through its high selectivity to ethylene, we propose that the industry will employ the use of heavier NGL feedstocks (propane and n-butane) for focused production of propylene and butadiene.

Ethylene Industry Review
"Frack to the Future"

The development of multiple shale resources has had a profound impact on the North American petrochemical industry. Given the disparity between domestic oil and natural gas prices, the increased production of natural gas liquids has not only yielded a renewed ethylene cost advantage, but has also changed the production dynamics in other petrochemicals. As ethane will continue to be leveraged through its high selectivity to ethylene, we propose that the industry will employ the use of heavier NGL feedstocks (propane and n-butane) for focused production of propylene and butadiene. This paper will begin with a brief look at the global olefin industry, followed by a more in-depth view of the effects of the shale gas boom on the outlook for the North American olefins industry.

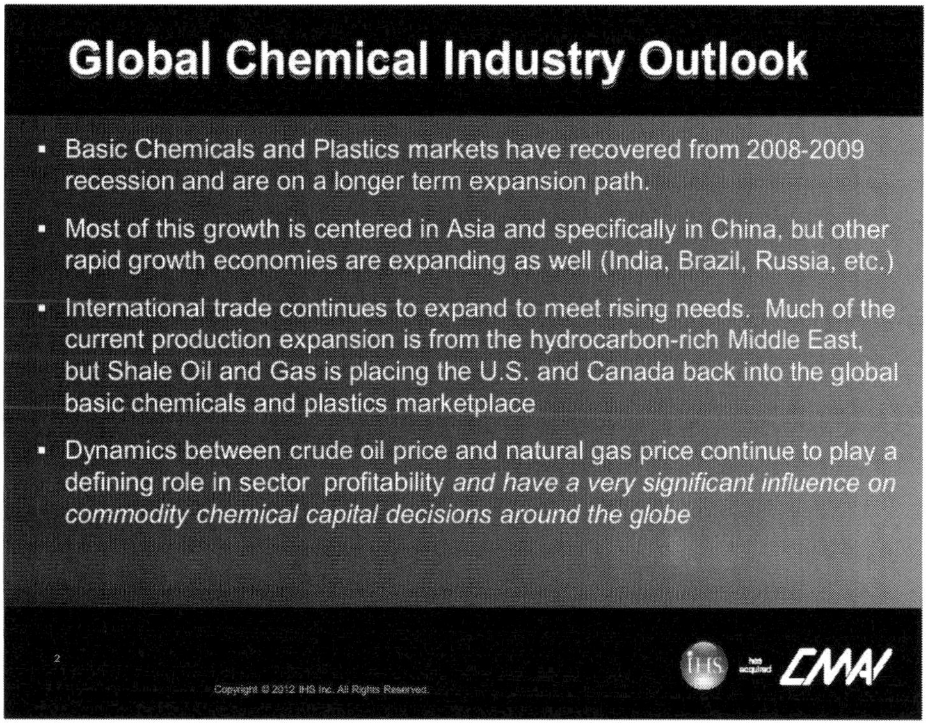

Basic Chemicals and Plastics markets have recovered from the 2008-2009 recession and are on a longer term expansion path. Petrochemical markets are growing at a pace that approaches 2 times that of the economy – these are key ingredients used to manufacture non-durable and durable goods that are vital to sustaining and improving our quality of life. Most of this growth is centered in Asia and specifically in China, but other rapid growth such as India, Brazil, and Russia economies are expanding as well. International trade continues to expand to meet rising needs. Much of the current production expansion is from the hydrocarbon-rich Middle East, but Shale Oil and Gas is placing the U.S. and Canada back into the global basic chemicals and plastics marketplace.

Dynamics between crude oil price and natural gas price continue to play a defining role in sector profitability and have a very significant influence on commodity chemical capital decisions around the globe.

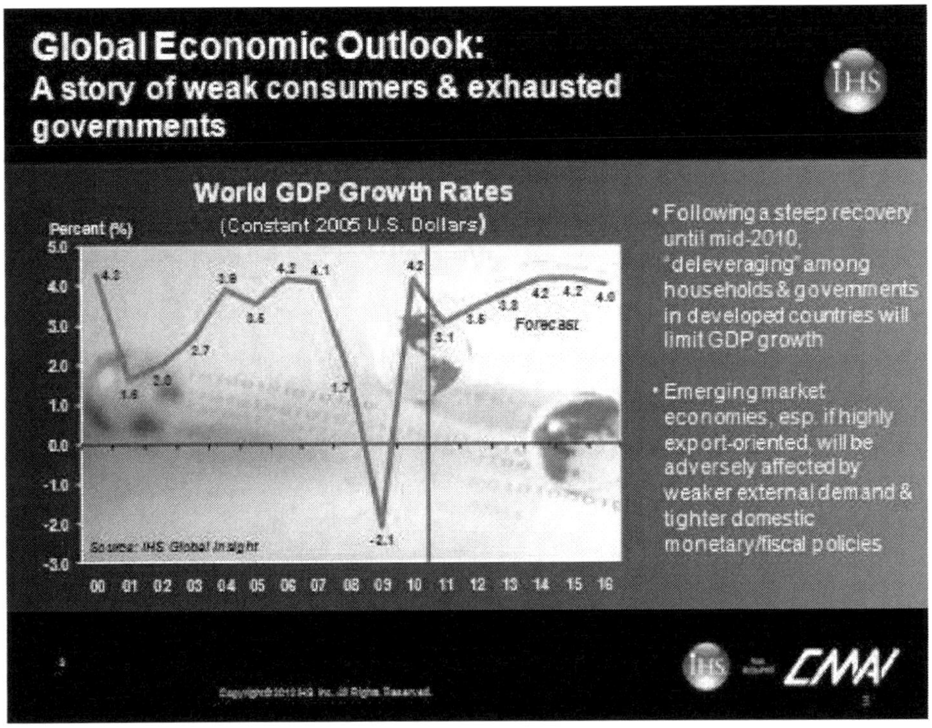

The Global Economic Outlook is a story of weak consumers and exhausted governments. Following a steep recovery until mid-2010, "deleveraging" among households and governments in developed countries will limit GDP growth. Emerging market economies, especially if highly export-oriented, will be adversely affected by weaker external demand and tighter domestic monetary/fiscal policies.

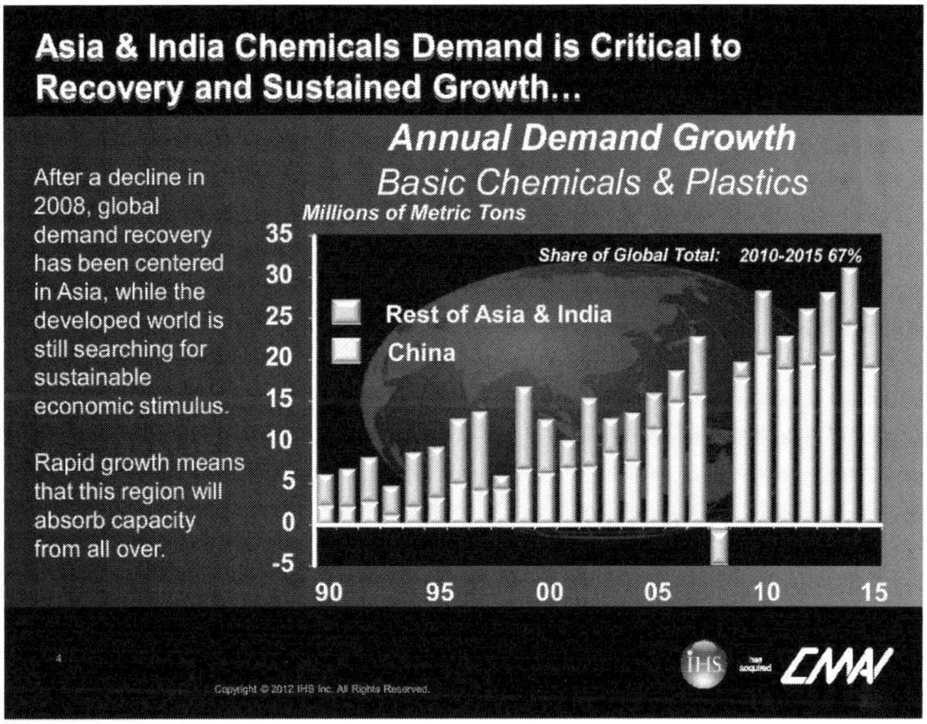

After a decline in 2008, global demand recovery has been centered in Asia, while the developed world is still searching for sustainable economic stimulus. In fact, Asia's share of global total demand growth for basic chemicals and plastics from 2010 through 2015 is expected to be 67 percent. This rapid growth means that the region will absorb capacity from all over the world.

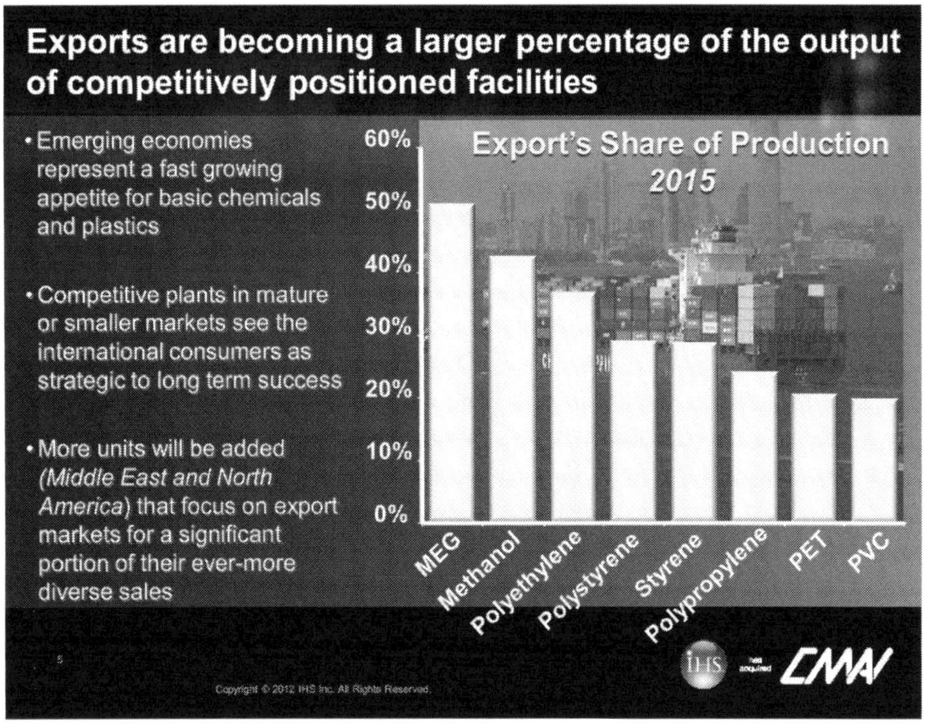

The volume of Basic Chemicals and Plastics international trade continues to climb. Trade is the perfect medicine for a mature, domestic market if it can develop a sustainable cost advantage. Exports are becoming a larger percentage of the output of competitively positioned facilities. Emerging economies represent a fast growing appetite for basic chemicals and plastics as more units will be added in both the Middle East and North America that focus on export markets for a significant portion of their ever-more diverse sales.

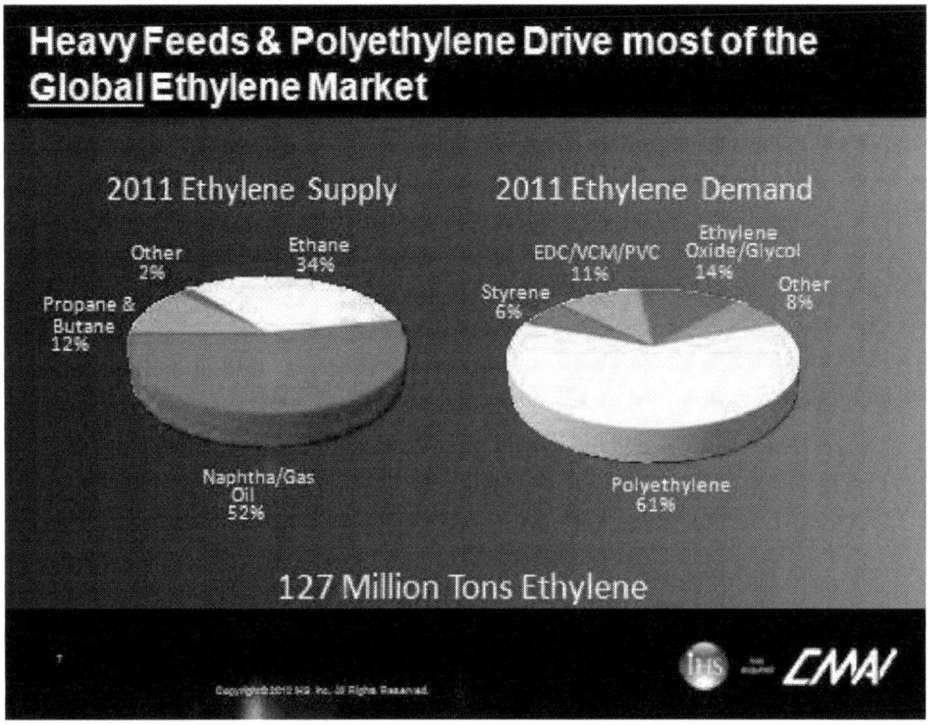

Heavy feeds and polyethylene drive most of the global ethylene market. Over half of global ethylene production is still based on naphtha feed. In 2011, the global light olefins industry produced an estimated 127 million tons of ethylene. The chart of ethylene production by feedstock indicates an estimated 52 percent of the ethylene produced in the world is based on naphtha/gas oil feedstock. Ethane cracking, which was by far the most competitive feedstock regardless of location during 2011 represents 34 percent of the global feed-slate but is forecast to increase over the next several years. The Asia/Pacific region (including India) will remain the largest producing region in the world.

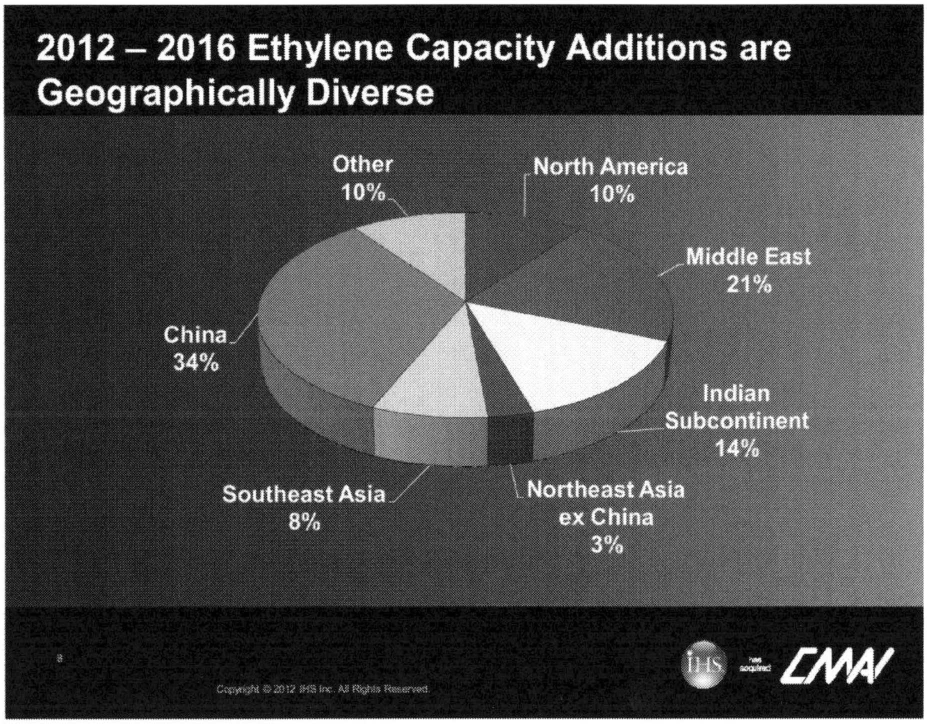

Future ethylene capacity additions are expected to be geographically diverse, with only 10% in North America through 2016. A significant portion of new capacity will be heavy feedstock based, which in today's energy environment may seem counter-intuitive. However, keep in mind the ethylene capacity additions during this time period are mostly in Asia, which is adding primarily heavy cracking capability. The majority of new Ethane cracker capacity in North America won't start up until after 2016.

While highly "advantaged" ethane cracking has been the primary driver for the rapid buildup of ethylene capacity in the Middle East region, we have recently seen the introduction of more feedstock diversity. Ethane has represented the dominate share of feedstock in the region and is forecast to remain so, but other feedstocks being utilized on new projects include light naphtha or condensate, propane and butane. This increased diversity in feedstock is being driven by limits on ethane supply as well as a desire to diversify the regional olefins product offering to other major value chains including propylene, butadiene and aromatics. By adding diversity of value chains for petrochemical derivatives, they help to achieve the over-riding objective of investments that lead to jobs creation.

With Saudi Arabia as the lead country in the region driving the vast majority of new capacity additions, this is where we see a significant shift of feedstock allocations for new projects. The new Saudi crackers are typically designed to operate with 50% of the ethylene produced from a heavier feed. The new project owners are offering broader derivative portfolio to win feedstock allocations, mainly based on co-product propylene, thus requiring the heavier feeds. Saudi Arabia is still investing up to the limits of its current ethane availability, though, and will continue to pursue ethylene investments.

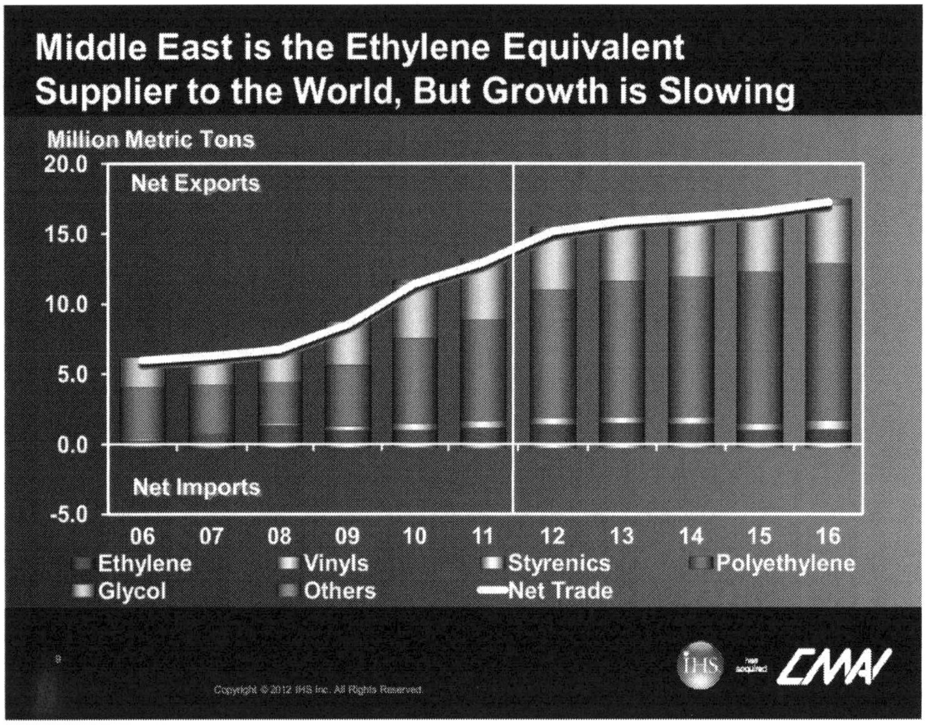

The Middle East is the ethylene equivalent supplier to the world, but growth is slowing especially in Iran due to the boycotts. Polyethylene and glycol account for nearly 90% of Middle East ethylene derivative exports and this is not expected to change over the next five years.

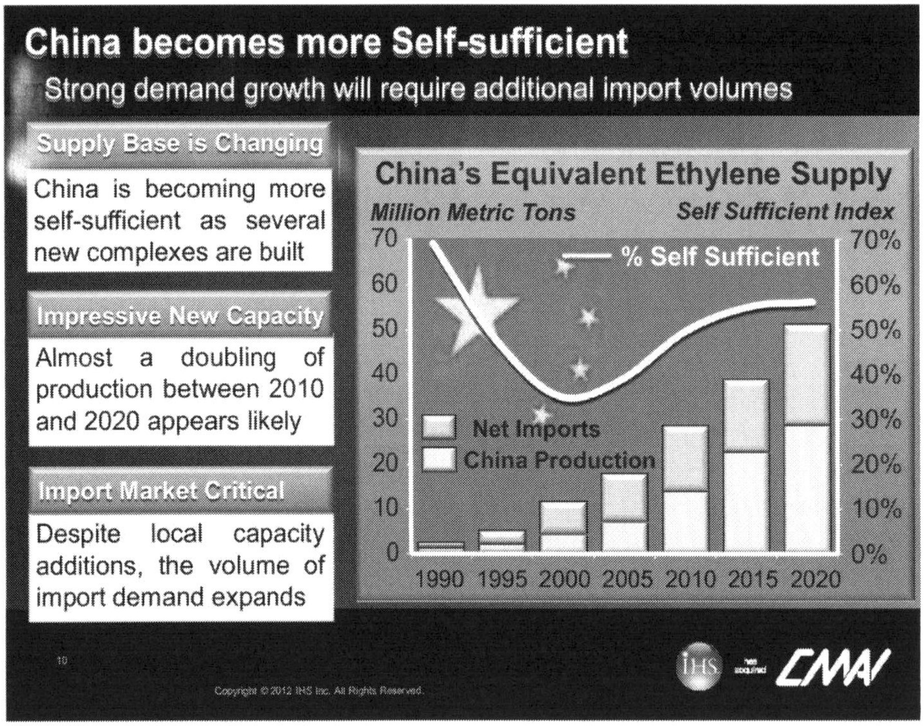

China, on the other end of the supply chain, is also driven by the economics of upstream integration and industrial development to keep self sufficiency levels of domestic production at reasonable levels (above 50% or higher). China is becoming more self-sufficient as several new complexes have been built and it appears likely that nearly a doubling of production will occur between 2010 and 2020. Despite these local capacity additions, the volume of import demand will continue to expand.

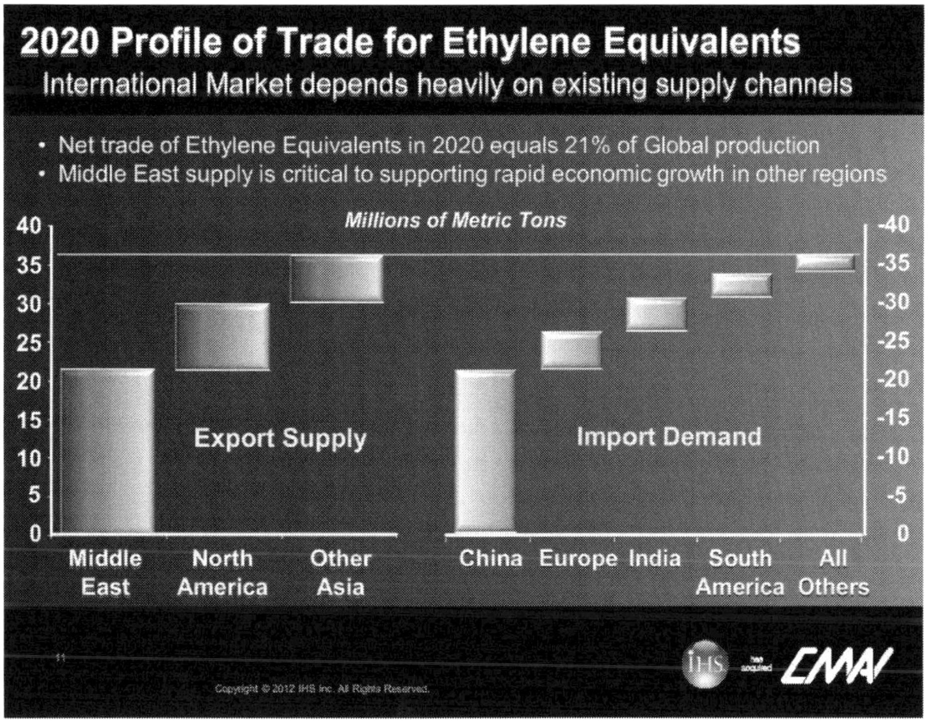

The low cost position of producers in the Middle East region will enable them to dominate global trade of ethylene derivatives as shown in this comparison of net equivalent ethylene exports. The international market depends heavily on existing supply channels for ethylene derivative trade. Net trade of ethylene equivalents in 2020 amounts to 21% of global production. Middle East supply is critical to supporting rapid economic growth in other regions, primarily China, and the Middle East will remain the largest and most important supplier of basic chemicals and plastics to world markets.

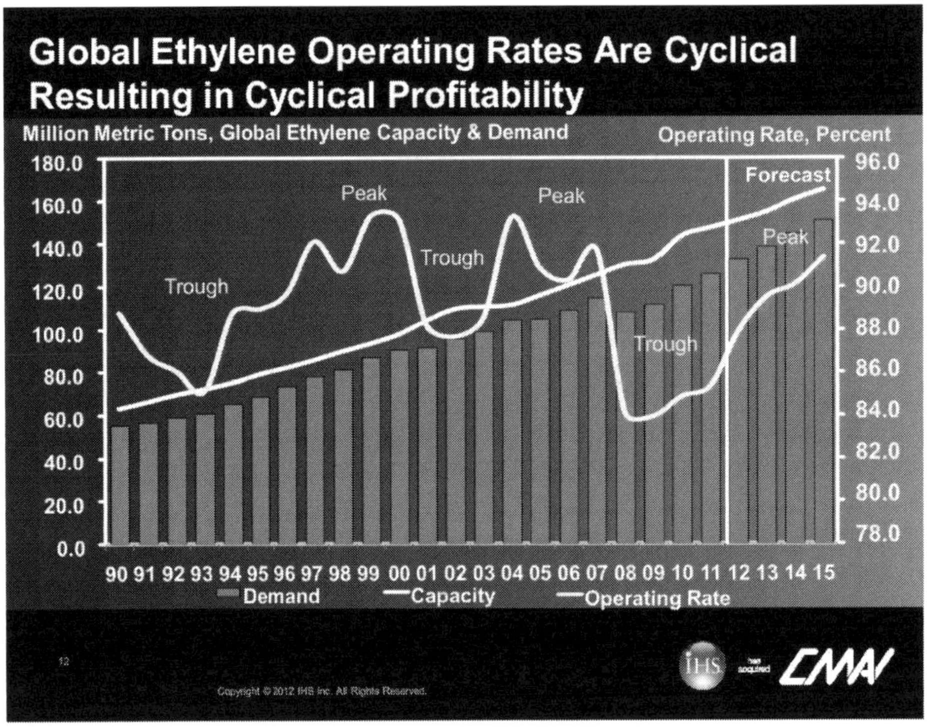

The ethylene supply and demand dynamics on a country basis are rolled up to provide regional and ultimately global views of the industry. One key aspect of the analysis is the measure of global capacity utilization. The utilization curve provides insights as to period of strength or weakness and helps to project the inflection points, where the industry transitions into peak or trough like conditions. From the historical perspective, global ethylene capacity utilization at or above 90 percent represents strength in the industry. During these periods we would expect to see strong, or "re-investment level", profitability (cash margins). Conversely, the periods of lower utilization are typically characterized by extremely low or certainly below "re-investment" cash margins.

Global ethylene operating rates are cyclical, resulting in cyclical profitability. The industry is now climbing out of a deep trough, with the next peak due in 2015.

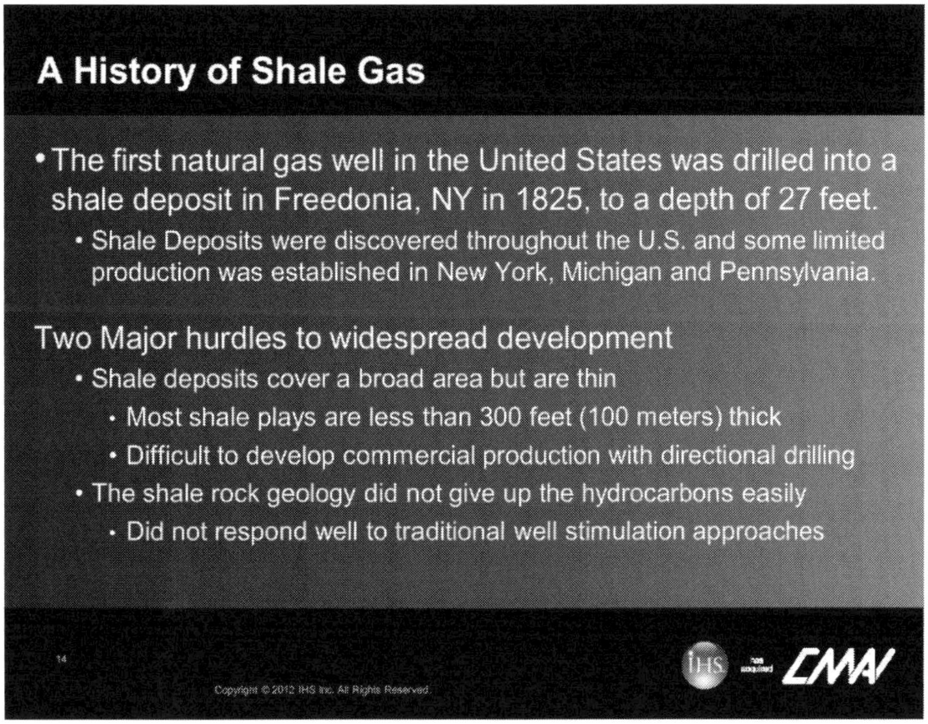

The first natural gas well in the United States was drilled into a shale deposit in Freedonia, NY in 1825, to a depth of 27 feet. Shale deposits were discovered throughout the U.S. and some limited production was established in New York, Michigan and Pennsylvania. There are two major hurdles to widespread development. Firstly, shale deposits cover a broad area but are thin. Most shale plays are less than 300 feet (100 meters) thick and it's difficult to develop commercial production with directional drilling. Secondly, the shale rock geology did not give up the hydrocarbons easily or respond well to traditional well stimulation approaches.

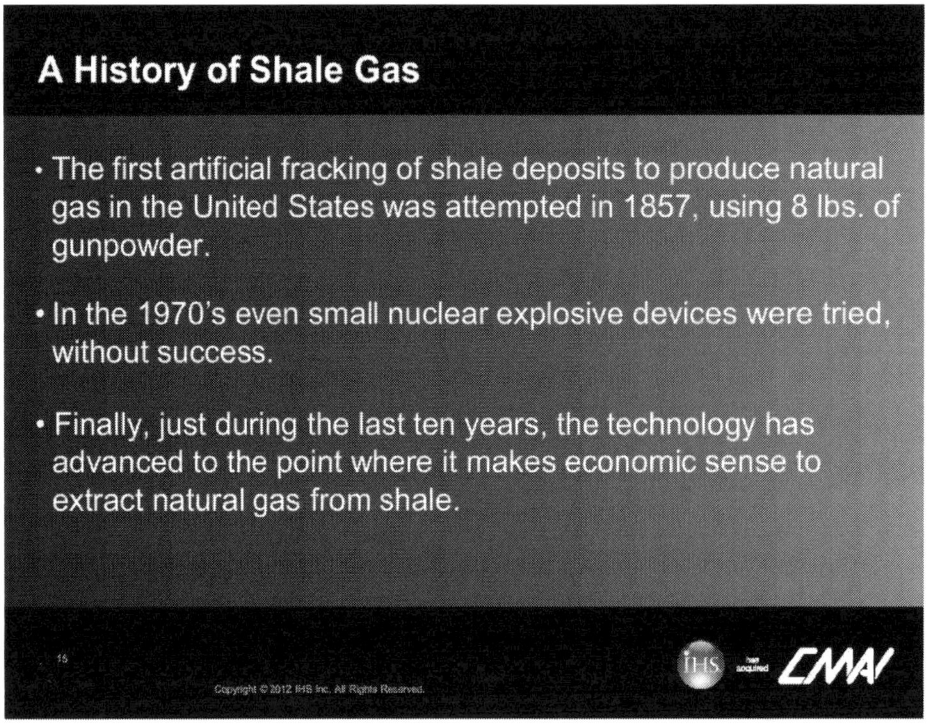

The first artificial fracking of shale deposits to produce natural gas was attempted in 1857, using 8 lbs. of gunpowder. In the 1970's even small nuclear explosive devices were tried, without success. Finally, just during the last ten years, the technology has advanced to the point where it makes economic sense to extract natural gas from shale.

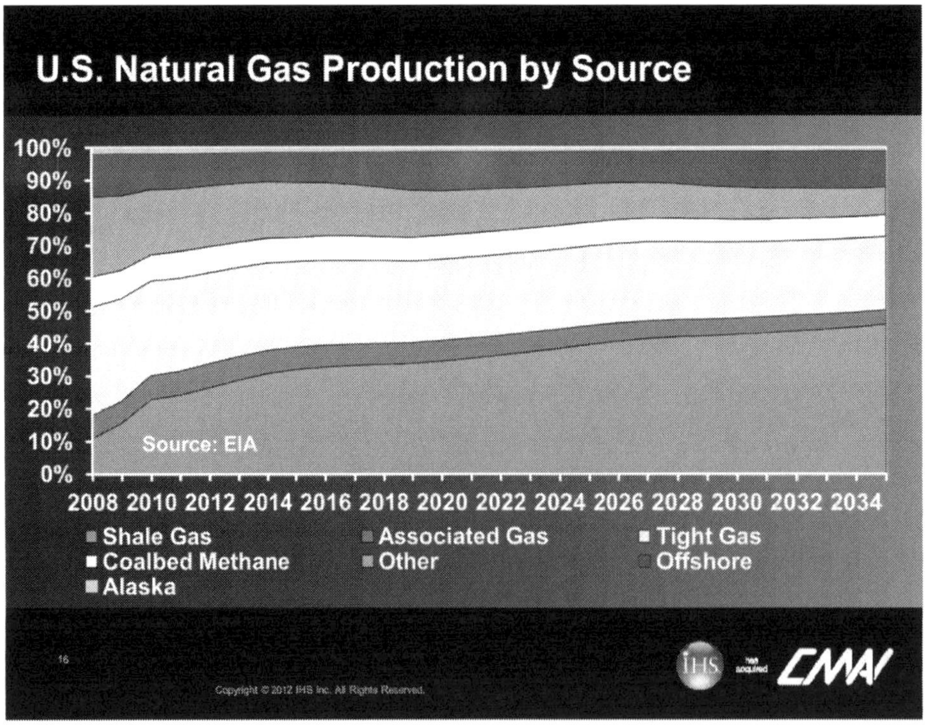

Shale gas has grown to about one quarter of total U.S. natural gas production currently, and is forecasted to continue to grow its share to over 40% by 2030.

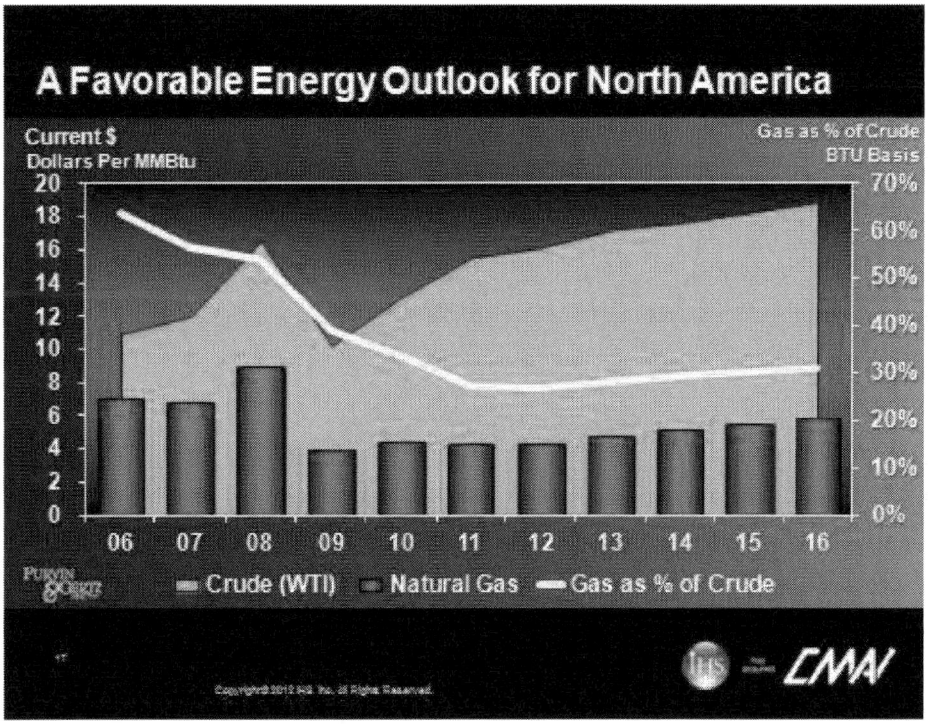

Extremely volatile and unpredictable energy markets have represented one of the greatest challenges to global olefins producers over the last several years. If you consider that crude oil prices were trading early in 2007 near $50 per barrel and peaked near $140 per barrel mid-2008 only to collapse to $40 per barrel by early 2009, and back to over $100 per barrel last year, you can understand why energy represents such a challenge. Natural gas markets have proven to be equally challenging to the North American producers. However in recent years, the lower value relative to crude oil has provided a competitive edge for the region once again. The value of natural gas in North America as a percentage of the BTU value of crude oil can provide a general measurement of the competitiveness of steam crackers in North America compared to naphtha crackers in other parts of the world. Since 2006, natural gas prices have been trading close to 50 percent of the value of crude oil on a BTU basis. In this type of energy environment, North America ethylene derivatives produced from ethane feedstock can be very competitive relative to derivatives produced from naphtha cracking.

Base energy dynamics favor U.S. Ethylene. The spread of non-conventional gas drilling in the United States has incentivized producers to go after wet-gas plays and treat natural gas as more of a by-product; this, along with an unseasonably warm winter, have combined to push natural gas prices below $3.00 for a sustained period of time. Should this continue, it could amplify the variable cost advantage garnered by the U.S. The more we develop our shale oil and gas reserves to generate fuel, the more attractive we become as a manufacturer of petrochemical products.

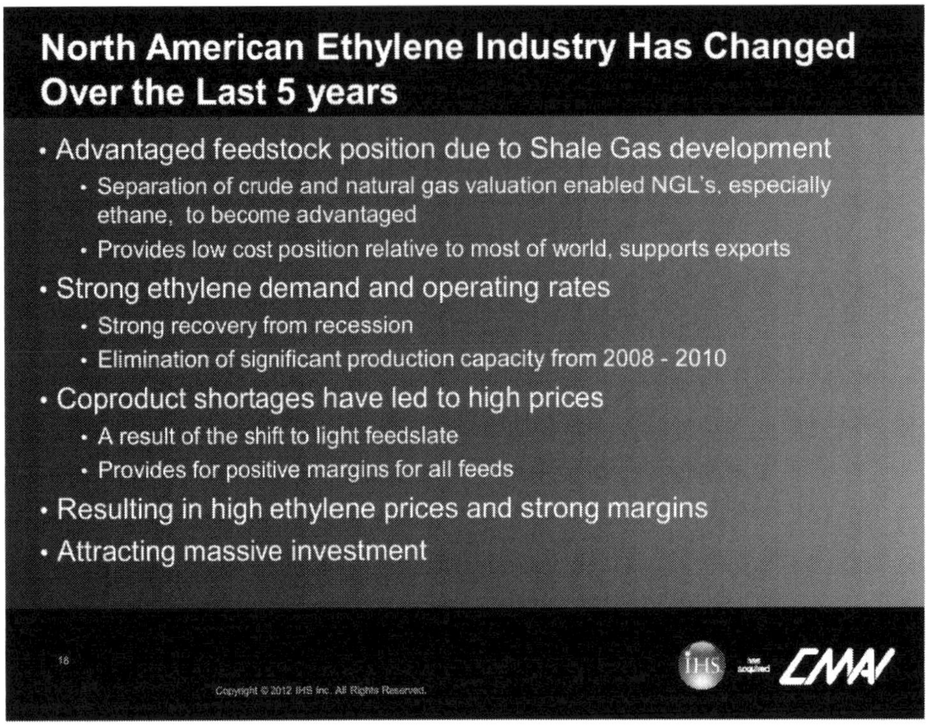

Five years ago the North American chemical industry was looking ugly! The global investment focus was on Asia & the Middle East and there were capacity rationalizations in North America. Canada was starting to see natural gas and ethylene plant feedstock production declines and Mexico had project Phoenix die and had not built a major chemical plant in 20 years. The United States had an oversupply of high cost chemical assets, with no feedstock advantage, as natural gas and crude oil were near BTU parity. Domestic demand was shrinking and the trend was to off-shoring of manufacturing.

But times have changed! North America now has an advantaged feedstock position due to Shale Gas development. The separation of crude and natural gas valuation enabled NGL's, especially ethane, will allow a low cost position relative to most of world which will support exports. Strong recovery from the recession will lead to strong ethylene demand and operating rates and the elimination of significant production capacity from 2008 – 2010. Co-product shortages have led to high prices as a result of the shift to light feedslate which provides more positive margins for all feeds. All this will result in high ethylene prices and strong margins which will attract massive investment.

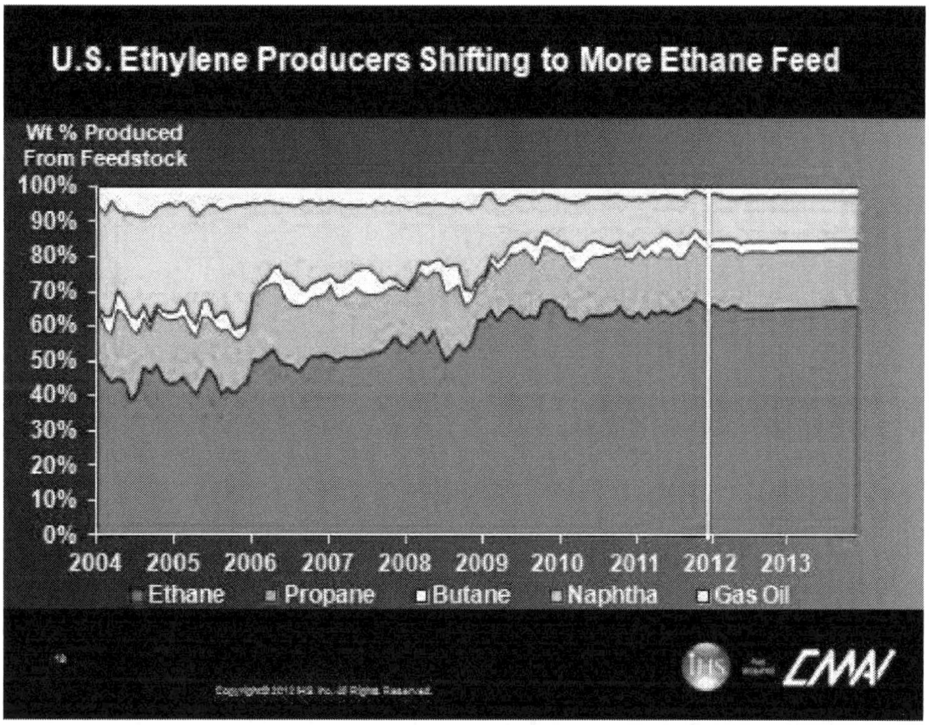

From a starting point that was already pretty light compared to Europe and Asia, the U.S. feed slate has gotten progressively lighter over time, so that now about 70 percent of U.S. ethylene is made from ethane and only about 10 percent is from naphtha.

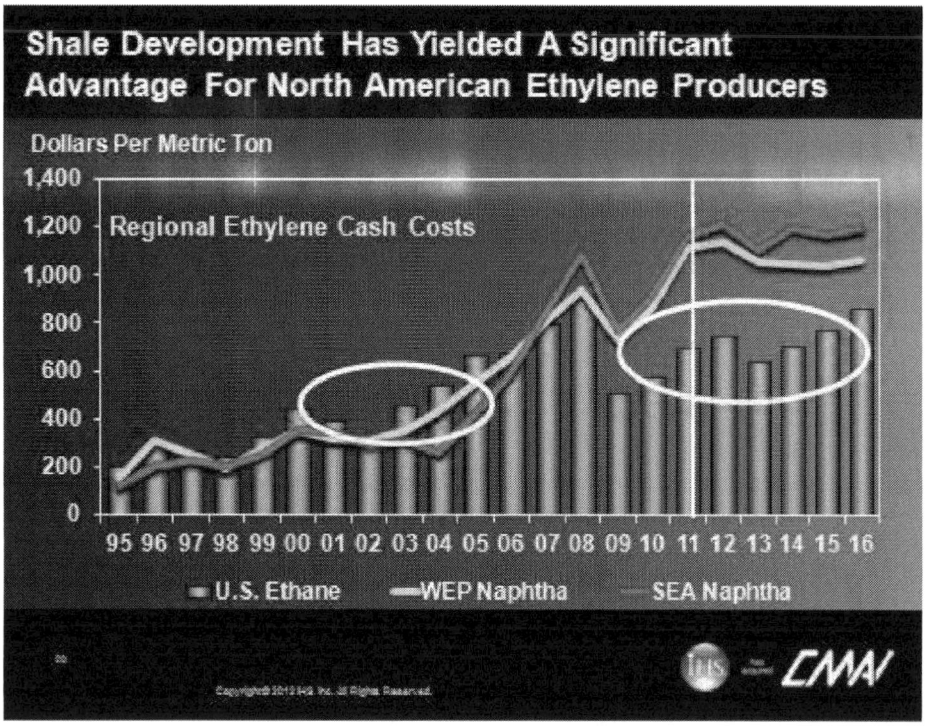

The increased ethane production from shale gas has yielded a significant advantage for North American ethylene producers versus those in Western Europe and Asia. Regional ethylene cash costs by feedstock are shown in the graphic from 1995 to 2016. Cash cost models include U.S. ethane, West Europe naphtha, and Southeast Asia naphtha. It is a combination of rising oil prices and lower natural gas prices (in North America) that create a sustainable advantage for ethane cracking in the U.S. compared to other regions (excluding the Middle East). This is the basis for a more optimistic outlook for ethylene derivative exports from the North America market versus Asia.

Rising supplies of NGLs in North America could prompt some increased use of LPG within the European feed slate as well, but the Middle Eastern feed slate will trend heavier for chemical diversification, so the advantaged US position appears to be sustainable.

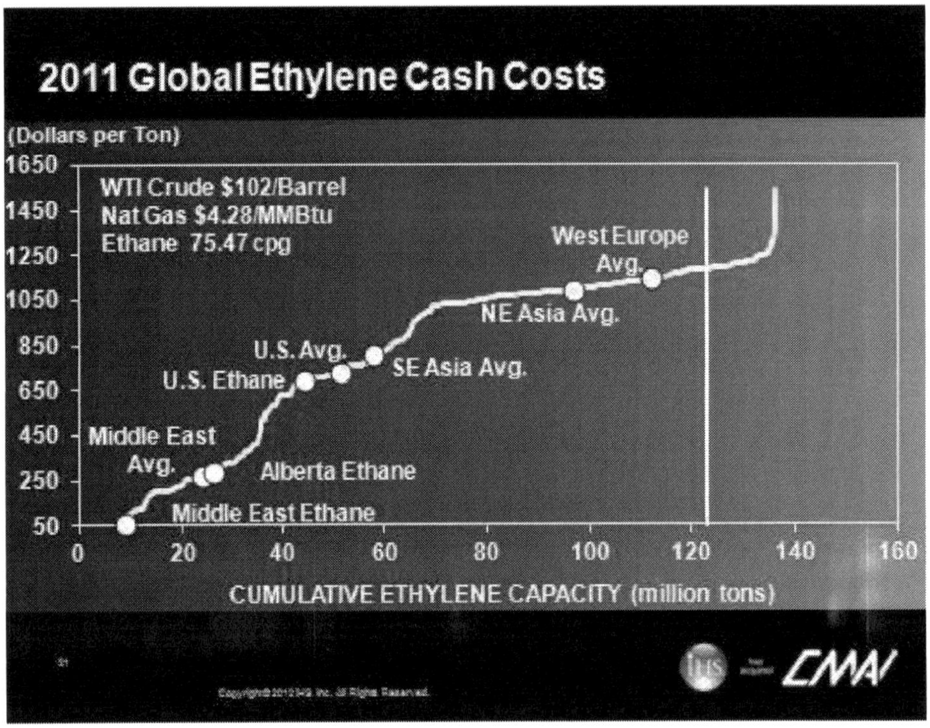

CMAI's cost curve analysis estimates the cash cost of production for every steam cracker site in the world and then aligns them in order of increasing cost on the y-axis and cumulative capacity on the x-axis. The general shape of the curve is broken down as follows: "advantaged feed areas" (far left); market based feeds (flat-middle); "isolated units" (far right). The regional points on the curve represent the weighted average for the region or country noted.

A review of the 2011 cost curve demonstrates the significant advantage that is created for the Middle East regional producers in a high energy environment. Compared to the flat portion of the curve, Middle East producers enjoyed an average ethylene cash cost advantage of more than $600 per ton. Other advantaged feedstock units that are present in the first quartile include units located in Western Canada, with Ethane crackers in the US and Southeast Asia located in the second quartile.

The global cash cost curve for ethylene used to be much flatter than it is now. When natural gas and crude oil were close in price on a BTU basis, the height of the curve was lower. As crude and natural gas prices separated, the height expanded. Over the last several years the cost of European and Asian naphtha-based ethylene production on the right end of the curve rose with crude oil and naphtha prices while North American ethane costs remained low, making the curve much more sloped than in the past. In fact, with the trend toward mixed feed crackers in the Middle East, Alberta ethane crackers are now competitive with the average Middle East cracker.

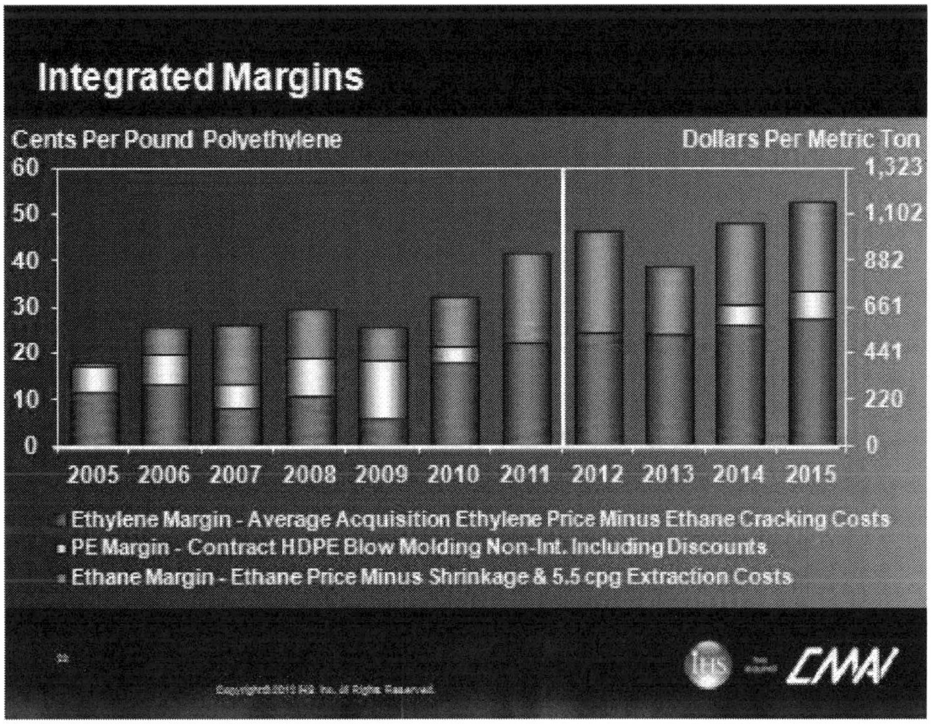

Reflecting the favorable cost position supported by the U.S. ethane advantage, the integrated chain margin outlook is very strong. With increasing supplies of ethane feed coming to market, the supply and demand balance appears to remain balanced, even with the additional ethylene capacity we are now forecasting. There is a great deal of profitability for all portions of the product chain, but primarily in ethane and ethylene. The economics of energy-derived raw materials represent 60% - 75% of total petrochemical production cost – and it drives capital and jobs decisions.

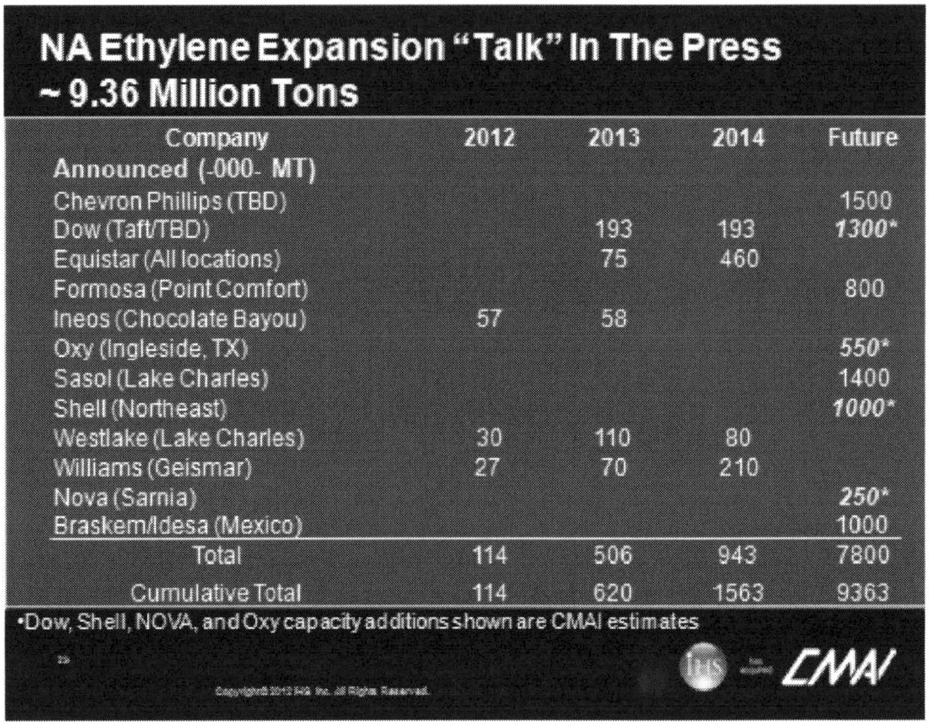

NA Ethylene Expansion "Talk" In The Press ~ 9.36 Million Tons

Company	2012	2013	2014	Future
Announced (-000- MT)				
Chevron Phillips (TBD)				1500
Dow (Taft/TBD)		193	193	*1300**
Equistar (All locations)		75	460	
Formosa (Point Comfort)				800
Ineos (Chocolate Bayou)	57	58		
Oxy (Ingleside, TX)				*550**
Sasol (Lake Charles)				1400
Shell (Northeast)				*1000**
Westlake (Lake Charles)	30	110	80	
Williams (Geismar)	27	70	210	
Nova (Sarnia)				*250**
Braskem/Idesa (Mexico)				1000
Total	114	506	943	7800
Cumulative Total	114	620	1563	9363

•Dow, Shell, NOVA, and Oxy capacity additions shown are CMAI estimates

This has resulted the announcement of a large number of new ethane-based capacity expansions in North America; in total over 9 million metric tons.

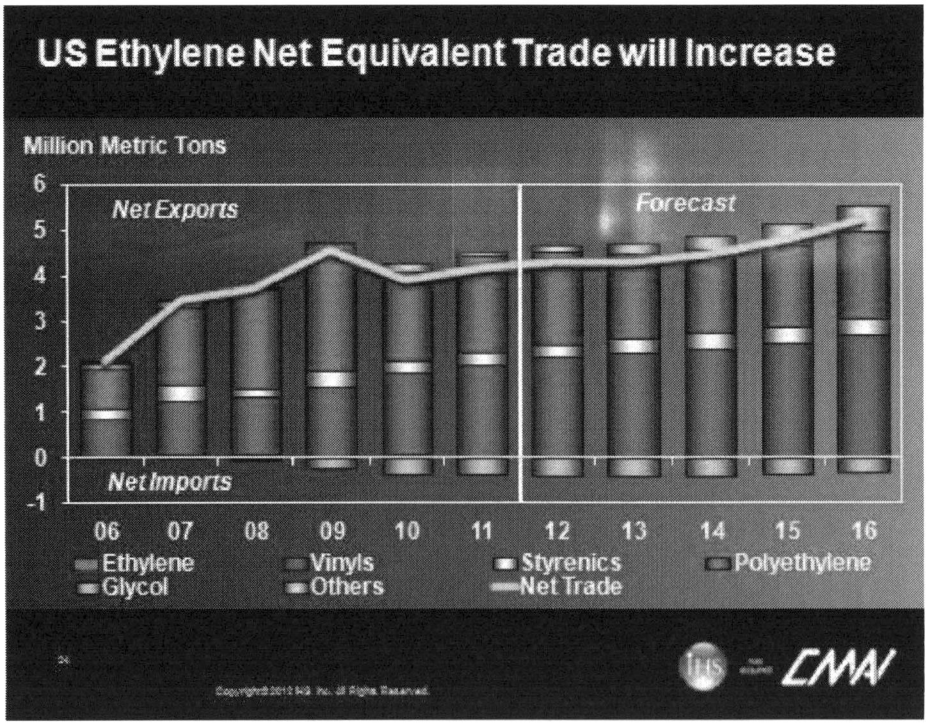

U.S. ethylene derivative exports are projected to increase from present levels, especially when the majority of the new capacity comes on stream after 2016.

Declining Co-products – Investing In Solutions!

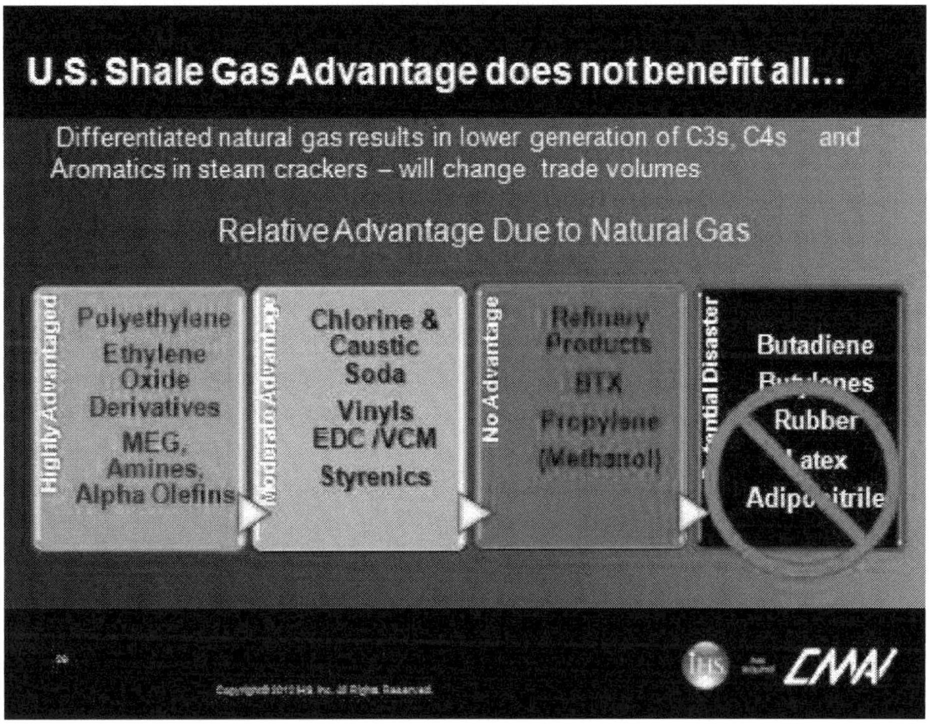

U.S. shale gas advantage does not benefit all products equally. Differentiated natural gas results in lower generation of C3s, C4s and Aromatics in steam crackers, which will change trade volumes.

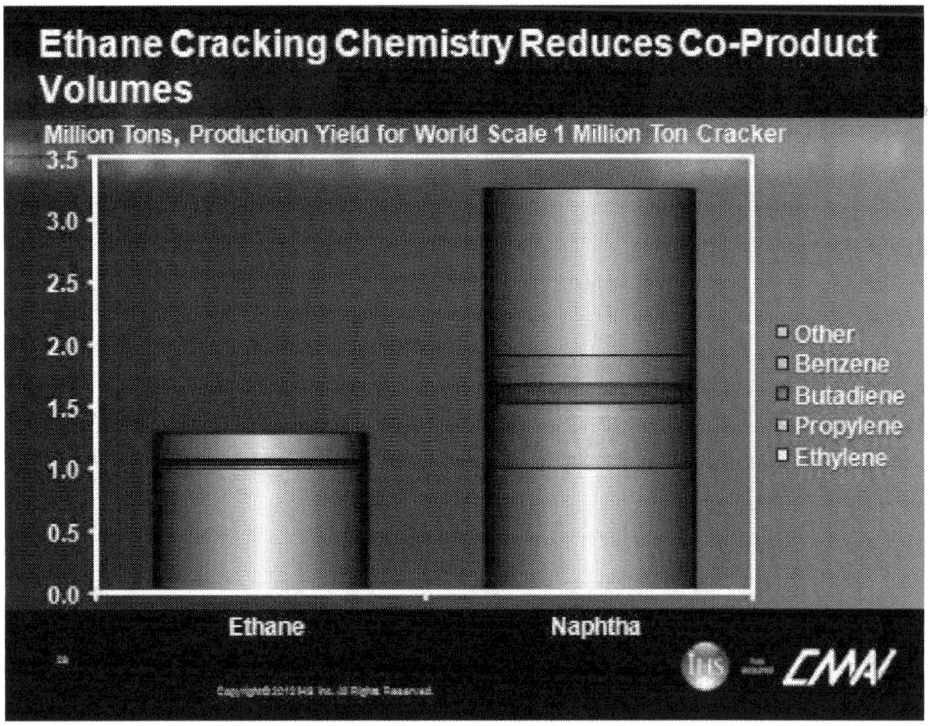

Ethane cracking chemistry reduces co-product volumes. Propylene and butadiene production is minimal in an ethane cracker, compared to a naphtha cracker.

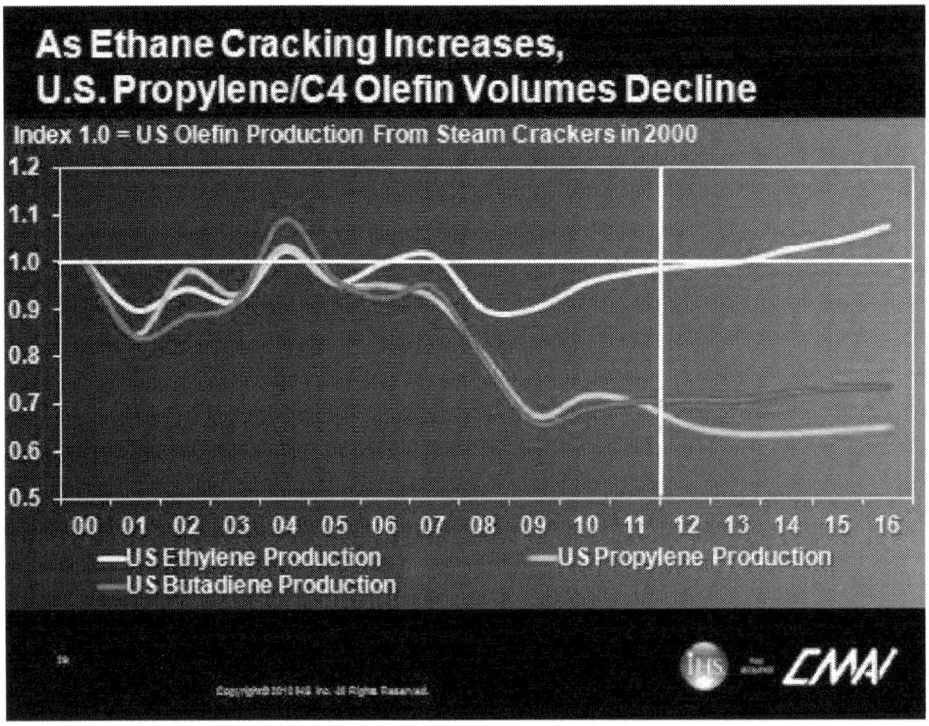

As ethane cracking increases, U.S. propylene/C4 olefin volumes decline fairly consistently.

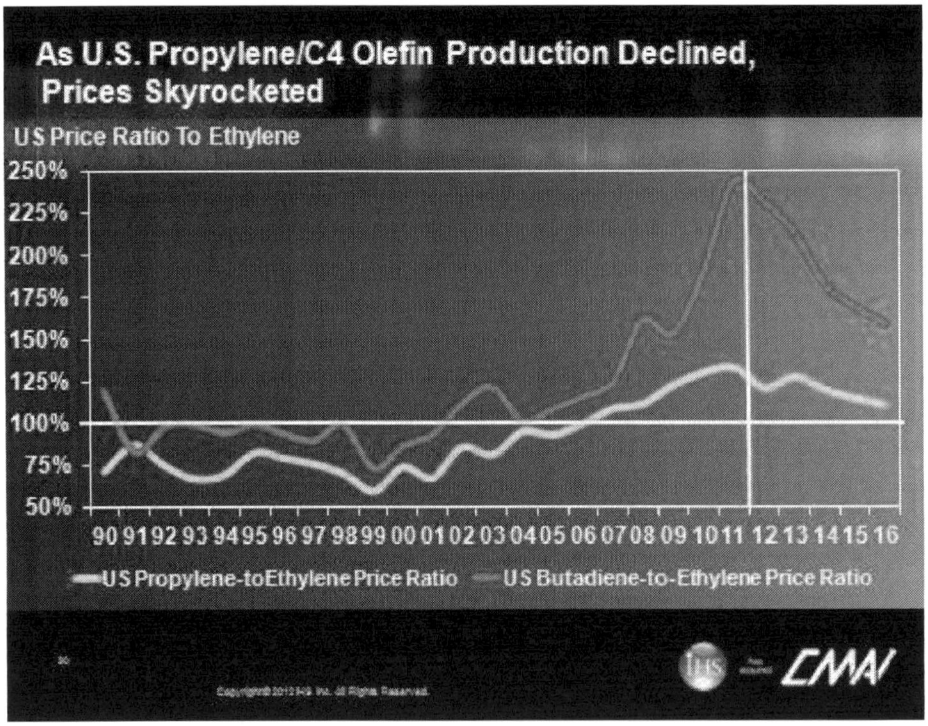

As U.S. propylene/C4 olefin production declined, prices skyrocketed. However, we believe on purpose production will help to alleviate this somewhat in the future, especially for propylene.

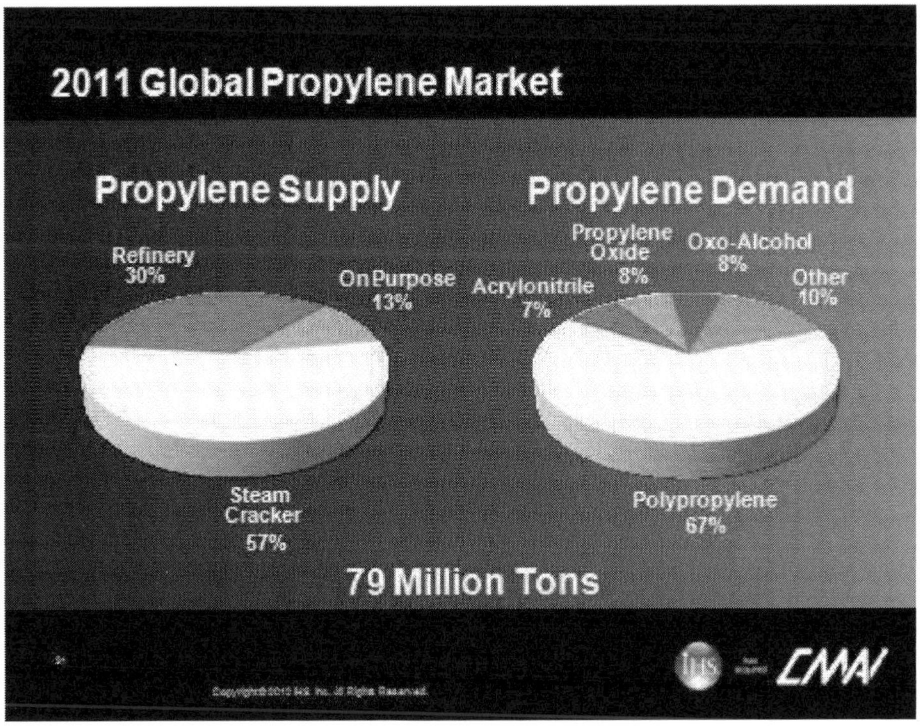

The high amount of heavy feedstock cracking in the world makes the steam crackers the primary source of propylene. CMAI estimates 57 percent of 2011 propylene production was sourced from steam crackers. About 30% comes from refineries. Steam crackers still produce the majority of propylene, which goes primarily into polypropylene.

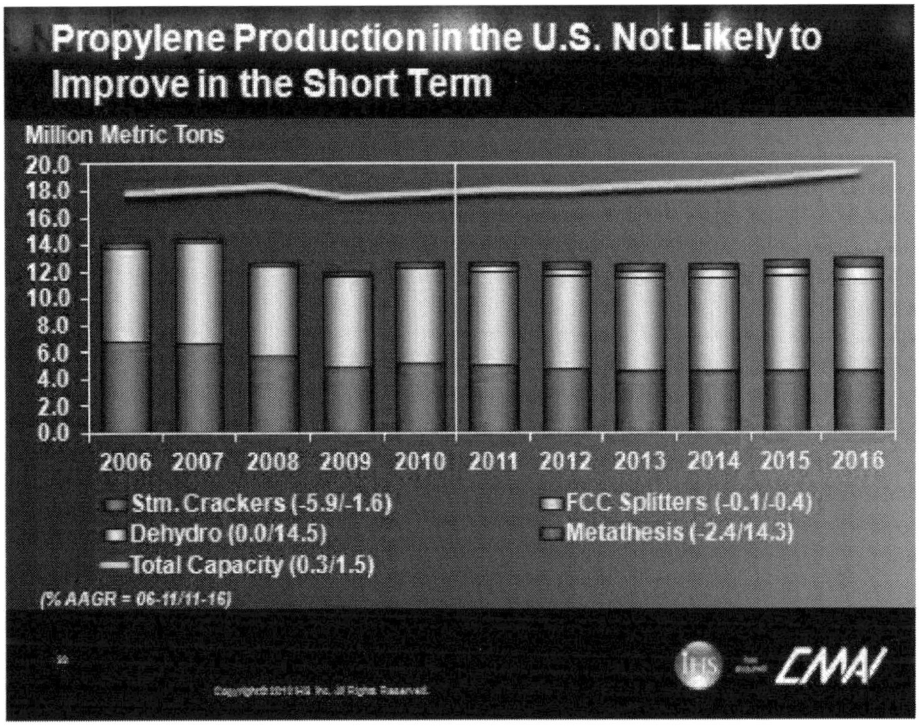

The lower amount of heavy feed cracking in the U.S. means that steam crackers cannot be the primary source of propylene. CMAI estimates about 40 percent of 2011 propylene production in the U.S. was sourced from steam crackers. Refineries now provide the majority, and on purpose units provide the balance of propylene supply in the region. Even with large increases in ethane based ethylene capacity, propylene production in the U.S. is not likely to improve.

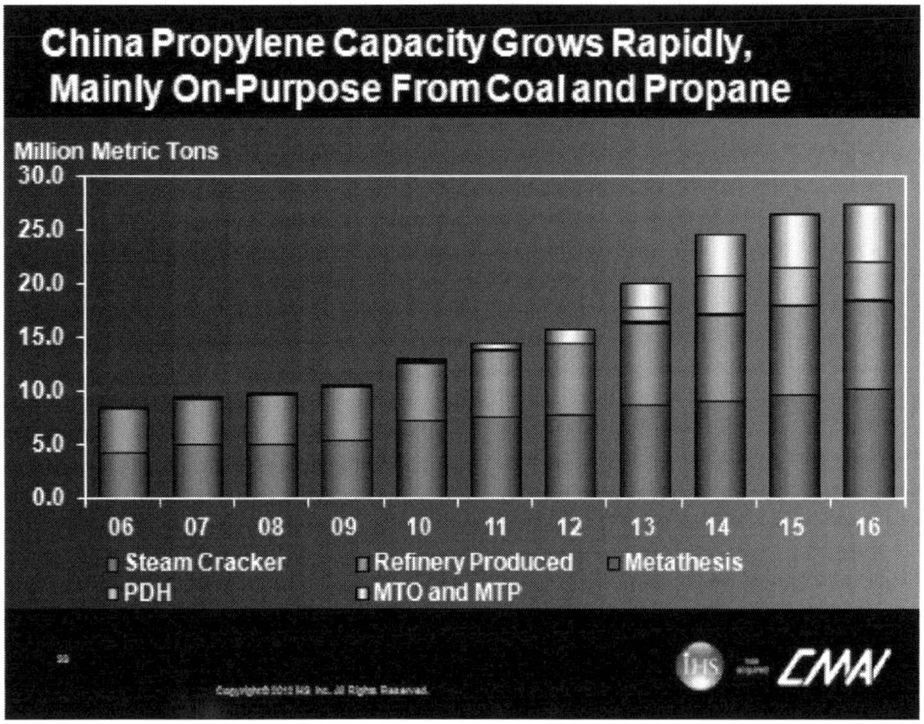

However, China propylene capacity is expected to grow rapidly, mainly through on-purpose production based on coal and propane feedstocks using methanol to olefins and propane dehydro processes.

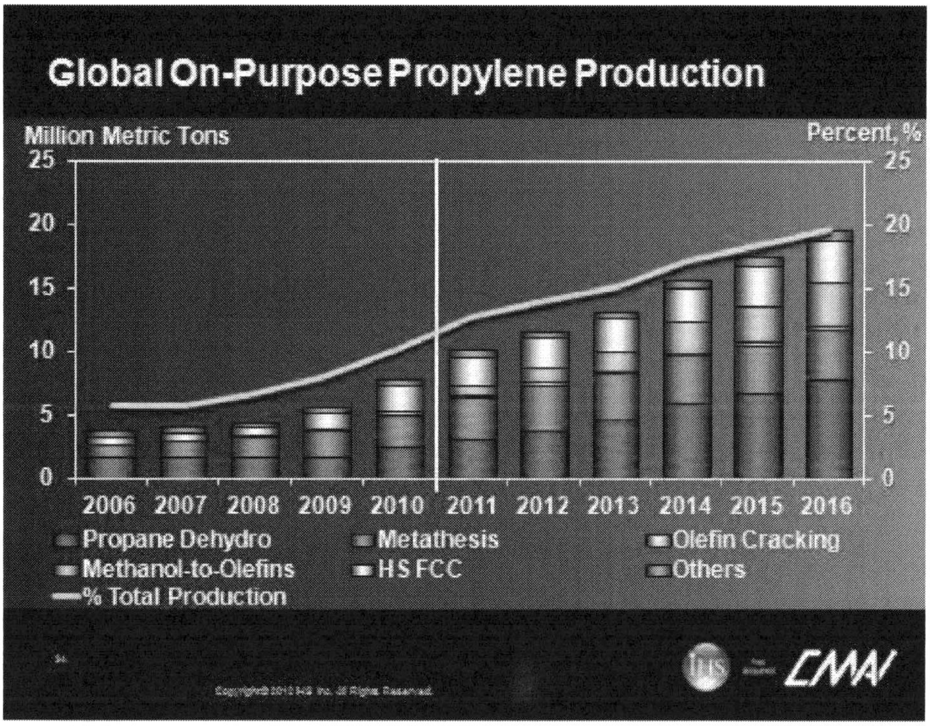

Global on-purpose propylene production is expected to increase from 13 percent of total production currently to 20 percent over the next 5 years.

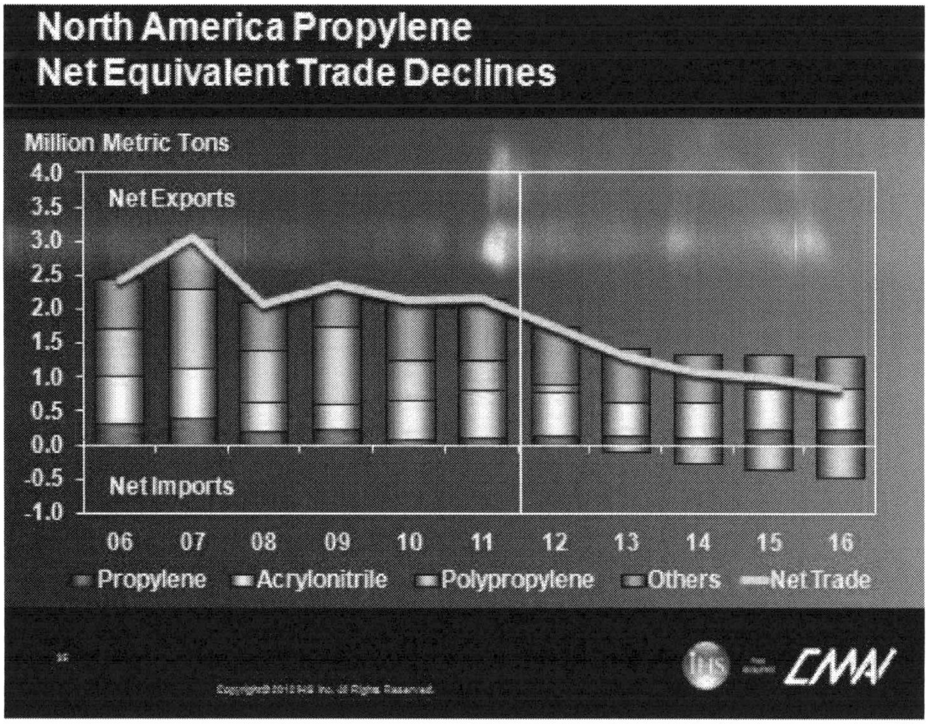

North American propylene net equivalent exports are expected to decline in the future, especially for polypropylene.

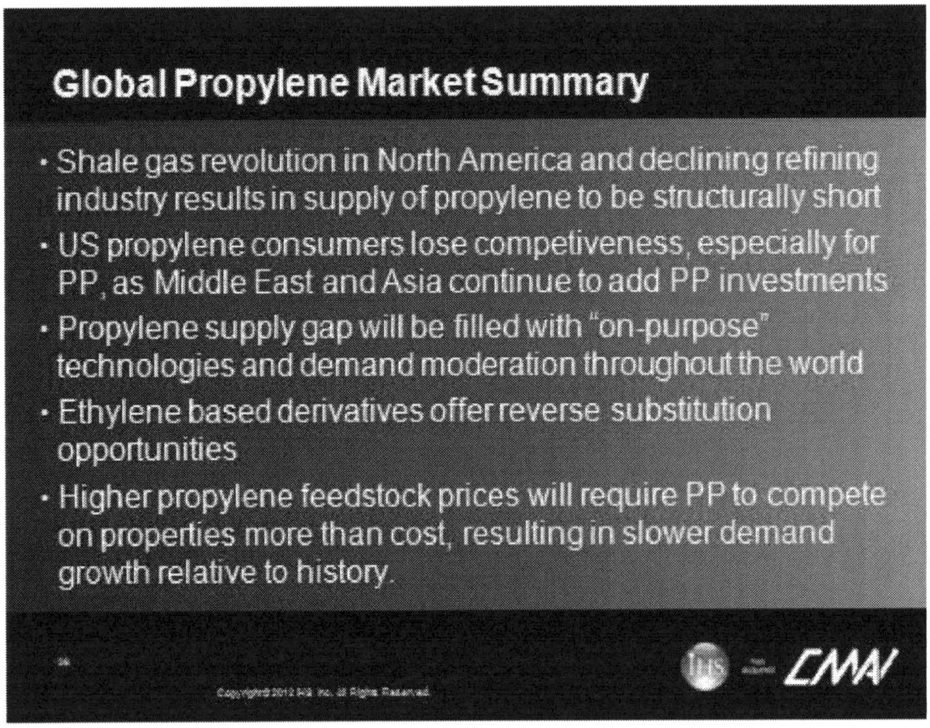

We expect further reduced U.S. steam cracker propylene supply as more ethane becomes available in 2012 and beyond. The domestic refining industry also faces challenges from additional ethanol blending, changing CAFÉ standards and alternative transportation fuels. Philadelphia is closing substantial capacity.

Imports of propylene, PP, and finished goods are looking to find their way into the North America market. These are needed to help rebalance the market, but entry barriers will have to be solved. Ethylene based derivatives offer reverse substitution opportunities.

A new propane dehydro (PDH) unit started production in 4Q 2010 and it is not likely to be the last on-purpose addition. There has been a recent Dow announcement for new units in 2015 and 2018 and LBI announced for another Metathesis unit.

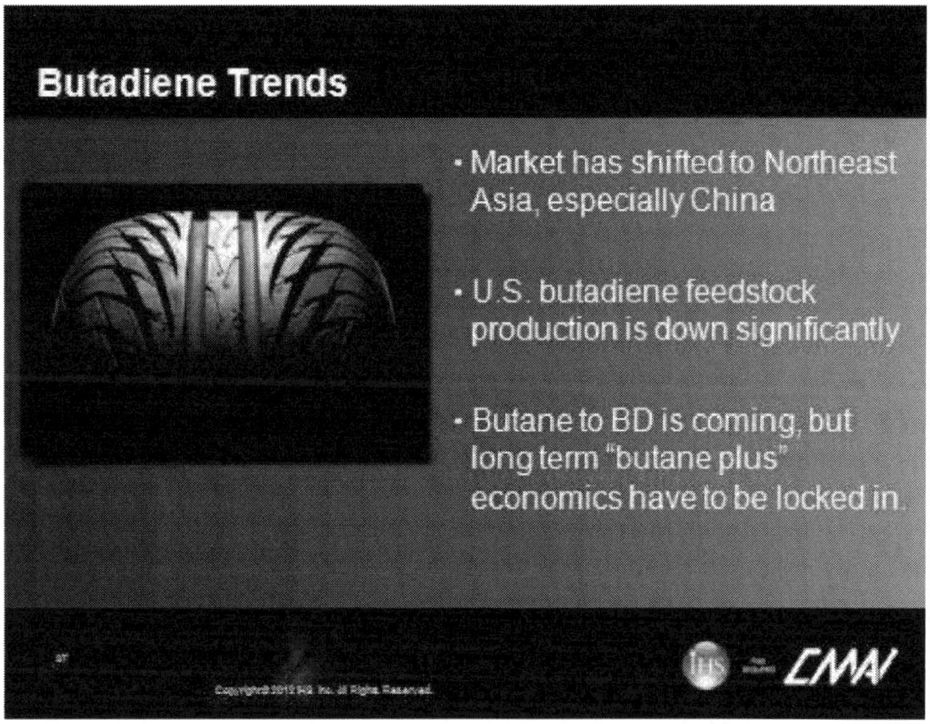

Butadiene faces similar problems as the market has shifted to Northeast Asia, especially China. U.S. butadiene feedstock production is down significantly. On-purpose butane to butadiene is coming, but long term the "butane plus" economics have to be locked in for projects to move forward. Asia currently dominates the global Butadiene market with most of the demand growth in China.

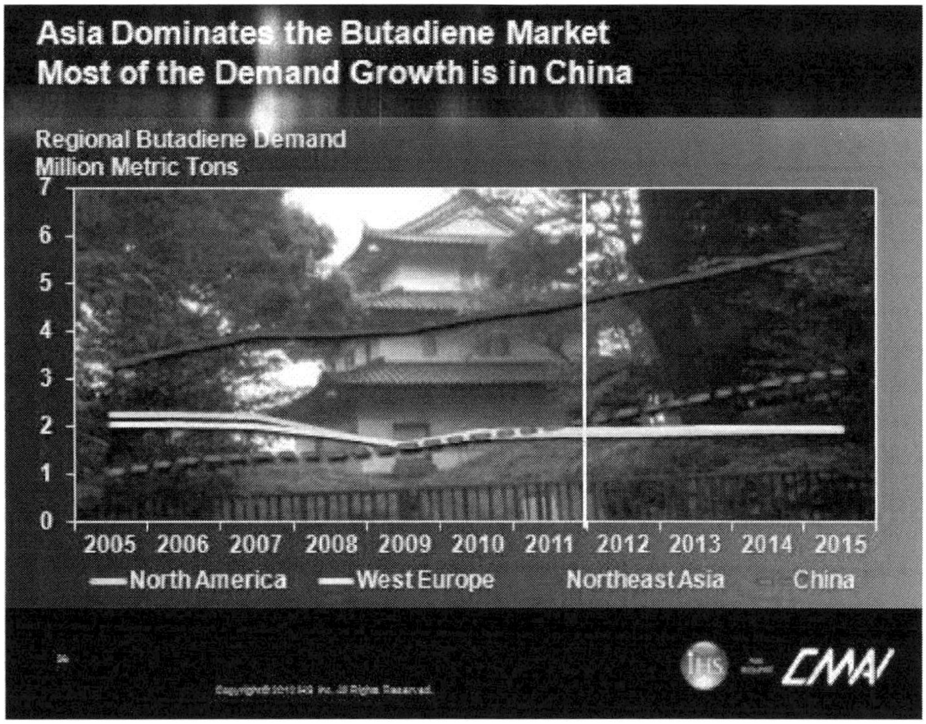

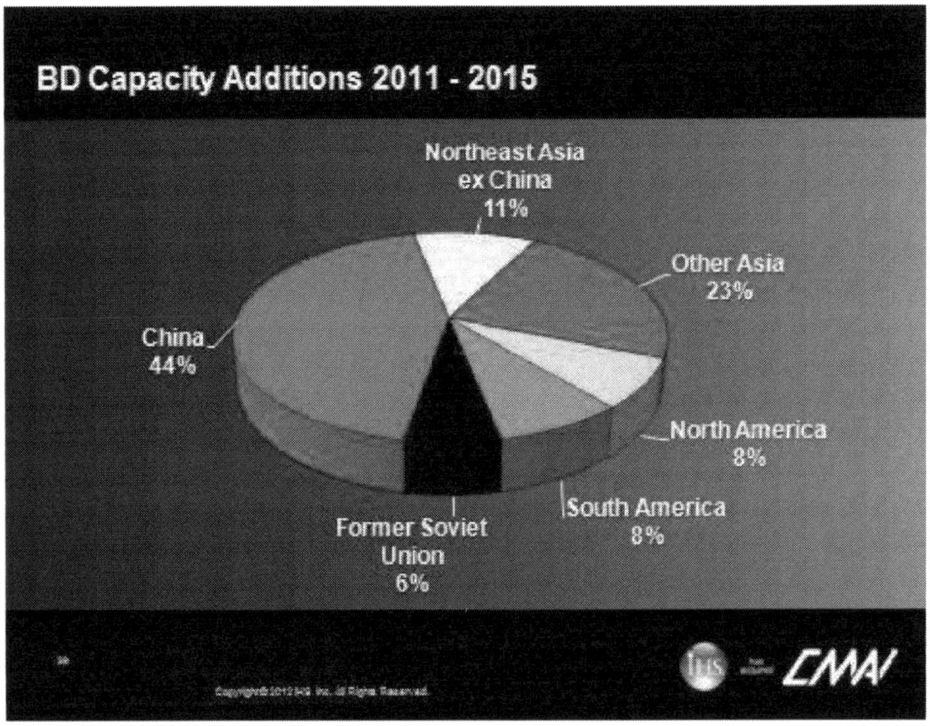

China also dominates future capacity growth.

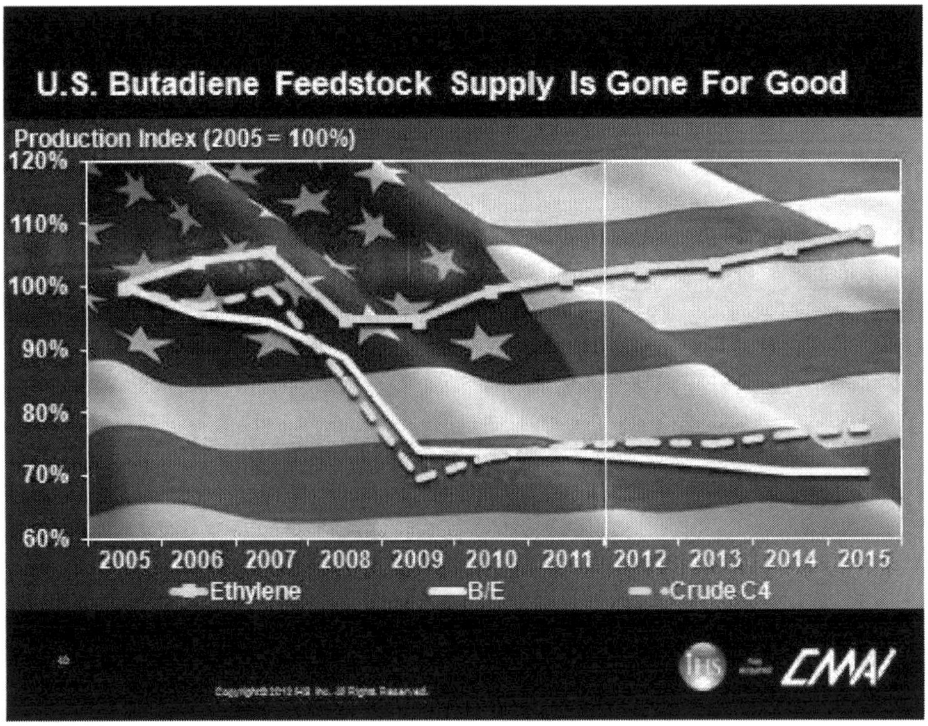

Due to the growth of ethane cracking, the domestic supply of butadiene feedstocks in the U.S. is probably gone for good.

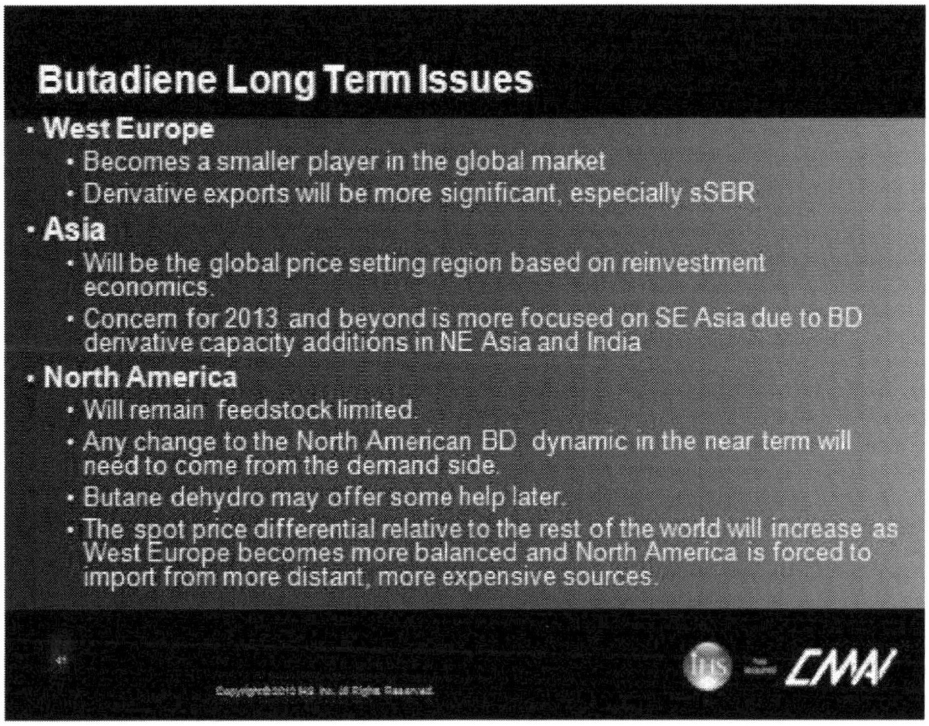

West Europe will become a smaller player in the global market as supply and demand of butadiene become balanced. Derivative exports will be more significant in the global market, especially solution styrene butadiene rubber.

Asia will be the global price setting region based on reinvestment economics. Market balance has improved with new 2010 capacity but concerns remain for next two years. Concern for 2013 and beyond is more focused on SE Asia due to BD derivative capacity additions in NE Asia and India.

North America will remain feedstock limited. Any change to the North American butadiene dynamic will likely have to come from the demand side. Butane dehydro may offer some help later. The spot price differential relative to the rest of the world will increase as West Europe becomes more balanced and North America is forced to import from more distant, more expensive on a delivered basis, sources.

Summary and Conclusions

Gas-based feedstock advantage over crude oil-based naphtha will be leveraged in North American ethylene and derivatives, as well as in on-purpose co-product production for propylene and possibly butadiene. Incremental ethylene and polyethylene produced in Asia from naphtha will set prices. U.S. ethylene producers will become "price-takers" instead of "price makers" in global markets.

North American ethane to polyethylene chain will enjoy stronger and less cyclical margins as long as gas and crude stay disconnected. Investment focus returns to North America for a rebirth and more expansion of export-oriented ethane-based ethylene and derivative capacity. Investment in on-purpose production of co-products takes advantage of rising propane and butane supplies (and lower prices), also part of the shale gas story.

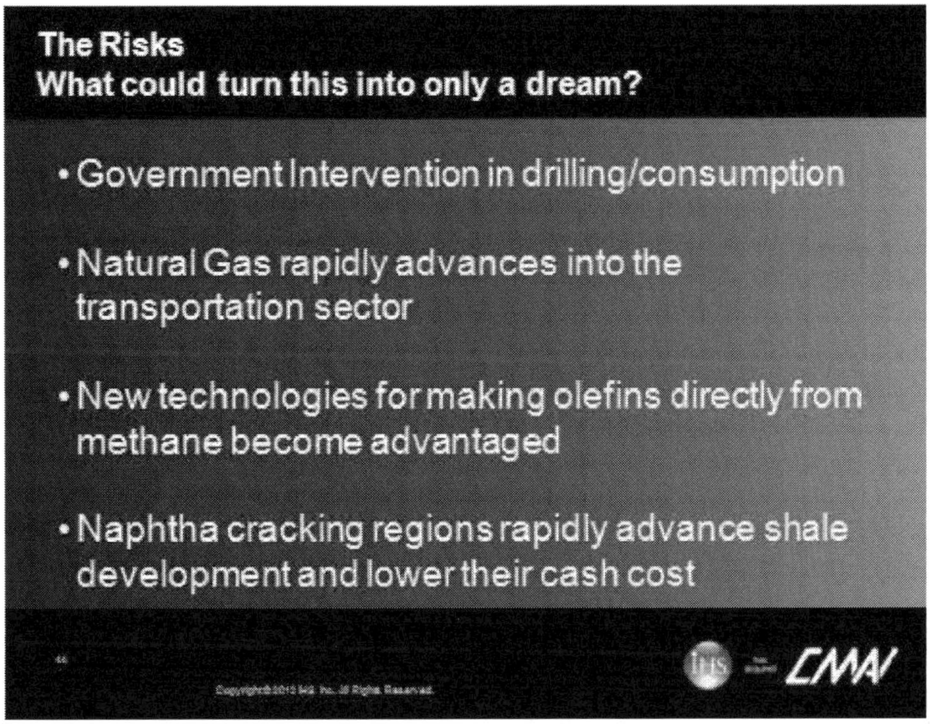

What are some of the downside risks to this forecast? Government intervention in drilling, such as fracking bans, could restrict the growth of shale gas production. On the other hand government incentives for consumption of natural gas versus coal and nuclear power could overwhelm our production capability. Natural Gas could rapidly advance into the transportation sector as a replacement for gasoline or diesel fuel. New technologies for making olefins directly from methane become advantaged. Naphtha cracking regions rapidly advance shale development and lower their cash cost.

In conclusion, global ethylene supplies will be sufficient to meet demand growth while global margins will be cyclical, but advantaged producers will enjoy extended strong margins. China is driving Asia market trends with strong capacity addition and demand growth. China coal based olefins production will be significant but still small relative to naphtha based capacity. The pace of Middle East ethylene capacity addition has eased and Middle East ethylene equivalent exports drive the market. U.S. ethylene producers were saved from a dire fate by low cost natural gas liquids, keeping ethylene chain margins strong throughout the forecast. U.S. ethylene derivative production growth will be driven by export opportunities.

PAPER 8b

Outlook for Ethylene Profitability

A Wall Street Perspective

Gregg Goodnight
UBS Equity Research
Houston, Texas

Prepared for Presentation at the 2012 Spring National Meeting
Houston, Texas, April 1-5, 2012
24th Annual Ethylene Producers Conference
Transformation of the Petrochemical Industry

Outlook for Ethylene Profitability

A Wall Street Perspective

Gregg Goodnight
UBS Equity Research
Houston, Texas

Abstract

Ethylene is a key building block for the global petrochemical industry and is used to produce about one-third of all chemical products. Therefore, its financial performance is often used as a proxy for the entire chemical industry. Ethylene and derivatives are largely commodities whose prices and margins are sensitive to global supply and demand. Thus, the global operating rate is a key performance indicator that requires special attention and forecasting to the best degree possible. Annual profit margins for ethylene can vary by over a factor of four from peak to trough industry conditions.

The UBS global equity research team analyses supply and demand for the global ethylene industry and publishes a report for its clients in the financial community. In this paper, we present our view of the global ethylene cycle and discuss the impact of ethane produced from shale gas formations as an ethylene cracker feedstock and how this impacts global competitiveness of the North American ethylene industry.

Introduction

We believe that average global ethylene operating rates in 2011 were near 85% of nominal capacity, well below the levels that have historically been associated with strong profitability. Indeed, margins from ethylene from the cracking of crude oil-based heavy feedstock were quite volatile and at times were low and even negative. However, North American producers with the ability to crack light feedstock such as ethane were at times enjoying margins commonly associated with peak global conditions. Plentiful ethane and other NGL's produced from shale gas deposits resulted in relatively cheap light feedstock for crackers and produced high margins and an economic bonanza for many North American producers. In 2011, the industry responded with new ethylene capacity announcements totaling over 15 billion pounds over the next six years. Will these additions result in a return to marginal economics for the industry?

Global Operating Rates and the Ethylene Cycle

History has shown that global supply and demand for ethylene has rarely enjoyed a prolonged period of balance; undersupply or oversupply is the norm. In figure 1, we show global ethylene operating rates from 1985 to our expected conditions in 2017. These operating rates are shown on a nominal capacity basis, unadjusted for cracker unavailability. Historically, global operating rates have ranged from a high of 93% to a low of near 65%.

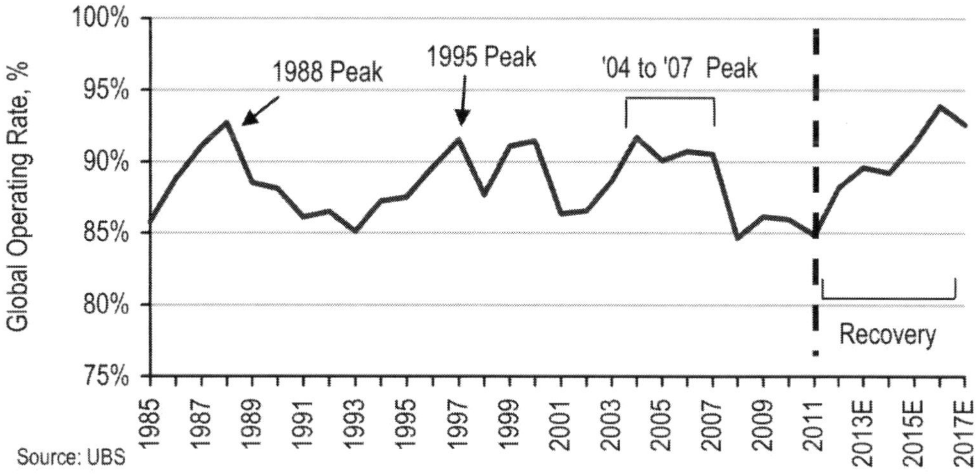

Figure 1: The Global Ethylene Cycle

The last cyclical peak for global ethylene operating rates was actually an unusual extended period of high operating rates from 2004 to 2007. During this period, industry profitability for ethylene was generally good in all geographies and for diverse sorts of feedstock. Top-of-cycle cash margins are achieved as global operating rates on an effective capacity basis surpassed 91%. Later in this paper, we will explain the relationship between effective operating rates and nominal operating rates.

The industry has also seen trough conditions, most notably in 2001 as a record amount of new global capacity additions of 7.1% was seen. More recently, global operating rates in 2008 were 84-85% of global capacity, due to moderate capacity additions and negative demand growth, and US margins dipped below $0.10 per pound and did not begin to recover until late 2009.

In the past, there appears to be a pattern of peak-to-peak and trough-to-trough durations of about seven years. We believe that there is a reason for the seven-year timing. Typically, it requires about three years to construct a new ethylene cracker after a project has been funded and the EPC (engineering, purchasing, and construction) contract is awarded. In the past, the industry had a tendency to approve large expenditures during peak financial conditions and delay expenditures during trough conditions. We believe that this tendency has again been seen as the number of new projects moving forward in 2008 to 2009 took a sharp decrease, setting the stage for a recovery of operating rates by 2013 to 2015. In retrospect, this industry behavior of project approval in good times and restraint in bad times has historically provided a mechanism, however inefficient, of matching supply and demand on a long-term basis.

Divergence of Light and Heavy Cracking Margins

Historically, the installed US ethylene capacity was typically based on a ratio of about 70% light feeds and 30% heavy cracking. Until recently, margins for light and heavy feedstock cracking have been very similar. In financial modeling, we used a 70/30 weighted average as a proxy for our estimate of US industry profitability. It was argued by some that heavy crackers had a slight edge in margins on a cash basis. However, beginning in early 2009, lower cost

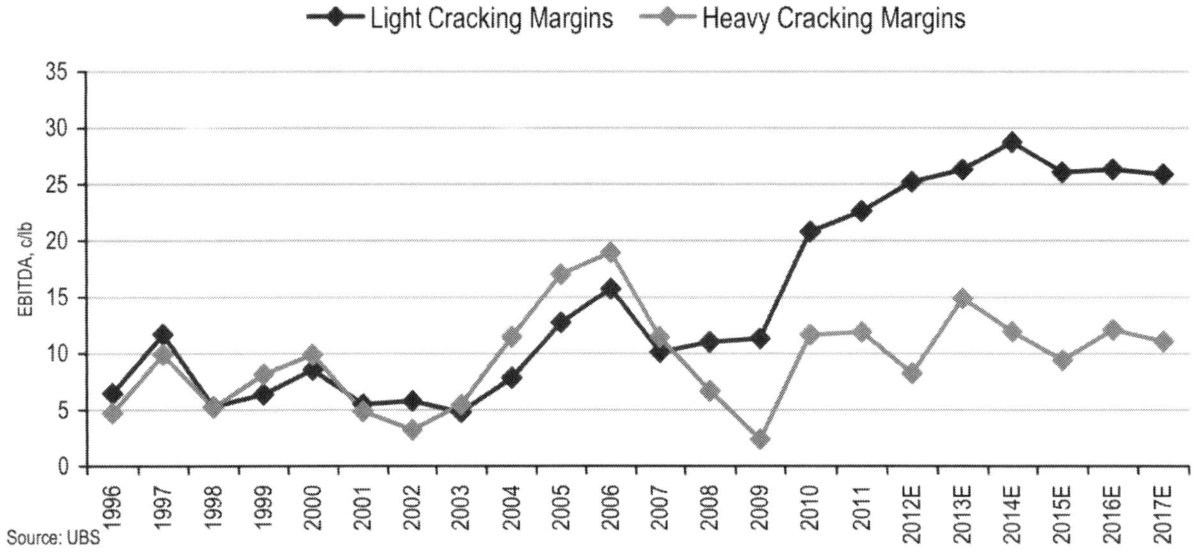

Figure 2: Cracker Margin Model for US

ethane associated with NGL's produced from shale gas began to give light crackers an edge, and producers responded by pushing feedstock slates to crack as light as possible. In addition, many US ethylene producers modified their crackers and brought in additional ethane supplies in "feedstock flexibility" projects. An output of our models is cash margins, or EBITDA per pound (earnings before interest, taxes, depreciation, and amortization) and EBIT per pound (operating income). We employ theses models to help forecast segment earnings for many of the commodity chemical companies that we follow, and in the past, the models have bridged to the reported financial results and thus have been validated by results from actual crackers.

US Margins and Global Operating Rates

Although ethylene itself has limited international sales as a commodity, most of the derivatives of ethylene are readily transportable and are typically imported and exported. The majority of ethylene is converted into the various grades of polyethylene, ethylene glycol, EDC, VCM, VAM and other derivatives that are commonly traded. While international arbitrage opportunities sometimes exist, international trade provides a leveling mechanism to establish values for ethylene on a global basis. We believe that in the long run, margins for ethylene can be correlated to the average global operating rates for the industry. During periods of high demand and operating rates, the market tension is increased and producers have more pricing power than in times of plentiful supply.

In Figure 3, we show historic ethylene cash margins plotted against nominal global operating rates. The margins we show are based on an assumed 70% light cracking and 30% heavy cracking split. The operating rates are shown using nominal global capacity, without any allowance for cracker downtime. Historically, most margins fell within an area representing a variation of plus or minus $0.03/lb. In years where were significant cracker outages, margins tended to be above the typical range as defined by historic precedent. Most recently, the advantage that cheap ethane from North American shale gas has resulted in higher average margins than those that would be anticipated from the prevailing global operating rates.

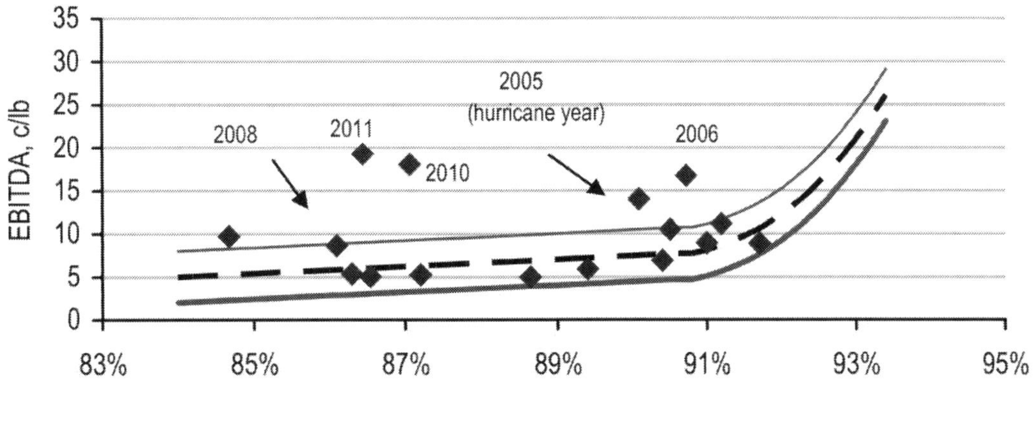

Source: UBS

Figure 3: Average US Cracker Margins vs. Nominal Global Operating Rate

Effective and Nominal Global Operating Rates

During times of unusually high downtime, availability of ethylene derivatives will tighten and the effective margins these products will command will be higher. In order to refine the global operating rate versus cash margin relationship for ethylene, we have studied the impact of cracker outages or unavailability on ethylene margin behavior. This unavailability can result from scheduled maintenance, or inadvertent downtime. In Figure 3, we show actual global cracker unavailability from 2007 to 2011 due to both planned maintenance outages and unplanned outages. The unavailability has ranged from 4.0% to 9.9% in the last five years. Some sources add this unavailability to the nominal operating rate to get an effective operating rate. However, we believe that this overestimates effective operating rate because there is a nominal amount of cracker maintenance downtime allotted for in the design of a typical cracker. We estimated that "normal" average cracker downtime or unavailability is 4.1% based on assumed average shutdown duration of 4 weeks every 3 years. We define the term "excess unavailability" as actual cracker unavailability minus the design unavailability.

	Cracker Unavailability		Global Operating Rate	
Year	Actual	Excess [1,2]	Nominal	Effective
2007A	4.0%	-0.1%	90.5%	90.4%
2008A	6.5%	2.4%	84.7%	87.1%
2009A	6.5%	2.4%	86.1%	88.5%
2010A	9.9%	5.8%	87.1%	92.9%
2011A	8.4%	4.3%	85.8%	90.1%
2012E	7.3%	3.2%	87.8%	91.0%
2013E	7.0%	2.9%	90.1%	93.0%

1) Excess unavailability is total downtime minus 4.1% standard downtime

2) Excludes market-related outages

Figure 4: Nominal and Effective Operating Rates

Source: Companies, CMAI, and UBS

In practice, we would expect that ethylene cash margin behavior would follow effective operating rate instead of nominal operating rate. In years of abnormally high cracker outages, the market should tighten. In Figure 4, we show recent cash margin behavior versus our calculation of effective operating rates. We note that in 2010 and 2011, global cracker outages have been relatively high and that this has tightened the market and improved cash margins. For 2012, our current expectation is for global maintenance outages to result in about 2.9% of cracker unavailability. Since it is still early in the year, the number of unplanned outages is still low.

Heavy Cracking Margins and Effective Operating Rates

Is indicated above, we traditionally have used a 70/30 weighted average of light to heavy cracking margin assumption as a proxy for US ethylene cash margins. Because light and heavy cracking margins were relatively close up until 2009 (Figure 2), this assumption was not critical to our analysis. However, as ethane prices fell due to plentiful supplies from shale gas formations, light margins soared and US light cracking margins exceeded historic norms. Recently, we have modified our operating rate versus margin analysis to include only heavy cracking margins, and these margins appear to still correlate well to effective global operating rates. We show this correlation in Figure 5.

It is important to understand that major ethylene producing regions such as Western Europe and most of Asia utilize heavy feeds for cracking. We have compared heavy cracking margins in different regions and have found reasonable agreement using our standard cracking economic models. Thus, when low global operating rates persist, we anticipate that all regions will demonstrate low margins.

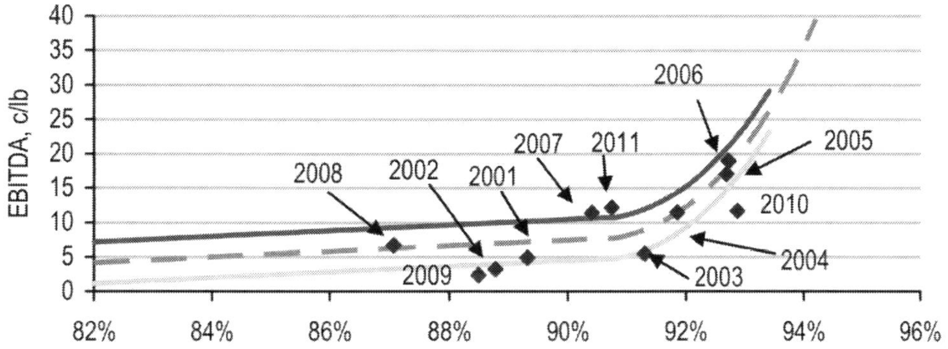

Source: UBS

Figure 5: Ethylene Margins and Effective Operating Rates

The North American Shale Gas Phenomenon

Beginning late in the 1990's, Mitchell Energy pioneered a new approach to the production of natural gas and natural gas liquids from shale formations. Key component of this technology included horizontal drilling techniques and special formation fracturing technology. Thus, natural gas wells of high productivity were developed from gas deposits that had previously been considered inaccessible. In the US, the major shale gas areas include the Barnett and Eagle Ford formations in Texas, the Granite Wash formation in Texas/Oklahoma, the Haynesville/Bossier formation in Louisiana, and the Utica/New Albany and Marcellus formations in Pennsylvania in New York. In addition to high natural gas productivity per well, several formations produced exceptionally rich co-product streams of natural gas liquids more typically found with crude oil production rather than dry gas wells. Until 2008, US ethane production was in the range of 750,000 to 800,000 bbl per day and the cracking of natural gas liquids providing about 70% of feeds for US ethylene crackers. Recently, the US ethylene industry increased the percentage of ethylene produced from NGL's to near 85%. In 2011, US ethane supply was aided by the ramp-up in NGL transportation and fractionation capacity so that by November, the industry produced about 930,000 bbl per day of ethane. However, at the same time US crackers increased their ability to consume ethane to near 1,000,000 bbl per day of ethane through feedstock flexibility and debottlenecking projects. Currently, the supply and demand for NGL's in the US has been relatively balanced, and ethane pricing has been volatile and dependant on cracker outages, new fractionation and pipeline capacity additions and the volatile energy market. Going forward, it is our view that ethane prices will soften to reflect plentiful supply by 2013 and possibly earlier.

Ethane Pricing on an Energy Equivalent Basis

If we convert pricing of crude oil and ethane to a dollars-per-million BTU basis, we can plot the cost of ethane relative to natural gas as we show in Figure 6. In the US, ethane is sold on a spot basis at prices that are below crude oil, but above natural gas values. From years 2005 to 2007, ethane prices were highly correlated to crude oil with a correlation coefficient of 0.86. During that period, ethane prices were poorly correlated to natural gas prices with a correlation coefficient of only 0.2. However, by 2011, ethane began to track natural gas plus a transportation and fractionation premium, and ethane's price correlation to natural gas exceeded the correlation to oil (0.68 to 0.450).

We believe that the incremental availability of ethane in 2013-14 will push ethane prices towards natural gas pricing plus a premium for transport and fractionation of $0.10 to 0.15 per gallon. The current (2/14/12) UBS forecast for natural gas prices for the US for 2013-14 is $4.00 and $4.75 per million BTU. This would imply that if ethane prices push towards their natural gas limit, they will be near $0.45 to $0.50/gal in 2013-14.

Figure 6: Ethane, Gas, and Oil Prices on Energy-equivalent Basis

Global Ethylene Supply Growth, 2013-2014

Each year, we conduct a survey of the global ethylene industry to understand the likely projected supply and demand balance. This is done through a both a literature search and discussions with engineering and operating companies. Our capacity survey suggests that the capacity growth in 2012 and 2013 will average only 2.4%, well below the average capacity additions of 6.3% per year in 2010 and 2011. We believe that demand growth for ethylene will exceed this level of capacity additions and a period of rising operating rates is likely. For the period of 2012 through 2017, we project annual additions to average 4% of global capacity. For a detailed breakdown of our capacity additions summary, please refer to our published reports.

Of significant note is the regional pattern of ethylene capacity additions. Until recently, it was generally thought that no new North American grassroots crackers would be built in the foreseeable future, with flat domestic demand growth and high priced and limited feedstock. Since the beginning of 2011, cumulative new capacity addition announcements of over 15 billion pounds per year have been made. These consist of a wide variety of incremental expansions, greenfield projects, brownfield projects and flex-feed projects. In Figure 7, we show announced cumulative capacity by each geographic region. Asia continues to exhibit near-term capacity growth. However, capacity additions in the Middle East have slowed with higher priced gas and reduced availability of advantaged feedstock allocations.

In our supply forecast, we are currently including 4.1 million metric tons of ethylene capacity that is coal based installed in China by 2017. We believe that this MTO (methanol-to-olefins) technology has certain advantages such as limiting naphtha imports, the operating cost and capital costs have no clear advantage over shale gas-based NGL's. In addition, there are numerous shale gas deposits globally, most notably in Argentina, Poland, Western China and Australia that may prove more attractive for ethylene production. Except in the US, we do not include any new projects based on NGL's obtained from shale gas formations.

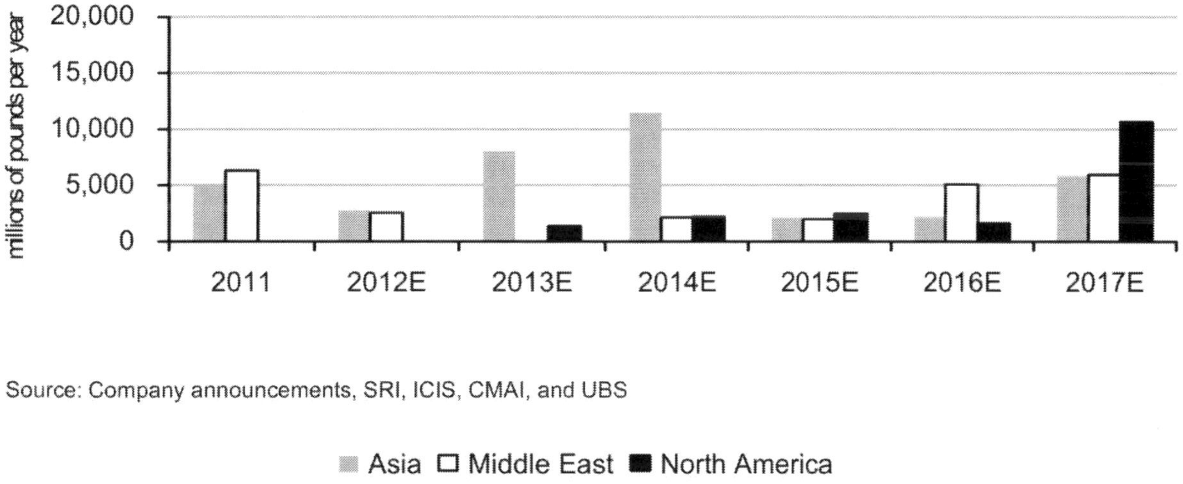

Figure 7: Geographic Distribution of Ethylene Capacity Additions

Global Trends in Ethylene

Our 2012 survey of global ethylene capacity has suggested the following trends for ethylene:

Iran – After the completion of four grass-roots crackers with a total capacity of 4 million MT of capacity late in the last decade (NPC#6 Amir Kabir and NPC#7 Marun, NPC#9 Arya Sasol and NPC#10 – Jam Petrochemical), Iran's new cracker additions have noticeably slowed. The full impact of this new capacity addition has yet to be felt with crackers seeing low effective operating rates and a host of issues related to derivative plants and exports.

China – After the addition of seven new crackers in 2009-10, 2012 will see incremental capacity additions. While a further six crackers are in construction or in late planning stages, China faces feedstock supply issues. There are perhaps 15 other crackers in planning the planning stage, and some will undoubtedly be built. Various Chinese companies including major coal companies are building coal-to-olefins plants. Our survey includes 10 plants being constructed by the end of the decade, and a further 10 to 15 plants are in the planning stage. However, we believe that additional units are less probable due to relatively high capital and operational cost.

Western Europe – There are no new crackers in our survey for Western Europe, although perhaps one or two projects could advance by 2016 to 2017. We have included in our survey the shutdown of three small crackers totaling 615,000 MTPA ethylene capacity. We feel that Europe will be increasingly a target market for imports from the Middle East and will endure relatively low utilization rates, on average.

Global Demand Growth, 2012-2017

Based on our GDP driven regional demand model, we forecast that global ethylene demand growth will average 4.7% per year in 2012 and 2013. This is significantly higher than our near-term capacity growth projection of 2.4% and suggests a tightening of the market for ethylene derivatives is likely in the near term. For the full period of 2012-2017, our demand growth assumption is 5.1%. Our assumption of global ethylene demand elasticity is 1.2x, in line with historical elasticity trend. On any given year, inventory adjustments in the supply chain can bias demand growth from one year to the next. The most recent example of this is in 2010, as global ethylene demand increased an estimated 9.1% over the trough year 2009.

Global Operating Rate Forecast, 2012-2017

As we show in Figure 1, our outlook for global ethylene operating rates is an improving trend in the next two years followed by an expended period of relatively high operating rates near about 90%. Traditionally, this operating rate has been the threshold for a greater expansion of margins, which we now believe applies generally to heavy cracking margins in all geographic regions. It is interesting to compare this expected period with the recent 2004 to 2007 period. In the past the industry has had peaking operating conditions, but these typically lasted for not much more than 6 to 12 months such as those in 1988 and 1995.

Margin Forecasts – 2012 and 2013 and Beyond

Our estimate of the average USGC ethylene EBITDA margin is $0.18 per pound in 2012, high by historic standards. This is based on a proxy 70/30 weighting of light and heavy cracking. Light crackers (ethane-based) margins are expected to average $0.24/lb for the year and heavy cracking margins are projected to average $0.06/lb according to our models. Heavy cracking margins will be a strong function of co-product pricing that has recently been quite volatile. However, our view is that demand for butadiene and other co-products will be firm enough to support margin contribution to ethylene. For individual crackers, margins may diverge from these results that are based on models representing typical industry conditions, cracker capacities and fixed cost burdens. For 2013, our forecast is for average margins to improve by $0.03/lb as increased availability of ethane from shale gas trims the average expected price of ethane feedstock.

Beyond 2013, we look for a tightening supply and demand balance globally. This will support margins for crackers using all feedstock. However, one thing that following this industry has taught us is that predicting future trends is a hazardous occupation!

Conclusion

We believe that the global ethylene industry is entering a period of improved global operating rates, and that this period could be sustained. North America is set to take advantage of low-cost ethane from shale gas. Newly announced North American based capacity will be built and will primarily rely on exports for fill up its plants. Capacity additions in the Middle East will still be significant, but will slow. Heavy crackers based in Asia and Western Europe will come under margin pressure and smaller units will likely be shut down, but existing larger crackers will be able to take advantage of higher co-product pricing in order to maintain positive margins.

References

1. Cash, A.; Goodnight G., "US Chemical Initiation". UBS, January 24, 2011

(8d) Panel Discussions

Sanjeev Kapur, Fluor Corporation, Sugar Land, TX and Gregory E. Yeo, ExxonMobil Chemical Company, Baytown, TX

Panel discussion will mostly focus on current business developments and drivers (e.g. Shale Gas related developments in US, Global dynamics and their impact on future of the business). This will also cover the impact of environmental issues and technology trends. The panelists include business leaders from major international companies, covering a wide spectrum of industry from different geographical regions.

Control and Optimization of Refrigeration Compressors Using Process Model Based Control

245888

Robert Moore & Maren Greseth, RMoore Controls, LLC

Introduction

Control of refrigeration compressors is important during each phase of operation. At steady state, optimal control of the refrigeration compressors can help maximize production rates and minimize horsepower. During startup and shutdown, good controls can assist the operator by managing the refrigeration compressor during transient conditions. This paper described a new design for controlling refrigeration compressors. Our Process Model Based Control design is based on the compressor specifications and gas physical properties to maintain a stable accurate distance from surge on each stage while optimize the suction pressure with changes in refrigeration load.

Compressor Surge

The flow through a centrifugal compressor stage will become unstable if the volumetric flow into the stage is insufficient. This condition, known as "surge", is disruptive to the process and can be damaging to the compressor. Surge can be avoided by maintaining an inlet volumetric flow rate (Q_S) in excess of compressor surge limit (Q_{surge}) by manipulating the False Load Valve. For each stage there is a Polytropic Head (H_P) vs. Inlet Volumetric Flow performance curve that indicates where the stage is expected to surge.

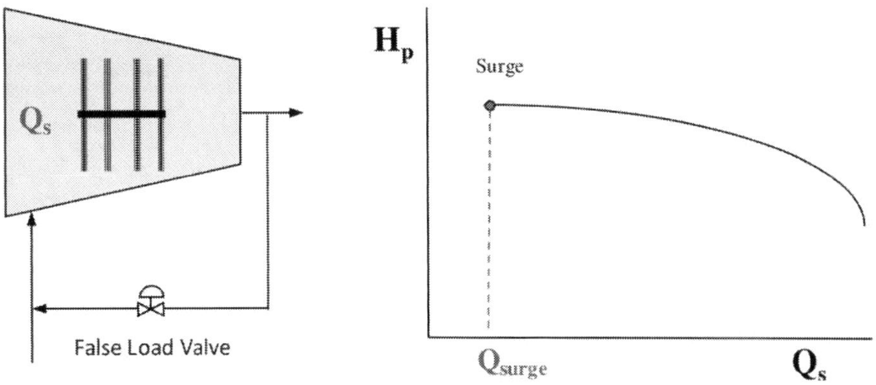

While avoiding surge is critical, it is important to avoid excess flow which will result in wasted power and reduced refrigeration capacity.

Surge Control – Current Practice

Single Stage

The objective of surge control is to maintain a desired "distance from surge". The standard approach is a design that is independent of gas composition. This is accomplished by making simplifying assumptions about the gas properties and in some cases transforming the fundamental performance relationships. With the standard design, "Distance from surge" is defined in terms of inlet and outlet conditions.

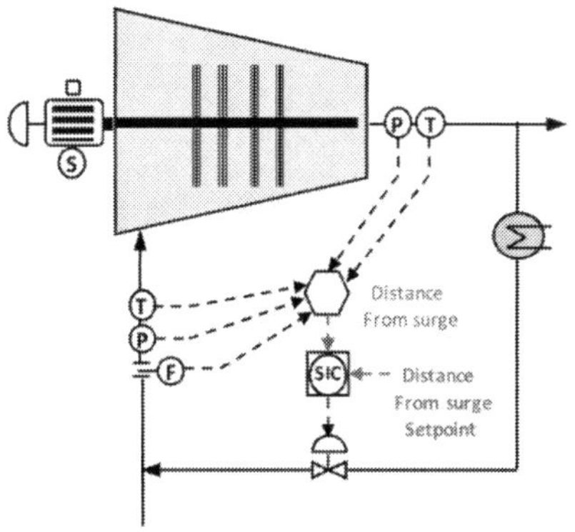

The concern with this approach is the error introduced by the simplifying assumptions and the designs dependence on discharge conditions. The compression ratio is flat near surge and so provides little information in resolving the distance from surge. Also the drop in compression ratio when surge occurs can cause an incorrect control response at surge. The consequence is that surge is more likely and when it does occurs it is made more severe by the control action.

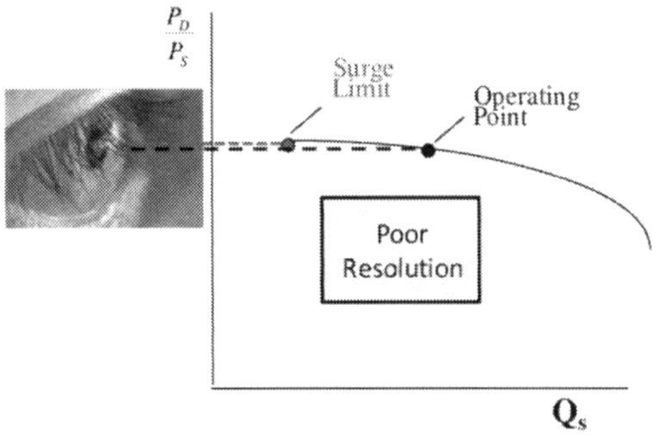

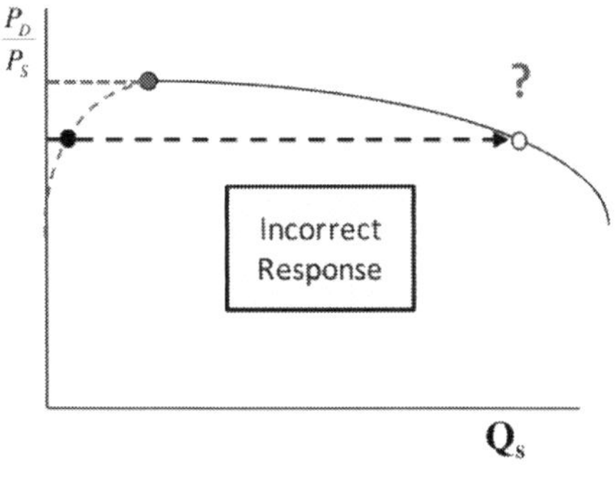

Refrigeration Compressors

The application of standard surge control to compressors with side stream feeds is made even more difficult by the presence of internal flows. As an illustration consider the difficulty of applying the standard design to the second stage of a refrigeration compressor:

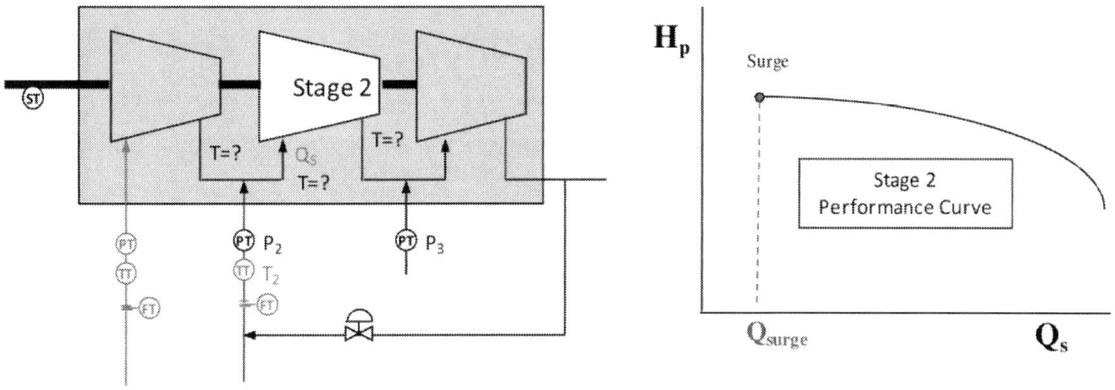

The measurements available to calculate "distance from surge" for the second stage as defined by the standard approach are its suction and discharge pressure. There is no measure of stage inlet flow, inlet temperature or discharge temperature available. The "fix" to the standard design is to make more simplifying assumptions and add additional tuning constants. The result is additional error in the estimation of distance from surge, more difficult tuning, increased likelihood of surge, and the eventual use of setpoints that are "safe" but result in wasted power and reduced refrigeration capacity.

Surge Control – Using Process Model Based Control

Single Stage

Our approach to compressor control in general can be illustrated with a single stage compression of a gas of known composition. We use the physical properties of the gas along with conventional instrumentation to calculate and control inlet volumetric flow.

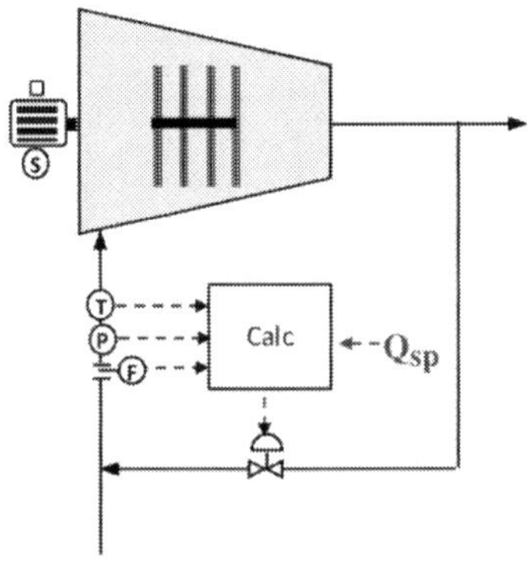

Accuracy is high because this approach requires few simplifying assumptions and does not rely on discharge conditions. This approach also provides excellent resolution and the correct response at surge. The result is that surge is less likely to occur but if surge does occur the control action helps minimize the severity.

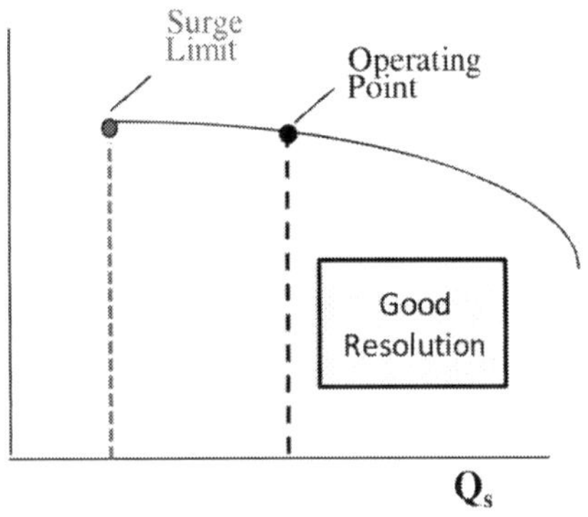

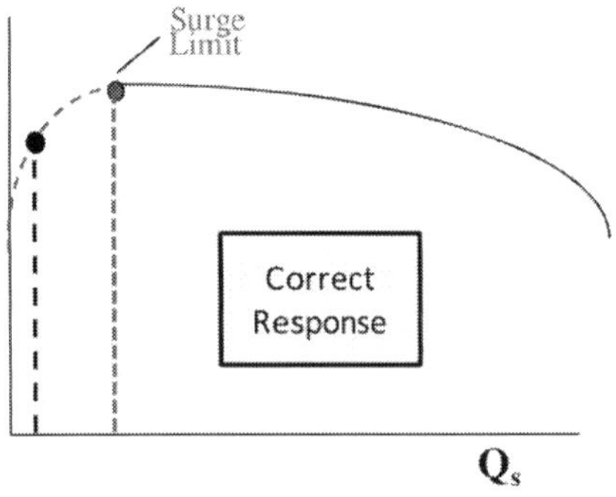

Refrigeration Compressors – Surge Control

The value of our PMBC approach is most pronounced when applied to multistage refrigeration machines. Our solution employs a proprietary 1st principles thermodynamic/aerodynamic model that relates internal conditions to external measurements. The model is easily configured to accurately estimate the internal conditions of any pure component refrigeration system. With this model we calculate the side stream feed flow rates needed to maintain the desired distance from surge at each stage

Refrigeration Compressors – Suction Pressure Optimization

Our 1st Principle modeling approach leads to the next logical step of optimizing the compressor speed based on refrigeration load. Using a simple model of the most constrained refrigeration exchangers our system will minimize the power used in delivering the required refrigeration load.

Implementation

By its nature, Process Model Based Control is a solution engineered to meet the needs of a particular process not a generic tool to be applied to a variety of processes. As a result, most of the work is already done! All we need is to get the basic process information in order to configure it for your particular refrigeration system.

Steps Required to Implement Surge Control

- Gather Process Information: Compressor specs, Flow meter specs, Instrument Tags
- Develop configuration files (offsite)
- Validate Model
- Modify PLC (minor)
- Commission Surge Control
- Tune Flow Controllers
- Validate Surge Points (not necessary but recommended)

Steps Required to Implement Suction Pressure Optimizer

- Review P&IDs
- Gather detailed exchanger information
- Develop configuration file (offsite)
- Commission Optimizer

Results

The result will be a system that

- **Is Easily Implemented**
  Configuration is done off site. Validation is conducted in a background mode. Implementation primarily involves tuning the flow controllers. The optional surge tests needed to verify surge points should be a relative "non event" due to the stable nature of our control design.

- **Maintains accurate distance from surge on all stages**
  The model provides the accurate estimates of internal conditions required to maintain the desired distance from surge on all stages

- **Maximizes refrigeration capacity / Olefins production**
  The stable nature of the control design allows operations closer to surge, the result of which is increased refrigeration capacity which can translate into increased olefins production.

- **Minimizes Power**
  The combination of our surge control and speed optimizer results in minimized power requirements at all refrigeration loads.

- **Is Easily Maintained**
  You should expect a system that is easy to maintain! A need for maintenance of the surge controller would probably indicate instrument error, a change in the compressor, or a change in the refrigerant composition.

Basic Care

Presented at AICHE, EPC meeting in Houston, April 2012

Gampa Bhat
Gampa Bhat & Associates, LLC
April,2, 2012

1

Table of Contents

2

Introduction:

What is Basic Care:

A Comprehensive Operation, Maintenance and Health Management System for Large Critical Compressor Trains in Petrochemical Industries. It also includes plans to address both near term (while the unit is still operating) and long term (during shut down) issues, to operate the plant with Safety and Reliability. Basic Care ownership stays with the operation.

Why it is necessary:

In today's world, the engineering tools and design, compressor venders manufacturing techniques and health monitoring system have led us to construct large size Ethylene plants. Though the capacity is important, the safety, reliability and environmental aspects of operation are more important to sustain profit or minimize production losses. The health monitoring system can send alarm notice of deviation but what do you do with the alarms or too many alarms and how effectively address the issues in a timely fashion and who is the equipment owner need to be clearly designed and understood.

3

- Just an operating round or writing work requests are nor effective without prioritization and senior management engagement.
- The Basic Care management system is not another program but an embedded process which includes top down and bottom up organization people with clear responsibilities and accountabilities.
- Basic care does not add additional manpower, it prioritizes and focuses on how we treat large compressor trains.
- Basic Care brings Passion and Loving Care for the most valuable asset in Ethylene Crackers.

4

Case for Action:

- Before we create a Basic Care Frame Work and an organization Structure, we need to have operational and maintenance history, production loss date, failure history and a date tracking system.

- The data may be stored in operation log books, maintenance work orders, centralized computers or in engineers failure analysis memos.

- Form a small team of machinery engineers, operation and maintenance technicians, production engineers and headquarters analysts to dig the $$ losses and failure histories for the last 5-10 years.

- Present the data in "Pareto" charts showing operation, maintenance, spare parts, design defects and engineering misapplication format.

- Get an endorsement from the plant manager to form a Basic Care frame work and an organization structure.

- Experience shows operation, maintenance and spare parts related losses are much greater than the design defects or engineering misapplication.

5

Basic Care Elements:

- Operators Rounds
  - Daily

- Management Focus and Engagement
  - Weekly once, deck walk

- Maintenance and Spare parts review
  - Semi- annual
- Monthly Failure Analysis, RCFA, Loss data review and report preparation

- Stewardship (monthly)
  - Site Manager
  - Team members
  - HQ operation directors
  - Central engineering

6

Basic Care Frame Work& Organization Structure

- **Frame Work:**

  - Assign a senior level machinery engineer from central engineering to lead and form a team (creditability)

  - Team to include operation supervisors, technicians, maintenance technicians, site machinery and Instruments engineers, data analyst (working team)

  - 3-4 days offsite meetings

  - Review the existing operation and maintenance alarm histories, practices, loss data, procedures, gaps in the organization structures

  - Compare the data against the best run plant
  - Set the target for Run Length, Loss Reduction($$) or Production

7

- Team visits the site and the compressor decks. Look for housekeeping, broken gages, alarm panels, complexity of sealing and oil control system, steam and other leaks around the deck, and work order system

- Prepare a Roll-Out package "Custom" to each site or plant

- Roll-Out package will have site loss data, reliability data, previous failure history on when &why, spare parts management system, training, reporting structure, reliability traffic light system and simplified stewardship reports.

- Once the packages is accepted by the site, the central engineering machinery leader stays as an advisor.

8

Organization Structure: Basic Care Team

- Plant Manager – Sponsor
- Unit Operation Manager/ Supervisor- Lead and Maintenance
- 1-2 Unit Operating and Maintenance Technicians
- Site Machinery Engineer- Secretary
- Other Functional Engineers as needed (Part time)

9

Stewardship report:

Includes:

- Reliability Traffic Light
  - Green - OK
  - Yellow - Need Improvement
  - Red - Do something now
- Spare Parts Status
- Operating Procedures update
  - Training
  - Dry Run
  - Safe Park Procedures
- Near Term and Long Term Plans
  - Contingency Plans
  - Tools & Labor, Job Packs
  - Investment $$ and Plans
  - To obtain Funding
  - Production Loss Reduction or Gain

10

Conclusion:

- "Basic Care" is an embedded process to bring passion, ownership & responsibilities to the manufacturing organization.
- It engages Top down and bottom up organizations with clear responsibilities to obtain the targeted run length and loss reduction of the critical compressor trains.
- Please love your compressor trains.
  They also need TLC.

11

(38c) Operations Management Panel Discussion

Greg Berry, Mechanical, KBR, Houston, TX and Michael Sicker, Mitsubishi International Corporation, Houston, TX

The intent of this session is to present and develop a framework for management of rotating equipment within the plant. The papers and panel discussion will highlight such issues as the relevance of alarm management, provide commentary on the roles and responsibility of engineers, operators and management, and review the importance of preventive and predictive maintenance.

AIChE Paper Number: 39a

ACCEPTABLE OPERATING PARAMETERS FOR STEAM ASSISTED FLARES AT PURGE RATES

Caleb Hurd
Flare Design Engineer
Zeeco, Inc.

Tom Farmer
Flare Applications Engineer
Zeeco, Inc.

Prepared for Presentation at the 2012 Spring National Meeting
Houston, TX, April 1-5, 2012

ACCEPTABLE OPERATING PARAMETERS FOR STEAM ASSISTED FLARES AT PURGE RATES

Caleb Hurd
Flare Design Engineer
Zeeco, Inc.
Tom Farmer
Flare Applications Engineer
Zeeco, Inc.

Abstract:

Recent field observations by the Texas Commission on Environmental Equality (TCEQ) have revealed what appears to be poor flare destruction efficiency by some steam and air assisted flares when operating at turndown conditions. Using specialized infrared cameras, unburned hydrocarbons were detected coming from improperly operated flare systems. Despite these findings, the general belief among industry flare experts is that through proper operation of existing flares, the destruction efficiency can be very high, even at turndown rates. Meeting proper destruction efficiency requirements with existing flare systems at low gas flow rates requires more clear and detailed training for operations personnel at these facilities.

To help develop this training program, Zeeco performed extensive testing of steam assisted flares at the Zeeco Combustion Research and Test Facility in Broken Arrow, Oklahoma. The testing covered a wide range of operating conditions, specifically focusing on turndown conditions. Testing included emissions plume sampling, as well as real-time data from the same specialized infrared camera currently used by the TCEQ. Computational Fluid Dynamics (CFD) models of the testing generated and compared to the actual results in order to calibrate the prediction models. Through further research, it is hoped that the prediction models can then be used to determine if existing flares require any modification to meet current or possible new legislation. The focus of the testing was aimed at finding and setting the proper operating parameters for steam flares at turndown conditions to allow existing stacks to remain in operation with little or no modification, while still meeting the required destruction removal efficiencies (DRE). This paper discusses the approach and results of this testing, as well as recommendations for steam flare operation.

Objective

The primary objective of the steam flare research was to test current industry steam assisted flare tip designs for destruction and removal efficiency of hydrocarbons while firing at American Petroleum Institute (API) 521 recommended purge gas rates with steam injected at recommended cooling steam rates. A secondary goal was to develop a relationship between the flare tip design, the combined steam and purge gas lower heating value (LHV), and the hydrocarbon destruction and removal efficiency.

Overview and Background

In 1982, the Chemical Manufacturers Association began a comprehensive study on the efficiency of flare operations. This study reported that properly operated elevated, open flame flares had a destruction efficiency of 98%. Based on the results of this study, the United States Environmental Protection Agency (EPA) issued the 40 CFR 60.11 to 60.18 regulations and the US EPA AP 42 Chapter 13.5. Since 1982, these documents have formed the basis for permitting flare emissions.

During the fall of 2010, the University of Texas in conjunction with the TCEQ, performed tests to determine how air and steam assisted flares perform at turndown rates. The test results suggested that incorrectly designed or operated steam flares have the potential to over-aerate or quench the flame, reducing combustion and destruction efficiency. Rather than destroying hydrocarbons, improperly operated flares vent unburned gases.

In comparison to the TCEQ Flare Study Project, this steam flare testing was performed at API recommended purge rates, which were significantly less than TCEQ's previously reported turndown rates. The velocity seal, gas seal, and no seal purge rates given in ISO 23251/API Section 521 Fifth Edition were the testing focus. An outside contractor performed the emissions testing on the flare gas plume. A sample extraction device similar to that used in the TCEQ Flare Study Project was designed to be held by crane above the flare and pull a continuous sample for emissions monitoring.

Testing Facility and Instrumentation

All testing was performed at the Zeeco Combustion Research and Test Facility in Broken Arrow, Oklahoma and occurred from July 18–22, 2011. Ambient weather conditions could not be controlled, but weather data was recorded. Throughout the week of testing, the temperature ranged from 85 to 110°F and wind speeds ranged from 2 to 15 mph.
Monitoring devices consisted of a LSI FLIR GasFindIR camera, Sony DCR-SR42, and a Canon Powershot SX130 IS. These instruments were used to monitor the presence of unburned hydrocarbons, to detect the presence of a flame, and to provide still photos of the testing setup. Test facility technicians and engineers were responsible for controlling the gas and steam flow rates, operating video recorders, and positioning the sample extraction device.

A key element in the study was the ability to measure flare emissions from the flare plume at the exit of the tip. Using the TCEQ Flare Study Project as a guide, a sample probe extractor was designed and fabricated. The probe was 26'- 6" long and 12" in diameter, and consisted of an inlet bell, elbow, sampling section, and eductor (**Figure 1**). Near the inlet, a Vortab flow conditioner was inserted to mix the gas pulled from the plume before sampling. Several tests were conducted beforehand with nitrogen to guarantee that the flow conditioner created a homogenous mixture. The sampling section contained three ports into which a thermocouple, Pitot tube, and emissions sampling probe were inserted. This allowed for the opportunity to directly monitor temperature, calculate sample flow rates, and obtain a continuous emissions sample. The eductor was designed to create a vacuum for pulling the sample into the probe. Compressed air was connected via a 2" nozzle operated at 50 to 75 psig. This airflow created suction and induced effluent gas into the sample extraction probe (**Figure 1**).

Air Hygiene International, Inc., was selected to perform the emissions testing due to their extensive emissions experience, mobile capability, and ability to provide real time data. The emissions data was recorded on a Logic Beach Portable Data Logging System Hyperlogger, which retrieves calibrated electronic data every second.

The flare designs used throughout the testing were Zeeco's UFSC-36" and QFSC-36" Steam Assisted Flare Tips with three pilots. These models were selected because of their industry popularity and ability to represent a large portion of the tips used in actual applications. Steam tip drillings were designed to create the same total steam flow area regardless of QFSC or UFSC layout. The tip itself had four sight ports mounted vertically on the side to allow personnel to view inside the tip to check for burnback and center steam condensate at different levels. Additionally, to ensure that the gas entering the tip was evenly distributed, a gas distribution plate was installed at the base of the tip.

Figure 1: Sample extraction probe composed of inlet, sampling section, and eductor.

Natural gas and a mixture of natural gas and nitrogen were used as the purge gases for this research. The natural gas was issued from the local natural gas supply and the nitrogen was fed into the flare header from a nitrogen bank. Orifice runs were used to determine the flow rate of the gas and nitrogen, as well as the steam rates. For the nitrogen/gas mixtures, the nitrogen was introduced approximately 150′ upstream of the flare tip to allow proper mixing in the header.

To monitor flame patterns and detect burnback in the tip, 16 Type K extension wire thermocouples were tack-welded around the perimeter of the tip. These thermocouples were divided into three groups and mounted at three different height levels (**Figure 2**). Level one was 6″ below the top of the tip. Levels two and three were sequentially placed a foot lower from the preceding level. Level one consisted of six thermocouples and levels two and three of five each. Each thermocouple was given a specific label, and the temperature readings of all sixteen thermocouples were logged continuously throughout testing with two Pico TC-08 Thermocouple Data Loggers.

Figure 2: 36"QFSC Flare Tip with four sight ports and three layers of thermocouples to record tip temperatures.

Testing

Testing began on July 18, 2011, and spanned a period of five days. Ambient condition measurements were taken by Air Hygiene before testing began to be used as a control for background readings of carbon levels in the facility.

The LSI FLIR GasFindIR camera was used simultaneously with the extraction probe temperature readings and Air Hygiene's destruction and removal efficiency (DRE) data to properly position the sample extraction probe. The FLIR camera assisted in horizontal positioning of the probe by showing hydrocarbons. When the probe was operating within the flare plume and pulling a sample, hydrocarbons could be seen exiting the end of the eductor (**Figure 3**).

Figure 3: Image from the FLIR camera. Sample hydrocarbons can be seen exiting out the end of the educator.

Thermocouple data and Air Hygiene's calculated DRE values from the sample measurements were useful for proper vertical probe positioning. When the temperature was high and DRE values were low, it indicated that the probe was located in the flame where full combustion had not yet occurred. On the contrary, low temperatures and very high DRE readings—in addition with FLIR imaging—indicated that the probe was too high and located in the open atmosphere rather than within the gas plume.

Air Hygiene provided the average DRE reading for every five-second period, allowing for adjustments to be made during testing to optimize the location of the sample extraction probe.

Testing occurred in three phases.

Phase 1: Testing Per Procedure

Testing began according to the testing procedure as listed in Appendix C: Testing Procedure, with steam rates and purge rates chosen to meet API 521 specifications. For each purge rate (Velocity Seal, Gas Seal, and No Seal), the fuel flow rate remained constant and the compressed airflow through the eductor was kept constant by maintaining a 75 psig pressure. See Table 3, Purge Rates, in Appendix C.

- Velocity Seal

  The first purge rate tested was that recommended by API 521 for a 36" velocity seal (990 SCFH). Initially, pure natural gas was run through the flare with no steam assist, and the DRE was 98.52%. The center steam was then added and the flow varied. As soon as the center steam flow was introduced into the flare tip, the DRE dropped by approximately 28%. Center steam flow was then turned off and upper steam flow was then initiated and varied. For natural gas with upper steam only, the DRE remained stable around 90% DRE. The final testing at velocity seal gas rates included both upper and center steam assist. The center steam was held constant at 500 lb/hr while gradually decreasing the upper steam from 1655 lb/hr down to 0 lb/hr. Under combined steam assist flow conditions, the DRE averaged around 70%.

- Gas Seal

  Gas seal testing followed the same format as the velocity seal testing (gas only, gas and center steam only, gas and upper steam only, gas and both steam assists). The fuel flow rate remained constant at 250 SCFH of natural gas. Since the gas seal rate is lower than the velocity seal rate, the compressed air which creates the vacuum in the probe was reduced to 50 psig so the natural characteristics of the flame would not be disturbed. Similar to that observed under velocity seal conditions, DRE decreased with the addition of center steam, averaging 75%. Purge gas and upper steam only assist destruction and removal efficiencies ranged from 80.3 to 86.4%, while that of combined steam assist was around 80%.

- No Seal Purge

  No seal purge was the third fuel rate testing to take place. Fuel flow to the tip was set at 1992 SCFH. The no seal purge testing occurred in three stages (gas and upper steam only, gas and center steam only, and gas and both steam assists). The highest destruction and removal efficiency of the three stages tested occurred when center steam was turned off. The gas and upper steam DRE was between 86.1 and 96.6%. The gas and center steam testing showed scattered DRE values in the 70% range, and the combined steam assist DRE fluctuated around 74%.

- Velocity Seal with Natural Gas/Nitrogen

  Nitrogen was then mixed with natural gas to achieve an LHV of 300 Btu/SCF, and this mixed gas was used as the fuel for additional tests. Any addition of steam lowered the LHV of the combustion zone mixture below 300 Btu/SCF and respectively lowered the DRE. This testing and corresponding results followed the similar pattern as the previous testing.

For all purge gas flow rates, the addition of any steam lowered the DRE. For all three purge rates, the lowest DRE values were observed when the purge gas was combusted with only center steam while the highest destruction efficiencies occurred with just upper steam assist. When the two steam assist methods were combined, the DRE averaged between those values seen during the individual assist methods. Under the recommended gas purge and steam cooling rates, the DRE never met the desired 98% level.

Phase 2: Adjusting the Test Plan and Building a Hypothesis

Since the initial purge rate with steam assist testing did not produce the desired destruction and removal efficiencies, Phase 2 of testing commenced. During this phase, steam and purge gas rates were increased / decreased until DREs of 98% or greater were achieved. Several different combinations of fuel, upper steam, and lower steam were tested.

Testing began with fuel rates slightly higher than velocity seal rates, and were adjusted, along with steam rates, to try and map fuel/steam interactions relative to DREs. The natural gas flow rate ranged from 1992 to 7677 SCFH. All data points for Phase 2 can be found under Appendix B: Numerical Data points 57 to 88.

Data analysis suggested combustion DRE is closely related to the combined LHV of the steam and gas mixture at the flare tip exit, known as the Combustion Zone Net Heating Value (CZNHV). As the CZNHV increased, the DRE rose as well. This evidence indicated that DRE is a function of the CZNHV. To further refine this data correlation, gas and steam combinations were tested with a CZNHV to achieve 98% DRE in Phase 3.

Phase 3: Verifying LHV Hypothesis

After hypothesizing that CZNHV is a controlling factor in achieving desired destruction efficiencies of 98% or greater, testing was then initiated to validate this theory using 20 different fuel / steam flow rates.

Tests showed a connection between CZNHV and DRE. High DRE strongly corresponded with a high CZNHV. When CZNHV's spanned from 200-250 Btu/SCF, the DRE was typically between 93-97%. For 250-350 Btu/SCF, the DRE was 98-99%. See points 89-108 under Appendix B: Numerical Data. While test results indicated the steam assist method does impact the DRE, the CZNHV exerts considerably more influence over achieved DREs.

DRE Evaluation

Air Hygiene measured the emissions of CO, CO_2, NO_x, and THC. With their measurements, DRE was calculated based on the carbon content in the purge gas and the flare plume according to the DRE equation as specified in Appendix D.

Due to the low flow rates tested, visible flames were rare. In contrast to the TCEQ testing, no correlation was drawn between the flame color and DRE. All DRE conclusions were based on purge gas-steam ratios and emissions testing data.

Thermocouple Evaluation

The temperature on the top 2'-6" of each flare tip was monitored continuously throughout testing, allowing burnback to be detected within the flare tip as well as determine whether there was any correlation between temperature, probe positioning, and DRE.

Division and mapping of the 16 thermocouples throughout the five days of testing provided interesting results. High temperatures on lower levels of the tip indicated that combustion was occurring on the inside of the tip instead of localized on the upper perimeter. This phenomenon was observed during the first two days of testing at lower purge rates. As shown in (**Figure 4**), the temperatures recorded on Monday, show Level 2 temperatures consistently charted above Level 1, regardless of the purge rate being tested. A similar trend was seen on Tuesday (**Figure 5**).

The first two days of testing represent the initial testing (Phase 1) of the three API purge rates. As previously stated, the destruction and removal efficiencies during these tests were low.

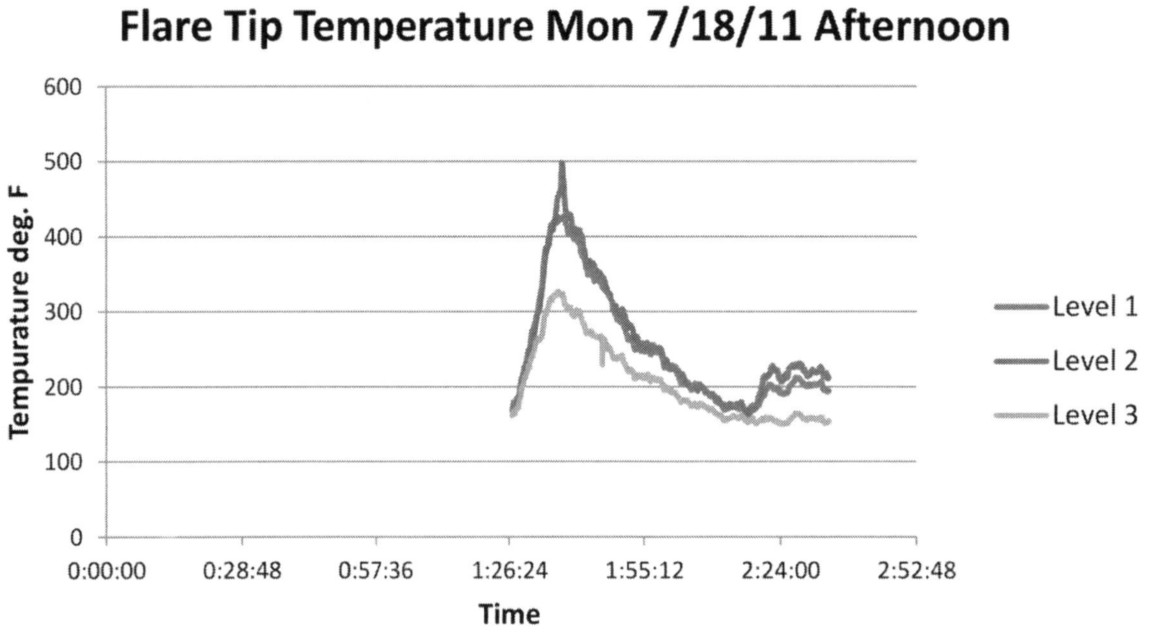

Figure 4: Average temperature readings for each level on flare tip on Monday. Level 2 is steadily greater than Level 1, indicating burnback in tip.

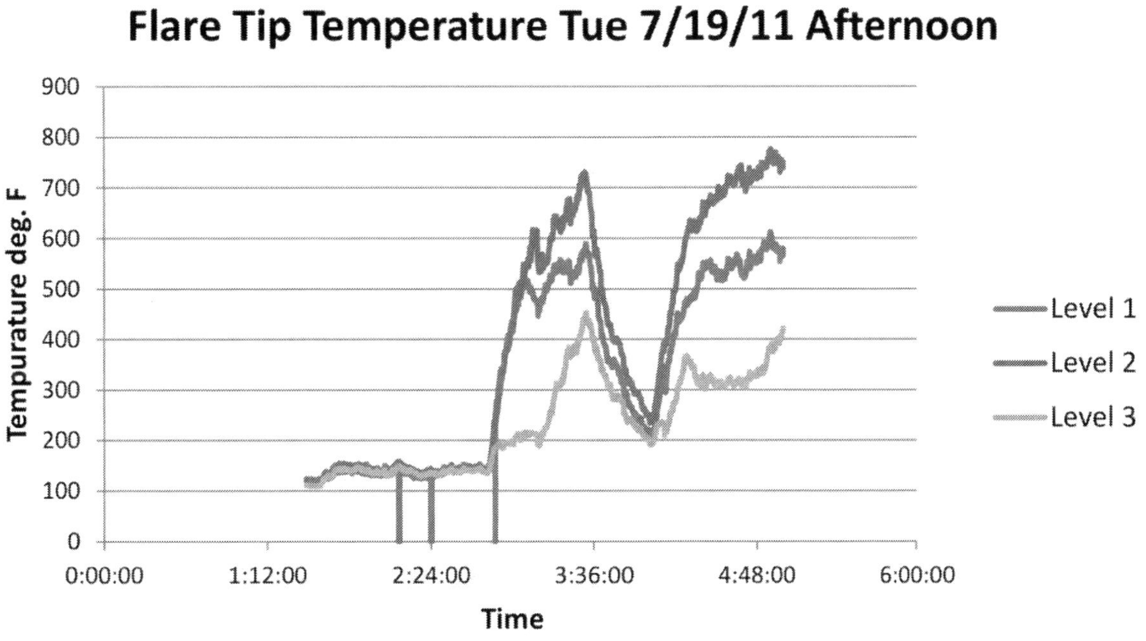

Figure 5: Average temperature readings for each level on flare tip on Tuesday. Level 2 is steadily greater than Level 3, indicating burnback in tip.

Phase 2 testing data suggested that proper flame retention ring burning leads to higher DRE values. Beginning on Wednesday, July 20, temperature readings indicated proper burning; Level 1 temperature was greater than Level 2 which was simultaneously greater than Level 3 (**Figures 6-7**). During the testing in which this temperature behavior was observed, a slight increase in DRE was also seen.

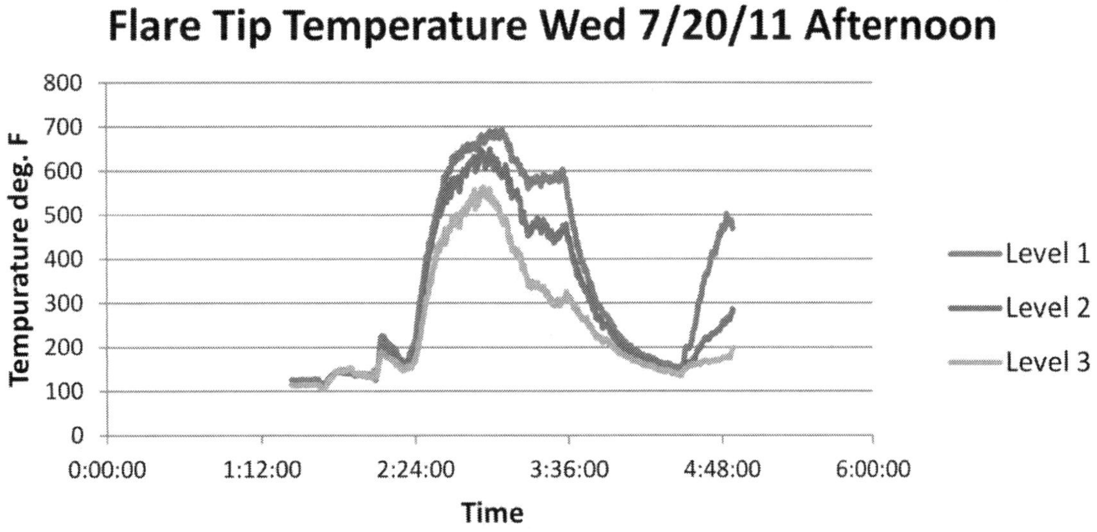

Figure 6: Phase 2 thermocouple data from Wednesday. Correct tip temperature sequence indicates no burnback in tip.

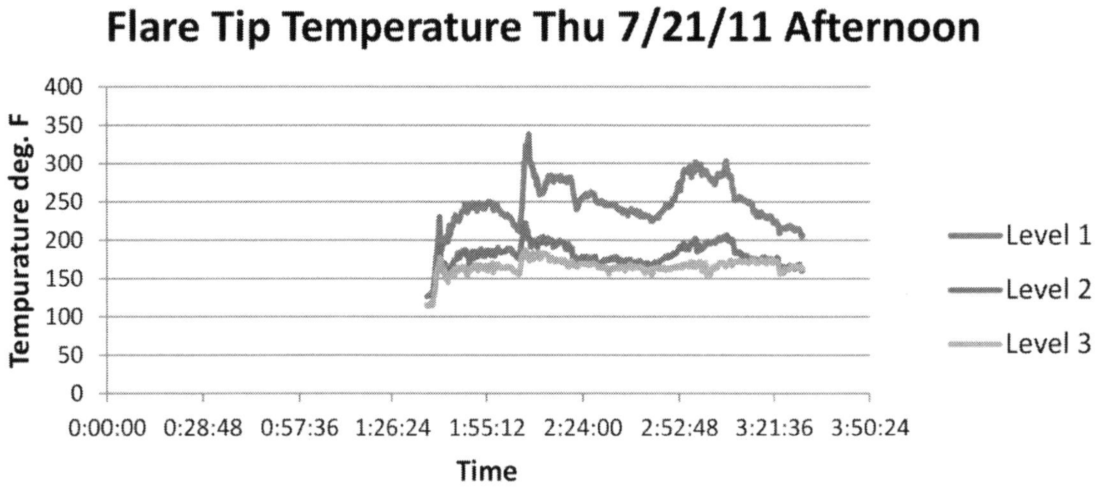

Figure 7: Phase 2 thermocouple data from Thursday. Correct tip temperature sequence indicates no burnback in tip.

During Phase 3 testing when high DRE standards were achieved, the thermocouple levels were all in order (highest to lowest temperature from top of the tip to bottom, respectively) (**Figure 8**). This phase of testing represented the point in time in which the combined LHV values were consistently high. The temperature range was typically 150-400°F and the DRE values 98-99%.

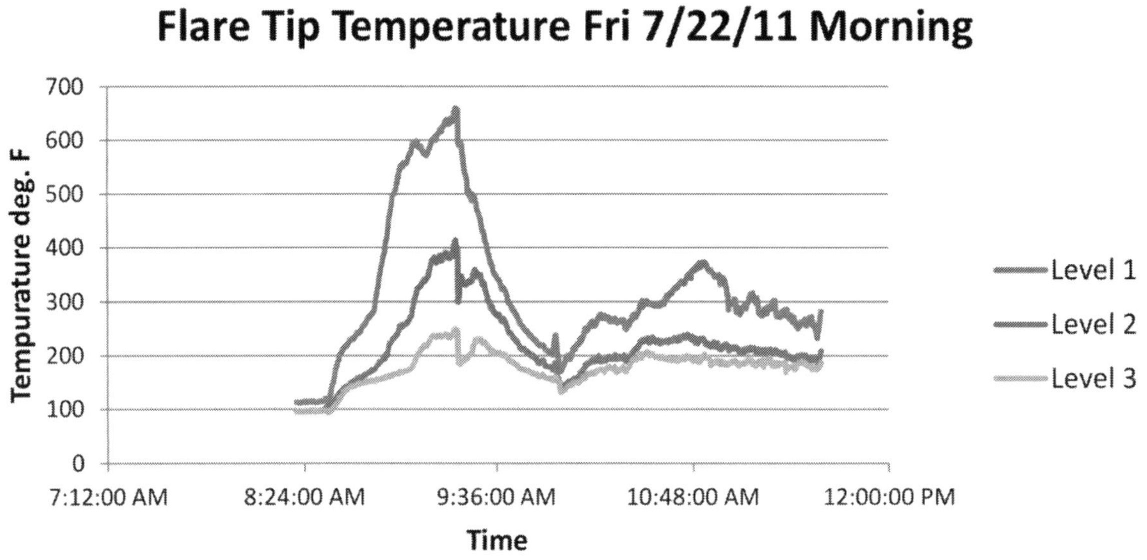

Figure 8: Phase 3 thermocouple data from Friday. Correct tip temperature sequence indicates no burnback in tip.

The best DRE values occurred when the flare tip temperature was between 150-400°F, no internal burning was occurring, and LHV's were high. From the thermocouple data, it was concluded that proper temperature arrangement is indicative of proper burning and directly relates to destruction and removal efficiency percentage.

Computational Fluid Dynamics

A secondary objective of this study was to generate, compare, and calibrate Computational Fluid Dynamics (CFD) models to the actual testing that occurred. If these models accurately represent real life testing, they could be used to determine what flare modifications are necessary to meet regulations.

A 3-D model of the flare tips that were tested was drafted and imported into the CFD software to calibrate the CFD model to the actual test results. A full analysis was then conducted considering environmental conditions, radioactive heat transfer, and combustion reactions. Two main test points were chosen for CFD analysis and modeled at varying wind velocities. Design specifications can be viewed below (**Figure 9**).

CFD Model Set Up

- 3-D Model created in Solid Works with 120' x 120' x 120' overall environment dimensions
- Uniform crosswind of 2 mph & 10 mph

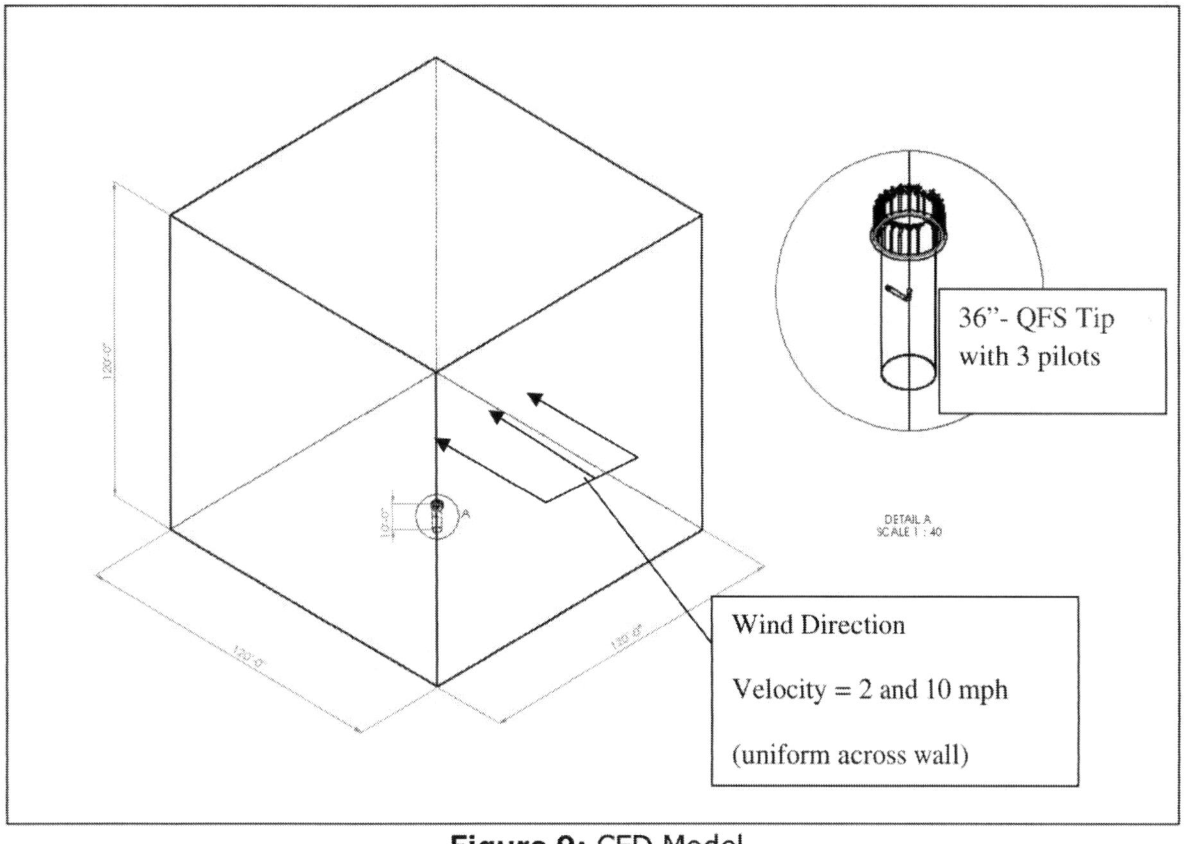

Figure 9: CFD Model

Flow Conditions Modeled in CFD

Test Point #14:
Gas Flow = 990 SCFH (100% CH_4)
Upper Steam Flow = 735 lbs/hr
Center Steam Flow = 500 lbs/hr
Test Point #107:
Gas Flow = 6648 SCFH (100% CH_4)
Upper Steam Flow = 530 lbs/hr
Center Steam Flow = 199 lbs/hr

Comparison with actual DRE results was the main method used to determine the accuracy of the CFD modeling. After the CFD design was complete, DRE was calculated through the method described below and compared with the experimental value determined through the outdoor testing.

The destruction efficiency for the CFD model was calculated as follows:

- Downwind wall of environment was separated out
- Mass flow through downwind wall was calculated (from CFD software)
- Mass fraction of components CH_4, CO_2, H_2O, & Air were also calculated from CFD software)
- Destruction efficiency of downwind wall equal to:

$$\textit{Destruction Efficiency} = \frac{MassFrac\ CH4\ x\ Mass\ flow\ Downwind}{Total\ CH4\ Mass\ Flow}$$

The model below (**Figure 10**) illustrates the flare tip in an open atmosphere to represent testing conditions. The red wall represents the sampling area; all gas species were measured to calculate destruction efficiency in this area.

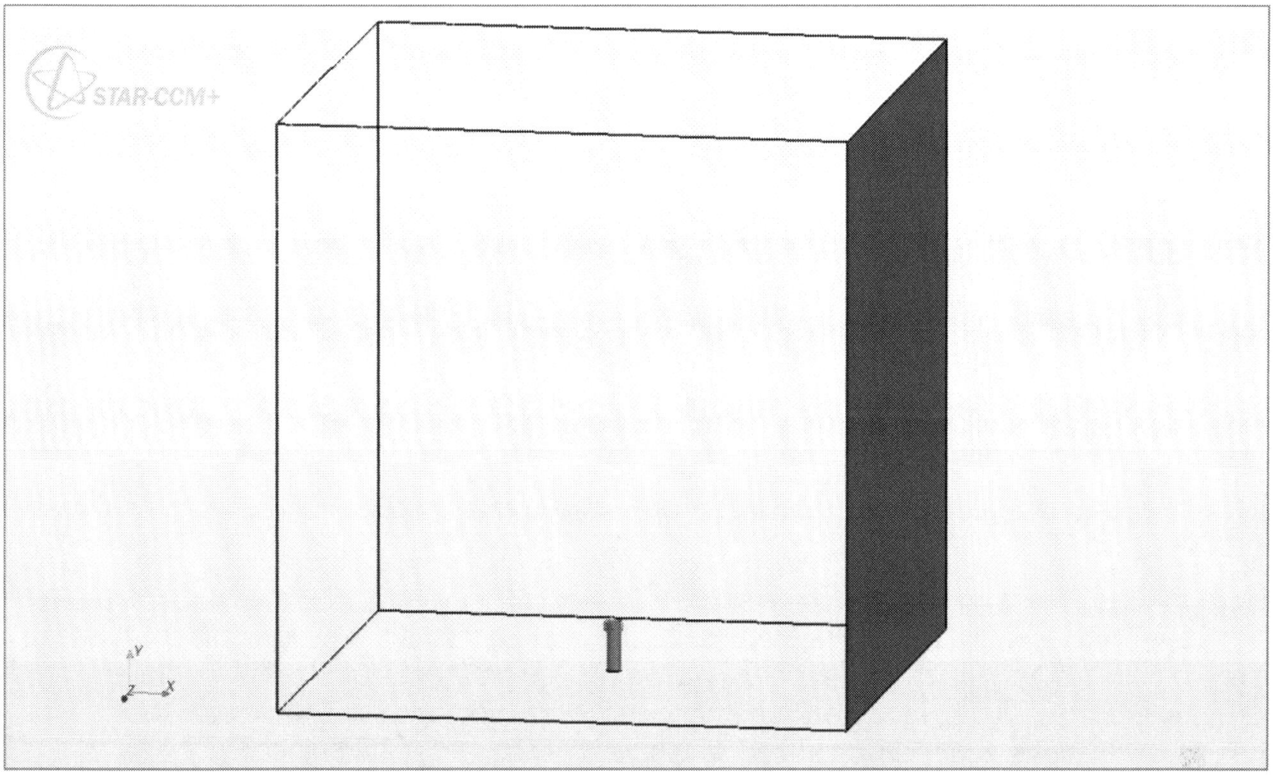

Figure 10: CFD model with red plane illustrating the point of destruction and reduction efficiency evaluation.

CFD Results and Comparison

Comparison of CFD results with actual testing data showed that CFD modeling can be used to analyze DRE in steam assisted flare tips. The data indicated that although the results of the CFD model showed a larger error at lower gas flow rates, modeling can be useful for predicting unburned hydrocarbons.

The results and comparative difference in DRE are detailed in Table 1 below.

Table 1: CFD Comparison Results

Test Point	**Fuel** (SCFH)	**Upper Steam** (lb/hr)	**Center Steam** (lb/hr)	**Combined LHV** (Btu/SCF)	**CFD Averaged DRE** (%)	**% Difference CFD to Test Data**
14	990	735	500	33.48	92.1	31.00
107	6648	530	199	275.99	98.2	0.30

Developing a link between CFD and testing through tailored combustion parameters will allow engineers to more accurately predict DRE in the future. Continued research will help to develop better ways of analyzing combustion efficiency through CFD.

Results and Conclusions

While constant steam injection (cooling steam) provides thermal protection to steam injection equipment, improper steam injection can create downsides. Though steam is an integral part of maintaining the integrity of flare components, improper use can become an impediment to the destruction and removal efficiency of hydrocarbons. Many factors must be considered when utilizing steam for combustion. From this research, it is understood that adequate combustion is not possible when using API 521 specified purge gas and current steam cooling rates. The test data advocates that the Combustion Zone Net Heating Value is a key component for complete combustion. Data consistently showed that the DRE achieved for a mixture followed the same trend as the CZNHV. It is important to note, however, that many Olefin plants have H2 in their gas. While any Hydrogen testing is outside the scope of this paper, it is noted that rules have been added to 40 CFR 60.18 (c)(3)(i) which indicate that hydrogen will help to improve combustion stability. From this information and other industry testing on Hydrogen flares, it is expected the any H2 in the gas stream will only help to improve the DRE while operating at a lower CZNHV.

Along with CZNHV, the temperature profile can be useful for determining whether proper combustion is occurring. Profiles with high temperatures along the perimeter and lower temperatures inside indicate little to no tip burnback and help increase DRE. The emissions calculations and thermocouple logs show that the DRE is influenced not only by the amount of steam introduced into the flare system, but also by the type of steam assist applied to the flare—a theory that is supported by CFD modeling developed in Phase 3. Achieving desired 98% DRE levels using center steam requires the highest CZNHV as it dominates the flow in the center of the tip as noted from the CFD model (**Figure 11**).

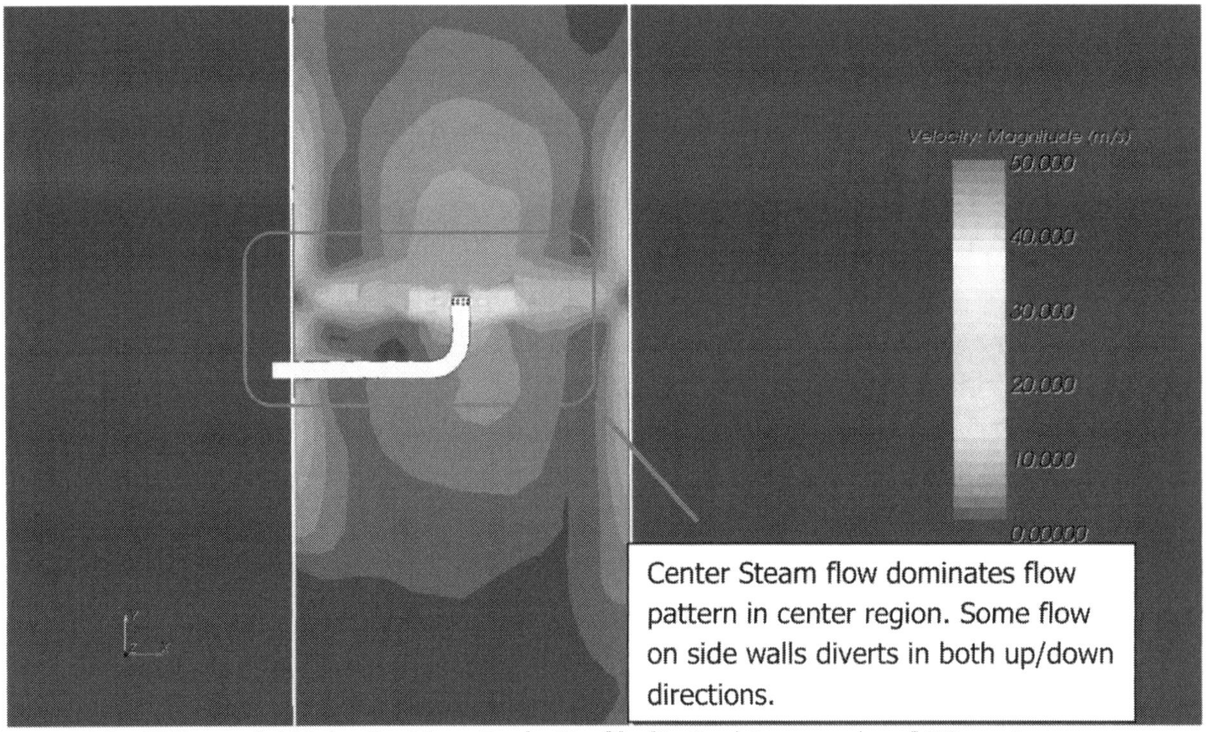

Figure 11: CFD model Velocity Magnitude Profile limited to a scale of 50 meters per second.

From this study, we are learning how to operate flares more efficiently as well as the effectiveness of CFD modeling in the design, evaluation, and operation of flares. Proper training for operators will be crucial as the flare industry moves forward.

While the focus of this testing is on flare tips with upper and center steam, it is expected that these results could roughly translate to high capacity inner steam tube type flare tips. The mixing point of the steam and gas for the inner tubes is at the exit point of the flare tip, so the steam injected into the tubes would likely have a similar affect to the DRE as the upper steam tested within this paper.

Recommendations

After analyzing the testing data, it was determined that in order to reach desirable DREs either fuel rates will have to be increased or applied cooling steam rates decreased. Minor reduction of steam rates is possible, however, the necessary reduction to the steam rates will likely result in condensate at the tip. As a result, an increase in the fuel flow rate will be required to raise the overall CZNHV. It is understood that sweep gases at many plants exceed the recommended API purge rates and is likely the most viable option. As cost of operation is a major factor in any facility, another solution is the use of nitrogen as a purge gas instead of increasing fuel gas as this may be a more economical decision.

Industrial recommendations are the following:

- A CZNHV of at least 225 Btu/SCF be maintained when using both upper and center steam. This can be accomplished by increasing the fuel flow rate or by decreasing the steam flow rate to ensure proper DRE rates.
    - The sweep gas rates of many facilities maybe enough to achieve this minimum recommended LHV value.
    - A drawback of increasing the fuel flow rate is that while the increased heating value of the fuel gas will help to improve the DRE of the flare, it may result in higher emissions if the flow of gas is greatly increased.
- Utilize nitrogen as a purge gas for emergency flares.
- Provide better insulation around steam line such that steam rates can be reduced without creating condensation.

Appendix A: Graphical Analysis

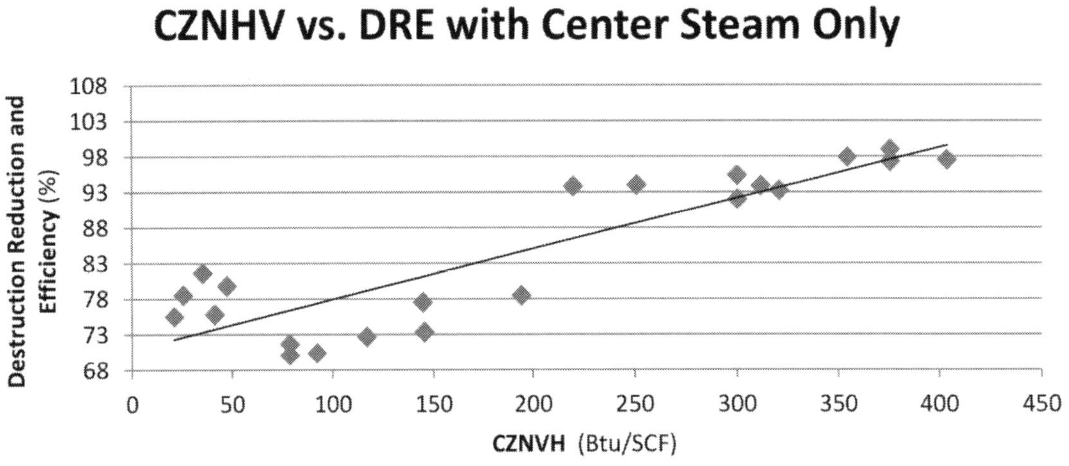

Figure 12: Combustion Zone Net Heating Value graphed against the destruction and reduction efficiency with the use of center steam only. The only points that have a DRE of 98% correspond to a combined LHV greater than 350 Btu/SCF.

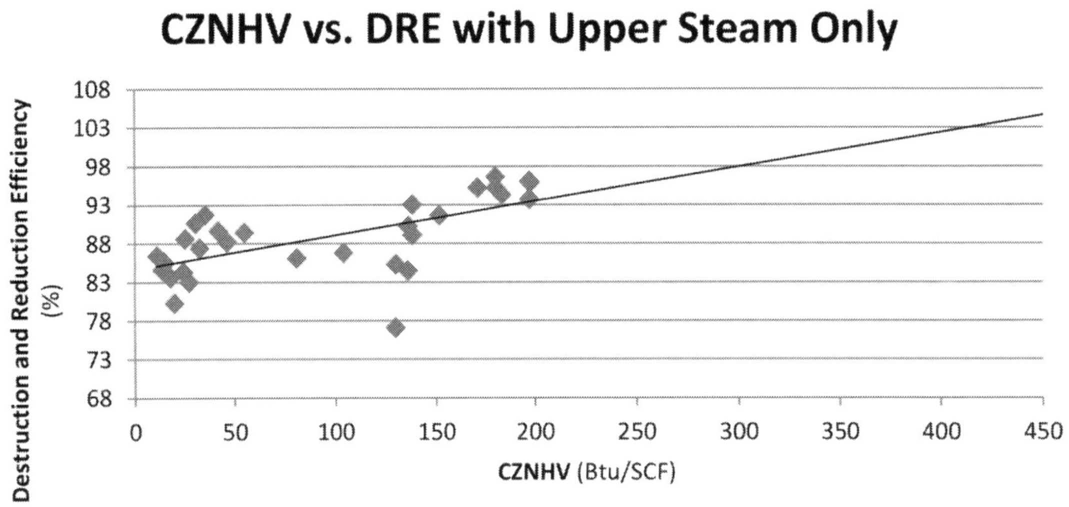

Figure 13: Combustion Zone Net Heating Value graphed against the destruction and reduction efficiency with the use of upper steam only. No points were tested that reached a DRE of 98%. Following the trend line, a DRE of 98% will occur with a combined LHV of 300 Btu/SCF.

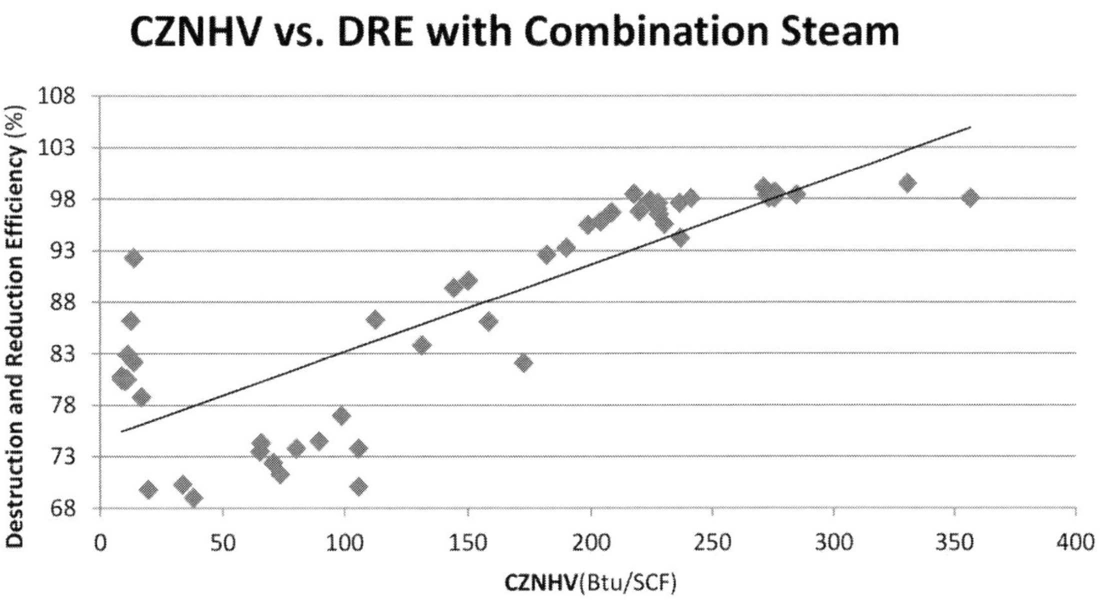

Figure 14: Combustion Zone Net Heating Value graphed against the destruction and reduction efficiency with the use of a combination of both center and upper steam. The only points that have a DRE of 98% correspond to a combined LHV greater than 220 Btu/SCF.

Appendix B: Numerical Data

Table 2: All Test Points Numerical Data

TEST POINT	FUEL (SCFH)	Upper Steam (lb/hr)	Center Steam (lb/hr)	Combined LHV (lb/hr)	Average DRE (%)
1	Ambient				
2	990	0	0	914	98.52
3	990	0	500	78	70.08
4	990	0	420	92	70.37
5	990	0	320	117	72.65
6	990	0	250	145	77.5
7	990	735	0	55	89.4
8	990	882	0	46	88.2
9	990	975	0	42	89.6
10	990	1170	0	35	91.7
11	990	1360	0	31	90.6
12	990	1655	0	25	88.6
13	990	1655	500	19	69.8
14	990	735	500	33	70.3
15	990	583	500	38	69
16	990	0	500	78	71.6
17	250	0	500	21	75.5
18	250	0	250	41	75.8
19	250	735	0	15	85.7
20	250	882	0	12	86.1
21	250	975	0	11	86.4
22	250	530	0	20	80.3
23	250	385	0	27	83
24	Ambient				
25	250	735	500	9	80.8
26	250	735	250	11	80.5
27	250	735	350	10	80.4
28	250	975	250	9	80.5
29	250	530	250	14	82.2
30	250	385	250	17	78.8
31	1992	735	0	104	86.8
32	1992	975	0	81	86.1
33	1992	530	0	138	89.1
34	1992	385	0	180	95.2
35	1992	385	0	180	96.6
36	1992	530	0	138	93

37	1992	0	500	145	73.3
38	1992	0	350	194	78.45
39	1992	0	250	251	94
40	1992	735	500	65	73.5
41	1992	735	350	73	71.3
42	1992	735	250	80	73.78
43	1992	530	250	99	77
44	1992	882	250	70	72.4
45	1992	975	250	65	74.3
46	990	0	500	26	78.5
47	990	0	350	35	81.6
48	990	0	250	47	79.8
49	990	975	0	14	84.6
50	990	882	0	15	85.5
51	990	735	0	18	83.6
52	990	530	0	24	84.3
53	990	385	0	33	87.4
54	990	735	500	11	82.9
55	990	735	350	12	86.2
56	990	735	250	14	92.3
57	2252	735	250	89	74.5
58	2716	735	250	106	70.1
59	2716	735	250	106	73.8
60	2716	735	0	136	84.5
61	2716	735	0	136	90.2
62	3890	735	0	183	94.3
63	3081	735	0	152	91.6
64	3081	530	0	198	95.9
65	3081	530	250	144	89.4
66	3081	530	214	150	90.1
67	3890	530	214	182	92.6
68	3890	530	175	190	93.3
69	1992	0	175	320	93.2
70	2573	0	175	376	97.3
71	2917	0	175	404	97.5
72	2573	735	0	130	77.1
73	2573	735	0	130	85.3
74	2573	530	0	171	95.2
75	3075	530	0	197	93.7
76	3075	530	0	197	96.1
77	2573	0	175	376	98.95
78	2573	0	250	300	95.35

79	2573	620	250	112	86.3
80	3075	620	250	131	83.8
81	3506	544	250	158	86.1
82	3889	544	250	172	82.1
83	7228	544	250	276	98.59
84	7228	735	250	236	97.59
85	7228	735	350	219	96.8
86	7677	735	350	230	95.57
87	5428	544	250	224	97.87
88	5428	544	175	241	98.07
89	16627	735	500	356	98.07
90	7677	735	500	208	96.7
91	7228	735	500	199	95.5
92	4330	421	199	227	96.5
93	3750	421	199	204	95.8
94	2005	210	95	217	98.48
95	634	0	95	220	93.8
96	1035	0	95	311	93.9
97	1268	0	95	354	97.9
98	5786	735	95	227	97
99	5786	735	95	227	97.58
100	7455	735	95	273	98.11
101	7455	918	95	237	94.2
102	7455	530	95	330	99.47
103	5550	530	95	271	99.11
104	5844	530	124	272	98.38
105	6232	530	124	285	98.39
106	6178	530	151	275	98.58
107	6648	530	199	276	98.54
108	7087	530	250	275	98.09

Appendix C: Testing Procedure

The procedure used for initial (Phase 1) testing was the following:

1. Install 36" tip with appropriate steam nozzles for a QFSC-36".
2. Turn on pilot.
3. Using natural gas (NG) as purge gas, bring the purge rate to 990 SCFH (36" Velocity Seal Purge).
4. Turn on lower steam. Regulate to a flow of 500 lb/hr.
5. Measure emission rates through sampling probe. Take video with FLIR Camera to help position probe and record sampling plume.
6. Vary steam rate to find maximum and minimum steam rates while still maintaining DRE ≥ 98%
7. Turn off lower steam. Turn on upper steam rate to 25 lb/hr per bull nose tip (975 lb/hr) (35 lb/hr per spider tip (735 lb/hr) for QFSC-36").
8. Repeat steps 5 and 6.
9. While upper steam is on, turn on lower steam to 500 lb/hr.
10. Repeat steps 5 and 6.
11. First purge rate is done. Repeat steps 4-10 for all NG purge rates listed in Table 3.
12. Change purge gas to natural gas/nitrogen (NG/N_2) mixture with LHV = 300 Btu/SCF. Repeat steps 4-11 with NG/N_2 Purge gas.
13. Change out steam tips to configuration of UFSC-36". Repeat steps 3-12. Step 4 can be neglected as center steam only points will not change.
14. Testing is finished.

Table 3: Purge Rates

	Velocity Seal	Gas Seal	No Seal	Velocity Seal	Gas Seal	No Seal
Purge Gas	NG	NG	NG	NG with N_2	NG with N_2	NG with N_2
Purge Rate (SCFH)	990	250	1992	990	250	2095

Appendix D: DRE Equation

$$DRE = 1 - \frac{mol\ \ THC_{out} measured}{\%\ of\ C\ in\ plume}$$

$$mol\ THC_{out}\ measured = \frac{ppm\ THC_{out}\ (mol\ sampled)}{1\ x\ 10^6}$$

$$\%\ of\ C\ in\ plume = \frac{C_{out}\ measured}{C_{in}}$$

$$C_{out}\ measured = 3\ mol\ THC_{out}\ +\ 1\ mol\ CO_{out}\ + 1\ mol\ CO_{2\ out}$$

$$C_{in} = mol\ C_{in} +\ 2\ mol\ C_{2\ in} + 3\ mol\ C_{3\ in} \ldots$$

$$DRE = 1 - \frac{total\ mol\ THC_{out}}{mol\ THC_{in}}$$

$$= 1 - \frac{mol\ \ THC_{out} measured}{(\%\ of\ C\ in\ plume) mol\ THC_{in}}$$

$$= 1 - \frac{ppm\ THC_{out}\ (mol\ sampled)}{(1\ x\ 10^6)(mol\ THC_{in})(\%\ of\ C\ in\ plume)}$$

$$= 1 - \frac{ppm\ THC_{out}\ (mol\ sampled)}{(1\ x\ 10^6)(mol\ THC_{in})(\frac{C_{out}\ measured}{C_{in}})}$$

$$= 1 - \frac{ppm\ THC_{out}\ (mol\ sampled)}{(1\ x\ 10^6)(mol\ THC_{in})(\frac{3\ mol\ THC_{out}\ +\ 1\ mol\ CO_{out}\ + 1\ mol\ CO_{2\ out}}{mol\ C_{in} +\ 2\ mol\ C_{2\ in} + 3\ mol\ C_{3\ in} \ldots})}$$

$$= 1 - \frac{ppm\ THC_{out}}{\left(\frac{3\ mol\ THC_{out}\ +\ 1\ mol\ CO_{out}\ + 1\ mol\ CO_{2\ out}}{mol\ C_{in} +\ 2\ mol\ C_{2\ in} + 3\ mol\ C_{3\ in} \ldots}\right)(mol\ THC_{in})}$$

$$DRE\ x\ 100 = \%\ DRE$$

where:

DRE = destruction and removal efficiency

Mol THC_{out} measured = total mol hydrocarbons measured in the plume sample

Mol THC_{in} measured = total mol hydrocarbons measured entering the flare

AIChE Paper Number: 39b

COMBUSTION OF MIXTURES: A MODIFIED IFC APPROACH

Gary Mueller
Global Discipline Lead - Air Quality
Shell Global Solutions (US) Inc.

Harshit Agrawal
HSSE Consultant
Shell Global Solutions (US) Inc.

Chris Caico
Consultant, Air Quality
Shell Global Solutions (US) Inc.

Ronald Olivacce
Laboratory Technician
Shell Global Solutions (US) Inc.

Prepared for Presentation at the 2012 Spring National Meeting
Houston, TX, April 1-5, 2012

COMBUSTION OF MIXTURES: A MODIFIED IFC APPROACH

Gary Mueller
Global Discipline Lead - Air Quality
Shell Global Solutions (US) Inc.
Harshit Agrawal
HSSE Consultant
Shell Global Solutions (US) Inc.
Chris Caico
Consultant, Air Quality
Shell Global Solutions (US) Inc.
Ronald Olivacce
Laboratory Technician
Shell Global Solutions (US) Inc.

Abstract:

While most studies on flares have been done with single hydrocarbon fuels or binary mixtures, the reality of most industrial flares is that they see different mixtures of many gases. Over the past five years the International Flare Consortium (IFC) has been conducting research on just this issue. In a recently published literature review, the IFC has explored approaches that have been used to describe the impact of vent gas composition and assist gas ratio on flare combustion efficiency. LeChatelier's Principle has been widely used to estimate the flammability limits (lower and upper) of mixtures of flammable gases. However, LeChatelier's Principle must be slightly modified when one or more of the gas components are inflammable or inert. This paper discusses some minor suggested modification of the reduced steam volume fraction parameter outlined in the IFC Literature Review, then discusses how this approach can be used to describe industrial flare operation, and the impact of assist gas ratio on the observed combustion efficiency. Finally, this paper will present the analysis of combustion efficiency data collected from seven recent flare test campaigns and illustrate the ability of this modified IFC RSVF method to describe the results obtained. A hypothesis will be offered as to how this approach might be used to predict the performance of other flares, given additional data.

While most studies on flares have been done with single hydrocarbon fuels or binary mixtures, the reality of most industrial flares is that they see different mixtures of many gases. Over the past five years the International Flare Consortium (IFC) has been conducting research on just this issue. In a recently published literature review, the IFC has explored approaches that have been used to describe the impact of vent gas composition on flare combustion efficiency [1]. LeChatelier's Principle has been widely used to estimate the flammability limits (lower and upper) of mixtures of flammable gases. However, LeChatelier's Principle must be slightly modified when one or more of the gas components are inflammable or inert. Equation (1) has been reproduced from the IFC literature review and displays the formula that can be used to estimate the flammability limit of a mixture of flammable gases with nitrogen, where nitrogen is the nth component of a mixture of flammable gases. Equation (1) is developed from the identity that for an inert, $F_{(i)} = \infty$, therefore the $x_{(n)}/F_{(n)}$ term in the sum goes to zero.

$$\frac{1}{F(mix)} = \sum_{i=1}^{n-1} x(i)\left(\frac{1}{F(i)}\right) \qquad \text{Equation (1)}$$

Other researchers have noted that not all inert species exhibit the same combustion behavior as nitrogen when mixed with flammable species. Zabetakis [2] has produced combustion flammability diagrams for many hydrocarbons, expressed as percent fuel versus percent inert for mixtures with air. Zabetakis has developed these diagrams for both nitrogen and carbon dioxide as the inert. **Figure 1** reproduces the combustion diagram for methane developed from Zabetakis' work. One can see that carbon dioxide has a more limiting impact on combustion than does nitrogen. The IFC literature review discusses the construction of simplified combustion diagrams for vent gases of varying compositions using Equation 1 to produce the combustion properties of a pseudo-fuel developed from the composition of the vent gas mixture of interest.

From **Figure 1**, one can see that the combustion diagram for any mixture of flammable species can be approximated by the trapezoid formed using the following computed properties of the vent gas mixture: LFL_{mix}, UFL_{mix}, I_{mix}^*, LFL_{mix}^*, and UFL_{mix}^*. For any vent gas composition, Equation (1) can be used to estimate LFL_{mix}, UFL_{mix}, LFL_{mix}^*, and UFL_{mix}^*. Using the IFC approach, outlined in their Literature Review, the stoichiometric air needed for combustion, λ, must also be estimated for any mixture to be able to construct the function C_{st} in **Figure 1**, which is the fuel concentration in a stoichiometric mixture of air in the pseudo-fuel/inert mixture. Additional mixing rule are needed to compute the values of λ_{mix} and I_{mix}^*. Equation (2) can be used to determine the stoichiometric quantity of air needed for any mixture of pure components.

$$\lambda_{mix} = \sum x(i)\lambda(i) \qquad \text{Equation (2)}$$

While Equation (2) was initially used by the author to also estimate I_{mix}^*, it was found to produce spurious results when the pseudo-fuel mixture contained large amounts of a second inert (e.g. nitrogen). Therefore, Equation 3, as recently proposed by Gogolek [3], has been used to estimate I_{mix}^* for all pseudo-fuel mixtures.

$$I(mix)* = \sum \frac{x(i)I(i)*}{F(i)*} F(mix) \qquad \text{Equation (3)}$$

One should note that Equation (3) will simplify to Equation (2) for vent gas mixtures that do not contain significant amounts of other inerts, and therefore these methods give similar results for all but a few limiting cases. In order to construct idealized flammability diagrams from the empirical pure component data, a unique value of I^* must be defined for each pure component. A way to uniquely define this value was not overtly presented in the IFC Literature Review. For this work the value of I^* has been uniquely defined as the value of the inert present at the point of tangency drawn from the origin of the flammability diagram to the flammability curve. Similarly, LFL^* and UFL^* represent the fuel concentration at the lower and upper intersection of a vertical line constructed at I^* with the flammability diagram. The utility of this definition will become more obvious later as we discuss flare operation and how flare operating points are represented on these idealized flammability diagrams for mixtures of vent gases (pseudo-fuels).

By constructing these idealized flammability diagrams, one can begin to assess the limiting quantity of assist gas that can be added to a given vent gas (pseudo fuel) mixture, such that combustion can no longer be supported (e.g. a nonflammable mixture results). Note that because each inert has a unique impact on flammability, these idealized flammability plots must be constructed for the predominant inert. For steam-assisted flares this inert is steam. The impact of smaller amounts of other inerts (e.g. nitrogen or carbon dioxide) can be incorporated by using Equation (1) to include them in the flammability limits calculated for the pseudo-fuel mixture (i.e. the flare vent gas). While Equation (1) holds for nitrogen as the inert, each x(i) term in Equation (1) must be modified to incorporate the impact of other non-nitrogen inerts in the vent gas on the flammability limit. Equation (4) illustrates the form of Equation (1) needed to account for non-nitrogen inerts[3].

$$\frac{1}{F(mix)} = \sum_{i=1}^{n-1} [x(i)/(1 + \sum_{k=inerts}(Ne - 1)x(k))] \left(\frac{1}{F(i)}\right) \qquad \text{Equation (4)}$$

For all of the flammability diagrams so constructed, the identity, 100= %fuel + %air + %inert, is valid. While some of the vent gases presented in the following data analysis did contain small quantities of carbon dioxide and water, the quantities of these other inerts were generally less than one volume percent in the vent gas. Therefore, in the subsequent data analysis presented here, Equation (1) was used, not the more generally applicable Equation (4).

While considerable empirical data exists for nitrogen and carbon dioxide as the inert gas [2,4], much less data exists for steam as the inert. While previous works are sufficient to demonstrate that the impact of steam as the inert will be somewhere between nitrogen and carbon dioxide, a unique expression for estimating where steam falls on this continuum has not been definitively defined. Based upon discussions with IFC principal investigator Dr. Peter Gogolek, the impact of steam on flammability has been estimated here by using an I_{steam}^* value that was estimated by giving the I_{N2}^* a 30% weight and I_{CO2}^* value a 70% weight (e.g. the I_{steam}^* value is closer to that of CO_2 than it is to N_2).

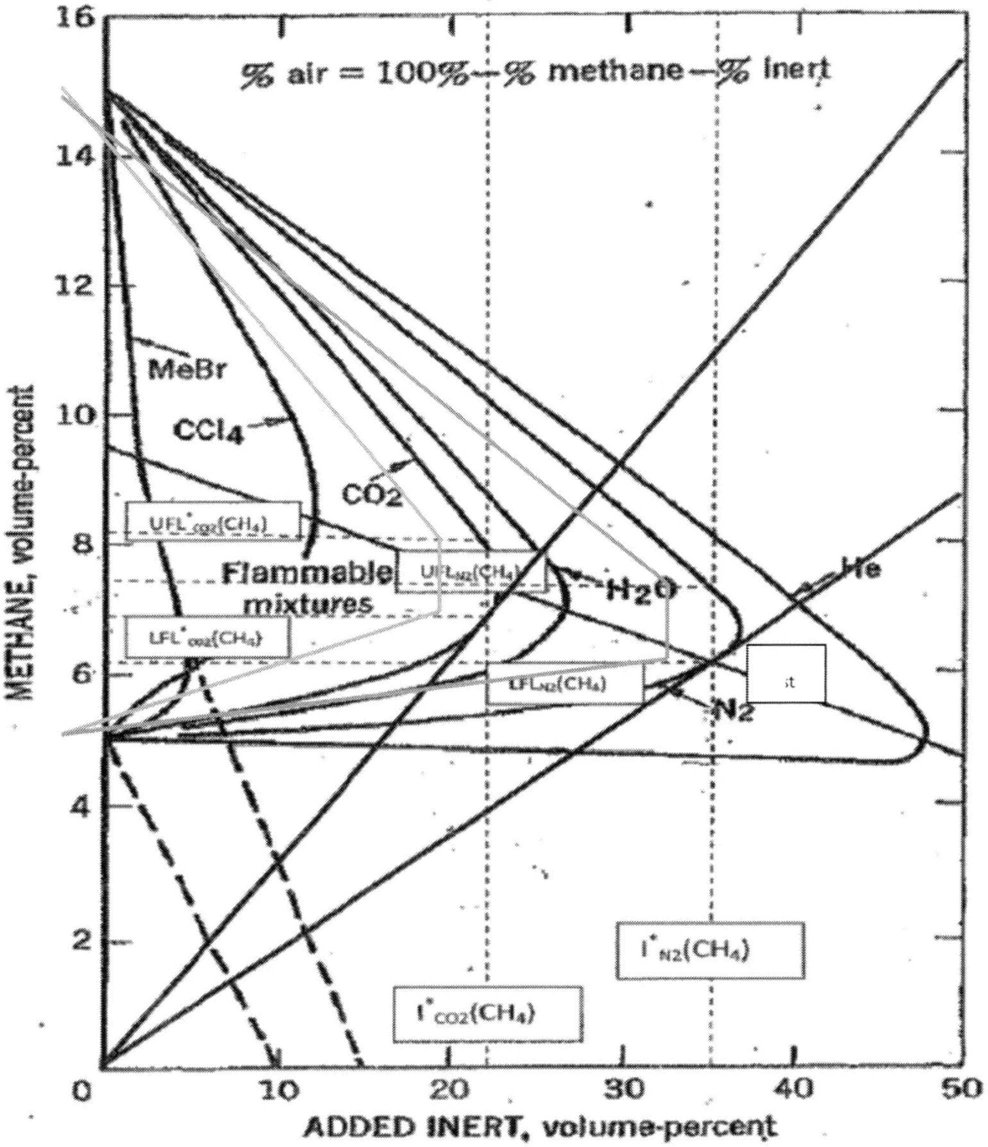

Figure 1: Estimated Flammability Diagram for Methane with Nitrogen and Carbon Dioxide[1]

Values for LFL_i, UFL_i, I_i^*, LFL_i^*, and UFL_i^* for pure component hydrocarbons were estimated from the work of Zabetakis[2] and have been tabulated in **Table 1**. Using these concepts, flammability diagrams for two limiting vent gas compositions encountered during recent flare testing were developed. **Figure 2** represents a high hydrogen content vent gas (pseudo-fuel); while **Figure 3** represents a high nitrogen vent gas. One can also see from **Figures 2 and 3** that the different vent gases represented have considerably different critical steam quantities.

Table 1: Values of I*, LFL*, and UFL* estimated from Zabetakis[2]

Component	I_{N2}^*	I_{CO2}^*	Est. I_{H2O}^*	LFL_{N2}^*	LFL_{CO2}^*	Est. LFL_{H2O}^*	UFL_{N2}^*	UFL_{CO2}^*	Est. UFL_{H2O}^*
Hydrogen	69.5	52.8	57.8	4.1	5.1	4.8	7.3	12.5	11.0
Carbon monoxide	55.0	36.9	42.4	13.4	17.6	16.3	18.1	28.3	25.2
Methane	35.0	22.0	25.9	6.2	6.7	6.6	7.6	8.2	8.0
Ethane	42.1	30.1	33.7	3.3	4.1	3.9	4.1	4.6	4.5
Propane	39.4	26.7	30.5	2.7	3.5	3.3	3.5	4.4	4.1
Butanes	36.5	25.7	28.9	2.2	2.7	2.6	3.4	3.9	3.7
Pentanes	40.5	24.8	29.5	1.9	2.1	2.1	2.7	4.0	3.6
Hexanes	40.0	26.4	30.5	1.6	1.9	1.8	2.7	2.9	2.8
Hexane+	36.8	25.7	29.0	0.7	0.8	0.8	1.3	1.4	1.3
Ethylene	46.9	31.0	35.8	3.4	2.8	3.0	5.9	8.4	7.7
Propylene	39.6	26.0	30.1	2.5	3.3	3.1	3.9	4.7	4.5
Butenes	41.0	28.4	32.1	2.2	2.6	2.5	3.5	4.0	3.8
Pentenes	42.4	28.7	32.8	2.0	2.3	2.2	2.9	3.3	3.1
Acetylene	64.7	48.9	53.6	2.6	2.6	2.6	4.0	14.7	11.4
Benzene	39.1	24.6	29.0	1.5	2.0	1.9	3.1	4.0	3.8
Hydrogen sulfide		25.3	28.9		5.3	5.1		12.9	12.0

The IFC Literature Review defines the critical steam volume fraction, X^*, as the intersection of the stoichiometric air line with the vertical I_{mix}^* section of the flammability diagram. This critical steam volume fraction was defined as the steam at this point divided by the sum of the steam plus fuel at this point, which is represented by the intersection of the light green line and the purple line in **Figure 2**. Mathematically, X^* can be expressed by Equation (5) below:

$$X^* = \frac{(1+\lambda)I^*}{(\lambda I^*)} \qquad \text{Equation (5)}$$

However, as seen in **Figures 2 and 3**, there exist steam volume fractions greater than this X* point that will still pass through the flammability diagram and therefore still support combustion. While this was not immediately apparent in the flare test data obtained from our initial test results, discussion with others who were also doing testing indicated that this conservative estimate of the critical steam volume fraction could indeed represent a reasonable number of operating conditions were high combustion efficiencies might exist. As a result, a Modified IFC steam volume fraction was proposed for this work.

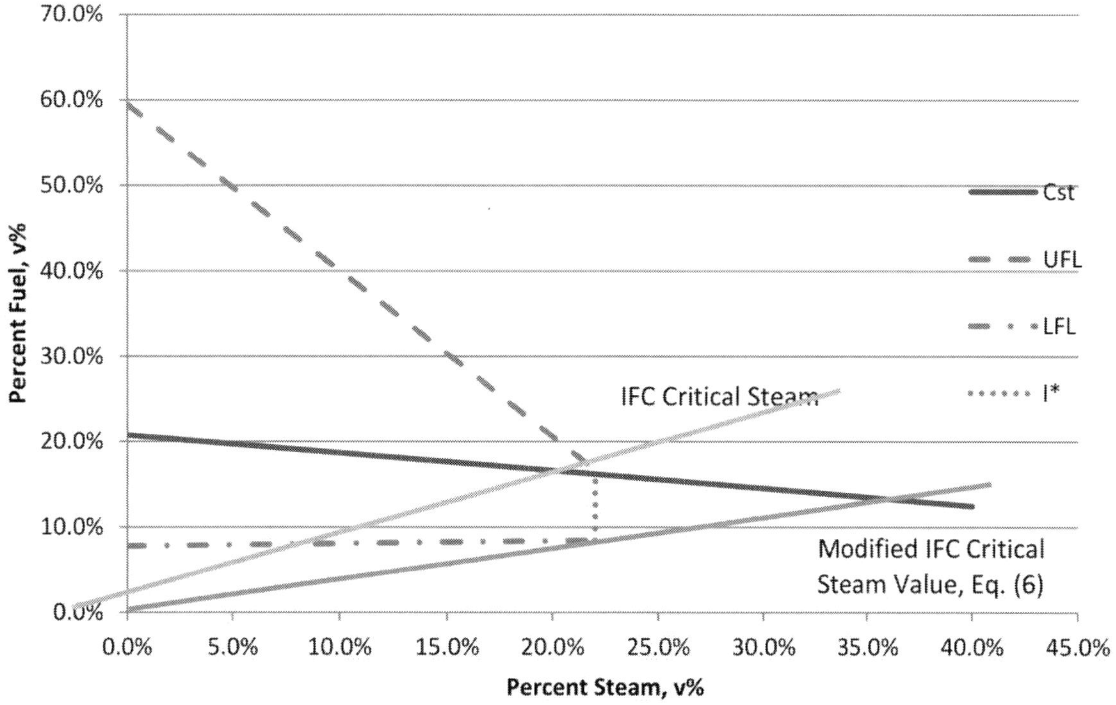

Figure 2: Estimated Flammability Diagram for High N_2 Flare Vent Gas

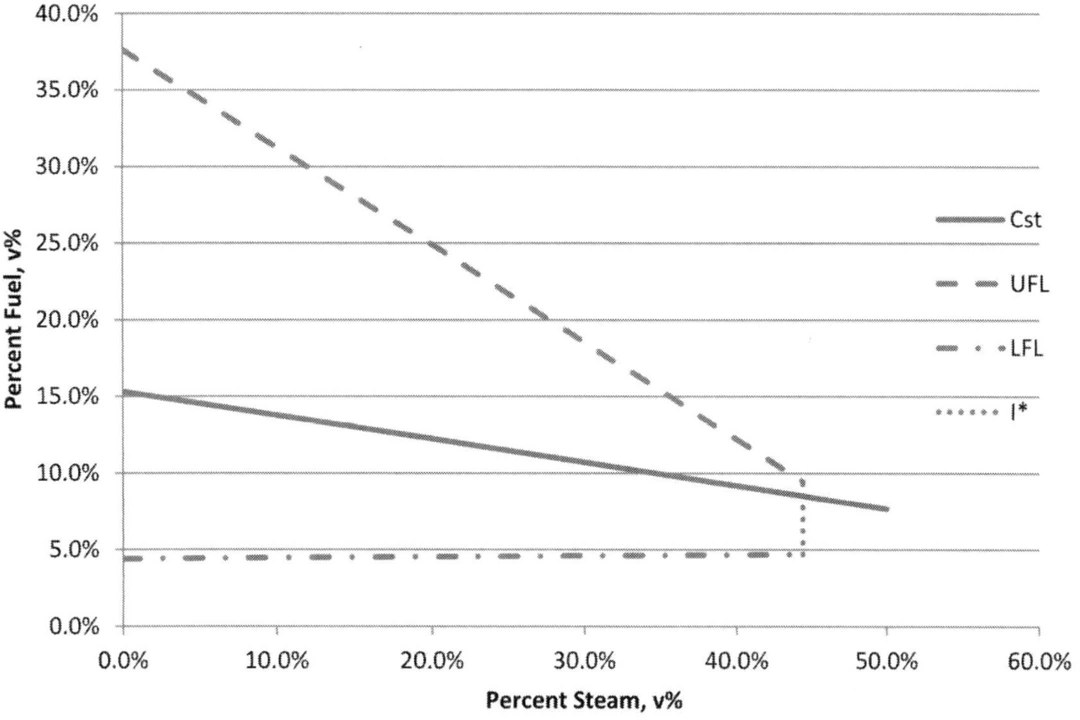

Figure 3: Estimated Flammability Diagram for High H_2 Flare Vent Gas

This new limiting steam value is not represented by the intersection of the vertical I_{mix}^* section with the stiochiometric air line, but rather the tangent to the LFL_{mix}^* point on the flammability diagram drawn from the origin. Recall that I_{mix}^* was uniquely defined in this Modified IFC theory by the tangent so drawn. This critical steam volume fraction is also represented on **Figure 2** as the dark green line. One can see that this formulation more closely defines the true limiting quantity of inert that will still support combustion. This modified critical steam volume fraction has been defined as X'^* and can be mathematically described by Equation (6) below:

$$X'^* = \frac{I(mix)*}{(I(mix)*+F(mix)*)} \qquad \text{Equation (6)}$$

In discussing flare operation, it is useful to expand these flammability diagrams to show the entire range of pseudo-fuel, steam, and air mixtures. This has been done for a given vent gas composition in **Figure 4**. One should note that the line from 0% steam, 100% fuel to 100% steam, 0% fuel represents the line of all possible mixtures of steam and fuel with no air mixed in. In essence, for flares that have center steam addition, the amount of center steam addition represents a unique point on this line and is indicative of the conditions that exist inside the flare stack just prior to exiting the flare tip. If a line is drawn from any point on the fuel/steam line to the origin, this line will represent all possible mixtures of air once the vent gas leaves the flare stack. The point on the fuel/steam line physically represent the point in space where the vent gas leaves the flare stack. If one then draws a line from this point to the origin, this line now represent all possible combination of the combustion zone gas with air. The point on the fuel /steam line representing the composition just prior to exiting the tip and the origin representing the point of infinite dilution with air. By definition, if this line crosses the flammability diagram, then there should be a point in space where combustion should take place if an ignition source exists. The existence of a point in space where combustion can occur does not tell us whether the flame is stable, blown off the stack, or smokes, just that combustion can occur. Obviously, the physical design of a flare will have an impact on the point in space where a combustible mixture is achieved and ultimately the stability of that flame.

Figure 4 can also be used to illustrate that not only is the amount of total steam added important, but also both the ratio of center steam to tip steam and the design of the tip steam injection ring can have an influence upon the limiting steam value where the flame is quenched. In **Figure 4**, the light blue line represents the potential dilution with air if only center steam is injected. Now let us assume that tip steam is also added. The yellow line in **Figure 4** represents an amount of total steam added to the flare above the critical steam value. It can be seen that this line does not cross the vent gas flammability diagram, and therefore combustion should not occur. However, if the amount of steam added in the yellow line case was not all added as center steam, but rather was split between center steam and tip steam, as different conclusion can be deduced. If one assumes that the center steam added is equivalent to the original light blue case, and tip steam is added to reach the total amount of steam added in the yellow case using an ideal 10:1 air to steam educator, then one obtains potential vent gas/steam/air mixtures represented by the medium blue line, which ultimately terminates at the yellow steam line. Clearly, this alternate method of adding the same quantity of steam to the flare leads to a different path where combustion could be supported even with a total steam addition rate beyond the critical value. The dark blue case presents the same center steam addition, but represent a less efficient 3:1 tip steam educator design. As can be noted, again the same total quantity of steam has been added, but the path taken by this design does not result in a vent gas/steam/air mixture that will support combustion.

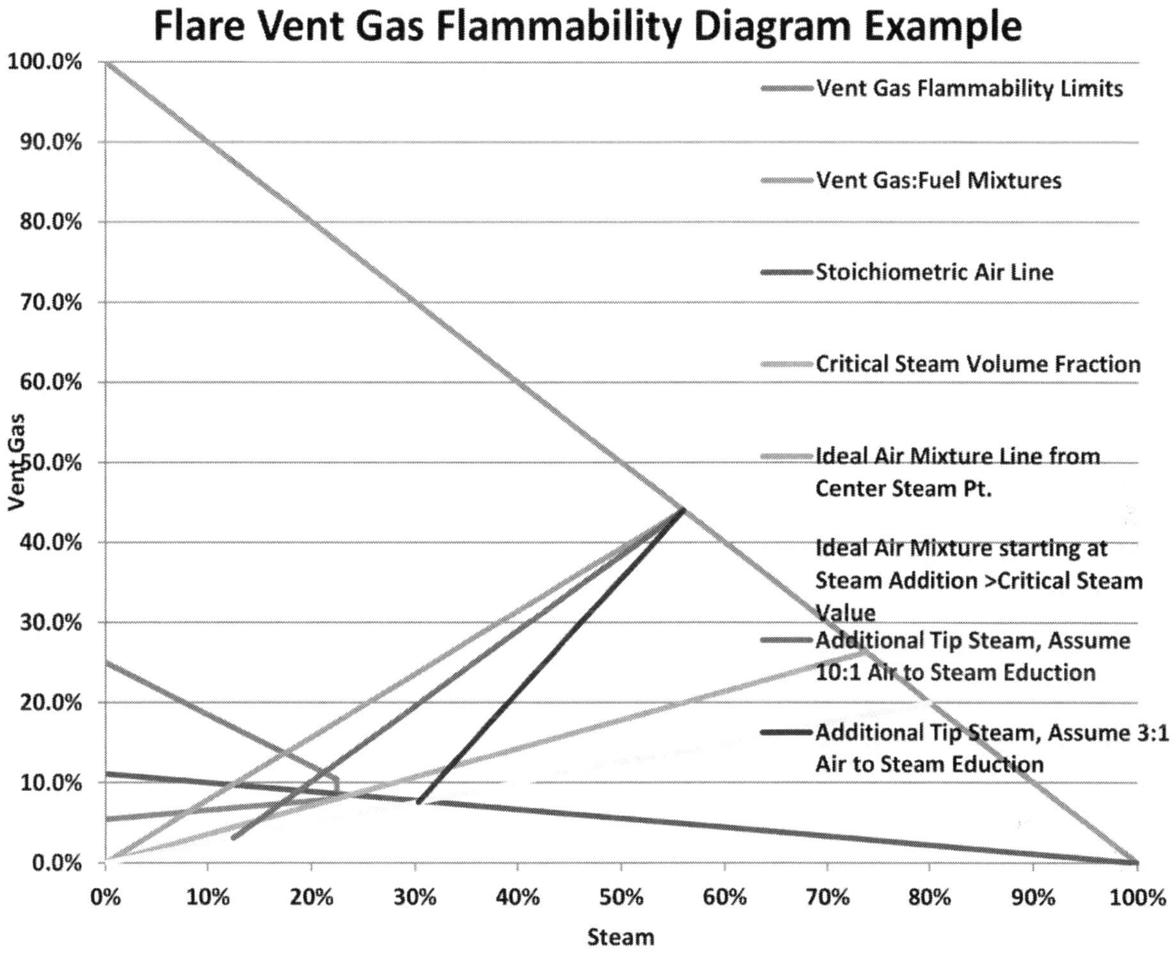

Figure 4: Flare Operating Conditions Expressed on Flammability Diagram

The IFC literature review further defined the dimensionless reduced steam volume fraction (RSVF) as the actual steam volume fraction, X= volume steam added to flare/(volume of fuel + volume of steam to flare) divided by the critical steam volume fraction, represented by Equation (5) in the IFC Literature Review, but by Equation (6) in this Modified IFC Approach. For a flare with center steam only combustion should not occur at an RSFV of 1.0. **Figure 5** displays combustion efficiency data obtained on 7 full scale flares over the past 2 years. The critical steam volume fractions have been determined according to Equation (6), and the observed combustion efficiency data have been plotted against the reduced steam volume fraction (RSVF).

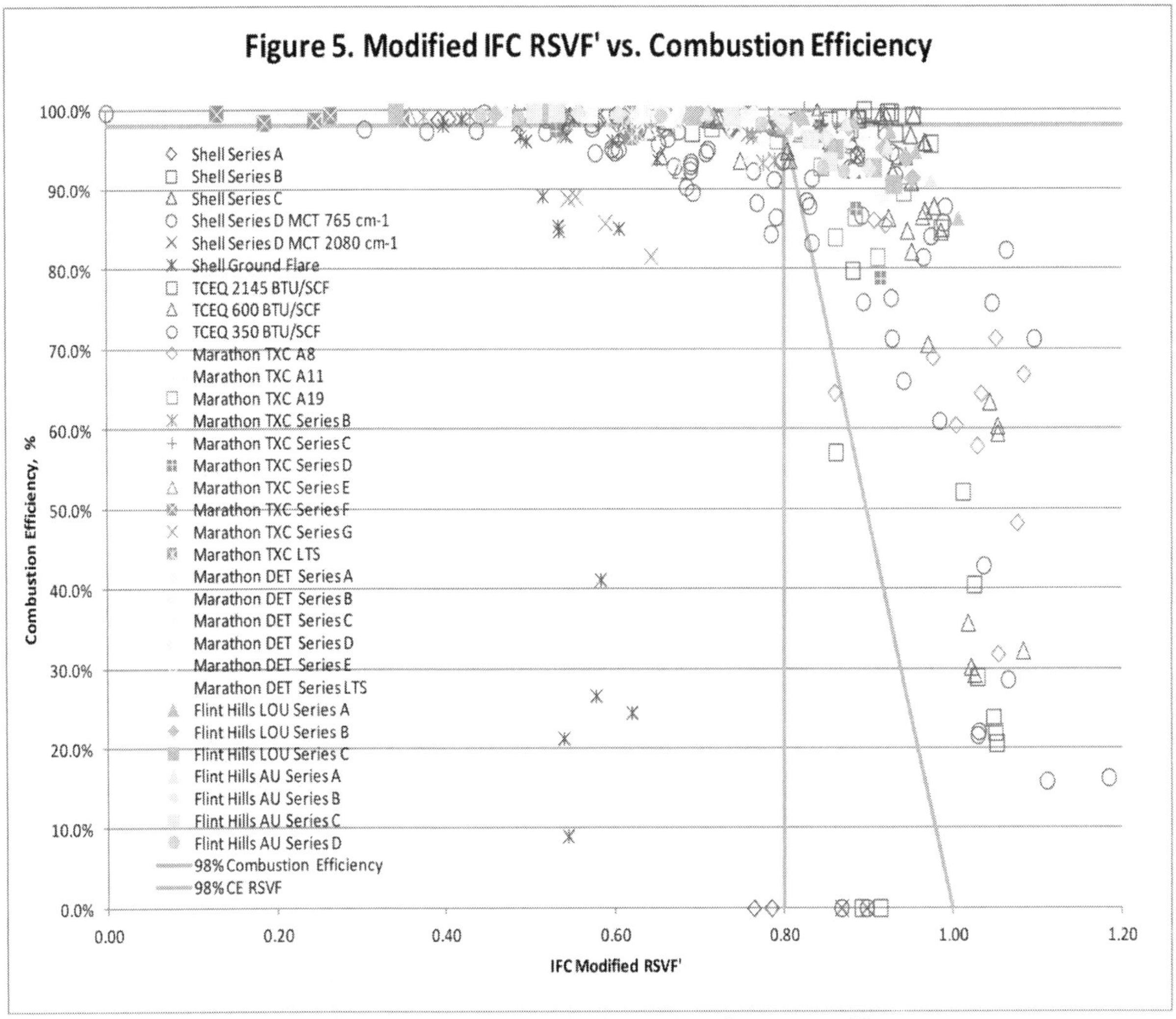

Figure 5: Modified IFC RSVF vs. Combustion Efficiency

Theoretically, the Modified IFC approach would say that above a RSVF of 1 no combustion should be observed. Clearly, while **Figure 5** indicates that combustion efficiency is indeed trending to zero as the RSVF approaches one, some measurable combustion efficiencies have been observed above a RSVF of 1. In this type of pooled, full-scale data there are many potential explanations for variation from the expected result, which include but are not limited to the following:

- Differences in the physical design of the flares (ground flare versus elevated, steam/air tubes versus tip steam, different ratios of center steam to tip steam)
- Different impacts of center steam versus tip steam, as previously illustrated in **Figure 4**
- Measurement uncertainty both in CE measurements at unstable combustion conditions, and steam and vent gas flow measurement at low flow conditions
- Uncertainty in pure component flammability data
- Temporal fluctuation in vent gas composition not picked up by online analyzers

While the ability of this Modified IFC approach to describe these full scale data is far from perfect, this approach does show some promise as a technique that can be used to estimate the critical RSVF (X'^*) and critical combustion zone net heating value (NHV_{CZG}) where a vent gas of a given composition will be extinguished.

Being able to determine the X'^* and NHV_{CZG} value where a given vent gas composition will be extinguished is valuable, of greater interest is a means to estimate the RSVF and NHV_{CZG} below which and above which, respectively, a combustion efficiency of 98% is maintained. From the combustion efficiency data obtained from the testing of any flare, an ample margin of safety parameter can be developed that when applied to the RSVF parameter would yield a limiting RSVF and NHV_{CZG} that ensures a combustion efficiency of greater than 98%. This factor can be estimated by plotting observed combustion efficiency data from a flare versus the Modified IFC reduced steam volume fraction (RSVF′ = X/X'^*), as illustrated in **Figure 5**. The $RSVF_{98\%}$ can be estimated by drawing a line from the theoretical point of flame extinguishment, (1.0, 0) on **Figure 5** to the last point on the test data that produced a reliable combustion efficiency of 98%. From the intersection of this line with the observed data, a vertical line can be dropped down to the RSVF axis. The intersection of this vertical line with the x-axis represents the margin of safety factor applied to the point of flame extinguishment that insures the desired combustion efficiency will is be maintained. For the data shown in **Figure 5** it is clear that this ample margin of safety factor is not necessarily the same value for each flare. However, as shown in **Figure 5**, an RSVF of 0.8 appears to be a reasonable approximation covering the bulk of the data presented in this paper. While beyond the scope of this paper, it is hypothesized that for flares of similar design this empirically derived maximum RSVF could be the same or similar.

Summary

Both vent gas composition and assist gas ratio can have an impact on the observed combustion efficiencies of flares. A modified International Flare Consortium (IFC) approach has been used to assess the critical limiting steam value that will support combustion for any given vent gas/assist gas mixture. This analysis approach has been used to analyze data from seven full-scale flare tests. While combustion efficiency trends to zero at the critical reduced steam volume fraction (RSVF) of 1.0 calculated from this Modified IFC approach for a variety of flares and vent gas compositions, some deviation from the ideal performance was note for a limited number of compositions and flare configurations. While several potential reasons for such deviations have been suggested, full rationalization of these discrepancies was beyond the scope of this work. An approach has also been outlined whereby the critical RSVF calculated from this approach can be used to define the maximum RSVF that can be employed to achieve a specified combustion efficiency. With the exception of one flare in this study, an RSVF of 0.8 gave reasonable assurance of achieving a combustion efficiency of 98%. Further data is required to assess how generally applicable this value is for other flares.

References

1. Gogolek, P., et. al. *"Emissions from Elevated Flares – A survey of the Literature." submitted to* Progress in Energy and Combustion Science, Elsevier Ltd. (expected publication date early 2011).

2. Zabetakis, M.G. *"Flammability Characteristics of Combustible Gases and Vapors,"* Dept. of Interior Bureau of Mines Bulletin 627, (1965).

3. Personal communication Dr. Peter E. G. Gogolek, Scientist, CANMET Energy Technology Centre – Ottawa, Energy Technology and Programs Sector, Natural Resources Canada, Government of Canada, June 20, 2011

4. Molnarne,M., P. Mizsey, and V. Schroder, *"Flammability of gas mixtures Part 2: Influence of inert gases."* Journal of Hazardous Materials, **A121**, pp. 45-49.

AIChE Paper Number: 39c

FLARE SYSTEM INSPECTIONS FOR OLEFINS FACILITIES

James Cleavinger
Fixed Equipment Technical Group Leader
Flint Hills Resources

Keith Wade
North American Ethylene Market Manager
ZEECO, Inc.

Prepared for Presentation at the AIChE 2012 Spring National Meeting
Houston, TX, April 1-5, 2012

FLARE SYSTEM INSPECTIONS FOR OLEFINS FACILITIES

James Cleavinger
Fixed Equipment Technical Group Leader
Flint Hills Resources
Keith Wade
North American Ethylene Market Manager
ZEECO, Inc.

Abstract:

When evaluating the safety and environmental aspects of an olefins facility one must surely consider the flare. The flare is responsible for safely combusting enormous amounts of flammable material during plant upsets and it must perform its designed function in an environmentally acceptable manner. Tolerances are tight and regulations are strict. A flare operating at less than peak condition typically will not perform to the design criteria, creating safety and environmental concerns.

Regular and effective inspection of the flare is absolutely necessary to ensure the safety and compliance of the facility. This paper highlights components of a proper flare system inspection process. The paper will address two basic types of flare inspections: in-service and shutdown. Each type of inspection has its own challenges and advantages.

Flare System Inspection Challenges – An Introduction

Because of the critical role the ethylene unit emergency flare fills, it is one of the last unit process systems available for maintenance and one of the first required back in service after a unit turnaround. With such a short window for maintenance, the condition of the flare system must be understood prior to a unit turnaround. Flares can be inherently difficult to inspect because of the intense heat radiation and physical elevation. However, the flare is dedicated to the safety of the process equipment and the surrounding community and must operate as designed at all times. This paper will discuss the two different types of flare system inspections: in-service and out-of-service / turnaround, and keys to conducting them effectively to keep your flare operating as designed.

In-Service Inspection

Prior to the 1990s, the petrochemical industry standard for flare system maintenance scheduling was based upon time in service rather than detailed inspection. Unfortunately, this approach did not take into account the actual mechanical condition of the flare equipment at the minute level. If flare problems were evident during operation or visible from a ground level inspection, then maintenance time would be built into the turnaround schedule. Upon investigation during the outage, addressing the maintenance issues previously observed would typically reveal additional problems. Discovering maintenance issues in this way usually leads to extended downtime. Facilities were essentially forced to have various spare and often expensive flare parts on hand, or plan for a replacement of their flare tip at every outage to avoid extended downtime during turnarounds.

Alternatively, an in-service inspection from the air can detect problems in the flare system that cannot be identified from the ground. Correctly and completely identifying any flare system issues helps to define the turnaround scope of the flare early enough to procure the equipment and/or schedule the time needed for repairs during an outage. Each facility typically has criteria that determine the proper timing for an in-service inspection, though they are usually performed months in advance of a turnaround.

The main challenge during an in-service inspection of a flare system is getting to the correct vantage point to safely view the upper flare system components while the flare is in service. The original in-service inspection concept included the use of a professional photographer in a manned helicopter. This approach was limited by the direct industry knowledge of the photographer, because the only photos provided for review would be the ones the photographer decided to shoot. Knowing what to shoot to properly inspect a flare system was the key. Over time, the photographers became quite knowledgeable and provided superior quality inspection photos and video. This type of inspection required the owner to interpret the photos, usually in consultation with the Original Equipment Manufacturer (OEM).

A manned helicopter inspection can be performed when a flare unit is in stable operation and atmospheric conditions are favorable, i.e. little to no wind, and clear skies. Depending on the complexity of the flare system, this inspection can take 30 minutes to two hours of helicopter airtime. The inspection includes an evaluation of all photos and videos while the helicopter remains on-site to determine if further shooting is required. A formal flare inspection report is available within weeks and provides enough time for the maintenance groups to prepare for any upcoming plant turnarounds.

This particular type of inspection method was not universally accepted due to concerns of placing helicopter personnel in harm's way. Tight coordination between operations and maintenance is critical to minimize the exposure of the manned helicopter inspection team to potential hazards. The manned helicopter is equipped with a direct radio link to the plant's operations control room to communicate any plant upsets, allowing the helicopter adequate time to evacuate the flare area.

A second type of in-service flare inspection is inspection by a remote control helicopter. Early flare system inspections in the late 1990s and early 2000s using remote control helicopters were limited by the quality of the photographic equipment and the availability of experienced fliers. The helicopters are expensive and the concern over losing control on one while airborne over a process area requires a qualified flier. Today, photography and video capabilities have improved and more remote control fliers have acquired the necessary experience needed. Similar coordination between operations and maintenance is required for a remote control helicopter inspection. Real-time video is provided during the remote helicopter inspection process to determine if more in-depth or additional inspections are necessary.

The manned and remote control helicopter inspection methods are used to identify any issues within the flare system that may require maintenance but are not identifiable from ground level. Each facility has repair and replacement criteria to determine the proper timing for an in-service inspection and they are usually performed months in advance of a necessary plant turnaround.

During an in-service inspection, the main inspection area is the flare tip and upper structure. These areas are impossible to properly observe and inspect from the ground. The upper structure includes the ladders, platforms, flare stack, supporting structure, guy wire lugs and connections, aviation lighting fixtures and cabling. Any ancillary flare components such as the exterior of the molecular seal, steam piping, pilot gas piping, ignition gas piping and any retractable systems like the pilots, thermocouples, and aviation lights should also be inspected. The longest lead item is typically the flare tip itself and therefore, it should be the primary focus of the in-service inspection. The flare tip inspection checks for any indication of material failures including welds on steam manifolds, welds on the flare tip body, the muffler, the pilots and the internal steam/air tubes. Visual indication of internal burning could indicate a steam/air tube internal weld failure, insufficient center steam, or capping of the flare. Flare capping is the result of a large amount of steam being supplied to the upper steam ring while not enough steam is provided to the lower steam ring. The visual indication will be evidenced by discoloration of the flare body and/or muffler area and can be seen as a "hot spot" in low lighting. Repeated operation of the flare with internal burning will severely damage the flare body and internal steam / air tubes.

In-service inspections should also carefully observe the pilot gas connection to the pilot gas manifold. This area is prone to piping failure if too much pipe stress is introduced onto the pilot mixer. Piping supports, guides, and hangers need to be inspected for proper alignment and bolting. This is important on the pilot gas, ignition lines, and the steam lines. In the steam piping, the heat-affected components of the flare tip can be prone to weld failure caused by a lack of cooling steam or water hammer from wet steam. Steam piping insulations are a commonly overlooked area and should be considered when conducting the inspection process

If the flare system was supplied with ladders and platforms, they need to be visually inspected during the in-service inspection so that problems can be identified and corrected before personnel are allowed access. If the system is a structure-supported flare, the structure itself should be inspected for damage or missing bolting. In a guy wire supported system, the attachments to the flare stack and the lugs should be observed in the inspection. In a self-supported flare system, any flanged connections in the flare stack should be inspected. Visual inspection of the flare stack itself for indications of corrosion or internal burning will help determine the turnaround maintenance scope and ensure that parts and scheduling are accurate. A careful and complete in-service inspection reduces downtime during turnarounds by identifying long lead-time items and accurately forecasting the timeline necessary for flare system maintenance during the turnaround.

Out-of-Service or Turnaround Inspection

Even with the improvements in the photography and video capabilities of the remote control helicopters used for in-service inspections, there are still areas of the flare system that cannot be inspected until the flare is out of service. While the external components of the flare system can be inspected fairly well with the in-service inspection, accurate inspection of the internal components requires the flare system to be out of service and cleared for inspection. API standards 510 and 570 should be used as criteria in developing out of service inspection details.

Components such as the flare system knock out drum, molecular seal, and liquid seal drum must be internally inspected while the flare system is out of service and cleared. The OEM specifications require internal components inspections while the flare is out of service. The steam/air tubes, and the knock out and liquid seal drums should be examined for any weld failures, fouling, and internal corrosion issues. The results of each of these internal inspections can significantly impact a flare system outage and result in additional plant downtime.

Summary

Flare systems fill a critical role in the safe and environmentally compliant operation of an ethylene plant and therefore must operate as designed. Unfortunately, the useful life and performance of a flare are impacted by unexpected problems within the ethylene unit operations or supporting utility system. A combination of inspection approaches should be utilized in order to most effectively maintain and efficiently operate a large ethylene flare. In-service inspections are the right tool to help identify problems, schedule parts and prepare a turnaround scope of internal inspection or repair for a flare. The in-service inspection should be evaluated in conjunction with the flare OEM or other flare suppliers for an unbiased opinion. Some concerns raised during the in-service inspection may require more immediate attention while others identified can help define the scheduled turnaround scope and ensure needed parts are on-hand in advance.

Out-of-service inspection of the flare system will often yield unforeseen additions to the planned scope of work, but is necessary for proper inspection of internal engineered components. Some of these issues may be able to be deferred until the next scheduled maintenance window, while others will require immediate attention. Using both in-service and out-of-service inspections to proactively identify problems and plan necessary maintenance based on actual observed conditions reduces unplanned downtime and potential system failures.

AIChE Paper Number: 39d

PERFORMANCE, MAINTENANCE AND ENVIRONMENTAL BENEFITS OF AN INTEGRAL SCR

Stephen De Haan
Vice President Olefins
Lummus Technology

Thomas Garza, P.E.
Senior Fixed Equipment Specialist
Chevron Phillips Chemical Company LP

Jacob Hilbrich
Process Engineer
Chevron Phillips Chemical Company LP

Prepared for Presentation at the 2012 Spring National Meeting
Houston, TX, April 1 - 5, 2012

PERFORMANCE, MAINTENANCE AND ENVIRONMENTAL BENEFITS OF AN INTEGRAL SCR

Stephen De Haan
Vice President Olefins
Lummus Technology
Thomas Garza, P.E.
Senior Fixed Equipment Specialist
Chevron Phillips Chemical Company LP
Jacob Hilbrich
Process Engineer
Chevron Phillips Chemical Company LP

Abstract: Selective Catalytic Reduction (SCR) involves injecting a NOx reducing chemical, such as aqueous ammonia, into an exhaust stream in the presence of a catalyst within a specific temperature window of approximately 650°F to 1100°F. To reduce NOx emissions from an ethylene cracker, Chevron Phillips Chemical Company LP retrofitted new convection sections on 13 existing ethylene furnaces, which incorporated both integral steam superheating coils and integral SCR's for NOx removal.

This paper expands on the information presented at the 2008 AIChE Ethylene Producer's Conference titled "Cracking Heater Convection Retrofit with Integral SCR." The subject of this paper will discuss the long term environmental benefits, performance, and maintenance of the installation of an integral SCR.

Introduction

Cedar Bayou Olefins Plant implemented an integral SCR for NOx removal on each of its existing 13 cracking furnaces, beginning in 3Q 2003 and completed in June 2004. The 2008 AIChE Ethylene Producer's Conference paper "Cracking Heater Convection Retrofit with Integral SCR" discussed the engineering design, implementation, and initial operation of the integral SCR. This paper expands on the information presented in 2008 and focuses on the long term environmental benefits, performance, and maintenance of the installation of an integral SCR.

Environmental Benefits

Starting in 3Q 2003, a retrofit of the convection sections of all 13 cracking furnaces was performed. The existing convection modules were replaced, steam superheating capability was added, and an SCR for NOx removal was included. A simplified diagram of a furnace convection section with the integral SCR is shown in Figure 1.

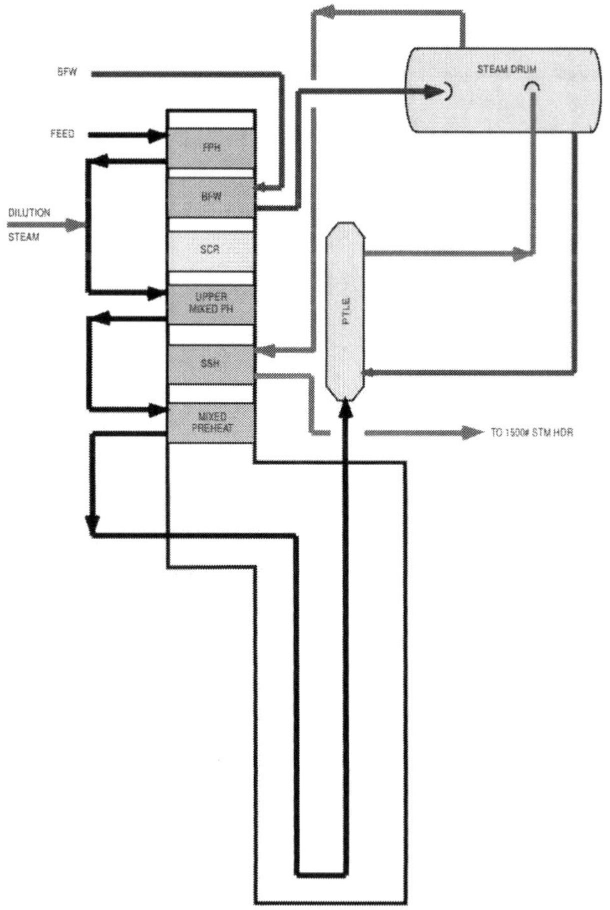

Figure 1 – Integrated SCR: Simplified Diagram

The selected SCR catalyst, Haldor Topsoe DNX-930, uses a corrugated fiber-reinforced titanium-dioxide (TIO_3) carrier. The center is homogeneous and impregnated with the SCR-active components: tungsten trioxide (WO_3), and vanadium pentoxide (V_2O_5), see Figure 2. This catalyst is considered a medium temperature type and operates best at approximately 750°F. The SCR was therefore added above the upper mixed preheat to assure that the flue gas temperature was sufficient.

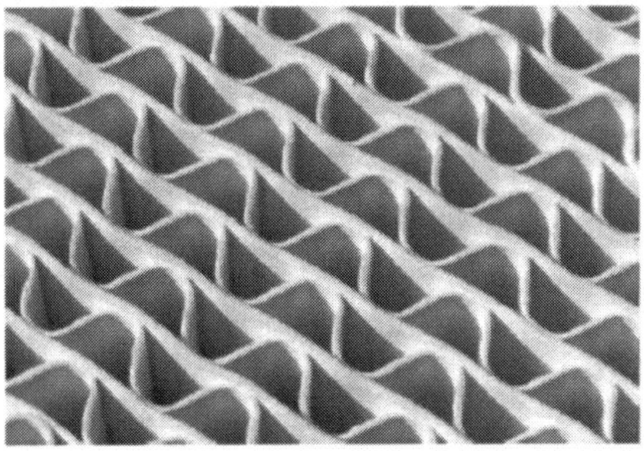

Figure 2 – Haldor Topse Catalyst DNX-930

Performance

The original goal for the integral SCR was to remove NOx to a concentration below 10 ppm in the stack, while maintaining an ammonia slip, in the stack, also below 10 ppm. The NOx concentration that enters the SCR is generally between 80-100 ppm. With over eight years in operation, the cracking heaters have continuously achieved the reduction goal. In addition, since steam superheating was added to the convection sections, the original separate steam superheater was idled. The NOx generated in the steam superheater was therefore eliminated.

Figure 3 shows the total tons of NOx created by the 13 cracking heaters, as well as the lb NOx/mmbtu from 2008 to 2011.

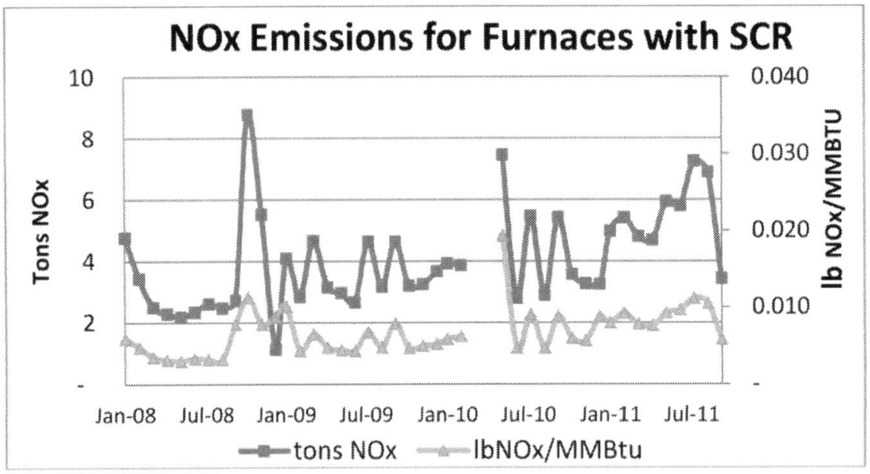

Figure 3 – Environmental Benefits of Integrated SCR's

The recycle ethane furnace (BA-113), as previously presented via the 2008 AIChE Ethylene Producer's Conference paper "Cracking Heater Convection Retrofit with Integral SCR" , was the first to be converted. As of January 2012, the initial catalyst charge has been on-line for almost eight and half years, which has well exceeded the three year guaranteed life of the catalyst. The performance of the recycle ethane furnace is shown in Figure 4, which shows the single digit ppm levels of both NOx and NH_3 slip in the stack.

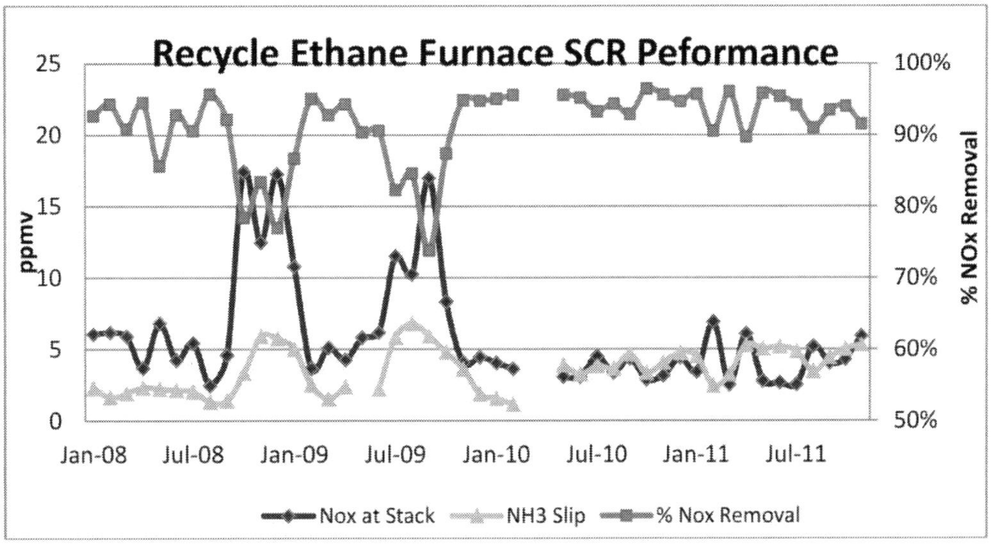

Figure 4 – Recycle Ethane Furnace SCR Performance at Eight Years in Service

Similar results have been achieved for furnaces BA101 through BA112. Figure 5 shows the performance of these furnaces and the single digit ppm levels of both NOx and NH_3 slip in the stack.

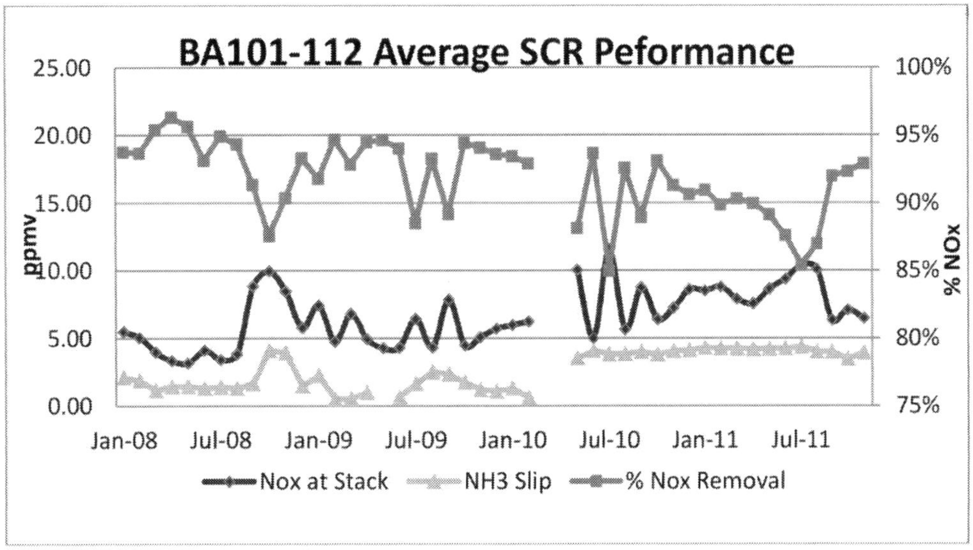

Figure 5 – Furnaces BA-101 through BA-112 SCR Performance at Eight Years in Service

The convection section retrofit for the recycle ethane furnace resulted in reduction of the stack temperature from 490°F to 299°F. Comparable improvements, the stack outlet temperature of BA-101 through BA-112, from approximately 420°F to 275°F, were also achieved.

The only significant impact of the integrated SCR on furnace operation results from the temperature of the flue gas entering the catalyst. Generally, the furnaces are not limited by the operating temperature of the SCR; however, the maximum operating temperature can be reached during operation. The maximum temperature for the inlet of the SCR is 825°F. There are times, most frequently when operating at minimum feed rate and high conversion, that this can be the furnace limitation. Furthermore, afterburning, in the convection section caused by low excess oxygen in the radiant section, can lead to high temperatures at the SCR bed. These high temperatures can limit the furnace feed and/or conversion by not allowing additional firing until the temperature of the flue gas entering the SCR is reduced. Adjusting excess air in the radiant section and/or balancing feed and conversion to reduce fired duty reduces the SCR temperature and removes the limit.

Maintenance

The maintenance of an integral SCR focuses on the catalyst, NH_3 injection, and instrumentation.

The major catalyst maintenance requirement is to vacuum any particles that can collect on the catalyst. These particles are deposited during decoke operation, since these furnaces decoke into the firebox, (Figure 6). High flue gas temperatures and low SCR NOx removal efficiency can be an indication of a fouled SCR. The plugging that occurs can cause poor temperature distribution in the convection section and poor NOx reduction due to the channeling that occurs across the bed.

Each time a furnace is brought down for maintenance and the furnace box is entered, the SCR should be inspected and vacuumed, if necessary. The vacuuming can either be completed in place, or the beds removed, cleaned and reinstalled. If time permits or the convection section is being cleaned, it is generally better to fully remove the bed for cleaning and inspection; however, the vacuuming can be completed more quickly in place. The SCRs can operate about two years between cleanings. Furthermore, care must be taken when the SCR is installed to seal the area around the box and at the door of the SCR to limit channeling along the sides of the furnace box. Flue gas channeling that is caused by either fouling or a poor seal will negatively impact NOx reduction.

In addition to cleaning the catalyst, the NH_3 injection system must also be inspected when the furnace is shut down for maintenance. The ammonia injection nozzles can plug and cause poor NH_3 distribution across the SCR, which can lead to increased NOx and NH_3 concentration in the stack.

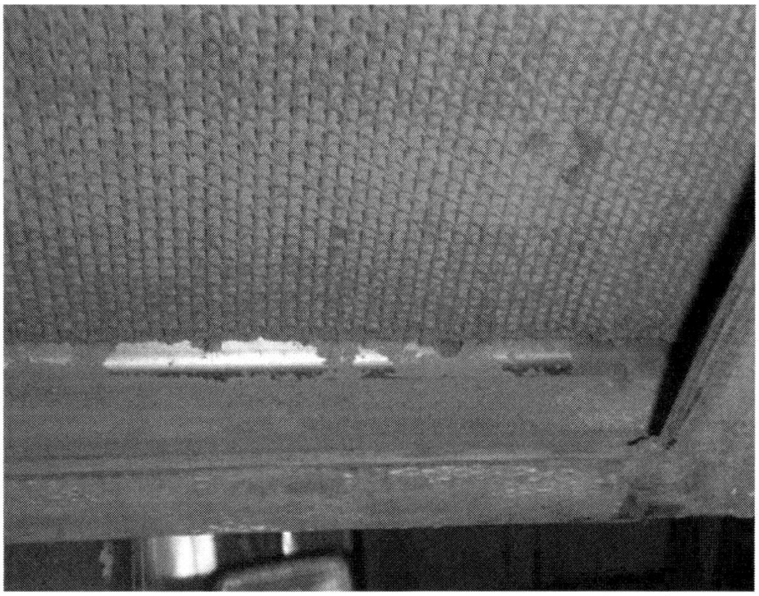
Figure 6 – Bottom of SCR bed prior to cleaning

The original guaranteed life expectancy of the SCR was three years. Eight of the 13 beds are original. Four furnaces have needed one replacement bed, and one furnace has required two replacements. The beds were replaced due to low NOx removal efficiency. The low efficiency can either be caused by poisoning, which is mostly due to chromium from the furnace tubes, or high temperatures during operation or decoke that can damage the SCR. The high temperatures may be caused by afterburning, as mentioned above. The SCR beds can be installed with coupons that can be used to test catalyst activation and help determine whether or not the bed needs to be changed. It is important to keep in mind that the NOx reduction efficiency can be significantly impacted by instrumentation and that replacing the catalyst bed is not always the solution for poor NOx reduction.

The instrumentation is critical to the success of the integral SCR. Not surprisingly, the NOx analyzer itself can impact the success of the integral SCR; however, several other measurements can also impact the NOx reduction. Among these, the calculation for ammonia slip is the most significant. The NH_3 flow control system on the furnace is automatically controlled and increases NH_3 flow to achieve a desired NOx emission of less than 10ppm without violating a NH_3 slip greater than 10ppm.

One way to monitor ammonia slip -- the amount of ammonia emitted through the stack -- is by calculating the value. The calculation is based upon the ammonia flow and concentration, the NOx concentration upstream and downstream of the SCR, the fuel gas flow and composition, and a correction factor that is the ratio of the measured slip to the calculated slip based on stack sampling. If any one of these values is not accurate, the ammonia slip calculation will be affected, and therefore can impact the NOx removal. If the calculated ammonia slip is erroneously high, the ammonia flow will be reduced and NOx will increase in the stack to stay below the NH_3 slip limit of 10 ppmv. For more information on the calculation of ammonia slip, consult TCEQ Air Regulations Chapter 117.8130. Ammonia Monitoring.

Results

As with any operating unit, there are issues that arise and require attention. As mentioned, catalyst and instrumentation maintenance are required and operational limitations can exist when operating an integral SCR; however, the performance of the convection sections and integral SCR installations has been extremely successful for the eight years since installation. Ethylene furnace NOx reduction has averaged 92% from 2008 to 2011, with eight original beds still in operation, as of December 2011. This design has achieved a NOx emission specification that other technologies cannot achieve.

Acknowledgements

The authors would like to thank the significant contributors to this paper:

- Dennis Salbilla of Haldor Topsoe Inc.
- Phil Crepinsek of Chevron Phillips Chemical Company.

References:

1. Karrs, M., P. Crepinsek, "**Cracking Heater Convection Retrofit with Integral SCR,**" Proceedings of the 20th Ethylene Producers Conference, Vol. 17, pp. 276-287, American Institute of Chemical Engineers, New York, NY (2008)

AIChE Paper Number: 39e

CONSIDERATIONS FOR FLARE GAS RECOVERY UNIT DESIGN AND OPERATION IN AN OLEFINS PLANT

John Swanson
Reliability Center Manager
Flint Hills Resources

Prepared for Presentation at the 2012 Spring National Meeting
Houston, TX, April 1-5, 2012

CONSIDERATIONS FOR FLARE GAS RECOVERY UNIT DESIGN AND OPERATION IN AN OLEFINS PLANT

John Swanson
Reliability Center Manager
Flint Hills Resources

Abstract:

In 2011, Flint Hills Resources installed a flare gas recovery unit at its Port Arthur Olefins plant. This paper will discuss items that had to be considered to apply flare gas recovery in an olefins plant. The items discussed will include: how to get accurate flare flow and composition data to size the unit, how to manage intermittent flare streams with high volume and high nitrogen content, controls for compressor staging and de-staging, handling of hydrocarbon liquids that condense during compression, and lessons learned during startup and operation.

Introduction

Flare Gas Recovery is a system of compressors, seal drums, knockout drums, and coolers designed to recover gas from a flare header and put it to an alternative use, normally as a component that is blended into the plant's fuel system. Flare gas recovery has been installed in quite a few refineries, but adoption into olefins plants has been much slower because the olefins plant offers some challenges to the designers that are not found in a refinery.

In 2011, Flint Hills Resources installed a Flare Gas Recovery Unit (FGRU) at its Port Arthur Olefins Plant. This paper will discuss items that had to be considered in the design and start-up of the FGRU.

Benefits of Flare Gas Recovery

The FGRU offers two benefits. First, emissions of Volatile Organic Compounds (VOCs) are reduced. This is because the generally accepted combustion efficiency for flares is approximately 98% versus the combustion efficiency in cracking furnaces of over 99.9%. Second, there are energy savings because the recovered gas replaced an equivalent amount of natural gas that otherwise would be purchased.

Design objectives

The FGRU design team worked against the following objectives:

1. The FGRU was expected to operate continuously between Light Olefin Unit (LOU) turnarounds. This meant that the team had to design the unit for a 5 to 7 year run.
2. The FGRU was designed to handle the normal day-to-day flare volume, with acceptable margin for flaring from maintenance activities and minor process upsets.
3. The FGRU designers had to consider that certain operations (dryer and reactor regenerations) put large quantities of inert gases into the flare.
4. The FGRU was not designed to handle the large flows that might occur during a major process upset, such as a cracked gas compressor trip.

In addition to the considerations listed above the team had to make the following decisions:

1. What type of compressors (liquid ring, reciprocating, etc.)?
2. What size and how many compressors to use?
3. How to handle liquid hydrocarbons and water that condensed during compression?
4. What second order effects, if any, would there be from connecting two previously independent flare systems?

Compressor Selection and Sizing

The team selected a contractor (John Zink) and the compressor type (liquid-ring) based on the positive experience of a similar system that had been installed at the FHR Corpus Christi Refinery. Although the prior experience with the technology at Corpus Christi was very helpful, it did not answer the question of what size and how many compressors were required.

Each of the flares at the facility is equipped with a GE Panametrics flow meter. In addition, a radioactive tracer study was conducted to validate the flow meters. The radioactive tracer study was conducted by TracerCo and it involved placing detectors at several points on the flare line, injecting a small amount of a radioactive tracer, and measuring the time that the tracer pulse took to travel between the detectors. The tracer study confirmed that data from the Panametrics meters was accurate for the FGRU design work.

The team analyzed 12 months of data at each flare and decided upon three 1000-acfm compressors. The normal base load is approximately 1½ compressors worth. A third compressor was added to allow flexibility for compressor maintenance and for increases due to process upsets and/or maintenance activities in the unit.

Liquid Hydrocarbons

Liquid-ring compressors use water to make the seal in the compressor and the FGRU designs all account for handling of this water. The process simulations showed that there would not be any free hydrocarbon in the water. However, plant experience led us to believe that there could be cases where liquid hydrocarbon would need to be addressed. The solution was to install a weir in the inlet knockout drum, with a separate draw off and pumps for the oil.

Specifying the right pump to handle a stream that did not appear in process simulations was a challenge for the design engineers. The challenge was eventually solved by telling the design engineers which make and model of pump to use. The team used nitrogen operated diaphragm pumps.

Inert Gases

Olefins plants perform operations on a periodic basis that put large quantities of nitrogen into the flare header. Regeneration of the propylene product dryers is a good example. Introducing large quantities of nitrogen into the plant fuel system would adversely affect the cracking furnaces.

The team designed a system that would allow the nitrogen rich vent gas from the propylene dryers to bypass the FGRU, while keeping the base-load flare gas going to the FGRU. The system consists of a 2nd seal drum, in parallel with the main seal drum. During dryer regenerations the nitrogen rich off-gas from the dryers is diverted to the flare through the 2nd seal drum, while the base-load flare gas continues to go to the FGRU.

Connecting Flare Systems

The team was concerned that connecting two independent flare headers through the FGRU might provide an opportunity for a high pressure event in one of the flare headers to pressure up the other flare header. This concern was mitigated in several ways: 1) Each flare system has its own water seal drum that will relieve to a flare if the pressure in the header gets above the seal depth. 2) The only connection between the flare headers is at the suction drum of the FGRU. Each line was fitted with an air actuated block valve that will stop flow to the FGRU if the header pressure goes above the desired value. 3) The lines from the flare headers to the FGRU are much smaller than the main flare headers. This minimizes interaction between the two flare headers.

Design for Unit Reliability

The objective of the design team was to have two compressors available 100% of the time, with the only scheduled outages being during LOU turnarounds. This meant that equipment that might need to be serviced more often than every 5 – 7 years would have to be spared. This includes the liquid ring compressors, demineralized water supply pumps, recovered oil pumps, and service water pumps. In addition, the fin-fans used to cool the service water were built in two banks so that one of them could be isolated as necessary.

Operational and Maintenance Issues

The FGRU was commissioned in July of 2011 and it has proven to be a reliable unit. The only issue of any significance had to do with some fouling that we observed in the gas recycle valve. The liquid ring compressors are constant volume machines, and a recycle valve is used to spill gas from the discharge to the suction in order to keep the compressors loaded. In some situations the recycle valve would fully open and would not recycle enough to keep the compressors from pulling the suction drum pressure below the low-pressure trip point. The valve was removed from service and disassembled. Flow through the valve had been obstructed by a hard material that had blocked some of the slots in the valve trim. We believe that this material was dissolved solids that were left behind as entrained liquid/mist in the recycle gas was flashed across the recycle valve. We have initiated a monitoring program on the recycle valve to ensure that fouling in the valve is detected before it becomes an operational concern.

Conclusion

Flare gas recovery can be successfully installed in an olefins plant, as long as the plant has a destination for the recovered gas, and the designers can account for large quantities of inert gases that come from dryer and reactor regenerations.

Flare gas recovery offers several benefits: 1) Reduced emissions of VOCs, due to the high combustion efficiency of furnaces as compared to flares, and 2) Energy savings, as the recovered gas can be used to replace purchased fuel.

AIChE Paper Number: 39f

OPERATION AND MAINTENANCE OF ENVIRONMENTALLY CRITICAL PROCESS ANALYZERS

Naresh Sethi
Engineering Group Leader Process Analyzers
Shell Oil Products Company LLC

Prepared for Presentation at the 2012 Spring National Meeting
Houston, TX, April 1-5, 2012

OPERATION AND MAINTENANCE OF ENVIRONMENTALLY CRITICAL PROCESS ANALYZERS

Naresh Sethi
Engineering Group Leader Process Analyzers
Shell Oil Products Company LLC

Abstract:

A key distinction between process control analyzers and environmentally critical analyzers is that while operation and maintenance of process control analyzers is entirely within the discretion of the plant management, specific rules mandate installation, calibration, operation, maintenance, quality assurance/quality control (QA/QC) and record keeping requirements for environmentally critical analyzers. These rules mandate specific daily, quarterly and yearly activities to ensure analyzer data quality and procedural requirements to take corrective actions when an environmentally critical analyzer is found to be malfunctioning.

Here, we discuss some of these rules in relation to the technology, operation, and maintenance of two types of environmentally critical process analyzers: CEMS (Continuous Emissions Monitoring Systems) that are used to monitor NOx, SOx and CO emissions from furnace stacks, and HRVOC GC (Gas Chromatograph) designed to monitor VOC (Volatile Organic Compound) emissions from Cooling Water Towers and Flares. Organizational capabilities and analyzer maintenance best practices to maintain high reliability of these environmentally critical analyzers to ensure compliance with applicable rules will also be discussed.

Introduction

Process analyzers are ubiquitous in any large petrochemical manufacturing plant. Process analyzers are on-stream instruments that provide real-time analytical data on process streams to control the process to ensure safe, reliable, and efficient operation. For most manufacturing plants process analyzers are also mandated license-to-operate requirements for compliance with federal and local environmental rules to monitor and quantify air pollutant emission from various sources. Mandated process analyzers used to measure air pollution emissions from emission sources such as furnace stacks, flares and, cooling water towers are referred to here as environmentally critical process analyzers. The focus of this paper is analyzers that measure concentrations of specific pollutants: NOx, SOx, CO and VOCs. A subset of VOC components; Ethylene, Propylene, Butenes, and Butadienes, are also referred to as Highly Reactive VOCs or HRVOCs.

Environmentally Critical Process Analyzers

Environmentally critical process analyzers referred to here are the CEMS and the HRVOC systems. These analyzers measure concentration of air pollutants in the gases that are vented to the atmosphere from manufacturing process units. CEMS analyzers are generally well known systems that have been in use in manufacturing sites to measure pollutants in flue gases from furnace vent stacks. CEMS analyzer systems are typically used to measure concentration of air pollutant such as NOx (Nitrogen Oxides), SOx (Sulfur Oxides), and CO (Carbon Monoxide). HRVOC analyzers, which are relatively new, were mandated by new regulations by Texas Commission on Environmental Quality (TCEQ) in early 2000s requiring manufacturing sites to monitor emission of Volatile Organic Compounds (VOCs) from cooling water tower exchanger system and flares. Light olefin components of the VOCs are referred to as Highly Reactive VOCs and hence the acronym HRVOC. HRVOC analyzers are generally not as well known as the HRVOC regulations are specific to manufacturing sites located only in Houston/Galveston and surrounding areas of Texas gulf coast[(1)].

An analyzer is an integral part of air pollutant monitoring scheme that also includes other instrumentation such as flow meters, temperature transmitters, and pressure transmitters. Air pollutant emissions are reported in total mass emission (lbs) and also rate of emission e.g. in units of pounds per hour (lbs/hr). The analyzer system is designed to continuously measure the concentration of the pollutants in the process stream. The analyzer output is combined with data from other instrumentation such as a flow meter to calculate total mass of pollutant emitted to the atmosphere and its emission rate. Various emission control regulations are based upon either the total emissions or the emission rate or a combination of the two depending upon specific pollutant and the operating permit.

CEMS and HRVOC analyzers are complex engineered systems. The entire system includes the analytical instrument that is used to measure concentration of air pollutant constituents and a sampling system designed to continuously provide a fresh and representative process stream sample to the analyzer. The sampling system is designed to extract a few micro liters of sample from the process stream that has been cleaned and is representative of the entire process stream. Reliability of the entire analyzer system is critically dependent on the reliability of the sampling system. The following are brief schematic descriptions of the CEMS and HRVOC analyzer systems.

CEMS Analyzers

Figure 1 shows a schematic of a typical CEMS analyzer system designed to measure concentration of NOx, SOx and CO pollutants along with Oxygen. The pollutant concentration as measured must be corrected for diluents effects of Oxygen and water vapors. The pollutant concentration data is reported in moisture free (dry) and O2 free basis. Oxygen is typically measured by an analyzer, but water content can be either measured using an additional analyzer or the sample system designed to remove all the moisture from the sample prior to introducing it into the analyzer. In the schematic shown, moisture content is removed, i.e., sample is dried before introduction to the analyzer.

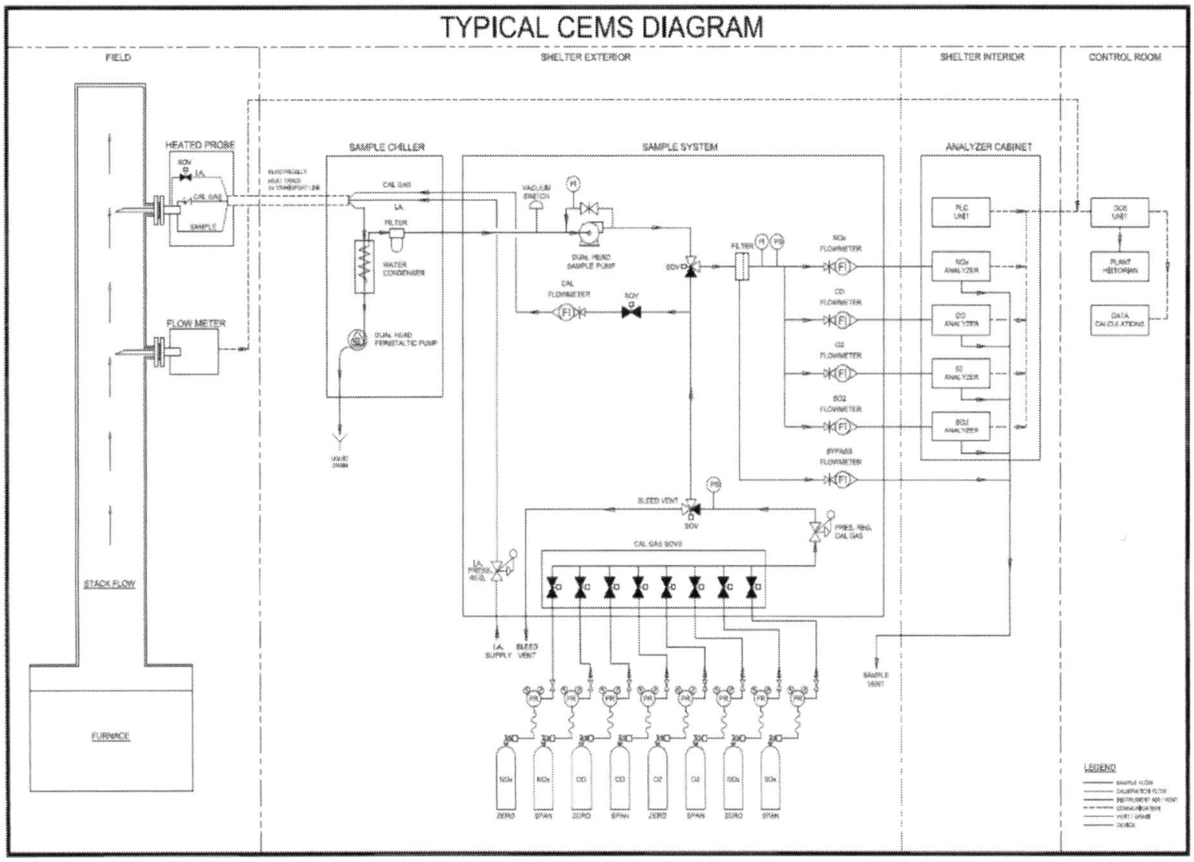

Figure 1: A typical CEMS engineering schematic.

The schematic in **Figure 1** shows a typical CEMS system as four connected sections. A sample probe or quill is inserted into the stack in path of flue gas flow. Extracted sample flows to the sample conditioning panel in heat traced tubing. Sample conditioning involves cooling the sample below the water dew point to remove moisture. Dry sample is then passed through filters to remove any particulate matter before introduction of the sample to the analyzer. The entire system is designed to ensure that a continuous sample flow is maintained from the sample probe to the analyzer to the vent.

The analyzer shown in **Figure 1** consists of four separate benches or detectors specific to measuring the four constituents: NOx, O2, CO and SOx. Each analyzer detector is calibrated using gas samples of known constituent concentration and adjusting the response of the detector when the gas sample is introduced to match its output to the known concentration. Typically, all detectors are calibrated at two concentrations; at zero and at span which is selected to be near the maximum expected detector response. A two level calibration ensures linearity of the detector response over the expected range of constituent concentration in the process stream. A PLC controller (shown as part of analyzer section) controls various valves and actuators to automatically switch from process sample to cylinder gases when needed to calibrate/validate the analyzer. Cylinder gases may also be passed though the entire sampling system including the sample probes before introducing in the analyzer so that the cylinder gas traverses the same path as the actual process sample. This is used to validate the entire analyzer system.

The PLC is used to communicate analyzer measurement data to a DCS system to be visible to process unit operators. Additionally, the PLC would normally be programmed to monitor and communicate various analyzer system health check statuses, e.g., general analyzer fault, alarm statuses, loss of sample flow etc. The DCS system typically communicates with a computer that hosts a database to archive all the analyzer data output; shown here as plant historian computer. The CEMS analyzer data thus archived in the plant historian becomes the official record of environmental compliance. For the compliance record to be complete and valid, all sections of the system must be working. For instance, even if all the components of the sampling system, analyzer and DCS were to be working correctly, but for some reason data communication link between the DCS and the historian were not working and thus the system having lost the ability to archive the data would render the entire CEMS system as out-of-order for the duration until the data communication link can be restored.

A separate computing system may be used for data reduction and to compute emission parameters in real-time such as total emissions and emission rates and communicate the results to pertinent systems and personnel. Computing system can also be designed to monitor various analyzer system health check statuses and alert personnel of analyzer system malfunction to respond and take corrective action.

HRVOC Analyzer – Cooling Water Tower

HRVOC regulations require continuous monitoring of VOCs in the any cooling water heat exchanger system in Houston/Galveston area with circulation capacity of 8000 gallons/minute or more and that has the potential to emit a HRVOC compound in case of a leak in the exchanger system. The primary intent for the HRVOC regulations to require continuous monitoring is to detect leaks in the exchanger whereby HRVOC compounds would be emitted to the atmosphere.

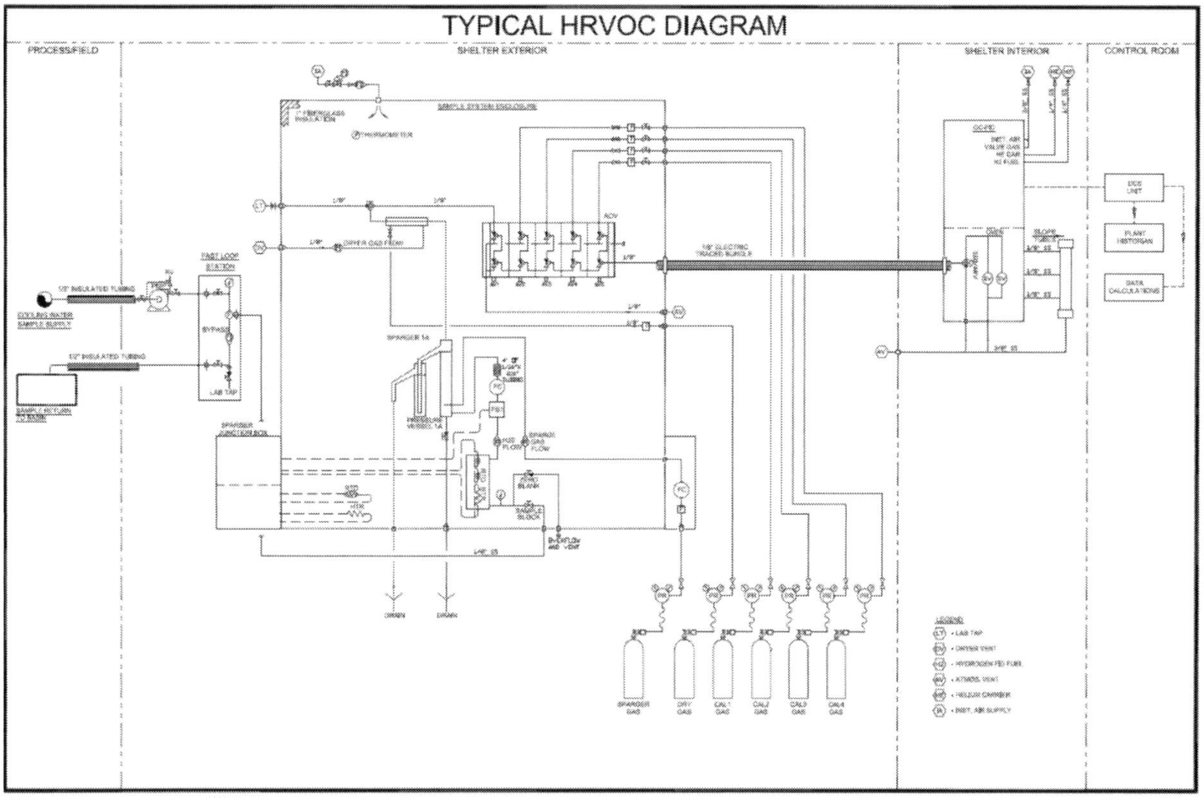

Figure 2: A typical Cooling Water Tower HRVOC engineering schematic.

Figure 2 shows a schematic of a typical engineering set up for a cooling water tower HRVOC analyzer system. The analyzer system is designed to measure VOCs in the circulating cooling water. The schematic shows four connected sections of the system wherein the water sample is extracted and flows to the sample conditioning system. The sample conditioning system contains a sparger to strip VOCs from the water. Bottled dry air or helium gas is used to bubble through the sparger system to strip entrained hydrocarbons in the water. Stripped hydrocarbons are then passed in to the analyzer for measurement. The analyzer in this case is typically a GC (Gas Chromatograph) that is designed to separate and measure concentration of hydrocarbon components present in the stripped gas.

The DCS, PLC and computing servers for HRVOC systems serve the same purpose as for the CEMS system.

HRVOC Analyzer – Flares

Flares in manufacturing sites in Houston/Galveston area that have a process stream routed to it containing potential HRVOCs is required to continuously monitor the composition of flare gas. A flare HRVOC analyzer system is designed to measure concentration of all constituents of the gas being sent to the flare. A schematic of the flare HRVOC analyzer system is not shown here, but it may be thought of as a combination of flue gas sample handling features of a CEMS system combined with a cooling water tower GC analyzer.

Flares are designed to combust hydrocarbons before exiting to the atmosphere. Ideally, therefore, there should be no VOC emissions from flare stack if complete hydrocarbon combustion is assumed. However, combustion is not assumed to be 100% and VOC total emissions and emissions rates are calculated based on flare destruction efficiency (DE). The DE determines % destruction (combustion) of various components during flaring.

DE is dependent upon two factors: Net Heating Value (NHV) of the flare gas and flare gas Exit Velocity (Ve). The NHV is expressed in units of Btu/scf and represents Btu's of heat release upon complete combustion of one standard cubic feet of flare gas. NHV of flare gas can be calculated from concentration of all constituents in the flare gas as measured by the HRVOC analyzer. Ve is expressed in unit of feet per second (ft/sec) and its value depends upon flare tip geometry and total flare gas flow rate. The following are two possible scenarios for DE as a factor of NHV and Ve:

1: DE is 99% for constituents lighter than Propane (includes HRVOC components Ethylene and Propylene), and DE is 98% for all other remaining constituents provided one of the following conditions is met:

NHV => 300 Btu/scf and Ve < 60 ft/sec;

OR

NHV > 1000 Btu/scf and 60 ft/sec <= Ve < 400 ft/sec;

2: DE is 93% for all constituents if the above conditions are not met.

Total emission and emission rate of a HRVOC component, e.g., Ethylene is calculated assuming that 99% of the Ethylene is destroyed and that only 1% of Ethylene in flare gas is emitted to the atmosphere. Note that if condition (1) is not met then a decrease in DE from 99% to 93% would mean 7 times increase in the calculated emission of Ethylene and Propylene HRVOCs to the atmosphere. For heavier components decrease in DE if condition (1) is not met would mean 3.5 times increase in reportable emissions rates. Generally, to ensure that condition (1) is not violated, natural gas is injected into the flare gas to keep the NHV of flare gas to be over 300 Btu/scf at all times.

Analyzers Operations And Maintenance

Operation and maintenance of CEMS and HRVOC analyzers present some special challenges that are normally not required of other process analyzers. The challenges stem from mandated requirements to follow a rigorous quality control and quality assurance (QA/QC) plans for analyzer maintenance to ensure analyzer system on-stream availability and its data accuracy. Analyzer QA protocols are designed to ensure that pollutant concentrations as measured by the analyzer are accurate and within specified control limits. The QC protocols include periodic analyzer system maintenance, routine system health check monitoring activities, and plans for corrective actions in case of malfunction to ensure that the analyzer system continues to operate within its specified control limits with minimum outage.

Operators of the CEMS and HRVOC systems must have a written QA/QC for analyzer systems that is approved by the executive director of the regulatory agency. The operating organization must develop, implement, and follow a written quality assurance plan (QAP) for monitoring including installation, calibration, operation and maintenance of the analyzer system. The rules governing analyzer specification and monitoring requirements are published in the federal register(2). The following are some of the QA/QC requirements relevant to the present discussion on operating and maintaining these analyzer systems.

- Following the initial installation or following any major changes to the analyzer, the system must be certified by an independent agency. During analyzer system certification, also referred to as the RATA (Relative Accuracy Test Audit) testing, analyzer system data accuracy is validated with reference to an EPA approved standard test method. Analyzer system cannot be used for environmental compliance monitoring until it is RATA certified. RATA may also be required periodically, e.g., once yearly, for some analyzer system.

- Each calendar quarter, a Cylinder Gas Audit (CGA) of analyzer data accuracy must be done. During the CGA the entire analyzer system, including sample conditioning, is verified for accurate operation. CGA is done by challenging the entire system with three different gases of known concentration of pollutants and diluents. Gas samples used for an audit must have been certified by comparison to the NIST standards. The difference between the analyzer response during the audit and actual gas concentration is used to validate the accuracy of the analyzer system.

- Analyzer calibration drift (CD) must be verified daily. Within each 24 hour period the analyzer must be challenged with gases with known concentration to verify analyzer data accuracy. CD is defined as the difference between the known concentration of the constituent and the analyzer reading. As shown in the schematics above, analyzer systems are designed to do the daily CD check of the analyzer automatically by switching from process stream to cylinder gases. The recorded CD must be evaluated and if found to be outside established control limits then immediate corrective action must be taken to bring the analyzer system back in control.

- Preventative maintenance activities to keep the analyzer system reliable must be carried out in timely manner. All aspects of maintenance, preventative or from corrective action, must be documented and all records maintained.

- Operating organizations are required to ensure that the analyzer system is operating in control and producing accurate compliance data minimum 95% of the time. Failure to keep the analyzer in control for minimum 95% of the time that the process unit is operating is an environmental compliance violation and reportable to the regulatory agency. The analyzer system is considered out-of-control if the daily CD checks fail, i.e., CD is outside specified control limit, or if any other part of the analyzer system malfunctions in a way that impacts analyzer data accuracy including, e.g., loss of data communication link between the analyzer and the DCS system even though there may be nothing wrong with the analyzer system.

- For each period that the analyzer system is considered out-of-control, the data from the analyzer cannot be used for emissions calculations. For every hour of out-of-control period, the data must be substituted with the highest hourly emissions data from the previous 24-hour period immediately prior to the start of out-of-control period. Thus poor analyzer reliability can directly lead to higher reportable emissions.

- For HRVOC analyzer systems specifically, samples must be collected for lab testing if out-of-control period exceeds specific number of hours and lab sampling and testing must continue till the analyzer system can be brought back in-control.

Organizational Capabilities And Tools

It should be clear from the above discussion on required maintenance, monitoring, and corrective action activities to operate compliance analyzer systems reliably that data monitoring tools and effective organizational capabilities are needed to ensure effective environmental compliance. The following are some of the key functional capabilities and tools that the operating organization can use to carry out all activities efficiently and cost effectively.

- Operating organizations must deploy sufficient number of technically competent workers on full time basis to carry out all required analyzer system repair activities, documentation and records retention. Technicians who work on these analyzer systems must be rigorously trained and be competent to carry out troubleshooting and repair in case of system malfunction in addition to routine preventative maintenance activities.

- Use of automated analyzer health monitoring tools is essential. For instance, tools that automatically evaluate the results of daily CD checks and are capable of electronically communicating the results can greatly simplify the organizational work-processes. Such tools can be used to automate a work-process that if carried out manually is very labor intensive and prone to errors. Such tools simplify the work-process by automatically singling out only those analyzer systems that may be having issue so that appropriate resource response may be directed to those. Thus responsible personnel can respond quickly to the analyzer system indicating malfunction and work to repair thus minimizing analyzer out-of-control period time. Also, use of automated tools ensures that daily CD which is a compliance record of analyzer data quality is captured and archived as a permanent record. Similar use of automated tools include smart diagnostic tools that continuously monitor various system health check parameters, e.g., alarm statuses, ensures that any present or emerging system malfunction would be quickly detected and communicated so that rapid corrective action can take place.

- Several vendors provide standalone compliance analyzers data acquisition systems that directly interface with the analyzer system and do several of the compliance calculations and other smart functions. Standalone systems work well when dealing with a few analyzer systems. However, as the number of analyzer systems increases this approach becomes increasingly costly to set up and maintain. Using standalone systems can also introduce a single point of failure mode in the analyzer system. Preference should be to always keep raw compliance data gathering and archiving separate from data computing and communication. This strategy eliminates single point of failure in the system since calculations can always be redone but raw compliance data once lost cannot be regenerated.

- Accurately tracking analyzer system out-of-control period is a critical function of the entire environmental compliance loop. Tracking analyzer system downtime often cannot be completely automated as manual review of the historical analyzer data is needed to accurately assess start and end of out-of-control periods. Organizations may need to assign a specialized role for accurately tracking and documenting analyzer system downtime.

Summary

Environmentally critical analyzers are complex engineered systems that, where applicable, are mandated license-to-operate requirements to comply with environmental emissions monitoring and reporting regulations. Operating organizations must dedicate resources and implement work-processes to ensure that installation, calibration and maintenance of these analyzers follow mandated QA/QC protocols. The QA/QC protocols are designed to ensure accuracy of the analyzer data output and to maximize analyzer on-stream availability. In this paper, specific mandated QA/QC requirements have been discussed in reference to two types of environmentally critical analyzers; CEMS used for measuring pollutant emissions from furnace stacks and HRVOC used for monitoring volatile organic compounds from cooling water towers and flares.

References

1. Texas Commission on Environmental Quality; Chapter 115 – Control of Air Pollution from Volatile Organic Compounds; Subchapter H: Highly-Reactive Volatile Organic Compounds Division1: Vent gas Control.
2. Environmental Protection Agency; Title 40 Code of Federal Regulations 60 (40 CFR 60) Appendix F – Quality Assurance Programs.

50a

LyondellBasell Industries multi-unit RTO implementation using combination of rigorous and simplified methodology

Sam Valleru

Principal Engineer – Real Time Optimization

Equistar Chemicals, LP

Thomas Kelly

Consulting Engineer – Real Time Optimization

Equistar Chemicals, LP

James Hackney

Senior Advisor – Advanced Process Control

Equistar Chemicals, LP

Prepared for Presentation at the 2012 Spring National Meeting

Houston, Texas, April 3, 2012

LyondellBasell Industries multi-unit RTO implementation using combination of rigorous and simplified methodology

Sam Valleru

Principal Engineer – Real Time Optimization

Equistar Chemicals, LP

Thomas Kelly

Consulting Engineer – Real time Optimization

Equistar Chemicals, LP

James Hackney

Senior Advisor – Advanced Process Control

Equistar Chemicals, LP

Abstract: Equistar Chemicals, LP implemented a Real Time Optimization project at its Channelview, Texas complex, which combines two olefins units and downstream recovery units into one RTO model. The project implementation is aimed at creating alignment between the long-term planning optimization models and the plant level controls and operation.

The multi-unit complex model utilizes a blend of rigorous and simplified methodologies at varying degrees. The underlying multivariable controller empirical models were utilized to maintain balance among RTO model consistency, speed and robustness. The model was built to optimize and enhance the overall profitability of the complex while traditional RTO approaches focus more on individual unit optimizations. This paper discusses the methodology, challenges and opportunities involved in the implementation and maintenance of the multi-unit RTO application.

Introduction

LyondellBasell Industries implemented Real Time Optimization project at the Channelview complex, Texas combining two olefins units and downstream recovery units into one RTO model. This paper reflects on the project evolution, model development, benefits and lessons learned.

Equistar Chemicals, LP is a wholly owned subsidiary of LyondellBasell Industries. Channelview site is one of the largest chemical complexes within the Equistar Chemicals. A multi-unit based RTO project was implemented using rigorous and simplified methodologies. The site is the home of two world scale olefins units, C4 recovery units, C5 recovery unit and Aromatics recovery unit; along with many other Intermediate plants and Derivatives plants. The scope of the RTO project and this paper is limited to the two olefins units, C4 recovery units, C5 recovery unit and Aromatics recovery unit.

Online optimization was implemented in Channelview Olefins units with individual online optimization applications dating back to 1989 using DMO® technology.[1] Later, these applications were upgraded to Aspen Plus RT-Opt® until the current project was implemented in 2008.[2] These separate real time optimization applications were limited to each unit with an optimization problem scope of the respective unit. The boundary of the each of the optimization problems was limited to mixed C4s, C5s and Aromatic streams leaving these units. The Olefins Optimization group develops individualized plans for each site. Channelview Olefins units are operated with flexible/multiple feedstocks. The choices of feedstocks are determined by maximizing an economic objective function that includes contributions from multiple plants. The central optimization problem is typically a linear problem. The individual unit optimization tools determine the optimal hydrocarbon feed flow rate and conversion/severity profiles with the given set of feedstocks and furnaces.

---

[1] Fatora III, F. C., G. B. Gochenour and D. N. Kelly, "Modeling Ethylene Plants for Real-Time Optimization Applications ," 5th Ethylene Producers Conference, Houston, 1993.

[2] Leggett, P. D., R. M. Stout and E. A. Teeter, "The Impact of Olefin Plant Revamps on Advanced Process Control and Real-Time Optimization," 12th Ethylene Producers Conference, Atlanta, 2000.

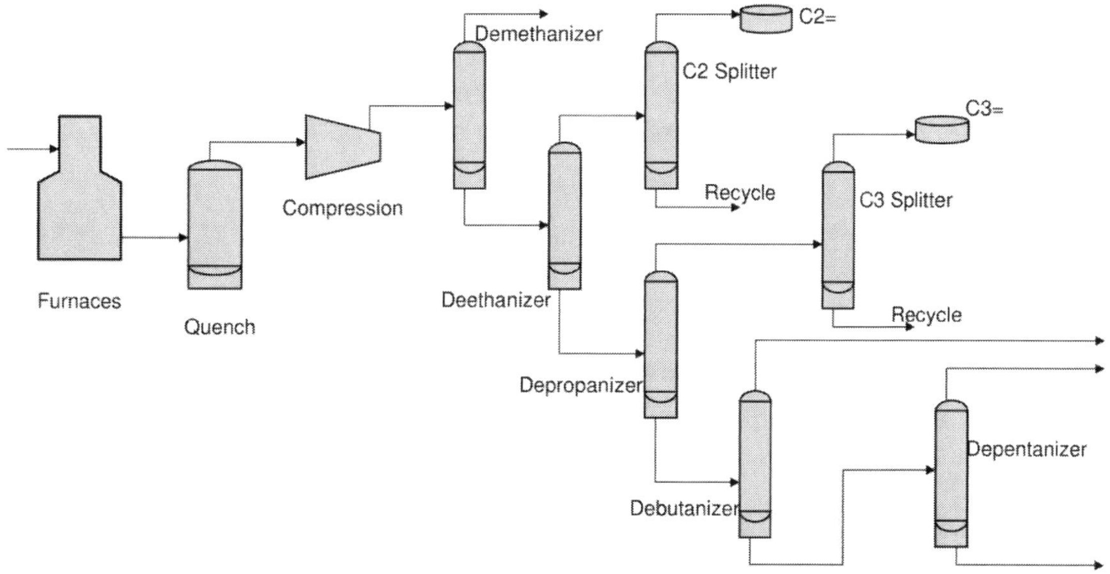

Figure 1. Simplified olefins unit diagram.

Project Scope Development

The current RTO project was developed as part of a centralized project to improve the overall enterprise optimization effort to streamline the optimization process from the corporate level to the plant level and to assure consistency in the process.[3] A core team of experts from the optimization group, process engineering, and process control engineering groups along with the experts from the individual sites identified the applications and work processes to achieve this goal. This Olefins Optimization Improvement Project identified the use of on-line optimizers as a key element in operating the plants to their optimum constraints. It was also identified that it would be beneficial for the RTO models to be similar in structure to that of the central optimization model. This will allow comparing and tuning of the models effective and efficient.

In this effort, the project team considered using a larger RTO model. The computing resources, model fidelity, and support requirements were important aspects considered at this point. A very large rigorous model may not be successful even if the computing power is available to solve the larger problem. The team considered using a simplified model where appropriate, based on the success of rigorous/simplified RTO application implementation at the company's Chocolate Bayou Olefins facility.[4] The team debated the pros and cons of a

[3] Hackney, J. E., "An Optimization Program in Lyondell," 20th Ethylene Producers Conference, New Orleans, 2008.

[4] Rickard, K. A., *et al.*, "Some Key Issues on a Successful Application of Real Time Optimization System to an Ethylene Plant," 19th Ethylene Producers Conference, Houston, 2007.

simplified approach as well as a rigorous approach. The choice of simplified versus rigorous approaches in developing these on-line optimization tools has a significant impact on the capital requirement for the project implementation and on the continuing support requirements to maintain the optimization tools.

Rigorous vs. Simplified

In order to support a selection process it is necessary to have an understanding of the technologies employed, the similarities and differences, and the ongoing support requirements required to develop and implement an on-line optimization system.

Both approaches:

- Utilize Aspen Online® to communicate with the historian to:
  - Retrieve plant operating parameters for the parameterization.
  - Return optimum set points to the control system.
- Utilize the Equation Oriented solution engine in Aspen Plus® to solve the model and perform the optimization.
- Utilize SPYRO® for the rigorous modeling of the cracking furnace coils.
- Use the same steps in execution of the optimization process
  - Evaluates process data from the historian to determine if the plant is at steady state. If the variability of the process data is too great, the optimization cycle is delayed until steady state is achieved.
  - Performs a parameterization to establish offsets to tune the model to match the operating conditions in the plant.
  - Solves an optimization case to determine the optimum operating conditions for the entire plant.
  - Repeats the steady state detection process to ensure the plant is still at steady state. A change to an unsteady condition during the period of time that the optimization model was solving invalidates the calculated optimum and prevents communication of the setpoints to the control system.
  - Applies the offsets from the parameterization and communicates the optimum setpoints to the control system for implementation.

The rigorous modeling approach:

- Utilizes a rigorous first principles model for the portion of the plant downstream of the furnaces.
- Includes optimization targets for:
  - Furnace severity targets.
  - Process gas compressor suction pressure targets.
  - Tower composition targets (product recovery).
  - Refrigeration system targets.

- Makes optimization decisions that include the tradeoffs of product recovery versus energy requirements.
- Rigorously models the non-linear behavior of process variables that are plant constraints.
- Relies on steady state detection of a large number of variables and therefore has a long execution time. Two and one half to three hours per complete cycle.
- Requires a high level of skilled technical support to sustain a high service factor.
- May require multiple initialization files to provide enough robustness to converge the model, especially during changes that involve substantial changes in the plant feedstocks.

The simplified modeling approach:

- Utilizes simple (linear) models to represent the portion of the plant downstream of the furnaces.
- Utilizes the projected steady state values from the Advanced Process Control (APC) system as the steady state values for parameterization.
- Is very robust and has a high service factor.
- Has fast execution time. Multiple optimization cycles per hour are possible.
- Relies extensively on the capability of the APC system to provide projected steady state operating parameters and operation against the true constraints of the plant.

The simplified modeling approach can be modified to include rigorous models of selected parts of the plant. Candidates for rigorous models could include the Process Gas Compressor for optimizing the suction pressure and selected fractionation towers for optimizing product recovery versus energy. The inclusion of the rigorous model of the Charge Gas Compressor provides the capability to determine the trade-offs of yield impacts versus energy in optimizing the plant. The limitation on expanding the portion of the plant where rigorous models are used is determined by the availability of projected steady state values of the operating variables in the APC models. If steady state projections are not available from the APC system then the validity of the optimization is compromised because data would be supplied to the model from multiple time domains.

Project Execution

The project scope was finalized after the engineering team made a decision on the scope of rigor and the scope/degree of simplification. The structure of the simplified model was decided to be same as the planning model.

AspenTech was selected to implement the RTO project. Several design issues were encountered as this is a creative way of implementing an RTO project. Multiple discussions were held among cross functional groups to solidify decisions on the optimization and interface philosophies. A few of those are described below.

As noted, the furnace radiant sections and the CGC suction pressure/chilling train sections of the model were rigorous while the downstream recovery sections were implemented as split factor models. Online analyzer values were used within olefins units to update the split factors on a real time basis. Also, underlying DMC controllers were mimicked in the RTO to represent the constraints of the olefins units. The C4 recovery units, C5 recovery unit and the Aromatics recovery units were represented with static split factors in line with the planning model. The scope of the RTO model is shown in Figure 2.

Hardware and software requirements were identified and acquired by the project team. The project utilized an RTO server and RTO Watch server and the associated AspenTech software.

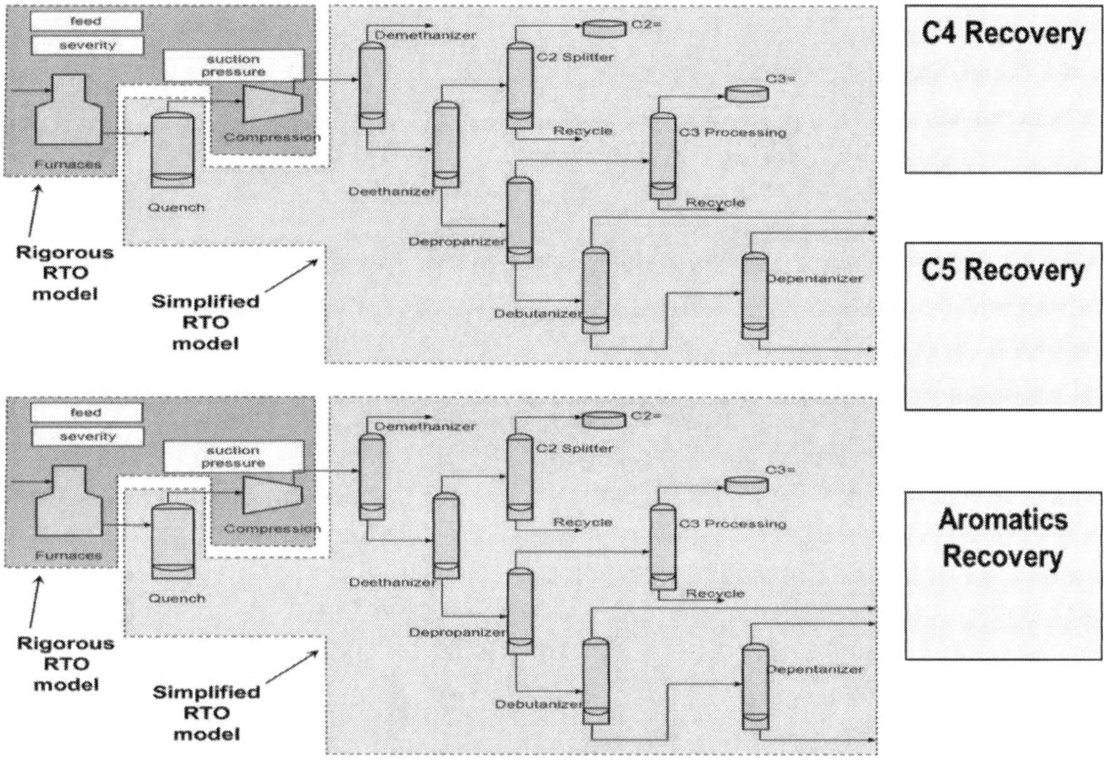

Figure 2. Site RTO scope.

Engineer interface process displays were developed to update the feedstock compositions. Several thousands of tags were created in the unit historian to implement a semi-automatic composition update system based on the types of feeds selected. The SPYRO feed component list was used as the basis for entering the components on the composition. A semi-automated system was developed to transfer and update the feeds/products/utilities prices. The same interface was also used to assign individual feed systems to a particular composition.

The operator interface included a web display which is similar in look to the existing multivariable controllers. Each olefins unit is assigned with an RTO ON/OFF switch that allows the optimization targets in that unit to be written by RTO if the switch is ON. Corresponding logic was implemented in the RTO to define the closed loop status of the RTO. Engineers and operators can view the status of the RTO using a custom display. The display indicates the switch status of each unit along with the steady state status information and the number of implementations statistics.

Continued Support

Originally the site RTO applications were set up to be maintained and managed at the site level by the unit RTO engineers. Over time this model evolved into the creation of a centralized highly skilled engineering support team. This evolution was driven by the engineer turnover at the sites and also the skill level requirements to maintain the RTO applications. Current support model still includes the site level engineering support personnel with varying skill levels. Roles and responsibilities were defined to address the engineer turnover at the sites to help manage the RTO site support.

Lessons Learned

The objective of alignment of the RTO model with the central optimization model was achieved after the completion of the project. The comprehensive model allowed the optimization of multiple units. The number of solutions implemented in a day has significantly increased compared individual rigorous model RTO implementations.

Multiple lessons were learned with many continuing challenges.

- Validation of the model with the plant current operations is very critical.
- The incorporation of the multivariable predictive controller (MVC) models in the simplified RTO model along with the traditional rigorous process modeling creates unique challenges. The degrees of freedom in the RTO model must be consistent with the MVC models to assure a consistent approach to plant constraints.
- The initial support model of unit engineer with partial RTO responsibility was found to be inadequate. The support model was expanded to include a larger central support role driving projects and model changes while the

unit engineer monitors the daily operation of the RTO, aligns with regional and company-wide planning, and validates unit constraints.

- The RTO benefit calculation needed to be more visible to the groups other than the RTO engineers.
- Any individual unit upset would limit the RTO uptime. With two olefins unit in one model, any unsteady operation in one unit will impact the overall RTO solution implementation.
- The RTO model simplification structure relies heavily on the quality of the underlying multivariable controller quality. This implies the need for immediate change management when model gains were modified in the multivariable controllers.

Conclusions

There are advantages in implementing comprehensive RTO models, provided the required support structure is in place. The effectiveness of the RTO implementations can only be measured in terms of how well RTO is able to operate against the unit constraints and how well the RTO implementations achieve the desired results. Consistent engineer review of the results and adjustment of the model to match the plant is essential for any RTO application to be successful.

AIChE Paper 50B

An Improved Approach for Online Optimization of Large-Scale Process Facilities

Kiran R. Sheth

R. Donald Bartusiak

Core Process Control

ExxonMobil Chemical Company

Baytown, Texas

Prepared for Presentation at the 2012 Spring National Meeting

Houston, Texas (April 02-05, 2012)

An Improved Approach for Online Optimization of Large-Scale Process Facilities

Kiran R. Sheth

R. Donald Bartusiak

Core Process Control

ExxonMobil Chemical Company

Baytown, TX

Abstract: Online close-loop optimization of large-scale facilities using nonlinear programming techniques has been practiced for more than 30 years. Traditional approaches are limited by the need to wait for the entire facility to reach steady-state for the purpose of model parameterization, and the computational effort required to solve the engineering principles model-based full-state nonlinear problem.

This paper describes a novel approach to decompose large-scale engineering models into smaller modules that can be parameterized asynchronously, and to reduce the order of the plant-wide model used for determining economically optimal targets. The paper also illustrates application of the approach to a feed-flexible ethylene plant.

1. *Introduction*

A primary goal of manufacturing facility operations is to run at or near the optimal conditions for as much of the available operating time as possible. A combination of regulatory controls, advanced multi-variable controls, and economic optimization are used to achieve this goal in an automated, close-loop fashion.

The primary objective of the regulatory controls is to reject disturbances that move the process away from its desired state (setpoints). Advanced multi-variable controls stabilize the interactive dynamic elements, and hold the unit at optimal operating conditions (targets). The optimal operating conditions are typically at the intersections of multiple throughput limiting constraints. Economic optimization determines the optimal operating conditions of the manufacturing facility to achieve the business objective (most commonly to improve the profitability), subject to operational constraints. This calculation requires mathematical description of the cause and effect relationships of the facility (model), an equation expressing the business objectives (objective function), and a software tool (optimizer). The economic optimum must be recalculated when profitability-affecting disturbances enter the system, including business condition changes (e.g. prices, supply of raw materials, demand of finished products). The term Real-Time Optimization (RTO) is used to describe when the economic optimum is calculated automatically, and implemented in a close-loop fashion via the advanced multi-variable and regulatory controls.

2. *Traditional Approach*

The traditional approach for RTO uses a steady-state, fundamental-principles model that represents the entire facility (Fatora and Ayala, 1992; Yoon et al., 1993; Krist et al., 1996). The model needs to be calibrated to the actual state of the plant (parameterization). Accordingly, parameterization or calibration of the model to the facility's operating conditions can only be performed when the entire facility is at a steady state. The optimizer uses the calibrated model to calculate the optimal targets. The targets are then sent to the control system for implementation. The facility responds to the new targets, and eventually reaches a new steady-state (line out period). Then, the RTO cycle is repeated. A conceptual execution cycle for traditional approach based RTO application is described in Figure 1.

Figure 1: Close-loop execution cycle for traditional approach based RTO application

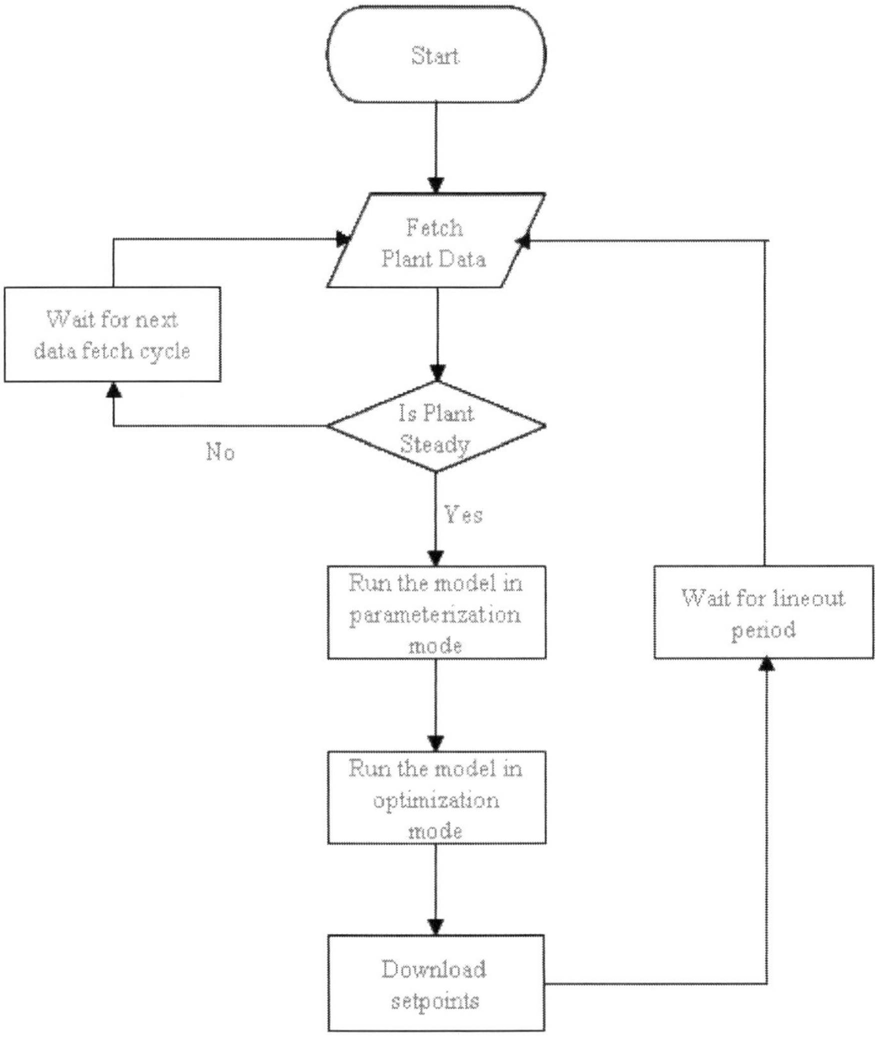

3. *Motivation for Change*

The challenges to effective deployment and sustainment of a RTO application include:

1. scope;
2. fidelity;
3. computational requirements;
4. frequency of re-optimization;
5. understandability; and
6. sustainability.

Regarding scope, the goal is to optimize as big a section of the facility as required to represent the business objectives. However, the state of available computing resources limits the scope of traditional RTO applications to a single-train ethylene unit or a refining unit (e.g. Crude, FCC, Reformer etc.). With increases in integrated and co-located manufacturing facilities (e.g. multiple ethylene units at same location), there is a need to expand scope of the optimization applications beyond single unit. Such expansion of scope often warrants careful consideration regarding model fidelity in order to keep computational requirements reasonable. Additionally, steady-state process models used in RTO applications require the entire facility within the application scope to be steady for accurate estimation of model parameters. The broader the scope of the application, the less frequent the facility will achieve the steady-state - reducing the frequency of re-optimization. And, last but not the least, the solution of non-linear program-based process models can also be difficult to understand because of the large number of unknown variables and equations within the model compared to the limited number of variables that are actually measured in the facility. Thus, trouble shooting the process model when it fails to converge can be very time consuming, if not futile. There is a need, therefore, to simplify the approach for optimizing a facility in order to address the challenges listed above.

The traditional approach involves the solution of a monolithic nonlinear model in parameterization and optimization modes. There are two basic mechanisms for improving the profitability of RTO applications. The first mechanism is to reduce the costs to develop the RTO model and commission the application. The second mechanism is to improve the sustainability of the application. The approach outlined in this paper addresses improvement of the sustainability of the RTO application.

The ExxonMobil Chemical Company (EMCC) approach specifically addresses the following sustainability needs:

<u>Steady-state limitations on frequency of RTO moves</u>: Traditional approach-based applications are limited to a maximum of eight to ten runs per day for a typical steam cracker. Most of the time this is not a practical limitation, but more frequent re-optimization is desirable after furnace transitions or refinery gas disturbances.

Engineering effort required for maintenance: Previous industry studies (Matthews et al., 2002) indicated that it takes 0.75 ± 0.20 dedicated engineering resources to maintain an olefins RTO application. A significant amount of this effort is spent in setting up and running offline cases, interpreting and explaining the results of offline case studies to Technical, Business Optimization, and Operations groups at site. While methods to interpret linear programs that are embedded in typical multi-variable advanced control applications (Peterson et al., 2011), diagnosing and troubleshooting of the nonlinear program-based RTO application requires a different skill set. Habitually untimely explanations are unsatisfactory to the site organization, and contribute to reduced utilization.

4. EMCC Approach

In the traditional approach, a rigorous, fundamental principles-based model of the entire facility is solved in parameterization and optimization modes using a nonlinear solver. The modular parameterization (MP) / sequential linear programming (SLP) approach involves two steps that we briefly describe in the following paragraphs (Sheth et al., 2011).

First, the model of the facility is segregated into two or more modules each containing a rigorous, fundamental principles-based model. Individual modules are solved in parameterization mode independently, and possibly asynchronously, of one another. Some inter-module data flow is required, for example to pass unmeasured stream composition information down the flow path. Subject to conditions on the timeliness of the inter-module data flow, the modules can be parameterized at higher frequencies than the traditional approach because of faster times to steady state of the smaller-scope modules.

Second, at the end of parameterization run, a reduced-order, linear model of each module is derived from its rigorous model. This is done by a matrix algebra operation that reduces the model from its full, state variable representation to a smaller input / output variable representation. This input / output model is linear, but its coefficients are updated with each parameterization module's execution, hence the designation, sequential linear programming optimization.

A reduced-order model of the entire facility is synthesized from the ensemble of reduced order models of individual modules. The reduced-order model of the entire facility is then used for optimizing the facility.

4.1 Modular Parameterization

Mathematical formulation for the traditional approach to RTO can be summarized as follows:

$$\begin{aligned} &\min \quad g(\mathbf{x}) && \textit{(Objective function)} \\ &\textit{subject to } \mathbf{0}=\mathbf{f}(\mathbf{x}) && \textit{(Model)} \\ &\qquad \mathbf{x}_{LB} \le \mathbf{x} \le \mathbf{x}_{UB} && \textit{(Bounds)} \end{aligned}$$

Where, $\mathbf{x} \in \{\mathbf{u}, \mathbf{p}, \mathbf{x_s}, \mathbf{y}\}$; $\mathbf{u}$ = input variables; $\mathbf{p}$ = model parameters; $\mathbf{x_s}$ = unmeasured state variables; $\mathbf{y}$ = output variables; scope of the model $\mathbf{f(x)}$ is facility-wide.

The objective for parameterization, or model calibration, phase is to find a set of model parameters $\mathbf{p}$, such that the difference between model outputs $\mathbf{y}$, and their plant measured value $\mathbf{y_m}$, is minimized. The objective for the optimization phase is to find a set of plant operating conditions $\{\mathbf{u,y}\}$, subject to operational constraints (bounds), such that plant profitability is maximized.

In the modular parameterization approach, the manufacturing facility is segregated into two or more modules with model descriptions $\mathbf{f_i(x)}$, such that in totality the $\mathbf{\Sigma f_i(x)}$ represent the model for entire manufacturing facility $\mathbf{f(x)}$. Subject to the scope of facility within the scope of the module reaching steady-state, models for each module are run in parameterization mode asynchronously of each other. Precision or fidelity of the model for each module is same as their counter-part in the traditional approach. However, the smaller scope of the facility within the scope of each module can reach steady-state at higher frequency than the entire facility, and hence, the execution frequency for parameterizing each module is higher than the same in the traditional approach.

Solving of a process model requires the computational algorithm to calculate output variables based on the value of input variables. Typical input variables include state of the material and/or energy feed streams and operating parameters. Typical output variables include state of the material and/or energy product streams. In a manufacturing facility, some material and/or energy streams may represent external feeds/products; the others may originate from one section of the facility and terminate into another (internal).

When a unified process model is used to describe the facility (traditional approach), the state of the internal material and/or energy stream is determined by solving the unit operation block where such streams originate. When the facility is represented by a group of modules (modular parameterization approach), the state of some material and/or energy streams can not be determined by a module, as the unit operation block where these streams originate may be outside the scope of the module. Instead, the state information for such material and/or energy streams must be provided in order to solve the module.

Thus, in modular parameterization approach, each module makes the state of its output streams available to other modules. The modules where such streams are input utilize the information to compute the state of their own output streams, and make it available for other modules to use. In the close-loop RTO application environment, these steps are repeated every execution cycle.

The optimization is performed using a derived, reduced-order model, and is described in the next section.

4.2 Sequential Linear Programming Optimization

For each modular parameterization module described by $\mathbf{0} = \mathbf{f}_i(\mathbf{x})$, the input variables, $\mathbf{u}_i$ and the output variables, $\mathbf{y}_i$ are identified. Typical input variables include optimization variables within the scope of the module, and variables describing state of the input streams to the module. Typical output variables include elements of the objective function and operational constraints that are within the scope of the module, and variables describing state of the output streams from the module.

With the input and output variables identified, a reduced-order linear model is derived by calculating the sensitivity (gain) matrix, $[\delta\mathbf{y}/\delta\mathbf{u}]_i$, from the Jacobian matrix, $[\delta\mathbf{f}/\delta\mathbf{x}]_i$, of each module. The model reduction is achieved by (a) linearizing the nonlinear model at the operating point, and (b) by reducing the full-state model to relevant input/output-only model. The reduced order model is accurate in the local region of the current operating point. The reduced order model is updated when each parameterization module runs as the operating point changes.

A reduced-order model of the entire facility (the SLP model) is assembled from the individual reduced-order models of the parameterization modules. This is a straightforward operation that only requires identification of the inter-module connecting streams. Once assembled, the SLP model is used to optimize the facility.

Note that the structure of the facility-wide, reduced-order model is fully defined, as soon as the input and output variables for each module are identified. Hence, the construction or assembly of facility-wide reduced-order model structure is a one-time, off-line activity. However, the coefficients of the module specific reduced-order model, are updated every execution cycle of individual modules as the operating point changes. Consequently, the coefficients of the facility-wide reduced-order model assembled from module-specific reduced-order models are updated as well.

Figure 2 represents conceptual structure for the facility-wide reduced order model, when the olefins manufacturing facility is segregated into a Hot-Ends (HE) module and a Cold-Ends (CE) module.

Figure 2: Conceptual structure for the reduced order model

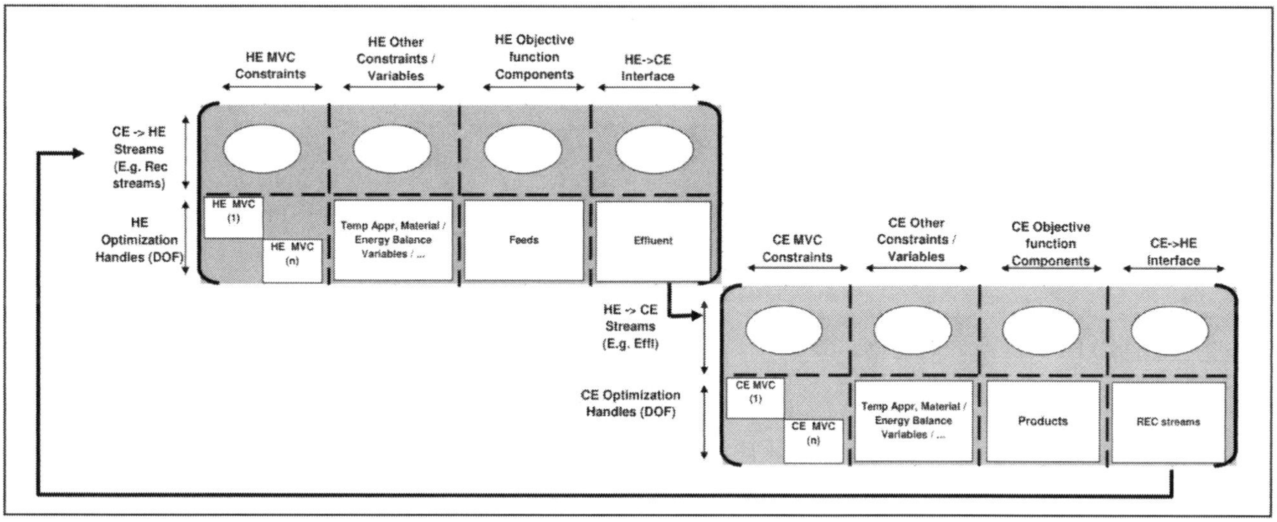

Figures 3a and 3b depict the conceptual execution cycle for parameterization and optimization steps for MP/SLP approach based RTO application.

Figure 3a: Parameterization cycle for MP/SLP approach based RTO application

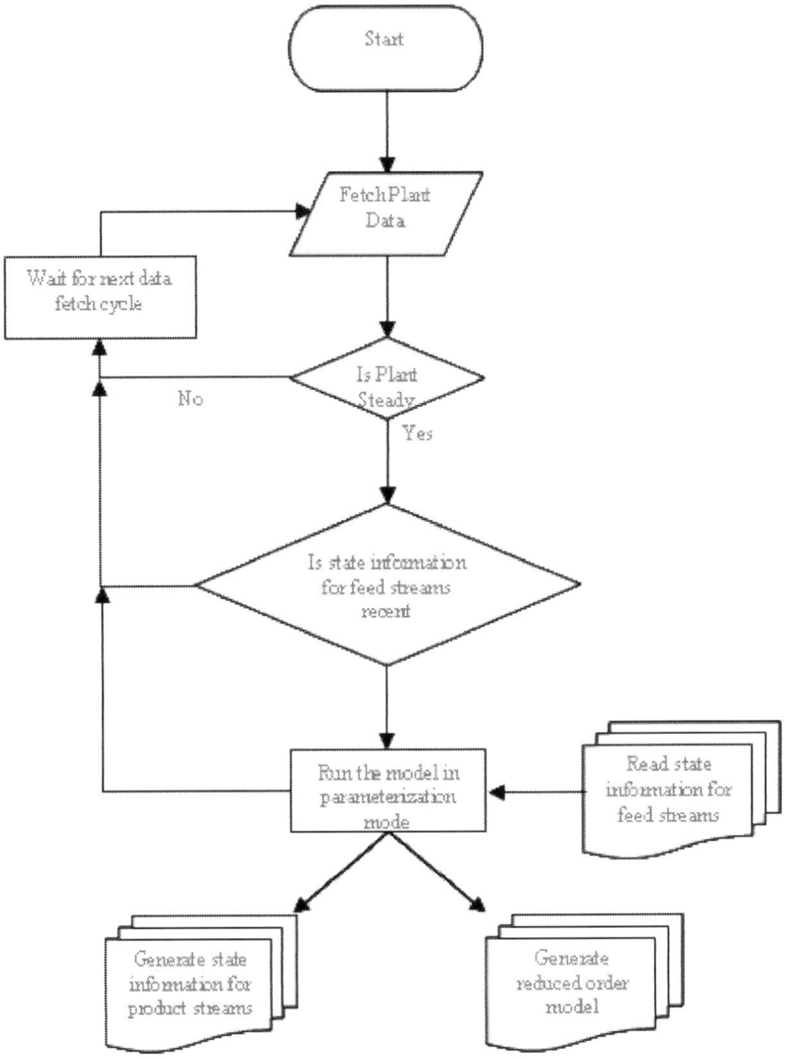

Figure 3b: Optimization cycle for MP/SLP approach based RTO application

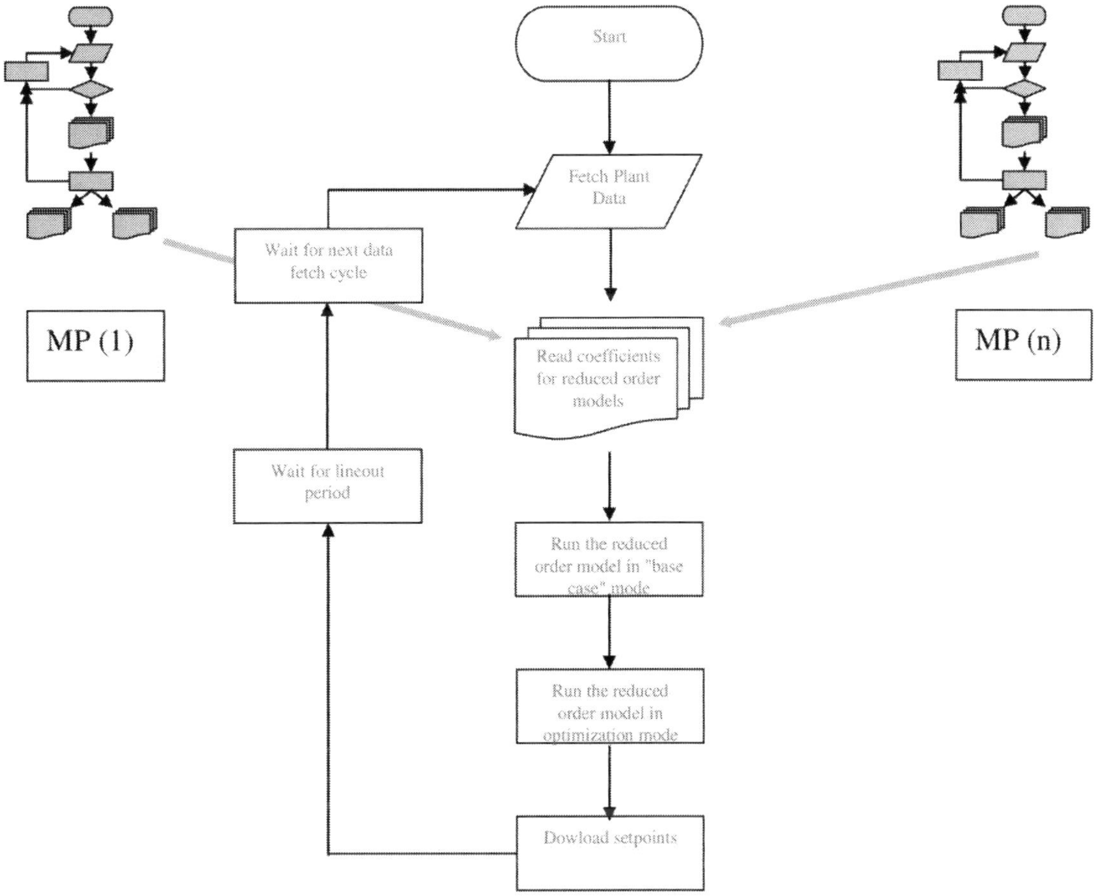

The SLP is constrained by step bounds that are tighter than what would be imposed with the traditional approach to prevent the SLP from moving too far from its region of validity and overshooting any internal optima. The SLP optimizer runs at a fixed, tunable frequency that is higher than the traditional approach-based optimization frequency, and higher than the time-to-steady-state for the facility. The SLP model is updated with latest sensitivity matrices generated by individual modules to maintain its local accuracy each time it runs.

When the steady-state model based optimizer is run at higher frequency than the time-to-steady-state of the facility, estimates for where the output variables will settle based on prior moves in input variables are required to ensure close-loop stability of the system. These estimates are provided by the underlying multi-variable advanced control application.

5. Test System to Validate MP/SLP Approach

To establish the feasibility of MP/SLP approach, an offline test system prototype was developed based on one of EMCC's feed-flex olefins manufacturing facilities. The facility consists of the feed header system, the furnaces, and the recovery section. Furnaces were modeled using EMCC's proprietary yield engine. The recovery section was modeled using rigorous process models such as flashes, towers, compressors, etc. Actual plant operation data (feed furnace line-ups, process values, bounds) were used in setting up the model.

The prototype consisted of 3 different components.

(1) A rigorous, engineering principles-based model of entire facility to optimize the facility using the traditional approach.

(2) Two rigorous, engineering principles-based modules (one for hot-ends, HE; and other for recovery section, CE) for parameterization, and a reduced order model of the facility (derived from rigorous models) for sequential linear programming based optimization.

(3) A virtual plant to implement optimization solution found using (2), so that the reduced order model in (2) could be updated, and the facility could be re-optimized in sequential steps.

Figure 4 compares the results obtained by optimizing the facility using traditional approach to results from using the MP/SLP approach. Each group on the X-axis represents a different component of the objective function (e.g. products, feeds, utilities, etc.). The Y-axis represents change in the component contribution ($/Hr) from the base case and the optimized case. Positive values for product components indicate higher revenue, and positive values for feed and utility components indicate reduced cost. A value of zero means no change from the base case.

There are several bars within a group on the X-axis. The rightmost bar is the result of optimizing the facility using traditional approach, and various bars on the left (within each group) represent the step-wise path to optimum using the MP/SLP approach. Note that the MP/SLP approach is <u>not</u> designed to give the ultimate optimum answer in one step. Instead, it is designed to lead to the optimum answer in multiple steps. If the two approaches were to give same optimum answer, trend of the bars on left should converge with the rightmost bar.

Based on the data presented below, we conclude that both approaches improve profitability mainly by increasing production of the same high-value products, and by increasing consumption of the same advantageous feeds. The differences in production and differences in feed and utilities consumption are virtually the same in both approaches.

Figure 4: SLP optimization converges to the NLP optimization (Objective function components)

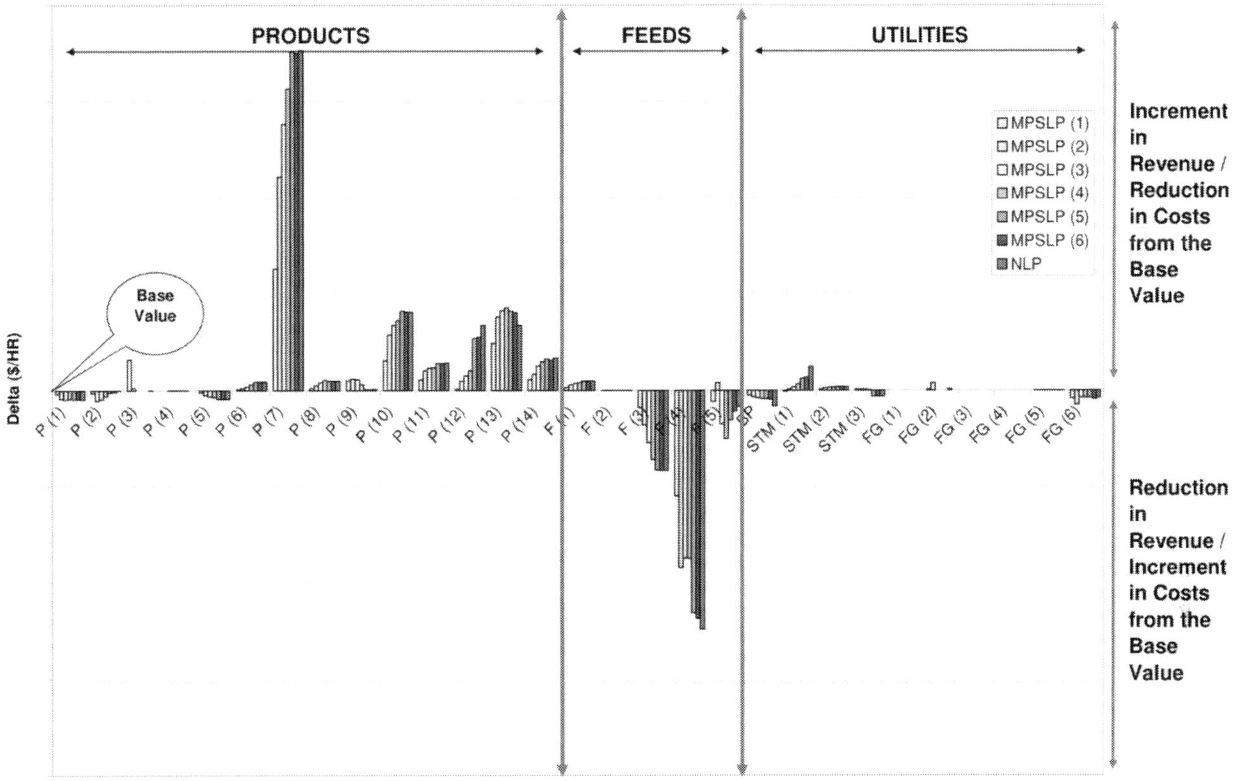

Similar analyses were also performed on the optimization variables, and operational constraints to confirm that both approaches converged to the same operating point.

6. Implementation at a Feed-Flex Olefins Manufacturing Facility

Upon confirming that MP/SLP approach can drive the manufacturing facility to the same optimal operating conditions as traditional approach in the offline environment, a second EMCC manufacturing facility was selected as the pilot site. This facility is highly feed-flexible, and can process feedstock ranging from lighter gas feeds to heavy liquid feeds. It also contains a multitude of steam-cracking furnaces, each with a multitude of convection and radiant sections. The feed header system is designed to facilitate cracking of several different feed streams in any of the on-stream furnaces. Cracked effluent is processed by a fairly complex separation train. The product mix consists of high value molecules ranging from Ethylene (C2=), Propylene (C3=) through Steam Cracked Gas Oil (SCGO) and Tar.

The advanced control layer for the facility consists of several Multi-Variable Controllers (MVCs). The scope of advanced control layer covers process units including the steam-cracking furnaces, primary fractionator, process gas compressor, several distillation towers, and C2= / C3= refrigeration compressors in the separation train.

The RTO application is configured by segregating the facility into two modules – a Hot-Ends (HE) module, and a Cold-Ends (CE) module[1].

Scope of the HE module includes the feed header system and all steam-cracking furnaces. Scope of the CE module includes primary fractionator, process gas compressor, all distillation towers, and the refrigeration compressors. In addition to the typical rigor practiced in modeling multi-stage compressors, multi-tray/multi-component distillation towers, and complex cold-box circuits, the multitude of convection and radiant sections within each furnace, are modeled individually in the RTO application.

The input variables for HE module comprise the furnace optimization variables and the state information for recycle feeds. The output variables for HE module comprise the furnace constraints, objective function elements such as import feeds and utilities, and the state information for furnace effluent stream.

The input variables for CE module comprise the compressor suction pressures, tower reflux and/or cut-points, and the state information for furnace effluent stream. The output variables for CE module comprise the product purity specifications, tower dPs, objective function elements such as high-value products and utilities, and the state information for recycle feeds.

The HE and CE modules execute asynchronously of each other as per the online execution cycle represented in Figure 3a. Mean time between successive HE module executions is 45 minutes, and mean time between successive CE module executions is 90 minutes. The facility-wide, reduced-order model (SLP) executes as per the online execution cycle represented in Figure 3b, and is configured to run every 60 minutes. As early as the traditional "open-loop commissioning" phase, benefits of implementing the MP/SLP approach were noticeable. Each module was able to execute at higher frequency and more robustly than one would expect from the traditional approach. Diagnosing convergence failure was much easier than the traditional approach. Optimization moves that often required offline case studies or sensitivity analysis were reasonably easy to understand and explain. Organizational acceptance of the application, though not effortless, was relatively painless.

In summary, benefits from implementation of the MP/SLP approach include:

- capability to increase scope representing large-scale manufacturing facility with high-fidelity model;
- reduced engineering cost for open-loop commissioning and maintenance, and increased on-stream factor due to modularity;

---

[1] Currently, the MP/SLP approach is being utilized for optimizing other ExxonMobil Chemical Company's ethylene manufacturing facilities using more than two MP modules.

- reduced engineering cost for close-loop commissioning and increased close-loop service factor due to improved understanding of optimizer moves and quicker turn-around time for offline case studies; and
- increased credits capture due to more frequent optimization and increased service factor.

7. Summary

A novel approach for real-time close-loop optimization of a large-scale manufacturing facility is presented. Two key elements of the approach include asynchronous parameterization of smaller-scale modules, and successive linear programming based optimization of a derived, reduced-order model. While driving the manufacturing facility to the same optimal operating conditions as traditional approach, the proposed approach offers additional benefits such as ability to increase the scope while retaining or increasing the fidelity, reduced engineering effort to commission and sustain the application, and increased credit capture. Benefits of the novel approach have been demonstrated by an on-stream application that has been in service for more than 5 years.

References

1. Fatora, F.C., and Ayala, J.S., "Successful Closed Loop Real-Time Optimization," Hydrocarbon Processing, June 1992.
2. Yoon, S.K., Kranz, R.A., Rao, A., and Stewart, W.B., "Global Real-Time Optimization of Flexible Feed Olefins Plant", Presented at the 5th Ethylene Producers Conference, Houston TX, March 1993.
3. Krist, J.H.A., Lapere, M.R., Wassink, S.G., Koolen, J.L.A., "System for Real Time Optimization", United States Patent #5486995, January 1996.
4. Matthews, W.E., Bartusiak, R.D., Hackney, J.E. and Jones, J.T., "Process Control Staffing Practices," Presented at the 14th Ethylene Producers Conference, New Orleans LA, March 2002.
5. Peterson, T.J., Punuru, A.R., Emigholz, E.F., Wang, R.K., Barrett-Payton, D., "Model Predictive Controller Solution Analysis Process", United States Patent # 7949417, May 2011.
6. Sheth, K.R., Bartusiak, R.D., and Fontaine, R.W., "Online Modular Parameterization and Successive Linear Programming for Improving Manufacturing Plant Operations", United States Patent # 8036759, October 2011.

50c

MPC REVAMP OF TWO NAPHTHA CRACKERS, LESSONS LEARNED AND RELEARNED

Vanessa Conz, APC Lead Control Engineer, Braskem
Suzana Fuchs, APC Controller Engineer, Braskem
Priscila Macedo Moura, Production Engineer, Braskem
Mark Darby, Consultant, CMiD Solutions
Doug Nicholson, Consultant, IPCOS (UK) Ltd

Prepared for Presentation at the 2012 Spring National Meeting
Houston, TX, April 1 – 5, 2012

MPC REVAMP OF TWO NAPHTHA CRACKERS, LESSONS LEARNED AND RELEARNED

Vanessa Conz, APC Lead Control Engineer, Braskem
Suzana Fuchs, APC Controller Engineer, Braskem
Priscila Macedo Moura, Production Engineer, Braskem
Mark Darby, Consultant, CMiD Solutions
Doug Nicholson, Consultant, IPCOS (UK) Ltd

Abstract

Braskem is the 8th largest petrochemical company in the world and the biggest producer of polymers in the Americas. The Braskem Triunfo site includes 2 Naphtha Crackers with a combined capacity of 1250 KTA ethylene and 685 KTA propylene. AspenTech DMCplus® controllers had been built for both plants over the past 10 years, but were not fully operational.

In 2010 a project was initiated to review the performance of the advanced control systems and improve their operability, service factor and economic impact. A further objective was to increase the knowledge of the site process control engineers. This paper gives an overview on the project, including the basis for deciding which areas to focus on, changes made at both the regulatory and advanced control level, and the results obtained. Changes in operational philosophies which improved process stability and efficiency are discussed. Techniques and tools used in the project are highlighted.

Overview

Braskem is a relatively new company, formed in 2002 through the merger of several petrochemical and chemical companies. The company has a focused growth strategy and has acquired companies in Brazil and, more recently, in the USA, Germany and other countries. Currently Braskem has industrial units in the four largest petrochemical centers in Brazil. Braskem utilizes advanced control as a means to reduce variable costs, increase energy savings and improve productivity in the company.

This paper describes a project covering two Braskem olefins units located at the same site in the south of Brazil. The site has three olefins plants and one aromatics unit. Two plants (Olefins 1 and Olefins 2) process naphtha. A third, plant uses ethanol as feed (Green Ethylene). The MPC project was on the naphtha based plants.

Olefins 1 started production in 1982. It is the largest ethylene producer in the site and has capacity to process 312 t/h naphtha. The plant has 12 naphtha furnaces, 2 recycle ethane furnaces. Olefins 1 normally operates with high feed rates typically limited by raw gas compressor constraints. It is the plant responsible for the pressure control of the site ethylene supply grid and is therefore affected by consumer demand variation. The base instrumentation is the oldest in the site with too many control loops in manual mode. These characteristics make Olefins 1 challenging for operators with a high number of interventions from control panel.

Olefins 2 started production in 1999 and has the capacity to process 180 t/h naphtha. The plant has 7 naphtha furnaces, one being a swing furnace that can be used on either plant, and 1 ethane furnace. It is more difficult for this plant to handle feed composition variability, as this unit processes more LGP and C4's along with naphtha.

Braskem owns several sites in Brazil. A particular challenge the company faces is job rotation of its people such as production and control engineers. New plant projects require experienced senior engineers to work on new projects. To succeed with MPC, it is well known that site engineers require the process and tools knowledge to maintain the installed base of applications. The use of projects to accelerate engineer training and learning is strategic to the company.

Both olefins plants already have MPC controllers installed. Olefins 1 has had controllers since 2004 and Olefins 2 since 2001. Table 1 lists the controllers that have been installed. After plant shutdowns and minor revamps, some of the controllers were not operational and others were not working well.

Table 1. Installed MPC controllers at the start of the project.

	MPC Controllers	Implementation Year
Olefins 1	Furnaces with O_2 optimization using draft pressure (14)	2007
	Deethanizer	2004
	C2 Splitter	2006
	MAPD Reactor	2004
	C3 Splitter (2 columns)	2004
	Debutanizer	2004
Olefins 2	Furnaces with O_2 optimization using dampers (8)	2001
	Primary Fractionator and Quench Tower (2)	2001
	High and low Pressure Depropanizer (2)	2001
	Deethanizer (1)	2001
	C2 Splitter (1)	2001
	MAPD Reactor (1)	2001
	C3 Splitter (1)	2001
	HPG Hydrogenation Reactor (1)	2001

Revamp Project

Motivation

The online factor in Olefins 1 was poor and the benefits in the existing MPCs were not being captured. The objective of the project was to improve the economic and qualitative benefits as well as increase on line factor. In Olefins 1 plant, for example, the variability of C2 Splitter impurities was high and the C3 Splitters had a large propylene loss (>20% mass). In Olefins 2, the goal was to remodel the parts of the controllers that were not working properly from the last shutdown in 2005. In addition, a parallel project installed new actuators on the secondary air dampers of the furnaces in order to control excess oxygen more efficiently. The project was an opportunity to review the furnace models and add the damper positions as additional MVs to the furnace controllers.

The site had just one engineer responsible for MPC for a number of years. In 2010, a trainee was hired to be trained as control engineer and a production engineer was assigned to the team to help the control engineers maintain controller on line factor.

The project was a good opportunity for engineers from the site to learn about MPC project methodology and develop expertise on advanced process control in order to handle future projects by themselves.

Approach

The scope of the project was challenging. There were many columns and furnaces involved and few engineers to execute the work. To speed things up, and to empower a team environment, the control group decided to hire advanced control consultants. The project was justified on economic return. Although some of the equipment in the scope had MPC controllers, certain controllers were selected to be remodeled due to their poor performance. The contract was made based on consultant hours in order to have flexibility in the project scope, e.g., in identifying benefits opportunities during the project. This approach requires discipline to calculate benefits and determine priorities.

Lessons Learned and Relearned

Organization

A multi-skilled project team was formed with control engineers, process engineers, production engineers, instrument technicians and operators. MPC consultants participated in all steps. DCS modifications were executed by the site control engineers. In addition, the site supply chain group provided important information about company optimization objectives and economic scenarios. The skills diversity of the group was essential to the success of the project and reinforced the conclusion that teamwork is very important for MPC projects.

The scope of the project initially involved regulatory control stabilization to avoid flooding of distillation columns, allow operation closer to products specifications and, particularly for Olefins 2, better energy performance on the cracking furnaces. These were the obvious opportunities in MPC. However, working with the consultants allowed a better understanding of the entire process and led to other opportunities as discussed below. In order to prioritize these new opportunities against the initial scope, it was important to measure the benefits on an ongoing basis. This also helped project promotion and acceptance.

To ensure the success of the project, it was crucial to monitor the results and performance in order to help the operators and engineers use the new capabilities in the controllers.

Solve the Correct Control Problem

Realizing the full benefits of the MPC applications required solving a control problem associated with the ethylene product header. A challenge to the Olefins 1 plant is the disturbances caused by disruptions in the demand of the ethylene customers, which historically had required overriding DMCplus targets or switching off the DMCplus controller. Solving this specific control problem reduced the impact of demand disruptions on both operations and the other MPC controls, resulting in less operator interaction, smoother operation, and higher online service factors.

Another important control problem is the DCS regulatory controls. Some MPC projects take the regulatory layer as they find it. Changing a DCS structure is hard work, operations must be convinced of the benefits, safety reviews must be carried out, drawings must be updated and console technicians trained. Nonetheless, change is often the right thing to do. Despite both Olefins 1 and 2 being well established ethylene units with functional multivariable controllers, several of the improvements in this paper are achieved by changing the regulatory control system. The correct regulatory structure can make plant operation and multivariable control better. As an example, using a pressure compensated temperature to control the tray temperature reduces the impact of pressure changes on the column. This simplifies the controller design making day to day performance better and long term support easier. Another example is reflux subcooling compensation in the DCS. In C3 Splitters this is a significant effect. If not corrected for, any independent variable that changes subcooling (i.e. pressure, cooling water temperature) will have a model to the overheads composition and column pressure drop.

Specific Improvements

Regulatory Design

Regulatory design, or redesign in our case, can have a significant impact on the control of the MPC. This includes linearization and improved rejection of disturbances due to the faster execution of DCS loops. Proper regulatory design can also increase the robustness of the overall system to model errors, simplify the controller design and better accommodate plant constraints. A summary of the redesigns follow.

C2 Splitter - The tray 13 temperature correlates very well with the bottoms ethylene composition. However, incorporating a temperature controller with the existing strategy of accumulator level control to reflux would result in an unstable control arrangement (i.e., there would be a snowball effect when the TC adds heat, the overhead accumulator LC adds reflux, etc.). Therefore an internal reflux controller was added which manipulates the side draw FC. Internal reflux is calculated as reflux flow minus side draw flow, based on setpoint values. As a result, incremental reflux changes translate into incremental side draw changes (at constant internal reflux). This control modification permits pressure compensate temperature (PCT) control of tray 13 via reboiler flow. The PCT controller has stabilized bottoms ethylene and the tower material balance (as seen in a more stable bottoms level).

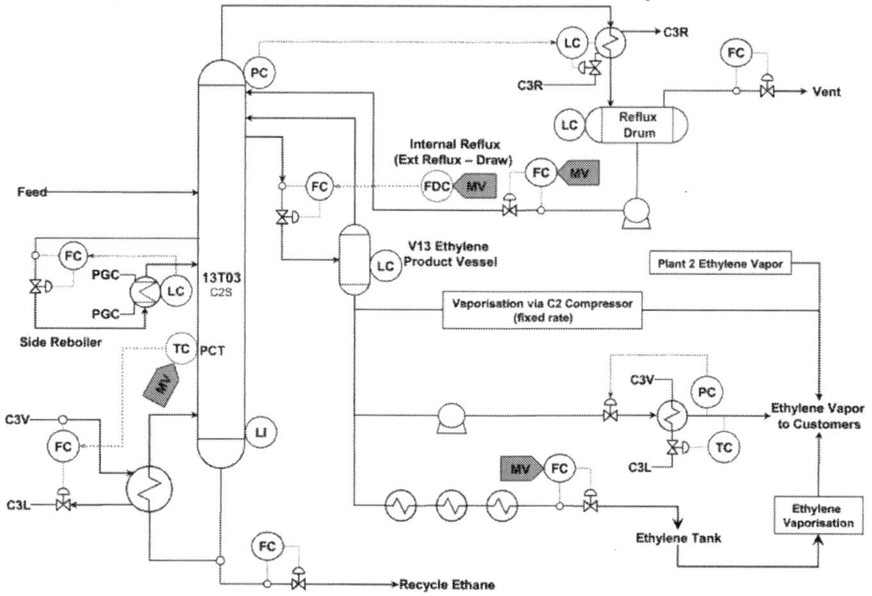

Figure 1. Revised C2 Splitter regulatory controls

Green Oil Tower – The original strategy had overhead accumulator level in cascade to reflux and bottoms level in cascade to reboiler quench water flow. Side draw was manipulated by a DCS-resident internal reflux flow controller (calculated as reflux minus side draw), with this strategy, the level controllers would fight when the FDC was saturated, leading to a violation in the column DP limit. The strategy was modified to have the level controller manipulate the C3 Splitter feed flow and the internal reflux controller manipulate reflux flow.

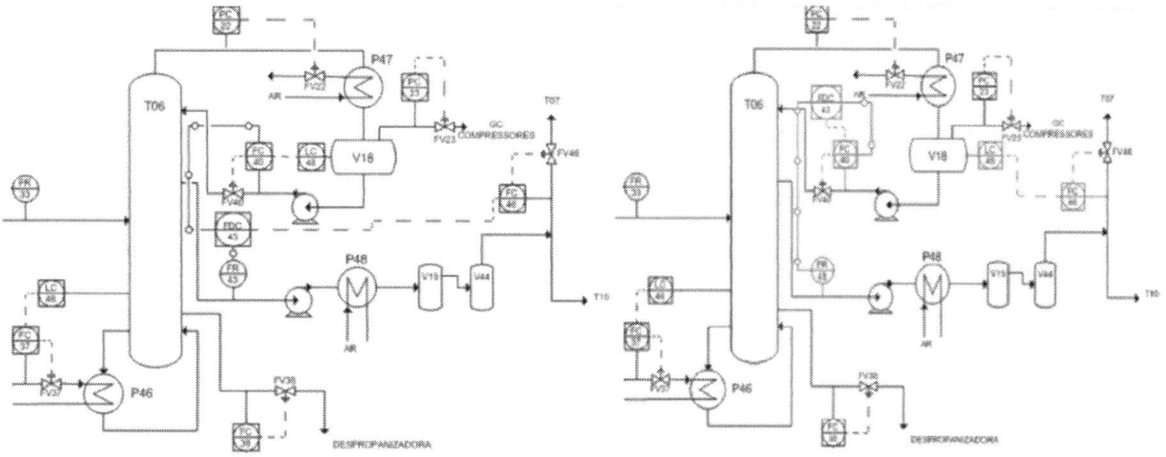

Figure 2. Revised Green Oil Column regulatory controls

C3 Splitter Plant 1 - The original strategy was a traditional design for high purity, small bottom product columns: bottoms level to reboiler quench water flow, with DMC manipulating reflux flow and bottoms flow. The overhead condenser is water cooled and, therefore, subject to ambient effects leading to disturbances in reflux temperature and, therefore, changes to internal reflux flow. Column DP is a constraint, resulting from resulting from tray downcomer liquid backup. Internal reflux control, calculated based on the difference between overhead and reflux temperature, was implemented in the DCS. A duty controller was incorporated into the bottoms LC to reboiler flow cascade. Although disturbances in quench water inlet temperature (due to furnace swaps) are realized relatively quickly in the bottoms level, addition of the duty controller showed definite improvement in providing tighter level control with less control action (and less interaction with the overhead system).

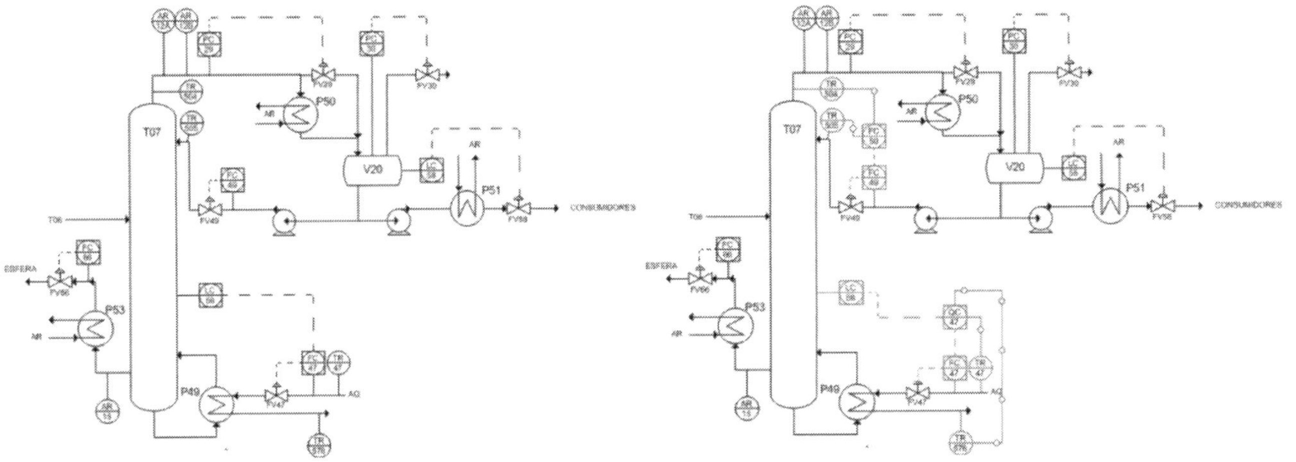

Figure 3. Plant 1 C3 Splitter revised regulatory controls

C3 Splitter Plant 2 - This large column is made up of an upper column and a lower column. There is a side reboiler at the bottom of the upper column heated by hot water. This exchanger reboils approximately 40% of the reflux. The original strategy was to fix the reboiler duty and control the level in the upper column bottoms by adjusting the flow to the lower column. However, even with a Q calculation on the reboiler duty, the duty of the reboiler was not constant when the water supply temperature changed. These errors in turn caused flow changes to the lower column and duty changes to the upper column, upsetting both columns unnecessarily. The lower column had a strict flooding limit so accurate control was important. The cure was simple. The base regulatory control strategy was altered to control the level by adjusting the reboiler duty and the accurate flow to the lower column was held fixed. This simple change corrected the inaccurate reboiler duty and allowed us to easily keep both columns steady during hot water temperature transients, this resulted in simpler models and a stronger controller.

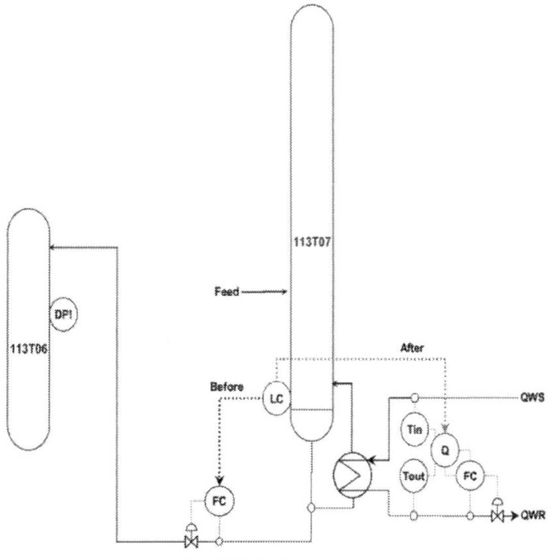

Figure 4. Plant 2 C3 Splitter Revised regulatory control

Plant 2 Furnace Upgrade

The primary purpose of the plant 2 furnace MPC upgrade was to include the revamped secondary air dampers to the existing furnace controllers. Inlet air dampers adjust the burner secondary air flow. These dampers are very important for effective air rate control; draft pressure alone is not adequate over the furnace operating range. The Plant 2 furnaces have four dampers, two on each side as shown in Figure 1.

Most of the current furnace controller models were judged satisfactory so response testing was limited to testing only variables that significantly changed the furnace air flow rate through the furnace. These are the four revamped inlet dampers and the furnace draft pressure. Testing of these variables was done manually over several days.

A challenge with all cracking furnace MPC controllers is ensuring the final control matrix is correctly collinear. This analysis is complicated by the physical arrangement of the furnace which means dependent variables (such as fuel gas pressure and valve position) in the radiant section will have a different gain ratio (by a factor of two) to variables such as radiant coil inlet temperature in the convection section.

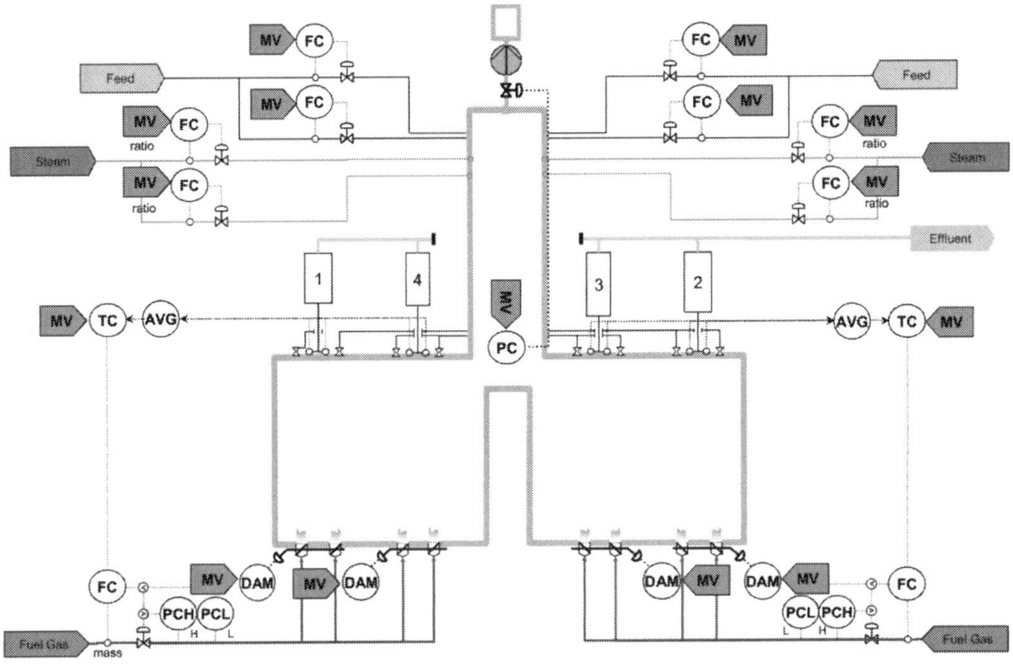

Figure 5. Plant 2 Furnace with revised inlet dampers

The identified model gains were modified using a custom Excel spreadsheet designed to find the minimum necessary changes to the identified gains to achieve colinearity. Using Excel allowed individual curves to be weighted according to how accurate each gain was. It also allowed the calculation process to be documented for further use if the furnaces are retested. The revised models are shown in Figure 6; final gain ratios are shown in Table 2.

The original furnace controller made use of a number of custom code calculations designed to inhibit the controller during certain conditions. For example, the controller would suppress movement of the draft pressure whenever the draft pressure was being reduced but not when it was being increased. At times this would cause oscillatory movement in draft pressure, moving slowly down and quickly up. Such calculations are usually applied after the initial controller is built in an effort to mitigate poor performance without retesting. These calculations are confusing and can only be implemented using a trial and error approach. For the revised furnace controller, all of these calculations were removed as they were not required with the revised controller model. Once the first furnace was completed, then all identical furnaces were changed to use the same controller model.

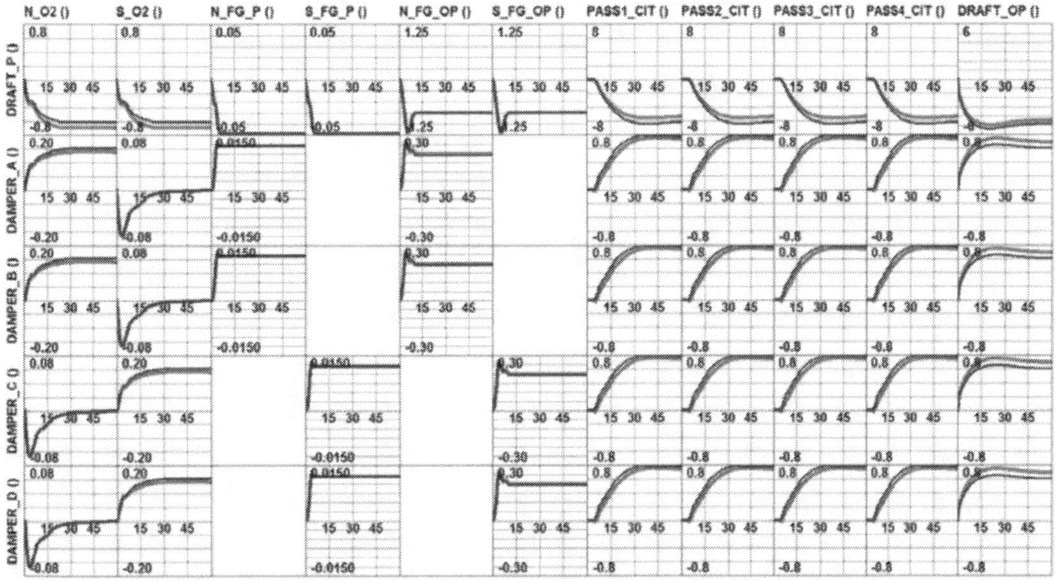

Figure 6. Plant 2 furnace air models: before (red) and after (blue)

Table 2. Plant 2 Furnace Air Final Gain Ratios

	N O2	S 02	N FG Pres	S FG Pres	N FG OP	S FG 0P	Pass 1 CIT	Pass 2 CIT	Pass3 CIT	Pass 4 CIT	Draft OP
Draft P	-0.6091	-0.6091	-0.0483	-0.0483	-0.7551	-0.7551	-6.1918	-6.1918	-6.1918	-6.1918	4.802
Damper A	0.15228		0.01208		0.18878		0.77398	0.77398	0.77398	0.77398	0.60025
Damper B	0.15228		0.01208		0.18878		0.77398	0.77398	0.77398	0.77398	0.60025
Damper C		0.15228		0.01208		0.18878	0.77398	0.77398	0.77398	0.77398	0.60025
Damper D		0.15228		0.01208		0.18878	0.77398	0.77398	0.77398	0.77398	0.60025
Ratio	**-4**	**-4**	**-4**	**-4**	**-4**	**-4**	**-8**	**-8**	**-8**	**-8**	**8**

C3 Splitters

All composition variables were modeled as steady-state variables, as opposed to ramp variables (as is often done to get a shorter controller horizon), to provide a consistent set of steady-state targets to the controller. The C3 Splitters can be constrained due to column DP or condenser limitations, both of which are relieved by reducing internal reflux. Note that a deviation above the upper propane impurity limit of the propylene product invariably results when reflux is decreased to control at one of these constraints due to the slower dynamic effect of the bottoms draw compared to the reflux. In Olefins 1, the DMCplus controller manipulated the feed rates to both C3 Splitters (and included the Green Oil Tower). This provides the future capability to adjust the feed split to balance propylene losses in the two Splitters when the models for the second Splitter (heat pump column) are added.

Ethylene Product Vessel V13

On Olefins 1, the side draw from the C2 Splitter goes to a separation vessel (V13) under flow control. From this vessel, the majority of the ethylene is vaporised to supply the site vapour customers. The pressure of the customer pipeline is controlled by adjusting the flow of ethylene from this vessel through 2 vaporisation routes. The remaining ethylene is refrigerated for cryogenic storage. This flow to storage consumes a significant amount of refrigeration duty.

Level control of this relatively small vessel was a constant challenge for the Olefins 1 operators due to customer demand changes, plant production changes and Olefins 2 production rate changes. This situation was improved by changing the C2 Splitter control scheme as described earlier and controlling both the C2 Splitter reflux drum level and the V13 product level, effectively doubling the size of the surge capacity. Previously the reflux drum had been under tight level control. The revised MPC controller was built to exploit the capacity of both vessels so that the flow of ethylene to storage was changed as slowly as possible, thereby minimising the impact of refrigeration capacity. At the start of the project, several of the operators believed that this level was impossible to control automatically as "everything affects it". The addition of the reflux drum capacity allowed the flow to storage to be automatically controlled during all but the most severe upsets. It is very popular with the operators.

Recommended Best Practices

PID Tuning

The DCS loops associated with the DMCplus controllers were tuned for disturbance rejection. For cases where this tuning resulted in significant overshoot for setpoint changes, the available DCS option of setpoint ramping was used to keep overshoot below 20%. For slower responding loops and levels, a tuning package [1] was used. Models for the tuner were developed either using DMCplus or by fitting data to simple transfer functions (e.g., dead-time plus ramp). The benefits of using a tuning package included:

- Ability to tune new loops with confidence (important for operator acceptance)
- Ability to achieve critically damped responses for level loops which were retained i.e., "not opened" for control in DMCplus; for example, the four cold box separator levels of the demethanizer feeds.

As an example, Figure 7 shows the open loop testing, resulting model, and closed-loop response for the C2 Splitter Pressure Compensated Temperature (PCT) PID controller.

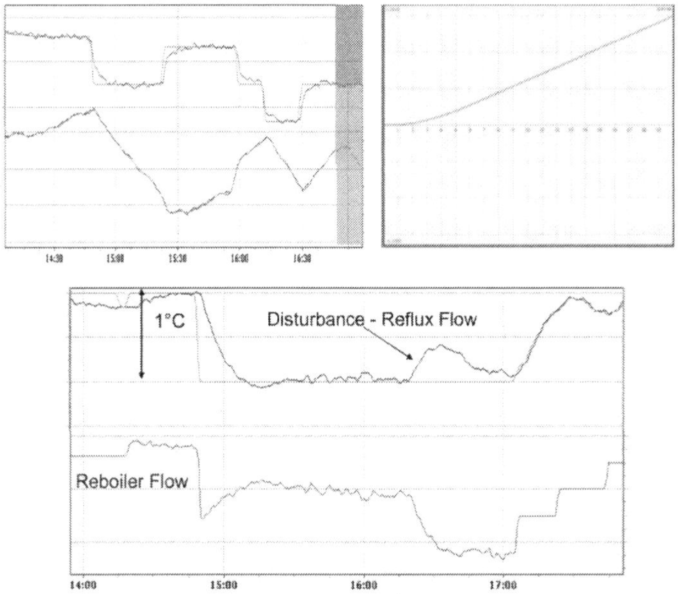

Figure 7. PID tuning of C2 Splitter tray temperature. Step test, integrating model, and closed-loop performance

The loss of downstream customers creates a significant disturbance to the plant and the refrigeration systems in particular. Many loops within the refrigeration system (compressor SIC, suction drum PICs and LICs of the various users) were retuned. It was important to verify tuning (and retune as necessary) during the hottest months of the year, when the system was most loaded.

Steady-State Simulation

Steady state simulations of high purity distillation columns such as C2 and C3 Splitters are quick to build and calibrate to observed data. We constructed models of the four high purity columns using the AspenTech AspenPlus steady state simulator. These models were used primarily to confirm the experimental steady state gains for the controller response models. High purity distillation processes such as C3 Splitters have a long settling time that can make the steady state gains difficult to estimate experimentally. Knowledge of the theoretical process gain is an advantage, especially for feedforward independent variables (such as feed composition) that are not easy to move. Figure 8 shows the estimated effect of feed propane composition on the overheads and bottoms composition of the Plant 2 C3 Splitter. Table 2 compares theoretical to experimental process gains.

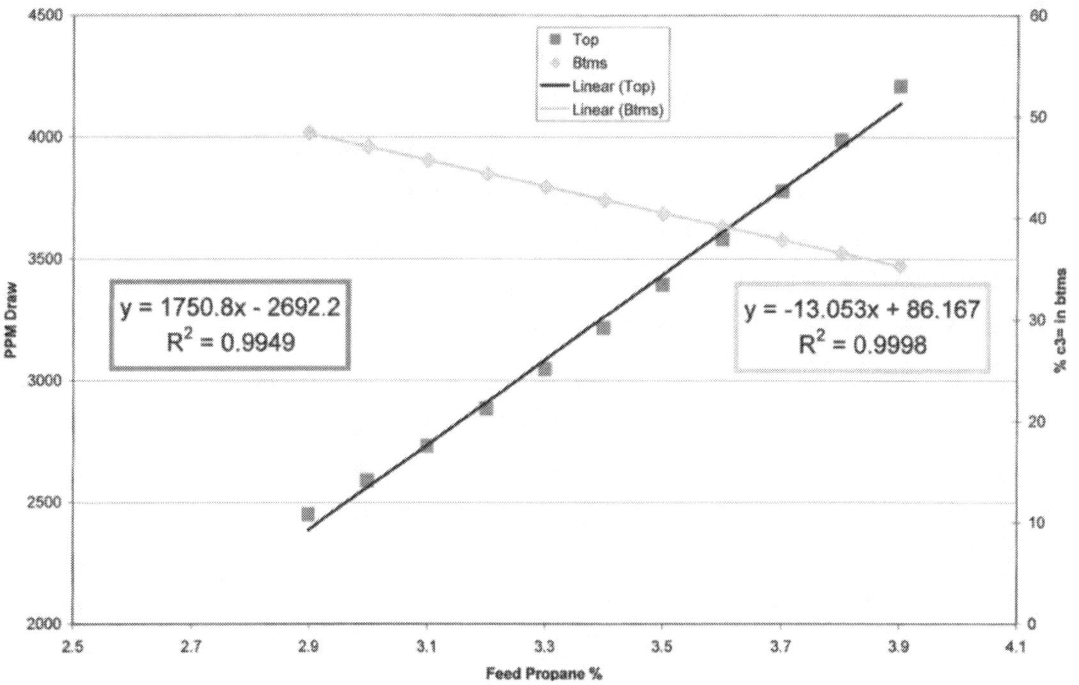

Figure 8. Simulated effect of feed composition on overhead and bottoms product compositions

Table 3. Simulation vs. experimental gains

Independent Variable	Overheads Propane in Propylene	
	Theoretical gain	Experimental Gain
Reflux (t/h)	24.5 ppm	21 ppm
Feed (t/h)	555 ppm	496 ppm
Feed Propane (Mole%)	1750 ppm	1400 ppm
Bottoms Flow (t/hr)(*)	-4670 ppm	-1572 ppm

(*) The error in the bottoms flow is likely due to truncation of the process response at the controller settling time interval of 20 hr. The actual response is about 48 hr.

The steady state models were also used to estimate several DCS constants used in the regulatory control system such as:

- Effect of reflux subcooling changes on calculated internal reflux
- Pressure compensated temperatures

Transforms

Transforms were used in several situations where significant nonlinearities were observed. A key reason for including the transforms was to provide stable, non oscillatory responses for situations where current operation might be far away from the steady-state targets (such as when a DMC is tuned on) and to provide for possibility of future operation at significantly different steady-state targets. All transforms used on the project were based on piece-wise linear transforms on either MVs or CVs based on the data. CV transforms were developed based on plots of predicted vs. actual and two iterations (typically) were used to develop the transform. An example, the C2 Splitter bottoms ethylene composition transform, is shown in Figure 9. Table 4 lists the key transforms developed on the project.

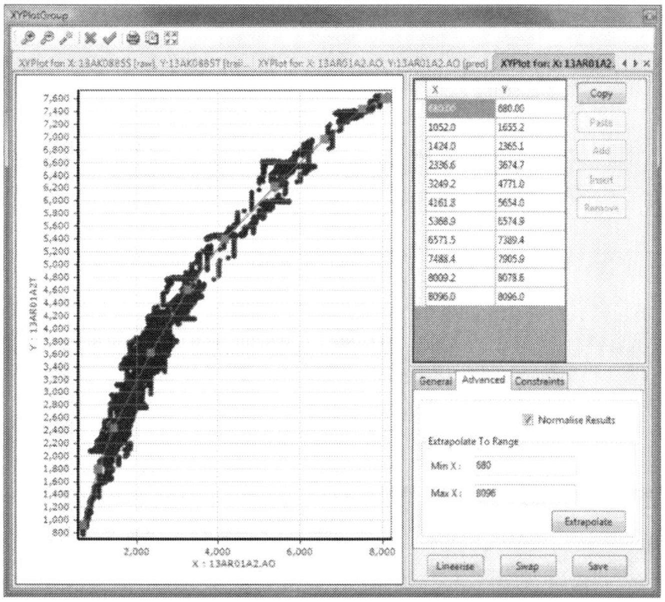

Figure 9. Transform for C2 Splitter bottoms ethylene. Analyzer: x axis, transformation: y Axis

Table 4. Controller transforms

Variable / Transform Type	Nonlinearity
Various controller OPs / CV	OP vs. flow nonlinearity
Demethanizer methane / CV	Gain increase with higher impurity
Demethanizer chiller level / MV	Dimensioning effect of chiller level on reflux flow with increasing chiller level*
C2 Splitter product ethane /CV	Gain increase with higher impurity
C2 Splitter bottoms ethylene /CV	Gain increase with higher impurity

Results

In the following, selected trends and statistics are presented to illustrate the improvements obtained with the revamped controllers, compared to previous operation. The data includes both normal (including constrained situations) and upset situations; no data has been excluded. Most plots show approximately 68 days of operation prior to the project and 47 days afterwards.

Prior to the project, Olefins 1 C3 Splitters operators would manipulate C3 Splitters feeds by changing PID setpoints. Including these feeds as manipulated variables in the APC controller, the parallel columns better absorbed feed and severity variations in furnaces. The number of interventions per day in the feed was reduced from 39 to practically zero. Figures 10 and 11 below shows feed of both Splitters before and after APC remodeling. C3 Splitter 1 is a conventional column and C3 Splitter 2 is a heat pumped column.

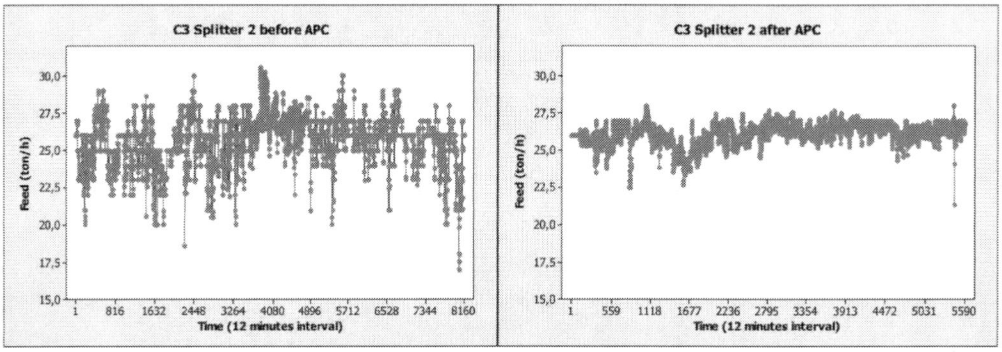

Figure 10. C3 Splitter 1 Feed

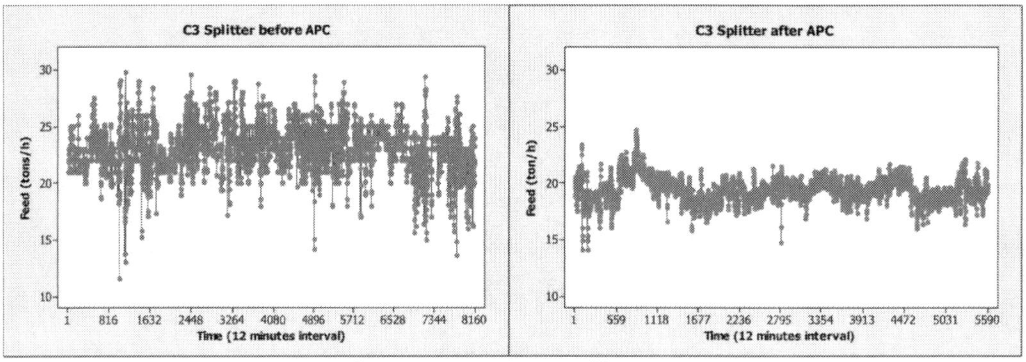

Figure 11. C3 Splitter 2 Feed

Another improvement in C3 Splitter 1 was the decreasing of variability in contaminants of propylene product. The regulatory control changes along with APC resulted in the variability reduction as shown in the Figure 12 below. The most important benefit in C3 Splitter 1 was the reduction of propylene loss in the column bottoms. The reduction in the average propylene losses was 5.2% compared to the baseline average (Figure 13). A significant part of the variability in the propylene loss in the bottoms is due to condenser limitations.

Flooding was an issue for C3 Splitter 1 in Olefins 1. Modifications in regulatory control such as including internal reflux control reduced the flooding which can be seen in the delta pressure - Figure 14. Standard deviations are shown in Table 5.

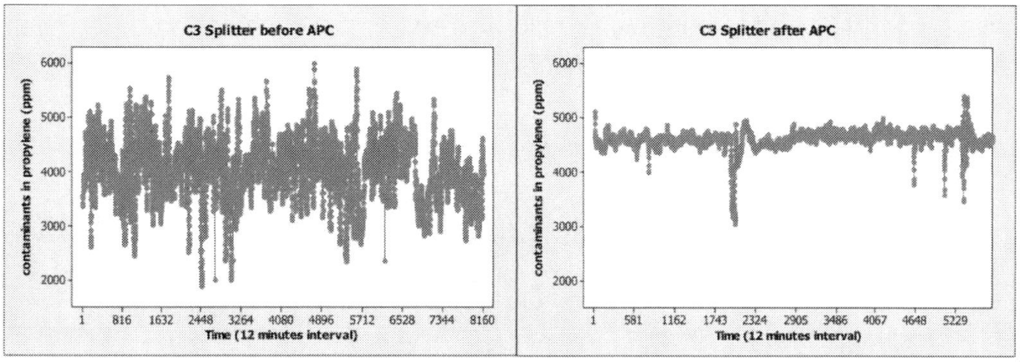

Figure 12. Contaminants in propylene product of C3

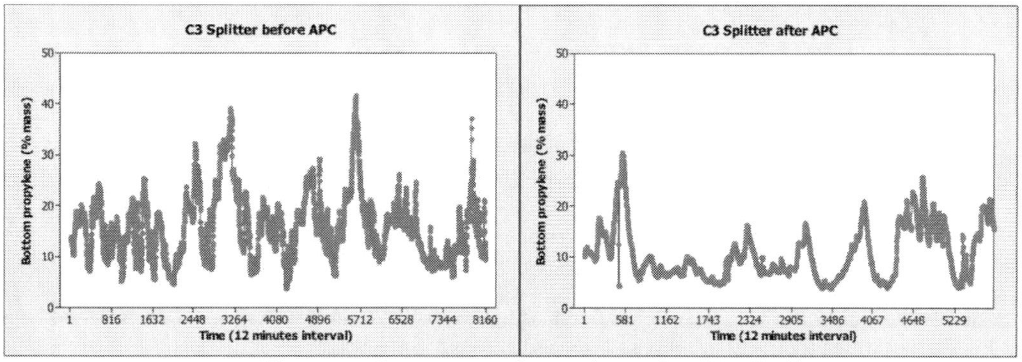

Figure 13. Propylene loss in C3 Splitter

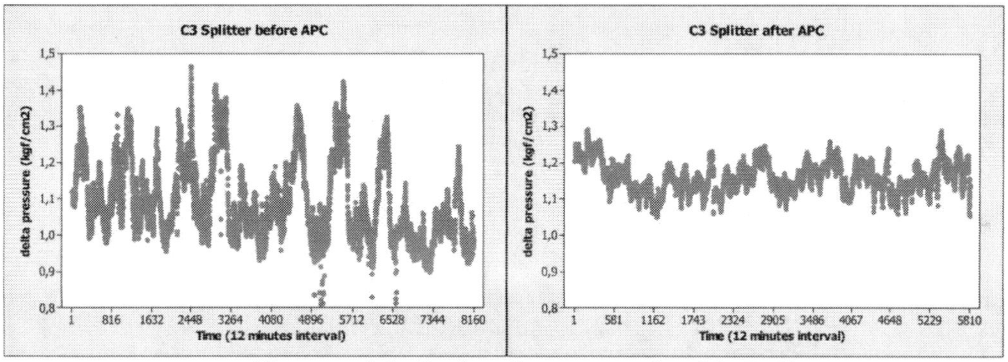

Figure 14. Delta Pressure in C3 Splitter

Table 5. C3 Splitter standard deviations - before and after results

Column	Variable	Before	After
C3 Splitter 1	Propylene in bottoms flow (%)	6.6	5.5
	Overhead contaminants (ppm)	570	200
	Delta pressure (kgf/cm2)	0.11	0.04
	Feed (ton/h)	1.86	1.03
C3 Splitter 2	Feed C3 Splitter 2 (ton/h)	2.16	0.73

As discussed previously, implementation of internal reflux was very important for APC success in the C2 Splitter. This approach stabilized the overhead and bottoms compositions and the result with the MPC was minor variability in the product ethylene contaminants and in the ethylene in bottoms ethane. These results are shown in Figures 15 and 16. Standard deviations are shown in Table 6.

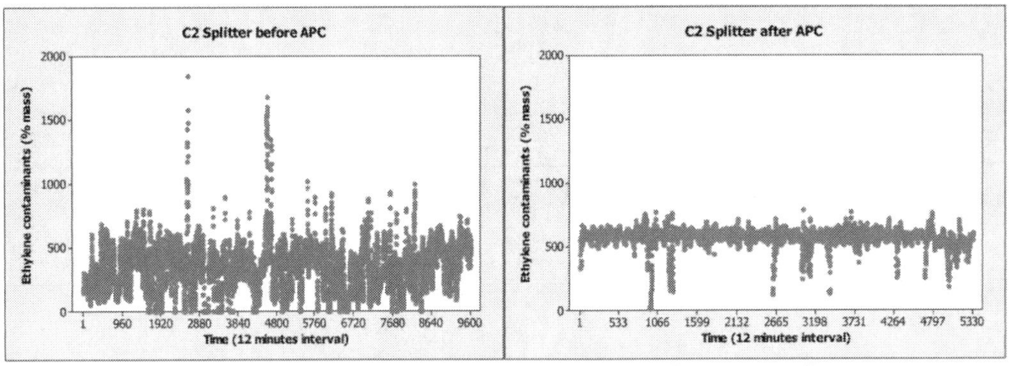

Figure 15. C2 Splitter overhead contaminants

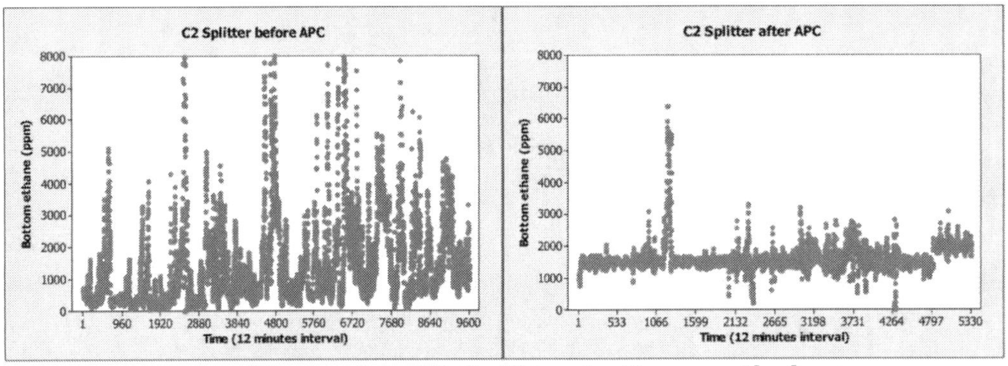

Figure 16. C2 Splitter bottoms ethylene

Table 6. C2 Splitter standard deviations - before and after results

Variable	Before	After
Overhead contaminants (ppm)	255	76
Ethylene in bottoms product (ppm)	1653	652

The new C2 Splitter controller significantly reduced the number of flow set point changes of ethylene to storage tank. The ethylene vessel product level was operating in manual mode and the number of interventions in flow to the tank per day was around 45. After the project this number is around 4.8. This result means the operators now have more time available to improve the process.

The ratio between ethylene flow to storage and C2 Splitter feed flow reduced by 26%. Sending ethylene to the storage tank is much less economic than directly to the product header. Sending ethylene to tank requires significant refrigeration duty. The C2 Splitter MPC helped to maximize flow to the customer header by better control of the product vessel level.

In addition, the overfilling of the product vessel is avoided. Figure 17 shows the behavior feed to flow to tank ratio before and after APC remodeling. Figure 18 shows the product level.

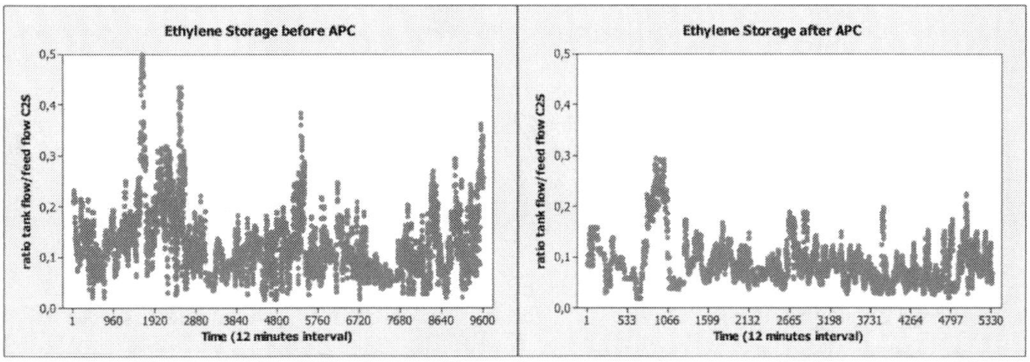

Figure 17. Ethylene flow to consumers and storage ratio

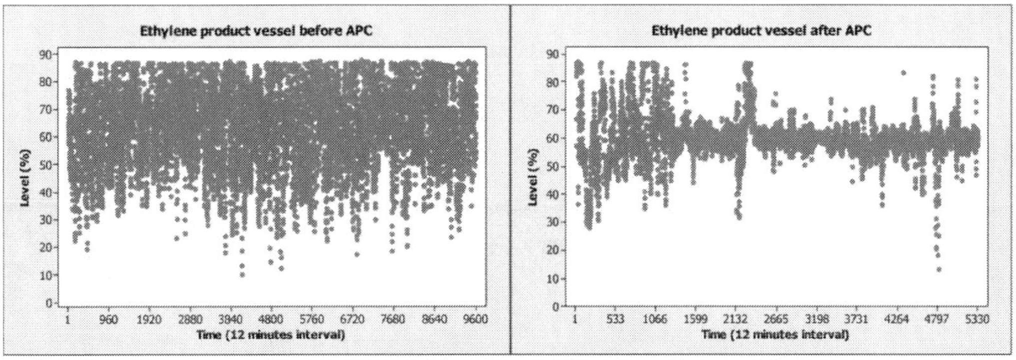

Figure 18. Ethylene Product Vessel level

Olefins 2 C3 Splitter Results - Due to a plant routine shutdown, runtime on the Olefins 2 C3 Splitter has been limited. However the first month of operation showed encouraging results – see Table 7.

Table 7. Olefins 2 C3 Splitter results

Variable	Average		Standard deviation	
	Before	Before	Before	After
OVHD contaminants (%)	0.350	0.408	0.067	0.045
Bottoms propylene (%)	15.3	6.46	11.80	4.28
Delta Press T06 (Kgf/cm^2)	0.308	0.313	0.018	0.005
Delta Press T07 (Kgf/cm^2)	0.732	0.723	0.019	0.016

Olefins 2 Furnaces - Table 8 shows the results in one furnace after remodeling and addition of new bottom dampers as manipulated variables to the furnace controller. The better air control allowed the minimization in excess air and consequently savings in fuel gas. Similar results have been obtained for the other six similar furnaces.

Table 8. Cracking Furnace results

CO_2 reduction	216 ton/month
Fuel gas reduction	29 ton/month

Next Steps

There are further opportunities in both olefins plants to improve process results through improved process control. A challenge for the Braskem control team is to maintain the revised controllers along with implementing new projects. Future work includes integrating the cold side controllers and implementing furnace severity control. Braskem also is preparing for the training of all operators, independently of the existence of projects, as new operators in the control panel are a frequent reality.

References

1. Dittmar, Rainer, Shabroz Gill, Harpreet Singh, and Mark Darby (2012), "Robust optimization based multi-loop PID controller tuning: A new tool and its industrial application," *Control Engineering Practice,* To be Published.

50d

On-line Energy Management System: Round 2

Brian Finnegan

Project Manager - Energy Management

Lyondell Chemical Company

David Chenevert

Principal Engineer & CVO Site Energy Champion

Equistar Chemicals, LP

Matt Michnovicz

Principal Engineer – Energy Management

Lyondell Chemical Company

Prepared for Presentation at the 2012 Spring National Meeting

Houston, Texas, April 3rd, 2012

On-line Energy Management System: Round 2

Brian Finnegan

Project Manager - Energy Management

Lyondell Chemical Company

David Chenevert

Principal Engineer & CVO Site Energy Champion

Equistar Chemicals, LP

Matt Michnovicz

Principal Engineer – Energy Management

Lyondell Chemical Company

Abstract: In 2008-9 LyondellBasell implemented an on-line, open-loop utility optimizer at one of its world-scale olefin sites. It's called our energy management system. While key project goals were met, the full potential of the system was not realized. After the system was brought on line, deficiencies were experienced in staffing for the system, the reliability of the models, and processes for maintaining and using the system. Additional opportunities were identified by analyzing the optimization actions. The deficiencies were addressed and additional opportunities were realized in a second round of engineering and commissioning.

This presentation discusses the identification and resolution of deficiencies that were encountered as well as the additional opportunities identified, including:

- Providing adequate staffing for design, start-up, and long-term maintenance
- Assigning credible limits within the models
- Assigning actionable recommendations to the appropriate personnel
- Ensuring reliable performance through the use of knowledgeable experts
- Providing APC to close the loop on rapidly changing variables

An explanation of how the changes are being incorporated into the company's third energy management system at another olefins site will also be discussed.

(Please note that this page – pg 2 - should consist of the title, author(s) and the abstract. The body of the paper should start on page 3)

On-line Energy Management System: Round 2

LyondellBasell is the world's third-largest independent chemical company. LyondellBasell's 2010 annual revenues were approximately $41 billion, and the company has more than 14,000 employees worldwide. Our vertically integrated facilities, broad product portfolio, manufacturing flexibility, superior technology base and reputation for operational excellence enable us to deliver exceptional value to our customers across the petrochemical chain – from refining to advanced product applications.

We manufacture products and develop technologies that improve the quality of life for people around the world. Our products are the basic building blocks used to manufacture countless everyday goods such as personal care products, fresh food packaging, lightweight plastics, construction materials, automotive components, durable textiles, medical applications, biofuels and many others. With the help of LyondellBasell materials, thousands of products are made safer, stronger, more affordable and more reliable.

Beginning in 2007, LyondellBasell initiated an energy reduction goal to reduce energy consumption across the company by 2% per year for the next five years. One tool chosen to work toward this goal was an on-line open loop utility optimizer. We refer to this type of tool as an Energy Management System (EMS). The EMS is an open-loop optimizer software that gathers real-time process and cost data and calculates an optimum way to deliver the energy required by a process. It then compares the actual quantities being consumed with the optimum quantities and presents them to a board operator as real time actionable opportunities.

Initial Projects

This EMS technology is not new but has not been implemented in real time at any LyondellBasell sites. A pilot project was envisioned for a small manufacturing site to validate the benefits of these types of projects. However, two large sites, one near

Houston in Channelview, Texas, and one near Rotterdam, in Botlek, NL, requested to be piloted. Management at both sites saw this as an opportunity to make significant progress on our corporate energy goal of reducing energy by 2% per year.

This EMS technology was initially viewed as an information tool rather than a control tool. For this reason the two projects started off as IT (Information Technology) led projects. One of the key requirements for a successful IT project is to develop clearly defined software requirements. This proved difficult because we did not have expertise with this technology, and we excluded the vendors that sold this technology from assisting in writing the requirements. We assembled a cross functional core implementation team to execute the project. The core team spent several months defining detailed requirements for a technology of limited familiarity. Ultimately, the project scope was reduced to just presenting actionable information to the operators, and most of the integration points that required use of our IT local area networks were eliminated. Responsibility for leading the execution of these projects was reassigned to the global energy management group. This allowed the pilot projects to be implemented without the level of detailed up-front software functionality definition required for IT-led projects. The core team evaluated different EMS technology vendors and selected one. There were not significant differences in the top vendors, so the project team selected the vendor deemed to have the most experience. The same vendor was selected for both sites, but two independent project teams were used for execution.

Both sites expanded their core project teams to involve more site personnel and support from central capital projects, IT and central control systems, along with the vendor selected. Led by the global energy management group, the systems were designed, configured, tested and implemented at both sites. Energy cost reduction opportunities were identified and captured during the design and development phases of both projects, even before the projects went online. Both projects were successfully executed on time and on budget. Both met their initial financial objectives. This presentation is not about the technology used but rather the internal and external human resources we utilized and how their use could be improved. We will discuss the personnel challenges the project teams faced and how we worked through these challenges.

The Botlek EMS project is viewed as successful and the plant continues to run near its optimal utility configuration. At present, the Botlek EMS occasionally identifies significant energy saving opportunities. It requires an ongoing effort for those responsible for the Botlek EMS to maintain operator vigilance since there are few opportunities to make changes when a site runs near its optimum.

The Channelview EMS is viewed as partially success. The project quickly paid for itself but when the optimizer ran into glitches, it was shut down for an extended period until repairs and improvements were implemented. The Channelview EMS initially did not achieve its full potential. The events that led up to the temporary shut down and the issues that have prohibited this tool from delivering its full value are noteworthy lessons for anyone implementing a similar EMS project. These lessons learned were integrated into the upgrade of the Channelview EMS and the new EMS project at the LaPorte, Texas site.

Throughout the EMS projects, we had excellent support from all levels of management. These projects were implemented over a period of time when our company was undergoing significant management changes due to entering and emerging from Chapter 11 bankruptcy. But the pressures of the bankruptcy caused frequent realignment of projects priority and core team members' duties. Our company's management continued to support these projects, but the impact of multiple management changes and fewer technical resources amplified what should have been minor delays into more significant issues. As the number of employees reduced, we lost project team members and the remaining resources were spread thinner. This also reduced the level of participation from key functional groups, requiring the site energy leads and energy management team to pick up more responsibility. The extent of personnel changes and reductions in technical resources at our Botlek site was less than the impact at the Channelview site and is a key reason the Botlek project realized more of its potential.

Project Differences

There were also significant differences and changes in the key technical resources provided by our implementation vendor. Both pilot projects started with very experienced vendor resources. We believe we had the two most experienced experts in the world on our teams, one on each project. Since these experts worked for the same company, we had an exchange of best practices. Further, each expert was backed up with additional highly skilled individuals. There was, however, a difference in the vendor support between the projects. The vendor resources that worked on the Botlek project were all located in the same building. The vendor resources for the Channelview project were spread around the western hemisphere. Several of the most experienced vendor resources for the Channelview project were also working on multiple other projects. They were swapped out with less experienced but more available resources as the project progressed. These changes caused significant delays, mainly to the vendor deliverables. In total, three vendor project managers were used to complete the projects. Also, because the Channelview vendor team was spread out between North and South America, the level of model testing on the vendor's system (using archived data) was insufficient to identify or resolve normal configuration problems that came up when the system was brought online in the production environment. To be fair, the Channelview EMS was significantly more complex than the Botlek EMS, but the changes and location of the vendor team compounded our problems.

After we went live with both projects, there were typical post go-live issues. These were anticipated, and we included provisions for post go-live support in our contracts and schedule. These problems were quickly identified and corrected by the Botlek team. Several of these required the vendor experts to become involved with the analysis and solutions. The Channelview EMS encountered similar post go-live issues but these issues did not get resolved in a timely manner. The delay in correcting these problems damaged the reputation of the system and contributed to it being shut down until the issues were resolved. One of the delays was caused by the key vendor expert resigning while no adequate back up was available. The loss of this key vendor resource strained the vendor capability, making it difficult to get someone with similar technical skills to assist in resolving our post go-live issues. Even with these problems, the Channelview EMS met its initial financial objective. The EMS was repaired but the following issues caused us to delay its restart until this system could be upgraded:

- The loss in EMS credibility with operators from being shut down.

- Poor reliability: a recurring issue with the optimizer allowing it to solve and calculate an optimum only about 80% of the time.
- Feedback from operators that some of the opportunities were not within their control
- Significant process changes due to a major plant turnaround.

Nonetheless, the success of these two pilot EMS projects resulted in the initiation of a third EMS at our LaPorte site. The LaPorte EMS project scope included resolution of the issues and an upgrade to the Channelview EMS, so that these solutions could be implemented both places. After evaluation, a different implementation vendor was selected for this EMS project. The selection of the implementation vendor was not biased towards the vendors' level of experience with an on-line utility optimizer since we had a good idea of what we needed. We wanted to expand the number of vendors that we felt were capable of implementing this type of technology. We used the same technology at Botlek and Channelview but do not believe this technology is any better or worse than its competitors. However, it is now more familiar to us.

Resolution of Issues

This implementation vendor was first used to upgrade the Channelview EMS and then to configure the LaPorte EMS.

- We created a development environment that was identical to the real time production environment but was isolated from it. This greatly improved the development and testing of the revised system at Channelview. The development environment allowed us to verify that the upgraded system was working the way we desired before production.
- The technical issue with the model not finding an optimum utility configuration 20% of the time was resolved by allowing the limits to change in proportion to the actuals values and limits after they were reconciled. This greatly reduced the number of times the optimizer could not converge on a solution. Optimized conditions are now identified +98% of the time.

- The issue of presenting operators with an energy saving opportunity that they couldn't implement was resolved. Operators can reclassify opportunities to other groups, with built-in justifications for repair or resolution.
- The Botlek plant had great success by using specialists instead of engineers as the first line of EMS support for board operators. This model was also implemented at Channelview.
- Awareness and support for realizing the EMS model opportunities has been partially addressed through improved communication. An automated email showing the value of each opportunity is sent regularly to key stakeholders. Managers also get an e-mail so they are aware of the monetary value of opportunities that could not be implemented. This also helps managers set priorities for opportunities that require maintenance work or plant outages.

We believe we now have a much better system. We continue to struggle with the priority of executing this system's opportunities with our operators, specialists, engineers and managers. These resources are extremely busy with important and urgent work. Saving energy is important but not always urgent, resulting in missed opportunities. Our experience showed the opportunities from the EMS become less valuable because the plant utilities are operated more efficiently; it thus became harder to keep the appropriate amount of focus on maintaining the EMS.

The Next Project

We have applied all of the lessons learned to the LaPorte EMS. As the time of this writing (January 2012), we believe this will be our most successful system. The LaPorte EMS is scheduled to go live in late March.

The Channelview project team had representation from local and central technical groups as well as site operations. However, ownership, daily use and maintenance of this EMS was assigned to people that were already fully committed to other work. The Channelview EMS did not reach its full potential because it was the third or fourth priority for several people who were engaged or utilized with their first and second priorities. The result is that driving gap closure and initiating needed model repairs was often delayed.

The upgraded Channelview EMS minimizes the time required to drive energy gap closures.

Key Learnings

Four key learnings were distilled from the first two EMS deployments and included in the La Porte project and the Channelview upgrades:

(1) Operator displays should only highlight gaps between actual and target performances that are in the operators' control. Gaps requiring other groups' intervention should be assigned to the other groups and the highlighted action removed from the operator displays.

(2) Site and vendor support should be locked in before deployment. Both site and vendor technical support staff turned over more quickly than expected. Trained, model versed support was not available quickly when EMS problems came up. With only one vendor available for support, we had a competitive disadvantage when seeking immediate help.

(3) EMS system performance should be distilled into easily accessible and actionable information. Analysis of EMS system performance was initially tedious. The supervisors and engineers responsible for checking on system uptime, recommendations and operator responses spent time paging through displays and logs every day to get this data.

(4) Training should be consistent and on time. Formal classroom training was originally planned but not conducted. As personnel numbers were reduced, training occurred on-shift, one-on-one as operations people were available during their shift.

Getting Back on Track

To get the Channelview EMS back on line, a small project was initiated to correct the model reliability, resolve the issues listed above and update the EMS for major capital improvements. These requirements were gathered, and a vendor was selected to implement the changes.

To ensure the LaPorte implementation was smoother than its predecessors, management and technical commitments were required prior to project approval. These were: sponsorship by the senior VP of manufacturing, support from the site manager and the energy group manager, an experienced site energy lead committed to the project, and buy-in by central and local control system personnel.

The key learnings and other opportunities were incorporated into the project design.

Opportunities were assigned to the appropriate people. Opportunities outside of operator control were easy to re-categorize and are not displayed on operator screens.

Each opportunity was tracked individually rather than in groups by process area or type of energy opportunity

Technical support was strengthened by including additional internal technical resources to the project team. Further, we contracted additional vendors to provide on-going technical support for the EMS as needed.

A server identical to the production server hosting the EMS system was purchased and installed for project development and testing. Because this system was not in production, it did not have the restrictions and change protocols the initial Channelview system had. So not only were problems or errors quickly resolved, but there were no translation errors developed due to differences between the development server and the production server.

New reports were automatically generated by the system, reducing the time needed by area personnel to see opportunities identified by the EMS. These reports show steam, electricity and fuel gas balances, overall and individual opportunities, and actions taken to reduce gaps.

Channelview

At the same time, the Channelview upgrade project was designed to resolve identified model errors and implement the same EMS upgrades incorporated in the La Porte system. Once the model was repaired and the upgrades were completed, operations personnel were reintroduced to the EMS. The Channelview upgrade project included some key learnings from the first implementation of the EMS model. The EMS is a supervisory system and can make recommendations as frequently as it runs. In our case, this is every 15 minutes. We found one system that the EMS recommended a change with every solution, which is too frequent for operations to respond. When a situation like this comes up, an alternate control solution should be chosen. In our case, a control systems engineer initiated a project to develop a stand-alone Advanced Process Control model for several steam exports with optimum targets that change frequently. These steam exports are handled by closed-loop controllers and capture considerable savings. The value of these particular controllers would not be recognized as quickly if the EMS were not deployed.

The Channelview EMS suffered from operators limiting its capabilities. Real optimum operation sometimes takes variables out of their previous ranges. If output variables are over-constrained, the optimizer can't do its work and benefits are limited. The project team worked with consultant resources and did additional training to overcome objections to movement outside the operators' comfort zone and achieve additional benefits.

Conclusions

This paper discussed the challenges we faced with these projects and covered the lessons we learned. The fact that we continue to implement these projects at appropriate sites is the best indication that these projects add value. We strongly feel that EMS (on-line utility optimizer) technology is worth pursuing and that it delivers value when implemented on complex utility systems. The more complex the utility system, the more applicable this technology is. Having said this, we believe the keys to successfully implementing future EMS projects are linked closely to how the projects are staffed after implementation. The technical issues with an EMS are much easier to resolve than the staffing issues.

Best Practices for EMS

1. Design the system to present opportunities to a board operator
    a. Keep it simple

    b. Avoid presenting opportunities that are not actionable
    c. Support the operator with a specialist, then engineers, then others
    d. Move rapidly changing opportunities to more automated systems.
2. Select a technology with multiple vendor experts that can support it
    a. The depth of local vendor experts is important
    b. An ongoing support arrangement should be set up to quickly resolve technical problems outside the level of in-house expertise.
3. Develop in-house technical support
    a. An EMS is a simplified RTO, and in-house control system engineers should become familiar with it
4. Have a test environment that represents the production environment as much as possible
    a. The EMS should be tested as an entire system that is as close to the production system as possible.
5. Keep engineers, supervisors and managers aware of opportunities the EMS is identifies
    a. Remind them that the EMS has returned the most value when it cannot find gaps between actual and optimum operation
    b. Make the value of the opportunity gaps transparent to all stakeholders

50e

Operator Training Simulators - Nova Chemicals' Experience for training, certification and other applications

Ron Besuijen

Technical Training Specialist

NOVA Chemicals

Roger Young

Technical Training Specialist

NOVA Chemicals

Michael Guaraldo

Project Manager

RSI Simcon

Prepared for Presentation at the 2012 Spring National Meeting

Houston, Texas, April 1 -5, 2012

Operator Training Simulators - Nova Chemicals' Experience for training, certification and other applications

Ron Besuijen

Technical Training Specialist

NOVA Chemicals

Roger Young

Technical Training Specialist

NOVA Chemicals

Michael Guaraldo

Project Manager

RSI Simcon

Abstract: The Nova Chemicals Site located in Joffre, AB Canada has numerous gas-cracking olefins units as well as down-stream polyolefins processing plants. Nova has been utilizing Operator Training Simulators at this site for over 12 years as part of a comprehensive training and certification program for its control room operators. This paper reviews how the simulators have been used as part of this program. In addition, it describes other ways that the simulators have contributed to the overall benefit of the petrochemical complex. The topics to be discussed are:

1. Current simulator models within Nova, history, and future development.
2. Training structure of simulator training.
3. Benefits of the simulator to the original startup of the E-3 Olefins unit.
4. Flaring reductions during startups and shutdowns of the Ethylene plants.
5. Development of trip applications for the Ethylene plants.
6. Simulator use for development of Polyethylene (Sclair technology) startup advanced control.
7. Additional simulator uses, including:
    a. Experimental Operations
    b. Testing of DCS software upgrades
    c. Network Security Testing
    d. Development of advanced applications
    e. Experion upgrades
    f. Incident Prevention
    g. Outside Operator training

Introduction

NOVA Chemicals operates three world-scale ethylene plants and two polyethylene plants in central Alberta, Canada. NOVA Chemicals is proud of its safety, operational and environmental performance and attributes a large part of our excellent record of accomplishment to the use of high fidelity, training simulators. The NOVA Joffre Simulator Training Program has been in existence since 1998, and serves four primary objectives:

- Panel Operator process training; normal operations thru all emergency conditions
- Startup and shutdown procedure validation
- Development and testing of experimental operations
- Development and testing of advanced optimization control strategies

The Joffre site currently has four, high fidelity, process simulator facilities, with one in development, and another in early project phase:

- Ethylene 1 – Cracking (2011) and Finishing (1998) models
- Ethylene 2 – Finishing (1999) with the Cracking model being developed.
- Ethylene 3 – Cracking and Finishing (2000)
- Polyethylene 2 – Process side modeled (2000)
- Manufacturing Infrastructure – Pipeline model being developed.
- Polyethylene 1 – 2013?

There are a number of factors that have contributed to the success of NOVA Joffre's Simulator Training Program, and among the most important is the continued support of senior Nova management.

About NOVA Joffre and the simulators

NOVA Chemicals operates three world-scale ethylene plants and two Poly-ethylene plants in central Alberta, Canada. Ethylene 1 (E1) has a capacity to produce 1.6 billion pounds per year and Ethylene 2 (E2) has a capacity to produce 1.8 billion pounds per year. The plants were started up in 1979 and 1984 respectfully. A third plant, Ethylene 3 (E3), came on line August 2000 with a capacity of 2.8 billion pounds per year. PolyEthylene 1 (PE1) produces 1.5 billion pounds per year. PolyEthylene 2 (PE2) produces 1.1 billion pounds per year.

The first simulator on site was located in E1 and contained the E1 Finishing and E2 Finishing models. The Finishing side was selected because of the high reliability of the Finishing side and the lack of opportunity for the panel Operator to maintain their skills.

A Cracking model has been developed for E1 and E2 has completed its own simulator facility and is developing a Cracking model to be completed by Sept. 2012. This has been done partly because the Furnace run times have increased from approx. 30 days to 300 days between decokes. This has led to the same issues of the Panel Operator not being able to gain experience and maintain their skills.

The NOVA Simulation System consists of the following:

- A training room that duplicates the actual control room
- An actual or emulated Distributed Control System (DCS)
- A process simulation

At NOVA Joffre, our training room utilizes actual Honeywell TDC3000 and Honeywell Experion furniture. The Honeywell DCS I/O communication is handled with a third-party product manufactured by Proconex Services. The plant simulations are built on RSI Simcon STS2000, and RSI Simcon INDISS platforms. Together, these three components serve to accurately duplicate the control room environment and realistic plant operations.

Control Room

Training Room

E2 Simulator

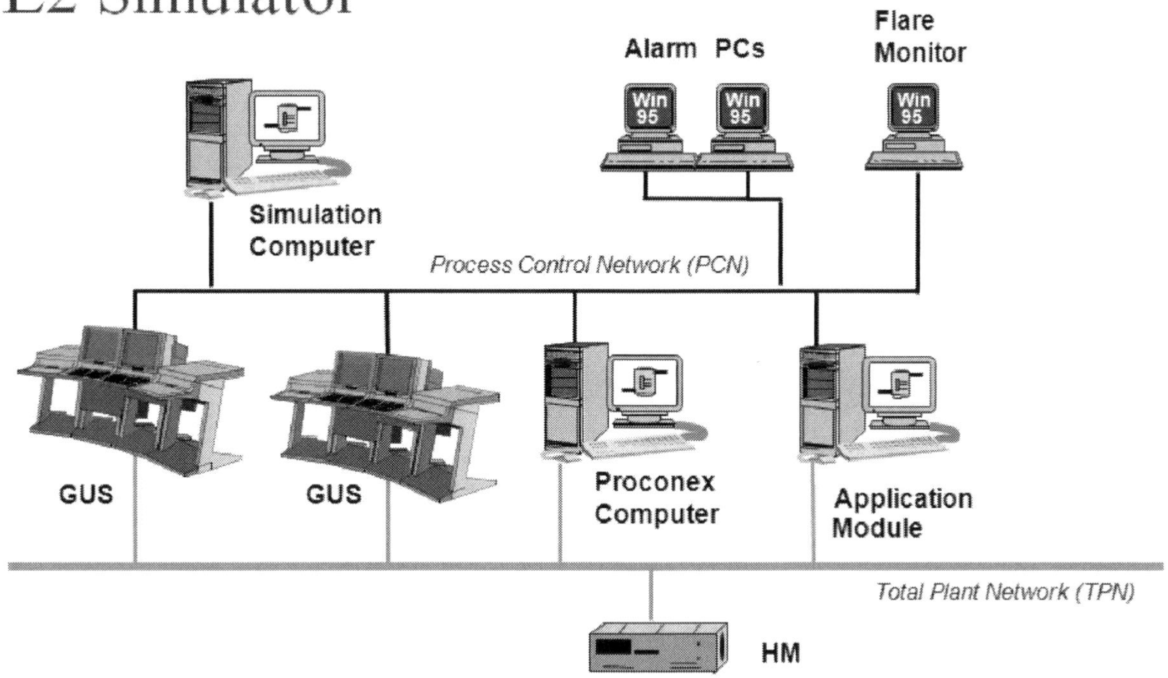

E2 DCS

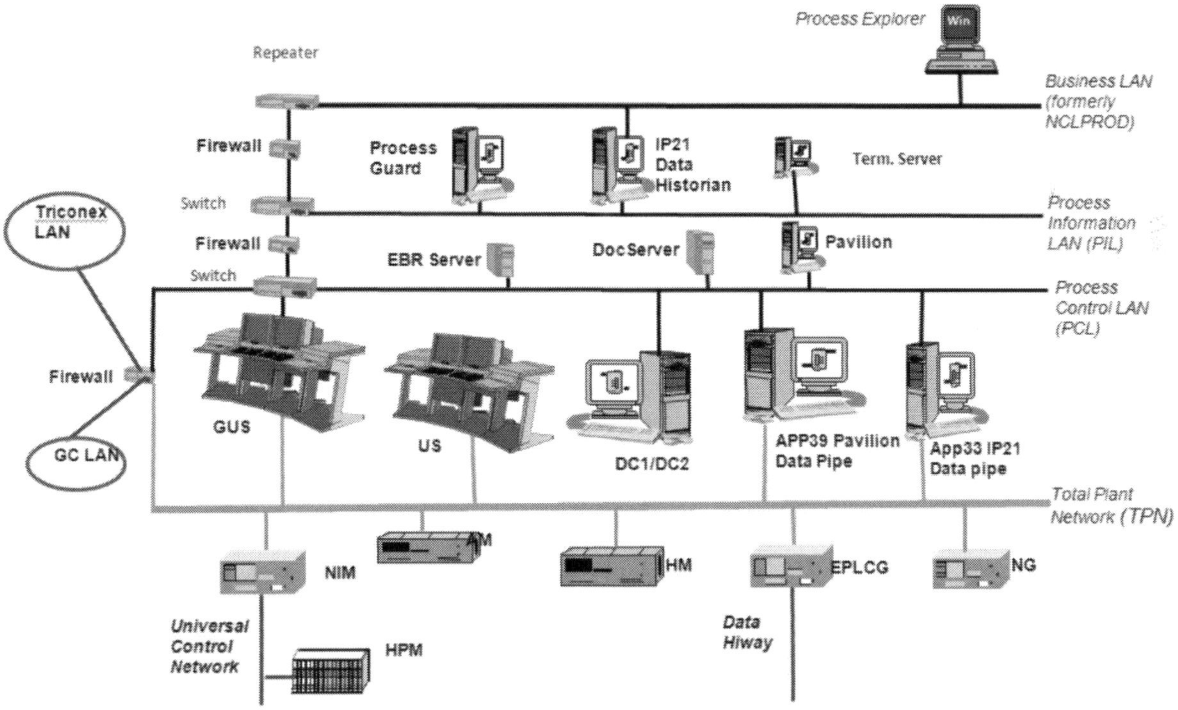

8. Ethylene 2 Training Structure

The following table gives an example of our sites Simulator Training Program. It outlines the structure of simulator training for the E2 Finishing model. Originally only 16 hours of refresher training were required. However E2 has a reliability of 99% on stream time and six years between plant shutdowns. This left little opportunity for the Panel Operators to maintain their abnormal situation management, and startup/shutdown skills. For this reason the simulator training was expanded to 24 hours.

Module	Duration *	Description
Orientation	1/2 hour	Overview of Simulator features (i.e. what the Simulator can or cannot do) and purpose of the training program. Also provides an opportunity for the Trainer and Trainee to discuss their training objectives.
Routine Duties	2 hrs	To provide Operators with the skills, knowledge and decision-making abilities required to perform normal routine duties including: initial checks, graphic navigation, process optimization, selector switches and alarm management.
Start-up/Shutdown	14 hrs	To provide Operators with the skills and knowledge necessary to safely and efficiently start-up every section of the Finishing end of the plant. This includes starting up all major pieces of equipment and making on-spec material.
Emergency Procedures & Trip Applications	24 hrs	At the end of this module, the Operator will be able to demonstrate proficiency in recovering the plant from major equipment trips and returning to a steady state condition. Some Emergency procedures will require the Operator to secure the plant to a safe condition. This will also include the use of automated trip applications that secure the Process.
Advanced Malfunction Training	16 hrs	To provide Operators with skills and knowledge necessary to "diagnose" and take "timely" corrective action to prevent a minor problem from becoming a major upset. A list of malfunctions has been developed and modeled from past incidents including situations such as transmitter drifts, reboiler fouling, etc. Operators develop troubleshooting skills and cognitive reasoning to

		ensure a good understanding of process impacts and key actions required to return the process to a stable condition.
Refresher Training	24 hrs	This is an annual program to "re-qualify" Finishing Panel Operators. The purpose is to ensure all necessary skills and knowledge to handle all nuisance and emergency trip situations are maintained. A plant start-up may also be planned for these individuals.

* Duration of training is dependent on the previous experience of the Operator.

3. E3 Startup Benefits

There were many benefits to having the simulator for the original startup of the E-3 Olefins unit. Although we had several experienced Ethylene Operators the new design was significantly different. The control system included an increased number of overrides and the safety system complexity was substantially increased from previous standards. The Simulator allowed Operations to become familiar with the process and control system before startup. Simulator training was mandatory for all panel operators.

Additional benefits:

a. To test the functionality of Safe View graphic interface. This was developed for E3.
b. Used for verifying Graphic layout for functionality and navigation. This helped reduce the number of display changes required by Operations for startup and upset conditions. This contributed to minimal changes required after startup.
c. The initial tuning of controllers was completed on the simulator and then transferred to the new control system.
d. Used to test Startup procedure and resulted in a significant change.
e. Process Safety Management changes increased the amount of logic required for Safety systems. These were validated on the simulator. This also gave Operations the opportunity to become familiar with them and the interfaces.
f. During the E3 simulator testing a significant number of ranges for transmitters and analyzers were adjusted. Engineering units were also corrected.

4. Flare Reductions

NOVA Chemicals continually strives to reduce loses to flaring during shutdown and startup conditions.

Although these initiatives were developed by many individuals in multiple disciplines the simulator helped Operations to accept and fine tune some large changes to the startup procedures. Operations are traditionally conservative in their approach to insure stability. The simulator has been an excellent tool to allow Operations to gain confidence in some of the large changes that were made and to validate concepts for flaring reduction.

Ethylene 1 & 2 Flare Reductions

In Ethylene 1 and 2, procedure changes reduced flaring during startup that has reduced flaring 200 – 300 Mg to 100 Mg. These improvements were achieved by starting up the Cracked Gas compressor at 20% - 30% of the traditional feed rates. Introducing Hydrogen to the Reactor at significantly lower mixed C2 flows also allowed the mixed C2 flow to be brought on spec. faster and the Reactor effluent directed to the C2 Splitter from the flare sooner. The challenge is to maintain steady feed flows to the Reactor to prevent reducing the space velocity to the point where channeling could occur and hot spots forming. This requires controlling some flows manually and good co-ordination between Panel Operators. The simulator has been a key tool that allowed the development of the skills required to succeed at reducing flare amounts. These skills would have taken much longer to develop relying only on actual plant startups.

Another initiative was to improve the E1 Propylene Refrigeration compressor startup. This system has multiple false load and liquid quench valves as well as dual drivers (a gas turbine and steam turbine) and startups on the system were prone to problems with compressor surges, high temperatures or trips due to high levels in knock out drums. Calculations were made for the optimal false load valves output to prevent the compressor from stonewalling. In the past up to 6 surges may be experienced during startup and this has been reduced to none. In both plants Panel Operators have practiced on the simulator and successfully started up the Propylene Refrigeration compressor for the first time with no surges.

The shutdown flaring has been reduced in E1 and E2 to 20 Mg from 100 Mg. These reductions have been achieved by keeping the Reactor on line as long as possible while plant rates are being reduced. The same issues apply with maintaining steady flows to insure sufficient space velocity through the Reactor. The Cracked Gas Compressor speed is controlled manually to prevent false load control problems which could result in surging.

E3 Flare reductions

In 2010 the E3 Operating Team received NOVA's "Excellence in Responsible Care" award from Nova's CEO, Randy Woelfel, for not only reducing flare emissions, but for reducing the number of community complaints of flaring noise and saving the company over $10 million in lost margins due to flaring over an eight year period.

The first attempts at reducing flaring were to feed the C2 Splitter with product that had a high amount of Acetylene. This prevented having to flare the Ethane which was approximately 40% of the flow to the C2 Splitter. Although it reduced total flare volumes it still required a large amount of flaring until the C2 Splitter Ethylene product cleaned up.

The next reduction was achieved by preventing flow to the Chilling Train by using N2 pumper trucks maintaining a higher pressure. This prevented Acetylene from contaminating the Chilling Train and The C2 Splitter. Flaring was still required while the Acetylene Reactor flow was established, brought on spec and feed could be introduced to the Chilling Train. This did allow for requiring only one flare stack which reduced noise to the community.

The next change to the Startup procedure was motivated by the presence of the N2 pumper and fired heater in the plant at arguably the most dangerous time of plant operation (Start-up). Also the dependency on outside contractor to supply of the service was always a concern.

The startup procedure was modeled and modified. During startup the initial plant feed rates are low resulting in low coil loading and reduced conversions. The Cracked Gas compressor is run at a reduced speed and the Process pressures are kept lower. Once the Reactor is on specification the process is brought up to pressure and the Deethanizers are brought fully on line. The plant is returned to normal operating conditions over about 60 minutes. As the plant is brought up to on line conditions the specifications for the various streams reach normal operating values.

Each change to the Startup procedure was first tested on the Simulator. The current procedure was tested for 1.5 years on the Simulator before it was implemented. This allowed problems to be identified and worked out by all the disciplines in a controlled fashion. All the Panel Operators had an opportunity to try and input on the new procedure. The E3 Simulator Trainer played a large role in resolving these issues and driving the process to change the procedure.

Development of trip applications for the Ethylene plants

Prior to the use of the simulator responses to plant upsets and Shutdowns were inconsistent and varied from team to team. The simulator allowed the opportunity to establish the best practice and to also develop automated isolation and upset response applications.

A plant shutdown in E2 requires 103 control moves by the Panel Operator. The trip application allowed the Panel Operator to complete the required isolations in a short period of time. This helped to retain inventories and prevent sending off spec. product to other areas of the process. This would leave the plant in a good state for startup. It also allows the Panel Operator time to address any other issues that may have caused the outage or were a result of the outage.

Because of the infrequency of plant outages it would have been very difficult to validate the application and correct any errors. This would have taken many years to accomplish without the use of the simulator. It also allowed Operations to become familiar with the application and become comfortable with using them.

E2 has developed the following trip applications:

- Propylene Refrigeration Compressor trip
- Cracked Gas Compressor trip
- Cracked Gas Compressor trip with an immediate restart
- Reactor Off spec.
- Ethylene Refrigeration Compressor trip

Polyethylene startup application

Simulator use for development of Polyethylene (Sclair technology) startup advanced control.

The following outlines how the Nova Chemicals Polyethylene 2 facility at the Joffre, Alberta site utilizes Dynamic Simulators to improve panel operator performance, and optimize the process.

The Polyethylene 2 facility, capable of producing one billion pounds per year of polyethylene from Nova Chemicals proprietary solution process, was a scale up from a pilot plant process. The facility was started up in 2001 after extensive work building a dynamic model of the process and training the panel operators.

All process panel related procedures were developed on the simulator based on operator knowledge and previous operating experience and use of the Process and Instrument Diagrams. Startup was very successful from a procedure and operator training viewpoint.

New panel operators are required to take 12 days of training on the simulator prior to attempting to qualify for the panel position. Once qualified they are required to complete 2 days of training to remain current in ASM (Abnormal Situation Response) and troubleshooting process issues.

One of the areas the simulator has been a great benefit to the Polyethylene 2 facility is in the development of computer applications to assist the operator in a controlled shutdown from or startup (startup) of polymer production. Before using applications, we found that the success of a startup or an efficient shutdown was largely dependent of the experience of the operator working the panel on that particular shift, due to the large number of steps required to complete either scenario (The startup procedure consists of about 100 steps to complete in one hour or less, the shutdown procedure consists of about 30 steps to be completed in 10 minutes).

We also knew that during each scenario that the risk of operator keyboard entry errors and stress increased substantially due to the high workload in a short timeframe, and for some operators there was potential to develop Repetitive Stress Injuries (RSI) in the wrist due to computer mousing activities during these events.

To reduce the variance between operators and the risk of entry errors or injuries, we developed a shutdown application and a startup application. Without the simulator to assist in development and debugging of the programs, it would have taken 2 years to develop each application in the plant and the increased the potential to produce additional offgrade or plant upsets. Using the simulator we were able to run and rerun each program over a two week period until we were confident that implementing the application would bring the benefits we were hoping to realize. Each panel operator was then trained on how to start and stop the application and monitor the process as the application completed each step.

Implementing the Shutdown and Startup applications seamless and the benefits were immediate. Keyboard entry errors during these times disappeared, and operator stress levels were reduced. The application also helps to ensure that critical steps are completed, thereby enhancing the safety of the shutdown or startup. The variance between experienced and new operators disappeared as well. For the startup application, the time to startup was reduced on average by 30 minutes each time it was used (a minimum of 12 times per year to switch between catalyst platforms), resulting in increased onstream time and production. Operator stress and workload has also been reduced and now they can effectively monitor the process and focus on the issue that caused them to shutdown or on getting the process on prime production when lighting off.

To further optimize and reduce downtime, we have recently begun testing of two new applications to help with reacting to the safe and efficient shutdown and restart of our large solution pump. In this case the testing can be done over two weeks to develop and debug in what would easily take more than 10 years without a simulator, as this occurrence may only occur once per year.

Again the same potential benefits can be realized in optimizing operator response regardless of experience, and reduce stress and workload so they can focus on the cause of the event, reduce downtime, and safely and efficiently get the plant into production again.

Additional Simulator uses

a. Experimental Operations

E1 was looking for ways to reduce flaring during Reactor Off spec. incidents. The past practice was to stop Feed forward from the Reactor outlet and flare. Plant rates have to be reduced by approximately 50% in a short period of time to insure complete combustion at the flare tip. One of the Operations Technicians had success with continuing feed forward and using the Off spec bullets to prevent flaring.

To gain acceptance from all of Operations the Simulator was used to test and develop a procedure to handle Acetylene Reactor Off spec scenarios. This has become part of the Simulator Training program in E1 with E2 investigating a similar strategy.

b. Testing of DCS software upgrades

The DCS software updates are tested on the Simulator. DCS network configuration changes are also tested on the simulator.

c. Network Security Testing

The E3 Simulator Network security is set up the same as the actual system. Any changes to Corporate Network Security model are first tested on the Simulator.

d. Development of advanced applications

The simulators have been used to test advanced control applications.

e. Experion upgrades

E1 is in the process of upgrading the DCS HMI to Experion. The simulator control stations were upgraded to Experion eight months before the plant upgrade. This has given Operations the opportunity to input on the layout and practice abnormal situations, Shutdowns and Startups with the new Graphics. The simulator has been used to verify correct tag data and allows the Graphic developers to test all the points on the simulator which will minimize the corrections required during the limited Shutdown window. A similar process will be used in E2.

E3 also upgraded to Experion although the graphics remained the same the E3 simulator was used to flush out any Graphics or Experion Alarming issues before going online. It was also used to give the Operators a live preview and hands on for the new Experion system.

f. Incident prevention

Occasionally problems are detected on the Simulator that would have been found in the real plant. This early detection has prevented incidents. One example is of a safety system upgrade on the E2 Acetylene Reactor. This was to add EBVs (emergency block valves) on the inlet and outlet. While training Operations on the new safety system changes it was noticed that when selecting the EBV the change zone would appear on the upper left of the screen. It required a direct entry of the desired output but did not require the Operator to use the Output change key. However it left the last tag in the detail display and if the Operator inadvertently used the output key it would make a change on that tag instead. On the simulator it resulted in a large flow change to the C2 Splitter feed and resulted in Offspec feed. Finding the problem on the Simulator first allowed a correction that prevented a similar problem from happening in the real plant.

g. Outside Operator training

To help train the outside Operators and to add the communication element to panel training we occasionally include the outside Operator in Simulator training. They will have a radio and respond to the locations they would in a normal upset. This gives the Panel Operator a better sense of the time required for outside Operator response. We have been given very positive comments on how helpful this has been for the outside Operators becoming used to abnormal situations.

Testimonials

"There are many abstract ideas that can be learned from a simulator that are hard to always pick up while working on and learning a new panel. Eg: moves on a Propylene Refrigeration machine that has its "fingers" everywhere in the finishing side of the plant, a small pressure move can really affect several exchangers and towers that will then cascade their affects to other parts of the plant. Doing little "experimental moves" on a live plant is not always possible, or in many ways a new finishing panel operator is timid about doing such moves no matter how much a trainer will encourage them to, but things like this can be done on a simulator. Not many types of plants match the complexity of an ethylene plant so this type of move can really throw a wave through things, which is why practice on a simulator can really help."
E1 Process Operator

"I find the simulator training very useful in the knowledge and confidence that it gives me in a real plant problem. I trained on the simulator before the plant was shut down in September, and was very confident when the plant was being brought down. Any problems that were experienced that day had already been dealt with in the training session. I felt more relaxed in dealing with the problems and didn't have any stressful moments when I did have problems.

I also had prior training on the startup of the finishing side of the plant, and found that was also very helpful when I came in to help start up after the shutdown.
I think the simulator has given me the resources to deal with most of the plant problems that would normally occur.
Being fairly new to the finishing panel, it is nice to be able to deal with problems in training before I have to deal with them in real life situations."
E2 Process Operator

"The simulator helps to give you the feel of how much of a change can be made. It helps to keep you familiar with trip procedures and the trip applications. It builds confidence in your ability to handle upset conditions especially when you do not work the panel very often."
E2 Process Operator

"I have benefitted from countless training opportunities to refresh my skills. I look forward to scheduled malfunction training days where ultimately my troubleshooting skills will be humbled, or the days on which I trial a new start-up procedure. I always leave feeling more prepared for any situation then when I arrived."
E3 Process Operator

Summary

The Simulator training programs is an integral part of Operations qualifications and ongoing skill development. The success of the simulator program has been due to ongoing support from Leadership to maintain the systems. Plant projects include updates to the Simulator This keeps the models updated along with changes to the facility. Any projects on site include updates to the simulator models. NOVA also dedicates a Training Specialist to each simulator facility to provide consistent training, to insure the system is maintained and the program followed. Simulators become expensive table ornaments without this commitment.

Acknowledgements:
George Dzyacky - RSI Simcon
Kevin Wilke - Technical Lead, NOVA Chemicals
Daryl Klaudt - E3 Technical Training Specialist, NOVA Chemicals
Mark Williamson - Team Leader Olefins, NOVA Chemicals
John Dennehy E3 Team Leader, NOVA Chemicals
Jamie Errington - Human Centered Solutions

The E2 Training Structure Table was updated from the paper "Process Simulation for Operations Training and Other benefits" by Keith Doe and Dave Coppard.

The following individuals were presented the Responsible Care award for reducing flaring in E3: Daryl Klaudt, Yost Kieboom, Kevin Wilke, John Dennehy, Mark Williamson, Ed Noble, Greg Heimsoth, Kaisa Nadeau, Nimish Rathod, David Woods and Ron Besuijen.

Testimonials: Clifford Roy, Bryan Fraser, Don Vanderlinden, Ed Noble.

50f

PRESERVING FEED COMPOSITION EFFECTS DOWNSTREAM OF AN OPEN LEVEL IN A DMCPLUS CONTROLLER

Matthew Fellows, P.Eng.
Process Automation Engineer
Dow Chemical Canada ULC

Rob Hoogerwerf
Advanced Control Subject Matter Expert
Dow Benelux B.V.

Rob Strashok
Advanced Control Technician
Dow Chemical Canada ULC

Luc Van Belleghem, P.Eng.
Lead Process Automation Engineer
Dow Chemical Canada ULC

Eduardo Arrieta
Technology Associate – Advanced Process Control
The Dow Chemical Company

Prepared for Presentation at the Year 2012 Spring National Meeting
Houston, TX., April 3 2012

PRESERVING FEED COMPOSITION EFFECTS DOWNSTREAM OF AN OPEN LEVEL IN A DMCPLUS CONTROLLER

Matthew Fellows
Lead Project Engineer
Dow Chemical Canada ULC
Rob Hoogerwerf
Advanced Control Subject Matter Expert
Dow Benelux B.V.
Rob Strashok
Advanced Control Technician
Dow Chemical Canada ULC
Luc Van Belleghem
Lead Automation Engineer
Dow Chemical Canada ULC
Eduardo Arrieta
Technology Associate – Advanced Process Control
The Dow Chemical Company

Abstract:

The FS-1 Fractionator is a natural gas fractionation unit which separates natural gas liquids (C3P) into propane, butane and C5P. The process consists of two distillation towers in series. DC3 is a depropanizer and DC4 is a debutanizer. DC3 is fed by three different feed pipelines, characterized by having fast and frequent changes to both flow and composition. The FS-1 Fractionator does not have significant feedstock or product storage, and plant feed rates are closely coupled to desired product flow rates sent by pipeline to customers. The product compositions ***and*** the product flows must be controlled to specified targets. The capability of meeting the production targets depends strongly on the compositions of the feed streams and the ability of the DMC controller to manipulate the ratio between the various feed sources.

A key design consideration is whether to open the DC3 level between DC3 and DC4. The advantage is that the delay and oscillation created by the level PID loop is removed and sharper, clearer models are created for DC4 CVs. However, opening the level breaks the relationship between the DC4 CVs and the feed compositions. The DC4 butane product flow would have models with the DC3 bottom flow, but not with the three different feed flows and their compositions.

The project team overcame this trade-off by opening the DC3 level, and re-building the feed composition effects around the open level. At steady-state, the problem can be treated as a mass/component balance. All propane that is added to the column must be removed as propane product or as an impurity in the butane product and so on. Steady state predictor models are created between the feed flows and compositions, with unity models to the product flows (ignore the effect of impurities). The DC3 bottom flow has a dynamic-only model to the butane product flow. The models are identified from open-loop data, but assembled in the way that we know they will perform in closed loop. The DC3 bottom flow is dynamically independent, but the need to control the DC3 bottom level within constraints means that it is not truly independent at steady state. The dynamic behaviour of the DC3 bottom flow must be accounted for, but the feedstock flows can be connected directly to the product flows at steady state, because the DC3 bottom level must remain in control.

Process Description

Introduction

The FS-1 Fractionator is a natural gas fractionation unit which separates natural gas liquids (C3P) into propane, butane, and pentane plus (C5P). It consists of two distillation columns in series, DC3, the depropanizer, and DC4, the debutanizer. DC3 is fed by three different feedstock pipelines, characterized by having fast and large changes in composition. DC4 is directly connected to DC3. Both propane and butane product are shipped by pipelines to customers, with only a small volume of surface storage available to buffer plant production to desired shipping rates.

Process Overview

Figure One shows the process overview of the FS-1 Fractionator. DC3 is fed by three feed sources. For FED1, no control is available for feed flow rate or composition, and both are treated as feed forwards. For FED2 and FED3, the flow rates, F2 and F3 are manipulated variables, and their compositions, X2, and X3 are feed forwards. The product flow rates, PR_C3 and PR_C4 are controlled variables, most strongly influenced by F2 and F3. The column product compositions, Y3_C4, Z3_C3, Y4_C4, and Z4_C, are controlled variables and are controlled most strongly by the manipulated variables of tower temperatures and reflux, T_TOP3, T_BOT3, F_RFX4, and T_BOT4 respectively. The level of DC3, LT3 must be controlled by the flow leaving DC3. The decision to open LT3, or leave in a regulatory controller is a key design decision for this controller and will be discussed further.

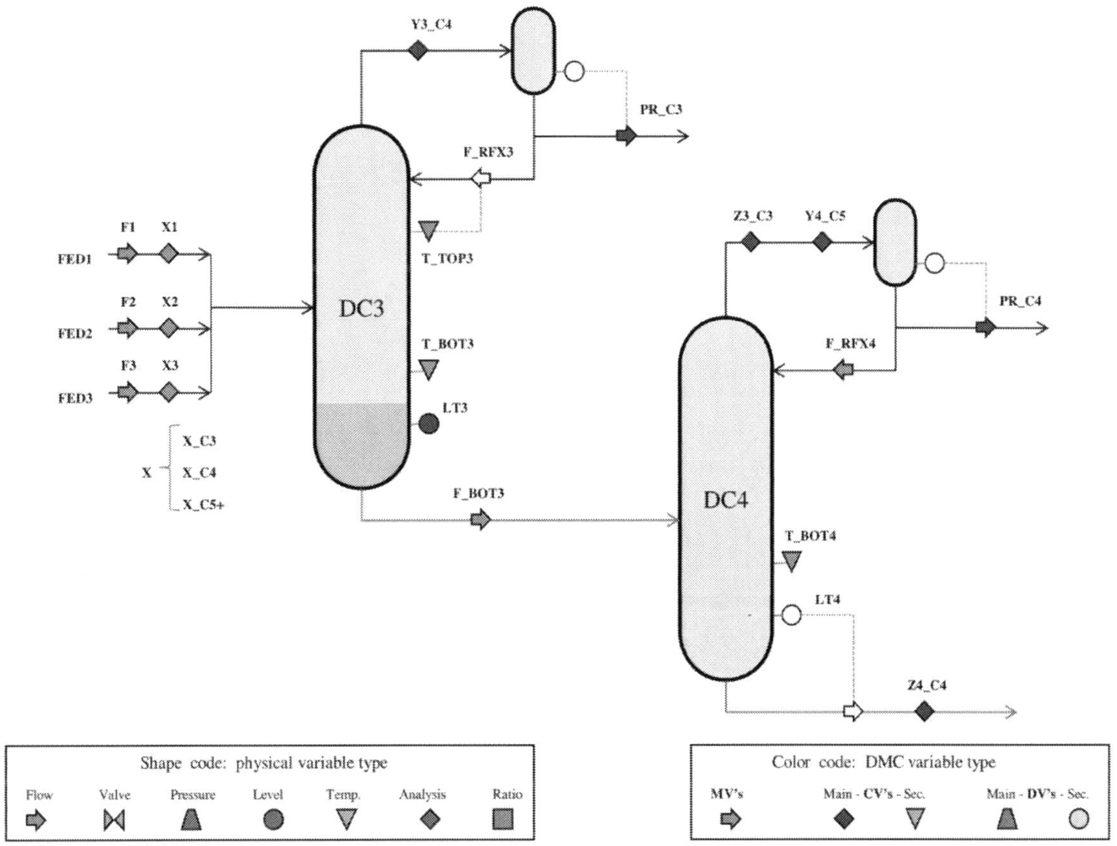

Figure One: FS-1 Fractionator Process Overview

Feed Characteristics

DC3 is fed from batch pipelines via FED1, and both the flow rate and composition are variable. DC3 receives feed also from two other pipelines via FED2 and FED3. The flow rate is controlled, but the compositions are variable. Figure Two shows the typical variation in feedstock flow and composition. The time scale of the plot grid is six hours.

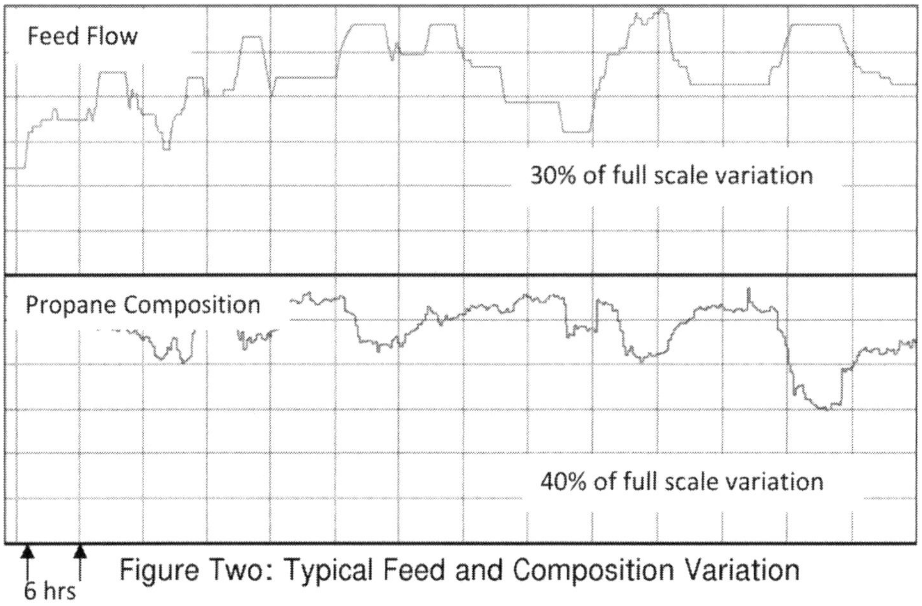

Figure Two: Typical Feed and Composition Variation

Controller Design

Key Design Decision – DC3 Level

A key design decision is whether or not to open the DC3 level in DMC. Opening the DC3 level presents several advantages. Opening the level removes delay and the waving "noise" created by a level PID loop, and it results in sharper, clearer models for DC4 CVs. Table One shows the typical construction of a model matrix for this application.

Indeps.		**Dependents**		
Name	**Type**	PR_C3	PR_C4	LT3
FED1	FF	G31		GB1
FED2	MV	G32		GB2
FED3	MV	G33		GB3
F_BOT3	MV		Gx	-GBF

Gain Multipliers	
G31 = a*X1_C3	GB1 = b*(1-X1_C3)
G32 = a*X2_C3	GB2 = b*(1-X2_C3)
G33 = a*X3_C3	GB3 = b*(1-X3_C3)

Table One: Typical Matrix Construction.

In this construction, the feeds, FED1, FED2 and FED3, have models with the propane product flow, PR_C3. The compositions are introduced via GMULTS, G31, G32 and G33. The level, LT3, has models with FED1, FED2, and FED3, as well as their compositions, with positive gains, and the flow out of DC3, F_BOT3, with negative gain. The primary disadvantage of this construction is that the butane production, PR_C4, only has a model with the flow out of DC3, F_BOT3. The butane production is influenced by the feed flows, and their compositions, but this relationship is broken with this construction.

In an open loop case, the typical matrix construction is correct. The flows into and out of DC3 level are independent. However, the level must still remain in control, within the time to steady state of the controller. Therefore, any changes in the feed flows must be passed through to the bottom flow, in order for the level to remain in control.

Table Two: Alternate Matrix Construction.

Indeps.		Dependents				
Name	Type	FD_C3_CV	FD_C4_CV	PR_C3	PR_C4	LT3
FED1	FF	G31	G41			GB1
FED2	MV	G32	G42			GB2
FED3	MV	G33	G43			GB3
FD_C3_DMV	DMV			1		
FD_C4_DMV	DMV				1	
F_BOT3	MV				0Gx	-GBF

Main Control Gains

Dynamic only model

Gain Multipliers		
G31 = a*X1_C3	G41 = c*X1_C4	GB1 = b*(1-X1_C3)
G32 = a*X2_C3	G42 = c*X2_C4	GB2 = b*(1-X2_C3)
G33 = a*X3_C3	G43 = c*X3_C4	GB3 = b*(1-X3_C3)

Table Two shows an alternate matrix construction. In this arrangement, two new CVs are added, FD_C3_CV, and FD_C4_CV, which represent the amount of propane and amount of butane fed to the system, respectively. The FD_C3_CV is created with models to FED1, FED2, FED3, and GMULTS with the feed propane compositions. The FD_C4_CV is created with models to FED1, FED2, FED3, and GMULTS with the feed butane compositions.

Two new independent variables can be created as conceptual/dummy MV's, FD_C3_DMV, and FD_C4_DMV. These new independent variables are equal to the new CVs, where FD_C3_DMV is equal to FD_C3_CV and FD_C4_DMV is equal to FD_C4_CV. Assuming that the DC3 level and the tower compositions remain constant within the time to steady state of the controller, then we can say that at steady state, the product flow of propane, PR_C3 is equal to the flow of propane component into the system, FD_C3_DMV, and the product flow of butane, PR_C4 is equal to the flow of butane component into the system, FD_C4_DMV, both represented with gains of 1. The bottom flow, F_BOT3, can only create dynamic disturbances in the DC3 level, represented by a dynamic only model, with gain 0Gx, because of the assumption that the DC3 level will remain balanced within the time to steady state of the controller.

Table Three: Alternate Matrix Construction with External Targets.

Indeps.		Dependents				
Name	Type	FD_C3_CV	FD_C4_CV	PR_C3	PR_C4	LT3
FED1	FF	G31	G41			GB1
FED2	MV	G32	G42			GB2
FED3	MV	G33	G43			GB3
FD_C3_DMV	DMV			1		
FD_C4_DMV	DMV				1	
F_BOT3	MV				0Gx	-GBF
		ET	ET	ET	ET	RP

FD_C3_CV ET = FD_C3_DMV SSMAN <-- PR_C3 ET
FD_C4_CV ET = FD_C4_DMV SSMAN <-- PR_C4 ET

In this arrangement, the desired Product Flows are controlled by the ***Component*** feed or inlet flows (DMV's), which in turn are controlled by the normal pipeline Feed Flows.

Table Three shows the desired control objectives of the controller. The product flow targets are controlled by the DMV component feed flows to external targets. The component feed flows are controlled by feed MVs with composition dependent gains. The external targets for FD_C3_CV and PR_C3 must be the same, as must the external targets for FD_C4_CV and PR_C4. This construction presents a transparency problem, because the production external targets will always appear to be satisfied, because the conceptual/dummy MVs can always give what is asked of them, regardless of whether or not this is actually possible.

Final Matrix Construction

Table Four: Final Matrix Construction

Indeps.		Dependents				
Name	Type	PR_C3_Pred	PR_C4_Pred	PR_C3	PR_C4	LT3
FED1	FF	G31	G41			GB1
FED2	MV	G32	G42			GB2
FED3	MV	G33	G43			GB3
FD_C3_FF	FF			1		
FD_C4_FF	FF				1	
F_BOT3	MV				0Gx	-GBF
		ET	ET	*Pred Only*		RP

(white)	Dyn Model with SS Gain
(light grey)	SS Step model
(dark grey)	Dynamic only model

PR_C3_Pred DEP = PR_C3 SSDEP (per Dynamic model with FD_C3_FF)
PR_C4_Pred DEP = PR_C4 SSDEP (per Dynamic model with FD_C4_FF)

The final construction of the matrix is shown in Table Four. Here, two feed-forwards are created for the C3 and C4 ***Component*** feed flows, FD_C3_FF, and FD_C4_FF, where FD_C3_FF = FED1*G31 + FED2*G32 + FED3*G33 and FD_C4_FF= FED1*G41 + FED2*G42 + FED3*G43. The feed forwards represent the flow of propane and butane into the system. The feed-forwards have a dynamic model with steady state gain of 1, which represents the time delay between the time product is added to the system and the time the product flows out of the system. PR_C3 and PR_C4 are the actual product flow rates. PR_C3_PRED and PR_C4_PRED are the predictions for those product flow rates, and their measurements (DEP) are equal to the Steady State targets (SSDEP) of PR_C3 and PR_C4 respectively. PR_C3_PRED and PR_C4_PRED have models with the feeds with gains equal to the feed compositions. This construction effectively controls the propane and butane flows into the system, and the predictor structure corrects for any mismatch between feed and product meters, composition changes in the product, and other mismatches.

This construction is equivalent to the Tray Temperature to overhead Analyzer predictors often used for distillation towers.

The external targets for the product flows must be ranked according to production priorities. It may not be possible to satisfy both depending on the feed compositions. Even if it is possible to satisfy both production targets, extreme swings in FED2, and FED3 may result if their compositions are too similar. Feed compositions must be monitored to turn off the secondary production target if needed.

Conclusion

Results

Figure Three shows the control of propane product with the original (non-DMC) control design. Figure Four shows the control of propane product after implementation of the final DMC design.

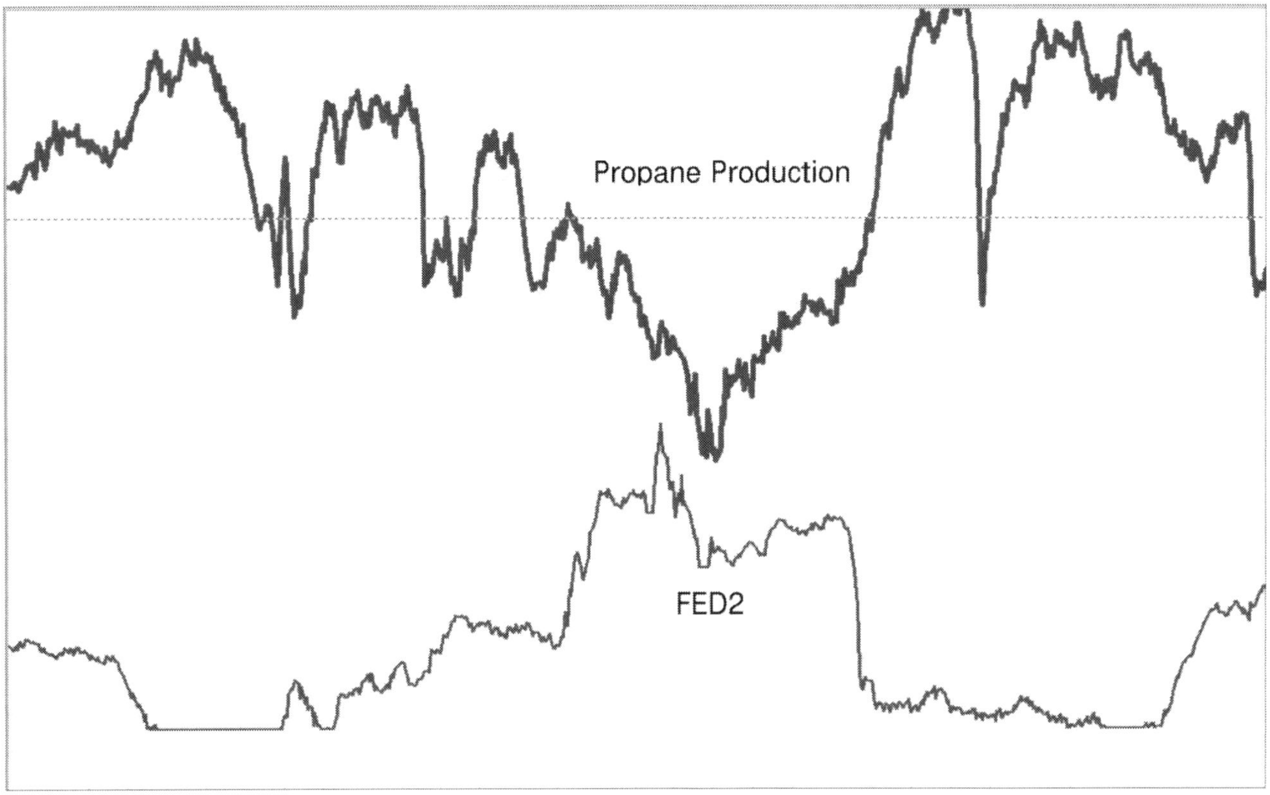

Figure Three: Propane Production with Original Control

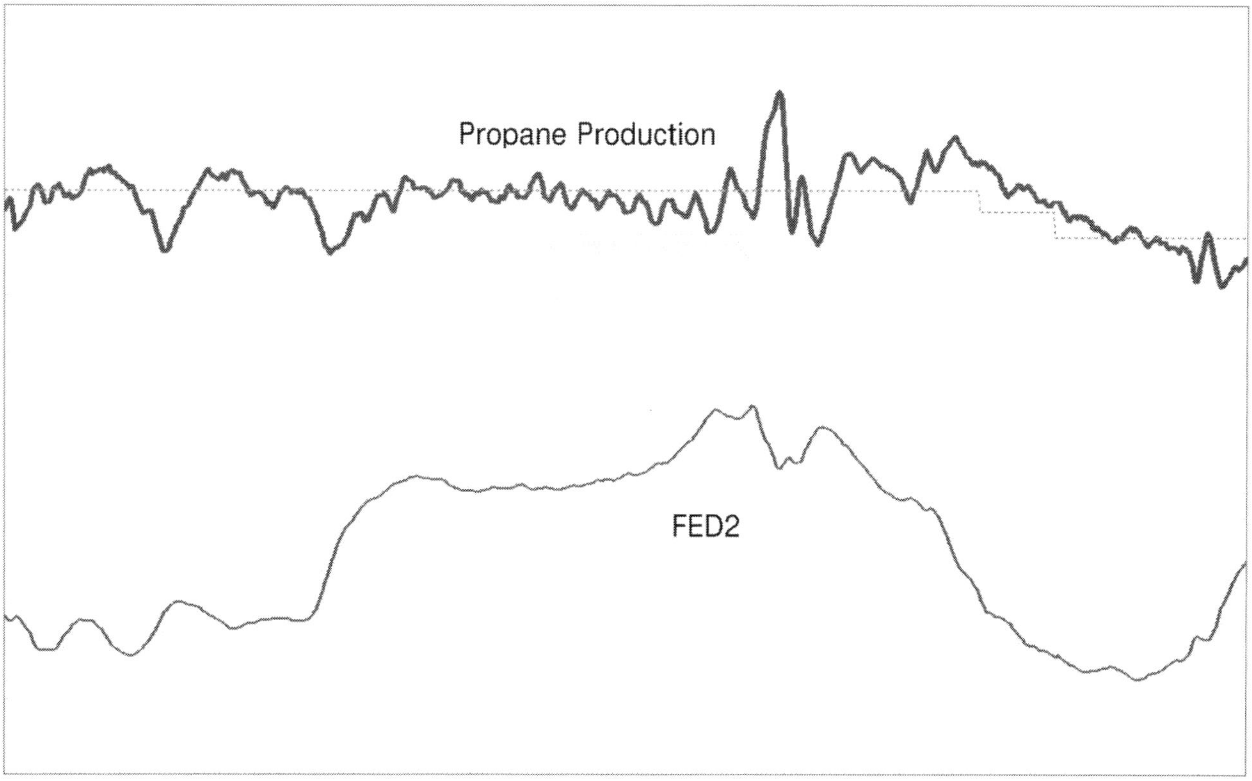

Figure Four: Propane Production with DMC Control

The original control design required the plant operator to set a total flow rate into DC3, controlled with FED2. The operator would observe the production of propane, and adjust the total flow rate. This design did not cope well with changes in feed composition. In the final DMC design, both flow and composition are taken into account; propane production is constantly monitored under closed loop control.

Conclusion

This controller is able to satisfy the objective of being able to control propane and butane production independently by considering the required closed loop performance of the controller. The models are indentified from open-loop data, and in the open loop system, the flows into and out of the DC3 level are independent, which breaks the relationship between butane production and the feed flows and compositions. However, considering the closed-loop objectives, and constraints, the DC3 level must remain quasi constant, therefore the flows into and out of the DC3 level must balance within the time to steady state of the controller, and therefore the butane production can be predicted with the feeds and their compositions. The open loop models are bent to match closed-loop performance. Understanding the control requirements and the physical process allowed the project to adapt open-loop models to meet closed-loop performance – which is the final and true objective for any controller.

AIChE Paper Number 51A

Natural Gas to Ethylene in One Step
Siluria Technologies OCM (Oxidative Coupling of Methane)

Sam Weinberger

Executive Advisor

Siluria Technologies

Erik Scher

Vice President R&D

Siluria Technologies

Rahul Iyer

Senior Director Corporate Development

Siluria Technologies

Prepared for Presentation at the 2012 Spring National Meeting

Houston, TX, April 12, 2012

Natural Gas to Ethylene in One Step
Siluria Technologies OCM (Oxidative Coupling of Methane)

Sam Weinberger
Executive Advisor
Siluria Technologies
Erik Scher
Vice President R&D
Siluria Technologies
Rahul Iyer
Senior Director Corporate Development
Siluria Technologies

Abstract: The oxidative coupling reaction of methane for direct, one-step conversion of C1's to C2's has been known for many years but never commercialized. Siluria Technologies has succeeded by bringing together 3 major scientific advances of the past decade that were unavailable to earlier researchers in this chemistry. The advances are (1) Biomaterial Templating - the use of genetically engineered biomaterials to influence the structure of inorganic catalysts grown on these templates, (2) Nanowire Catalyst Design - a synthetic method that permits the influencing of catalyst properties by changing the active sites, and which can be made under commercially scalable synthetic conditions, and (3) High-throughput Screening – that allows for measuring gas-to-gas reactions orders of magnitude faster than in traditional reactors.

In addition, all of Siluria's R&D and experimental work on the reaction has been limited to commercially viable process conditions and reactor configurations. Work in the 1980's and early 1990's on this highly exothermic reaction, typically using catalysts produced by conventional bulk methods, was limited to temperatures, pressures, or other process conditions that were outside of the practical engineering window for a commercial system.

The recent and rapid pace of natural gas discovery and development in North America and other regions provides a very strong business case for the deployment of this technology.

Introduction

Siluria Technologies, a small, venture-financed company in San Francisco is applying a novel materials production and discovery platform with the goal of creating a catalyst family and associated chemical process for the direct conversion of natural gas into the high value commodity intermediate, ethylene, thereby enabling low-cost and low-carbon production of materials, chemicals, and liquid fuels. The process, based on the oxidative coupling of methane ("OCM"), uses an engineered catalyst material capable of being deployed within existing industry production methods and infrastructure. The successful commercialization of such a process technology would have the potential to transform many industries by enabling the switch to an abundant and lower-cost feedstock. In addition, a novel pathway from methane to liquids could enable the economical use of stranded fossil and renewable methane.

This chemistry is not new, and it has been studied extensively, beginning in the 1980's by a number of well-known companies and universities. But it was never commercialized. Our current effort brings together three technologies that were previously unused on this problem - because they were not mature at that time. The new technologies brought to bear are key differences in Siluria's program compared to earlier work. But first, it's worthwhile to cover why this is a problem worth trying to solve.

Business Case

Methane, the predominant component of natural gas, provides an example of an attractive alternative feedstock due to its enormous supply, global availability and low cost. There are enormous proven reserves of methane widely distributed throughout the world. Current worldwide methane reserves are estimated at over 6,300 trillion ft3, which provides a 300-year supply based on current consumption rates.

Supplies of renewable methane and biosynthetic natural gas are being rapidly developed, driven by a range of factors including regulatory requirements. In many markets including the US, methane is inexpensive, priced at less than one-fourth the cost of crude oil on an energy-equivalent basis. Additionally, in these markets, methane's price is delinked from crude oil and trends predict even lower prices relative to that of oil, given the use of more productive exploration and production methods. For these reasons, methane represents an ideal alternative feedstock to crude oil fractions and NGLs from an economic, strategic, and environmental perspective.

Today, ethylene as an intermediate is currently limited in use to higher-value commodity chemicals because of the high cost and high value of the ethylene, which is linked to crude oil through worldwide naphtha cracking economics. By enabling the production of ethylene from methane, which is less than one-fourth the price of crude oil, the fundamental cost structure of both ethylene derived chemicals and liquid fuels can be dramatically improved.

Since the business case for the development and deployment of OCM is so compelling, why hasn't this been done already? As will be covered in a moment, it has been extensively researched by major energy and chemical companies, but never brought to market.

Technical Challenge

First let's look at the chemistry and why it is so challenging. Chemically, methane is an extremely stable and symmetric molecule. Its C-H bonds are very strong and it does not contain any functional group, magnetic moment, or polar distribution to facilitate chemical activation. As a result, methane activation (the splitting of the C-H bonds) requires extreme conditions such as high temperatures (>900°C), and/or the use of oxidation agents to complete conversion. Catalysts can also be used to reduce the extreme conditions required to activate the C-H bond provided they do so without forming a carbon oxygen bond.

Incumbent methane conversion methods are based on an indirect approach using high temperature reforming to synthesis gas (a mixture of H_2 and CO). Partial oxidation, catalyzed partial oxidation, and steam methane reforming are all commonly practiced methods for producing hydrogen-rich syngas. Depending on the downstream processes to convert syngas into a product the ratios of H_2 to CO must generally be adjusted to meet the stoichiometric requirements of the downstream process. The hydrocarbon products that result may have broad product distribution and therefore require significant additional refining/separations and further cost and energy input.

Thus, to better exploit existing natural gas resources, more efficient, economical and environmentally sound direct methods are required.

A promising alternative and efficient route to convert methane is the Oxidative Coupling of Methane (OCM). The OCM reaction is a direct reaction meaning that the desired and targeted products are formed during the reaction, and thus avoids the sequential steps to arrive at the desired product, in this case ethylene.

In OCM methane and oxygen react over a catalyst exothermically to form ethane and ethylene (C2) and small amounts of higher hydrocarbons (C3+ products).

$$2\,CH_4 + O_2 \rightarrow C_2H_4 + 2H_2O$$

The non-selective reactions are those that result in CO_2 and CO, and the key catalyst performance parameter is to maximize selectivity or maximize the coupling reaction over complete oxidation.

Motivated by the potential economic advantages of OCM, the OCM reaction has attracted significant attention since the fundamental work of Keller et al. in the early 1980's. This was followed by significant industrial and academic research into the early 1990's. Despite these efforts, no economically viable OCM process has been put into practice. The difficulties associated with the direct conversion of methane arise from both kinetics and thermodynamics. In particular, the complex nature of a homogeneous / heterogeneous reaction mechanism (involving both surface and gas phase reactions) imposes unique challenges to the catalyst requirements and function. For example, high temperatures are required for activation of methane, but at such conditions radical reactions in the gas phase are dominating favoring the formation of non-selective CO and CO2 byproducts and also potentially causing severe catalyst deactivation from decomposition or sintering. Additionally, the reaction is highly dependent on the local chemical and electronic environment of the surface oxygen sites of the catalyst, which places additional design constraints for an effective catalyst.

Siluria's Approach

Now let's look into the novel combination of work processes and approaches being used. The R&D approach is informed by the constraints of commercialization. The earlier literature placed the most emphasis on catalyst performance, that is, selectivity (the percent of the carbon in the converted methane which becomes heavier hydrocarbons) and single pass methane conversion. We believe that to focus on these parameters without constraining the other key aspects of commercial viability reduces the probability of commercial success. Let's look at these parameters in turn.

Gas hourly space velocity (GHSV): Prior catalysts were so inactive that large volumes of catalyst and large volume reactors would be required. Practical catalyst activity is such that the gas hourly space velocity, the volume of gas processed per volume of catalyst, needs to be orders of magnitude greater than many published bulk OCM catalysts.

Reaction temperature: The initiation and operating temperature is critical. Prior work required initiation at 800 to 900 C. Industrially practiced partial oxidation exothermic reactions typically run at temperatures much lower than that due to practical reactor constraints, usually below 600C.

Oxygen source: If using pure O2 as a feed, an air separation plant will be needed, a costly addition to any commercial system. Alternatively, by using air as the feed, the need for such a plant is eliminated, and in concert the nitrogen in the reactor helps to mitigate the exothermic temperature rise and safety concerns.

Pressure: Much previous research used methane partial pressures of less than 1 atm. This makes scaling difficult, and therefore we limit ourselves to catalysts that perform at greater than 1 atm pressure.

Catalyst volatility: Many previously reported high performing catalysts are unstable at the high temperatures previously required used, either because of loss of dopants, instability of the material, or loss of oxygen, etc.

Reactor type: For partial oxidation exothermic reactions, the commercially deployed realm of reactor types is typically limited to fixed bed and fluidized bed systems, typically running at temperatures less than 600 C.

To enable the discovery and development of commercially viable OCM catalysts with commercialization requirements defined as boundary conditions, we combine three highly innovative technologies, none of which were deployed in the 1980's or 1990's against OCM. The platform consists of the following elements: (1) "Bottom-up" synthesis of nanowire catalysts (2) Biomaterial templating and (3) High throughput screening.

Using the above three technology platform elements, Siluria developed a work flow to rapidly synthesize tens of thousands of catalysts and to screen them directly in the OCM reaction. The process flow starts with the fabrication of biomaterial templates using a phage well known in the pharmaceutical industry. The templates are then used as a seed layer to synthesize nanowire catalysts. The phage is removed from the nanowires and calcination is performed, leaving a dry powder comprised of inorganic nanowires without any biological component. Each and every nanowire catalyst sample made is then screened using a primary high-throughput screening tool to directly measure performance in the OCM reaction. The data from the primary screening tool is used to perform two functions. First, the best catalysts identified from are put through secondary screening in high throughput fixed-bed reactors, followed by process optimization in traditional fixed-bed reactors. Second, the correlation between the input parameters (synthetic inputs), and the output measurable (performance in the OCM reaction) is used to optimize top-performing catalysts by creating new catalyst libraries by varying the biological, synthetic, and conventional input parameters in the local phase-space surrounding the top catalysts. This Design of Experiments (DoE) methodology is then repeated to further refine catalyst performance.

As already covered, in order to meet all the OCM catalyst performance objectives required for commercially viable performance, experiments are conducted using a process window in which critical dimensions are fixed at commercially viable levels. Examples of these dimensions include; methane conversion, reactor type, catalyst stability, pressure, oxygen source, reaction temperature and GHSV. Then only the catalyst itself is varied within the above constraints to optimize the selectivity. This approach enables us to develop an OCM catalyst that is commercially viable.

As selectivity is improved to commercially viable levels, parallel work takes place on developing different catalyst compositions, morphologies and synthetic parameters, so as to mitigate the risk of catalysts failing to get through process engineering and optimization.

The fixed process window provides a foundation from which economic studies and an aggressive and ongoing demonstration and commercial engineering effort can be launched.

Commercial Engineering

In turning our catalyst system into a reactor system, we decided early on to go to market with a proven reactor system within its demonstrated commercial space of temperature and other factors. We felt that introducing a novel catalyst with a familiar reactor system was more likely to generate acceptance than a novel catalyst in a novel reactor configuration. In conjunction, and based upon an initial design basis, work on separations can begin. The separations system of course depends on what the end product and final application is, within the commodity chemicals, specialty chemicals, and fuels spaces.

Summary

The pursuit of a process to build longer-chain hydrocarbons from the single-carbon methane molecule has resulted in a long history of scientific and engineering heroics. Nevertheless, the value of methane is still largely based on its utility as a fuel, rather than its value as a feedstock for further value-added processing.

A recently published International Energy Agency report states, "While the world's natural gas endowment is known to be very large...global resources could be larger than currently estimated." Thus, the ability to create new and higher-value markets for natural gas is of significance. The opportunity to replace a fraction of the chemicals, materials, and liquid fuels currently produced from crude oil represents a chance to rapidly introduce a lower cost, lower carbon route to domestic production of these products. And, of course, a more economical pathway to liquids produced from stranded and otherwise underutilized sources of methane represents a potential game-changer on the supply-side of global energy markets.

A commercially viable OCM process can help achieve all three of the aforementioned goals. By combining the power of nanowire synthesis, with biomaterial templating, high throughput screening, and a commercialization driven catalyst R&D focus, new technological approaches have been brought to bear on a decades-old challenge and significant progress has been made

References

1. T. Ren, M. Patel, "Basic petrochemicals from natural gas, coal and biomass: Energy use and CO2 emissions", Res. Conserv. Recycl. 53, 513 (2009).
2. BP Statistical Review, June 2008.
3. Progressive Energy & CNG Services, "Bio-SNG Feasibility Study", November 2010.
4. Market research report from Nexant, Inc., "Ethylene" Sept. 2009.
5. A. Holmen, "Direct conversion of methane to fuels and chemicals", Cat. Today 142, 2 (2009).
6. P.L. Spath and D.C. Dayton, NREL Technical Report: Technical and Economic Assessment of Synthesis Gas to Fuels and Chemicals with Emphasis on the Potential for Biomass-Derived Syngas, NREL/TP-510-34929 (2003).
7. L. Mleczko, M. Baerns, "Catalytic oxidative coupling of methane – reactions engineering aspects and processing schemes", Fuel Process Tech 42, 217 (1995).
8. J.H. Lunsford, "The catalytic coupling of methane", Angew. Chem., Int. Ed. Engl. 34, 970 (1995).
9. G.E. Keller and M.M. Bhasin, "Synthesis of Ethylene via Oxidative Coupling of Methane", J. of Catal, 73, 9 (1982).
10. Fatih Birol et al, "International Energy Agency Special Report World Energy Outlook 2011", IEA, 48 (2011)

AIChE Paper Number 51B

Hydrogen Transport Membrane Technology for Ethylene Production

Uthamalingam (Balu) Balachandran
Materials Scientist
Argonne National Laboratory

Tae H. Lee
Materials Scientist
Argonne National Laboratory

Chan Y. Park
Materials Scientist
Argonne National Laboratory

Stephen E. Dorris
Materials Scientist
Argonne National Laboratory

Olutoye P. Akinjiola
Chemical Engineer
Psage Research, LLC

Prepared for Presentation at the 2012 Spring National Meeting

Houston, TX, April 1-5, 2012

HYDROGEN TRANSPORT MEMBRANE TECHNOLOGY FOR ETHYLENE PRODUCTION

Uthamalingam (Balu) Balachandran
Materials Scientist
Argonne National Laboratory

Tae H. Lee
Materials Scientist
Argonne National Laboratory

Chan Y. Park
Materials Scientist
Argonne National Laboratory

Stephen E. Dorris
Materials Scientist
Argonne National Laboratory

Olutoye P. Akinjiola
Chemical Engineer
Psage Research, LLC

Abstract: Ethylene, the largest volume commodity chemical manufactured today, is currently produced by steam pyrolysis of ethane. Thermodynamically limited in terms of its equilibrium conversion, this process requires substantial energy input. Our project offers a new approach for producing ethylene, wherein a hydrogen transport membrane (HTM) selectively removes hydrogen during the dehydrogenation of ethane, and thereby shifts the equilibrium toward ethylene production. Enhanced conversion beyond the thermodynamic limits by removing hydrogen provides higher conversion per pass, lowers the ethane recycle to the cracking unit, and reduces the number of necessary purification steps. Purification represents about 50% of the energy consumption and 70% of the cost in the energy-intensive pyrolysis approaches used today. Besides significantly reducing the energy costs, the new HTM approach yields savings in feedstock and capital costs. Preliminary laboratory experiments using small disk-type HTMs showed that the new approach can afford ethylene yields above the thermodynamic equilibrium limit at high ethane conversion. A mathematical model of an HTM membrane reactor used for ethane dehydrogenation was developed. The model-predicted ethylene yields were compared with experimental data from the membrane reactor. The results of our experimental work and modeling studies are presented in this paper.

Introduction

While ethylene is the largest commodity chemical, its production also requires the highest energy [1,2]. The conversion of hydrocarbons to ethylene is endothermic, and the one-pass conversion rate is limited by thermodynamic equilibrium to less than 30% at moderately high temperatures [3, 4]. Another limitation in conventional ethylene production is the high cost for separation/recycling unreacted raw material and product purification [5].

The advantages of membrane reactors over conventional packed reactors have been amply demonstrated in several experimental and theoretical/modeling studies [3,4,6,7]. The continuous removal of hydrogen byproduct provides enhanced conversion beyond that determined by static thermodynamic equilibrium until limitations imposed by the underlying kinetics of the reaction system and/or the catalyst becomes dominant. Compared to a conventional reactor, higher attainable conversion levels in a hydrogen transport membrane (HTM) reactor allow the process to operate at lower temperature, thus providing prolonged catalyst lifetime and lower production costs [5]. Furthermore, the membrane dehydrogenator offers the possibility of supplying hydrogen with higher purity than that supplied by a conventional reactor, and under less severe operational conditions.

The investigation explores the potential application of the HTM dehydrogenation reactor as a substitute for the conventional packed-bed reactor in an integrated ethylene production process. In this new approach [8-10], a dense HTM selectively removes hydrogen during the dehydrogenation of ethane, thereby shifting the equilibrium toward higher ethylene conversion (Fig. 1).

Preliminary laboratory experiments showed that this approach gives ethylene yields above the thermodynamic equilibrium limit, even at high ethane conversion [10]. Enhanced conversion beyond the thermodynamic limits by removing hydrogen provides higher conversion per pass, lowers the ethane recycle to the cracking unit, and reduces the number of necessary purification steps. Purification represents about 50% of the energy consumption and 70% of the cost in the energy-intensive pyrolysis approaches used today. Besides significantly reducing energy costs, the new approach should result in savings in feedstock and capital costs. A mathematical model is utilized to study the influence of various parameters and to identify the key parameters that are likely to affect the HTM membrane performance. The effects of isothermal and non-isothermal model assumptions are analyzed.

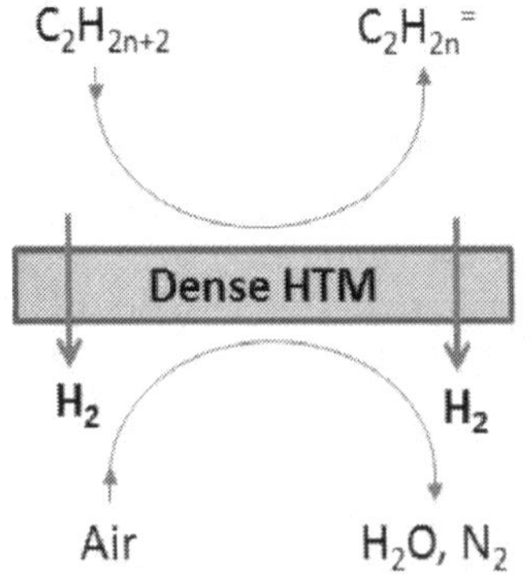

Fig. 1 Schematic of process for producing ethylene via dehydrogenation of ethane.

Experimental

A dense disk-type HTM in the form of a ceramic/metal composite was fabricated to produce ethylene by dehydrogenation of ethane [11,12]. The membrane for the dehydrogenation experiment contained 50 vol.% Pd with the balance being Y_2O_3-stabilized ZrO_2 (TZ-3Y). It was prepared from a mechanical mixture of Pd (Technic) and TZ-3Y (TOSOH) powders that were uniaxially pressed at ≈70 MPa in a stainless steel die. After sintering at 1400-1450°C for 5 h in air, the membrane had a thickness of ≈0.25 mm and a diameter of ≈20 mm. In the dehydrogenation experiment, ethane at ambient pressure was passed over one side of the HTM, while air flowed over the other side. The membrane was heated to ≈790-890°C. The hydrogen produced by dehydrogenation was removed and transported through the HTM to the air side of the membrane. The air provided the necessary driving force for transport of hydrogen through the HTM. The reaction between transported hydrogen and oxygen in air can provide some of the heat needed for the dehydrogenation reaction.

Because the ethylene production rate depends directly on the rate at which hydrogen is removed from the dehydrogenation zone, the HTM must deliver a high hydrogen flux for the new process to be effective. The hydrogen flux of HTMs was measured in tests using feed gases with known hydrogen concentrations. The HTMs were made as either a self-supporting sintered disk, like that in the dehydrogenation experiment, or a thin-film membrane supported by a porous ceramic substrate.

Thin-film membranes for hydrogen flux measurements were prepared by mixing palladium and TZ-3Y in a solution of α-terpineol and isopropyl alcohol with polyvinyl butyral as a binder and dibutyl phthalate as a plasticizer. A paste was prepared to give 60 vol.% Pd in the final thin film. The concentration of isopropyl alcohol was varied to control the viscosity of the paste. To prepare porous substrates for HTM thin films, alumina hydrate powder was mixed with carbon powder (0-30 wt.%) as a pore former, and then the powder mixture was cold isostatically pressed (≈70 MPa) in the shape of either a disk or a tube. The pressed specimen was then pre-sintered at 800-900°C in air for 5 h. Paste containing the membrane components was painted onto the surface of the pre-sintered porous substrate and then was sintered for 5 h in air at 1400-1450°C. Film thickness was ≈18 μm.

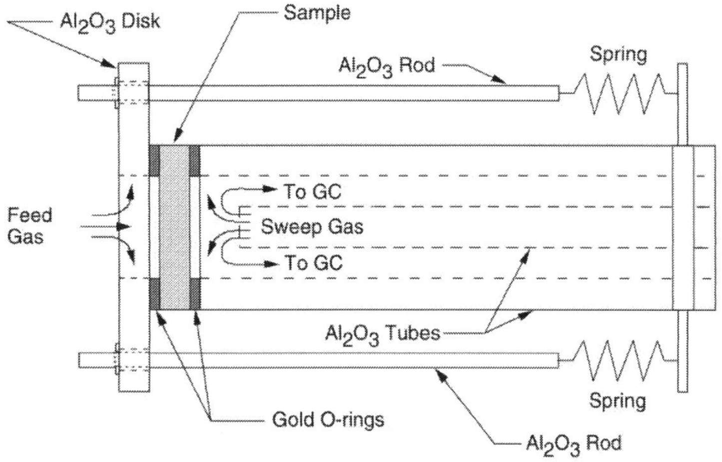

Fig. 2 Fixture for measuring hydrogen flux through HTM.

The hydrogen flux through an HTM sample was measured by affixing the sample to the Al_2O_3 tube in the assembly shown in Fig. 2. A seal formed during heating at 950°C, when spring-loaded rods squeezed a gold ring between the membrane and the Al_2O_3 tube. One side of the sample was purged with feed gas composed of 80-90 H_2/balance He, while the other side was purged by a sweep gas of 100 ppm H_2/balance N_2. All gases were at ambient pressure. The hydrogen flux was measured in the temperature range 400-900°C.

Gas flow rates were controlled with MKS mass flow controllers. A Hewlett-Packard 6890 gas chromatograph was used to measure the concentrations of He and H_2 in the sweep gas. The He concentration in the sweep gas was used to correct the hydrogen concentration in the sweep gas for leakage through the seal. The calculated hydrogen leakage rate was typically <10% of the total H_2 flux through the membrane.

Mathematical Model

A mathematical model was used to calculate hydrogen flux and permeation for the hollow-tube HTM reactor shown in Fig. 3. Dense membrane material is assumed to be coated on the outer surface of a porous ceramic tube. It is also assumed that reaction occurs on the surface of the catalyst, and the sweep gas flows through the inner core of the tube.

Fig. 3 Schematic of the HTM reactor.

Other model assumptions are:

1. Ethane-to-ethylene conversion selectivity is 100%.
2. Isobaric conditions exist inside and outside the reactor.
3. Pseudo-homogeneous conditions exist without axial dispersion.
4. Plug-flow behavior exists for the permeate and reaction zones.
5. Hydrogen diffusion in the membrane is the rate-determining step for hydrogen permeation.
6. Diffusion of reactants/products into/out of the catalyst surfaces is not rate limiting. This assumption is very critical, and it implies that the catalyst has been designed to overcome this possibility.

The expression for ethane dehydrogenation [13] is given by:

$$C_2H_6 \leftrightharpoons C_2H_4 + H_2 \quad (1)$$

$$r_{C2H6} = k_1[P_{C2H6} - \frac{P_{C2H4}P_{H2}}{K_1}] \quad (2)$$

where k_1 [kg-mol/(m-hr-atm)] is the reaction rate constant of catalyst and is given by:

$$k_1 = 1.64 * 10^{11} e^{\frac{-10378}{T}} (kmol\ hr^{-1} atm^{-1}) \quad (3)$$

T is the reaction temperature (K), and K_1 (atm) is the equilibrium constant [4], which is a function of temperature:

$$K_1 = 1.096 * 10^7 e^{\frac{-17300}{T}} (atm) \quad (4)$$

The membrane reactor model consists of stoichiometric mass balance equations that describe the steady-state spatial profile of ethane, ethylene, and hydrogen conversion along the tubular reactor's length. Standard mass-balance equations capturing the net rates of depletion and the flow rate for the components constitute the model equations.

The flow rate as a function of position is expressed as follows:

$$\frac{dF_{C2H6}}{dz} = \rho_B A_r \eta_r r_{C2H6} \quad (5)$$

where F_{C2H6} is the flow rate (kmol-hr^{-1}), ρ_B is the catalyst density (kg-m^{-3}), A_r is the reactor cross-sectional area (m^2), η_r is the reaction effectiveness factor, r_{C2H6} (kmol-hr^{-1}-m^{-3}) is the rate of reaction/production of each species per kg of catalyst, and z is an axial coordinate. The permeation rate of hydrogen through the Pd membrane is generally assumed to obey the half power law or the so-called Sievert's law [6,7,13-18], which states that H_2 permeation flux is inversely proportional to the thickness of the Pd layer, directly proportional to the difference of the square roots of the hydrogen partial pressures in the two sides of the membrane:

$$Flux = J_{H_2} = (\frac{Q_{Pd}}{\delta})(P_{H_2}^{0.5} - P_{H2,p}^{0.5}) \quad (6)$$

$$\frac{dF_{H2,p}}{dz} = J_{H_2} A_m / L = \eta_m (\frac{Q_{Pd} A_m}{\delta})(P_{H_2}^{0.5} - P_{H2,p}^{0.5}) \quad (7)$$

where:

P_{H2} = Partial pressure of H_2 in the retentate
$P_{H2,p}$ = Partial pressure of H_2 in the permeate
δ = Membrane thickness (μm)
Q_{Pd} = Permeability of Pd layer, kmol-μm/(hr-m^2-atm$^{0.5}$)
A_m = Membrane permeation area, m^2
J_{H_2}= Hydrogen permeation flux, kmol/(hr-m^2)
L=length of membrane (m)
η_m =Membrane effectiveness factor

Expressed in the Arrhenius form, the permeability is given by:

$$Q_{Pd} = Q_o e^{\frac{-E_a}{RT}} \quad (8)$$

where E_a is the permeation activation energy, and R is the universal gas constant. Thus, the hydrogen flux could also be expressed as:

$$Flux = J_{H_2} = (\frac{Q_o}{\delta}) e^{\frac{-E_a}{RT}} (P_{H_2}^{0.5} - P_{H2,p}^{0.5}) \quad (9)$$

The heat balance along the length of the reactor is:

$$\frac{dT_r}{dz} = \left(K_T(T_W - T_r)A_w/A_r - \rho_B \eta_r r_{C2H6} * \Delta H - J_{H_2} H_{H_2} A_m/A_r\right)/\left(u_s \rho_g c_p\right) \quad (10)$$

From the stoichiometry of the reaction in Equation 1, the flow rate of the reaction mixture, F_{Total}, is defined as:

$$F_{Total} = F_{H2}^{o} + F_{C2H4}^{o} + F_{C2H6}^{o}(1 + X_{C2H6}) \quad (11)$$

$$F_{Total} = F_{Total}^{o} + F_{C2H6}^{o} X_{C2H6}$$

X_{C2H6} is fractional conversion of ethane; F_i^0 is the initial flow of component i; F_{Total}^{o} is the initial total flow rate; F_{Total} is the instantaneous molar flow rate; K_T (kJ m^{-2} hr^{-1} T^{-1}) is the heat transfer coefficient between the wall and the reactor contents; T_w, T_r, and T_p are wall, reaction zone, and permeate zone temperatures, respectively. The heat of dehydrogenation is ΔH (kJ/kmol), H_{H_2} is the hydrogen enthalpy (kJ/kmol); ρ_g is density; c_p is the heat capacity; and u_s the superficial gas velocity of the reactant/product mixture, which is approximated as:

$$u_s = \left(\frac{RT}{P_T}\right) \frac{F_{Total}^{o} + F_{C2H6}^{o} X_{C2H6}}{A_r} = \left(\frac{RT}{P_T}\right)\left(\frac{F_{Total}}{A_r}\right) \quad (12)$$

where P_T is the total pressure in the reactor.

The temperature of the permeate zone is assumed to be constant because the fraction of H_2 permeate is very small relative to the sweep gas. Wall temperature is assumed to be fixed because the reactor is always immersed in a high-temperature radiant furnace. The dimensionless form of the set of equations reduces to:

$$X_{C2H6} = \frac{F_{C2H6}}{F_{C2H6}^{0}},\ X_{C2H4} = \frac{F_{C2H4}}{F_{C2H6}^{0}},\ X_{H2} = \frac{F_{H2}}{F_{C2H6}^{0}},\ Y_{H2} = \frac{F_{H2,p}}{F_{C2H6}^{0}} \quad (13)$$

$$\frac{dX_{C2H6}}{d\xi} = \rho_B A_r L \eta_c r_{C_2H_6}$$

$$\frac{dX_{H2}}{d\xi} = -\frac{dX_{C2H6}}{d\xi}$$

$$\frac{dX_{C2H4}}{d\xi} = -\frac{dX_{C2H6}}{d\xi}$$

$$\frac{dY_{H2}}{d\xi} = \left(\frac{Q_{Pd} A_m \eta_m}{\delta}\right)(P_{H2,r}^{0.5} - P_{H2,p}^{0.5})$$

$$\frac{dT_r^*}{d\xi} = \left(\frac{L}{u_s \rho_g c_p T_r}\right)\left(K_T(T_W - T_r)A_w/A_r - \rho_B \eta_r r_{C2H6} \Delta H - J_{H_2} H_{H_2} A_m/A_r\right)$$

where:

$$T_W^* = T_w/T_p$$

$$T_r^* = T_r/T_p$$

$$\xi = \frac{z}{L} \text{ dimensionless distance}$$

and the instantaneous partial pressure of the reactants and products are expressed by:

$$P_{C_2H_6} = (1 - X_{C_2H_6})P_T/\sigma \quad (14)$$
$$P_{C_2H_4} = (X^o_{C_2H_4} + X_{C_2H_6})P_T/\sigma$$
$$P_{H2} = (X^o_{C_2H_4} + X_{C_2H_6})P_T/\sigma$$
$$\sigma = (1 + X^o_{H_2} + X^o_{C_2H_4} + X_{C_2H_6} - Y_{H_2})$$

where X_{C2H4} is the fractional conversion of C_2H_6; Y_{H2} is the ratio of hydrogen flux to the initial ethane feed rate; and X_i^o (i=C_2H_4 and H_2) is the initial flow of each component into the reactor. The boundary conditions are:

$$at\ \xi = \frac{z}{L} = 0,\ X_{C_2H_6} = X^o_{C_2H_6}, X_{C_2H_4} = X^o_{C_2H_4},\ X_{H_2} = X^o_{H_2} \quad (15)$$

Results

Figure 4 shows the variation of hydrogen flux at 900°C with membrane thickness. The feed gas during these measurements was 80% H_2/balance He, and the sweep gas was N_2. As expected from Sievert's law (Equation 6), the hydrogen flux increases linearly with the inverse of membrane thickness. The results also suggested that the membrane's bulk properties control its flux in this thickness range. The HTM for the dehydrogenation experiment and the thickest membrane tested for Fig. 4 had a similar thickness (0.21-0.25 mm); therefore, their hydrogen transport properties are expected to be similar. Since the dehydrogenation experiment was conducted, the hydrogen flux for HTMs has been increased significantly by reducing their thickness, and corresponding increases in hydrogen flux have been observed.

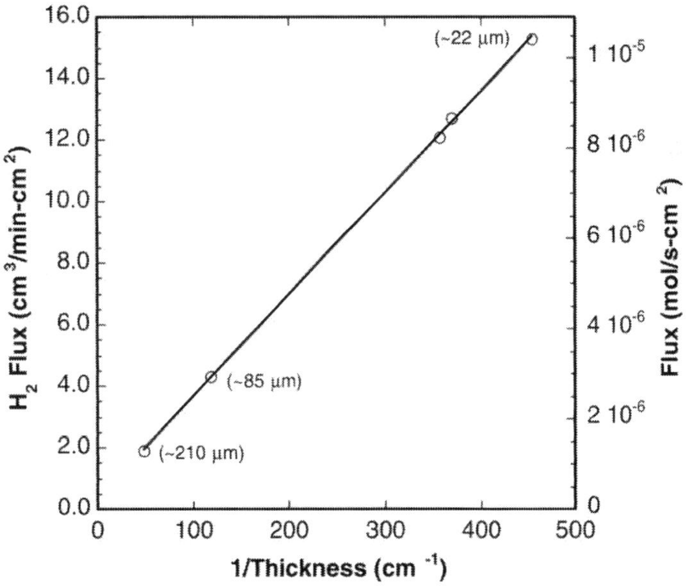

Fig. 4 Hydrogen flux at 900°C vs. inverse thickness of HTM. Feed was 80% H_2/balance He; sweep was N_2.

The data in Fig. 4 were obtained using gas flow rates of 150 cc/min with HTM membranes containing 50 vol.% Pd. Gas flow rates were not adjusted to keep the difference of the square roots of the hydrogen partial pressures in the two sides of the membranes constant when membranes of different thicknesses were tested. The data (Fig. 4) were obtained with self-supporting membranes that were prepared by polishing sintered disks. This fabrication technique is, of course, impractical for producing membranes with large surface areas, and thin samples made this way are very fragile. We thus sought to make thinner membranes with better mechanical integrity by a scalable method. To that end, we developed a paste-painting technique for fabricating dense thin-film membranes on porous supports [19].

A cross-sectional micrograph of a film made by paste-painting is shown in Fig. 5, where a dense, ≈18-μm-thick, HTM film is seen on top of a porous alumina substrate. Figure 6 shows the hydrogen flux profile with temperature for a ≈18-μm-thick HTM made by paste painting. The effective membrane permeation area is 1.2 cm^2, and porosity is ≈40%. The hydrogen flux data in Fig. 6 were obtained using feed and sweep gas flow rates of 500 cc/min. Gas flow rates have a pronounced effect on the hydrogen flux values for thinner membranes. Higher gas flow rates help to overcome the flux limitation resulting from concentration polarization in thinner membranes. Using feed gas of 90% H_2/balance He, with feed and sweep gas (N_2) flow rates of 150 cc/min, the flux at 900°C was ≈28 $cm^3/min\text{-}cm^2$. When the feed and sweep gas flow rates were increased to 500 cc/min, concentration polarization effects were reduced, and the hydrogen flux at 900°C (1/T = 0.00085 K^{-1}) increased from ≈28 to ≈52 $cm^3/min\text{-}cm^2$. The data in Fig. 6 are in good agreement with the predicted profile from Equation 9.

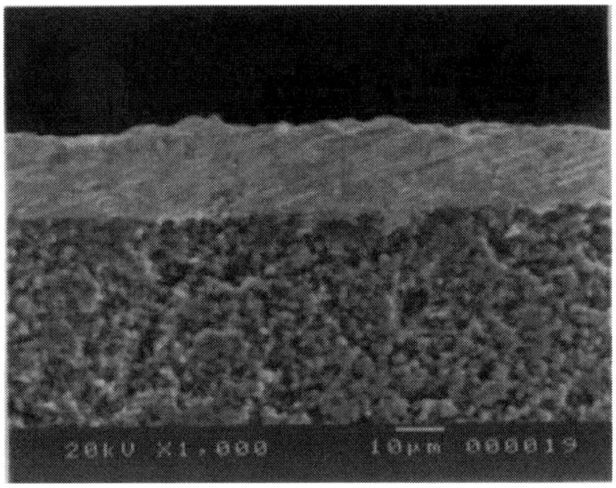

Fig. 5 Cross-sectional view of ≈18-μm-thick HTM film on porous alumina substrate.

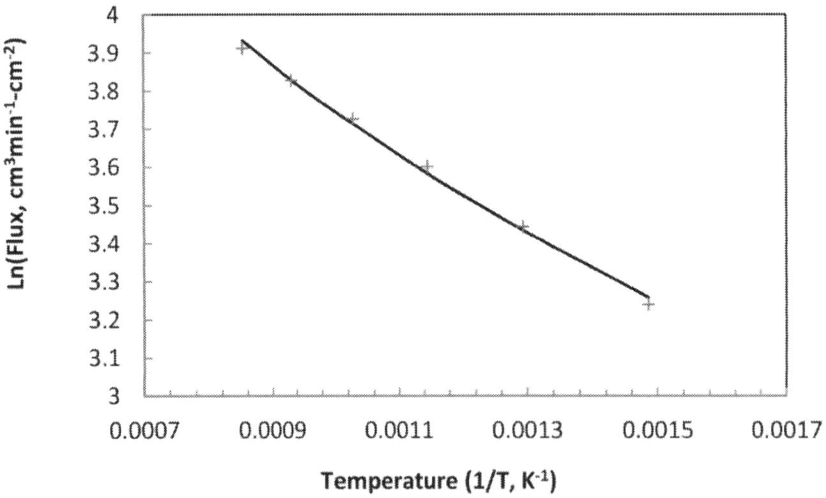

Fig. 6 Hydrogen flux vs temperature for ≈18-μm-thick HTM film on porous alumina. Feed (90% H_2/bal. He) and sweep (N_2) flowed at 500 cc/min.

From Equation 9 and regression analysis of the data, we derived a membrane permeability Q_o of 52.283 kmol-μm/(m^2-hr-atm$^{0.5}$) and an activation energy for the hydrogen permeation, E_a, of 8818 kJ/(kmol-K). These data compare well with other data on hydrogen permeation through palladium [14-18]. Thus, the permeation of Equation 9 can be expressed as:

$$Flux = J_{H_2} = 52.283\exp(-1.06/T)\,(P_{H_2}^{0.5} - P_{H2,p}^{0.5})/\delta \qquad (16)$$

Figure 7 shows the results from an ethane dehydrogenation experiment done at 790-890°C using a self-supporting HTM disk (thickness ≈0.25 mm) [10]. This experiment was conducted at BP (Naperville, IL). In this experiment, the hydrogen produced by dehydrogenation of ethane was removed and transported to the air side of the HTM, where it reacted with oxygen to form steam. The reaction between hydrogen and oxygen on the air side provided some of the energy needed for the dehydrogenation reaction and established a low partial pressure of hydrogen, which provided the driving force for hydrogen transport through the HTM. It was found that ethylene selectivity decreases with temperature, as would be expected, and that ethane pyrolysis is more susceptible to side reactions and even coking at higher temperatures.

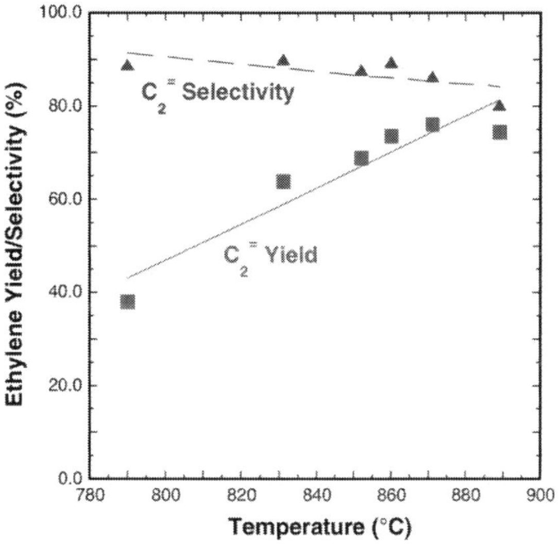

Fig. 7 Dehydrogenation of ethane at 790-890°C with 0.25-mm-thick HTM.

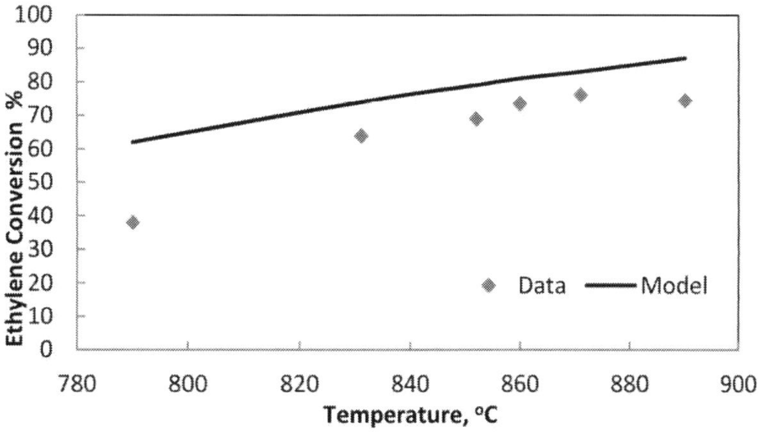

Fig. 8 Model-predicted ethylene yield variation with temperature.

The model-predicted ethylene yields are compared with experimental data from the membrane reactor in Fig. 8. The model results show the same trend as the data, the flux of hydrogen through the HTM enhanced ethylene yield beyond equilibrium conversions. Observed ethylene yields are lower than those predicted because the model did not account for selectivity or side reactions. Side reactions are unavoidable, and ethylene yield will invariably be lower than ethane conversion. This discrepancy is not a serious limitation to the model as a design guide because the ethylene selectivity also depends on choice of catalyst as well as the reaction conditions.

Figure 9 shows the model-predicted effect of the sweep gas flow rate relative to the feed flow on ethane conversion. Increased sweep rate lowers the partial pressure of hydrogen in the sweep gas and, consequently, increases the hydrogen partial pressure gradient across the membrane. Hence, the sweep rate will influence the hydrogen flux via changes in the differential partial pressure according to equation 6. The curves in Fig. 9 suggest that ethane conversion will increase with increase in the sweep rate. The results suggest that the sweep rate is a design and operating HTM reactor parameter that can be used to improve the reactor's performance.

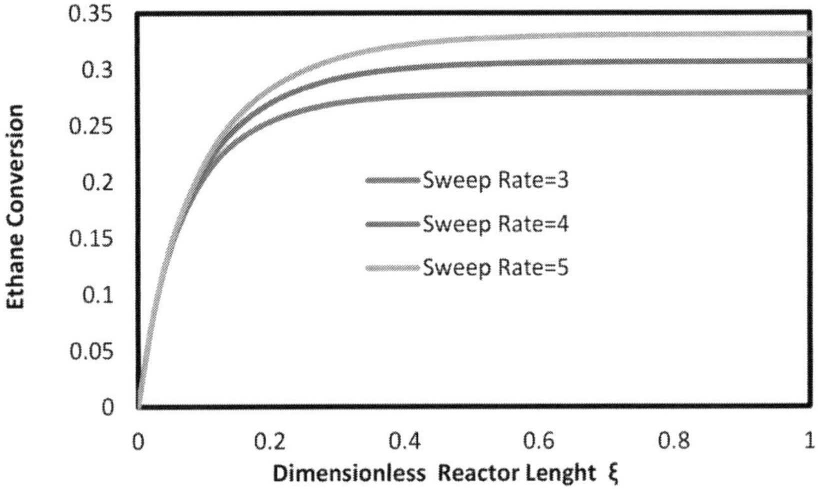

Fig. 9 Effects of sweep rate on ethane conversion at T=600°C.

Conclusions

We have demonstrated a new process whereby an HTM reactor is used to enhance ethane dehydrogenation to ethylene beyond equilibrium-limited values. The improved ethylene conversion per pass will reduce any feedstock recycling that may be needed to effect high overall production. In addition, pure hydrogen is produced upstream and under reaction conditions in the HTM reactor, unlike in the conventional fixed-bed reactors where hydrogen is obtained by product purification under energy-intensive cryogenic conditions. This new process will, therefore, reduce energy costs and overall costs, as well as greenhouse gases. Furthermore, the HTM reactor will allow ethane dehydrogenation to be performed at a lower temperature, which would suppress unwanted side reactions and prolong membrane life. The proposed mathematical model follows the observed trend of experimental data, and it could be used to investigate the parametric effects of the design variables of both the dehydrogenation reaction and the HTM reactor.

Acknowledgments

Work supported by the U.S. Department of Energy, Advanced Manufacturing Office (this Office was formerly known as the Office of Industrial Technologies Program) and Office of Fossil Energy, National Energy Technology Laboratory's Hydrogen and Syngas Technology Program, under Contract DE-AC02-06CH11357.

References

1. Chemical Bandwidth Study Phase I Exergy Analysis: A Powerful Tool for Identifying Process Inefficiencies in the U.S. Chemical Industry, U.S. Department of Energy Report (2004).
2. Chemical Bandwidth Study Phase II Exergy Analysis: A Powerful Tool for Identifying Process Inefficiencies in the U.S. Chemical Industry, U.S. Department of Energy Report (2006).
3. Wallsgrove, C., "Energy Philosophy in Prospective Petrochemical Projects", Proc. of the Sixteenth National Industrial Energy Technology Conf., Houston, TX, April 13-14, 1994.
4. Gobina E., Hou K., and Hughes R., "Ethane dehydrogenation in a catalytic membrane reactor coupled with a reactive sweep gas", Chem. Eng. Sci., **50**, 23, 1995.
5. Champagnie A. M., Tsotsis T. T., Minet R. G. and Wagner E. I., "The study of ethane dehydrogenation in a catalytic membrane reactor" J. Catalysis, **134**, 713, 1992.
6. Chiappetta C., Barbieri G., and Drioli E., "Effect of the pressure increasing and design parameters on Pd/Ag–based membrane reactor performance", Chem. Eng. Trans., **17**, 81, 2009.
7. Ayturk M. E., Kazantzis N. K., and Ma Y. H., "Modeling and performance assessment of Pd- and Pd/Au catalytic membrane reactors for hydrogen production", Energy Environ. Sci., **2**, 430, 2009.
8. Balachandran U., "Hydrogen transport membranes for dehydrogenation reactions", U.-S. Patent 7,329,791, issued Feb. 12, 2008.
9. Balachandran U., Lee T. H., Dorris S. E., Udovich C. A., Scouten C. G., and Marshall C. L., "Novel membrane technology for green ethylene production", Proc. of Clean Technology Conference, Boston, MA, June 1-5, 2008.
10. Balachandran U., Lee T. H., Dorris S. E., Udovich C. A., and Marshall C. L., "Hydrogen transport membrane technology for ethylene production", AIChE Paper Number 75b, Ethylene Producers' Conference, 2009 AIChE Spring National Meeting, Tampa, FL, April 26-30, 2009.
11. Dorris S. E., Lee T. H., and Balachandran U., "Metal/ceramic composites with high hydrogen permeability", U.-S. Patent 6,569,226, issued May 27, 2003.
12. Balachandran U., Lee T. H., Chen L., Song S. J., Picciolo J. J., and Dorris S. E., "Hydrogen separation by dense cermet membranes", Fuel, **85**, 150, 2006.
13. Abashar M. E. E., and Al-Rabiah A.A., "Production of ethylene and cyclohexane in a catalytic membrane reactor", Chem. Eng. Process., **44**, 1188, 2005.
14. Koffler S. A., Hudson J. B., and Ansell G. S., "Hydrogen permeation through alpha-palladium", Trans. Metall. Soc. AIME, **245**, 1735, 1969.
15. Davis W., "Diffusion of Hydrogen through Palladium", USAEC Report KAPL-1227, U.S. Atomic Energy Commission, 1954.
16. Balovnev Y. A., "Diffusion of hydrogen in palladium", Russ. J. Phys. Chem. **48**, 409, 1974.

17. Toda G., "Rate of permeation and diffusion coefficient of hydrogen through palladium", J. Res. Inst. Catal. Hokkaido Univ., Japan, **6**, 13, 1958.
18. Morreale B. D., Ciocco M. V., Howard B. H., Killmeyer R. P., Cugini A. V., and Enick R. M., "The permeability of hydrogen in bulk palladium at elevated temperatures and pressures", J. Membr. Sci., **241**, 219, 2004.
19. Balachandran U., Lee T. H., Chen L., Song S. J., and Dorris S. E., "Dense cermet membranes for hydrogen separation from mixed gas streams", Proc. 23rd Annual Intl. Pittsburgh Coal Conf., Pittsburgh, PA, Sept. 25-28, 2006.

Paper 51C

MICROWAVE ENHANCED CRACKING OF ETHANE FOR ETHYLENE PRODUCTION

Morgana Fall
Operations Engineer
Ceralink Inc.

Eric Wagner
Director Technology Development
Technip

Frank Cabe
Technician
Ceralink Inc.

Gary Eagleson
Senior Process Engineer
Technip

James Carnahan
Chief Scientist
Edison Laboratories

Ravi Lal
Commercial Vice President, Ethylene
Technip

Holly Shulman
Materials Scientist
Ceralink Inc.

246098

Prepared for Presentation at the 2012 Spring National Meeting
Houston, Texas, April 3, 2012

Microwave Enhanced Cracking of Ethane for Ethylene Production.

Morgana Fall
Operations Engineer
Ceralink Inc.
Eric Wagner
Director Technology Development
Technip
Frank Cabe
Technician
Ceralink Inc.
Gary Eagleson
Senior Process Engineer
Tecnhnip
James Carnahan
Chief Scientist
Edison Laboratories
Ravi Lal
Commercial Vice President, Ethylene
Technip
Holly Shulman
Materials Scientist
Ceralink Inc.

Abstract: A laboratory-scale microwave cracking system was built and tested to explore the feasibility of direct hydrocarbon cracking from ethane. The microwave system operated at 2.45 GHz in a single mode microwave applicator. The gas stream was a mixture of ethane and argon, which was introduced into the microwave zone through a quartz tube reactor. Samples were collected from the product stream exiting the hot zone and analyzed using a gas chromatograph. Results indicated that it was possible to crack ethane and obtain ethylene with high product conversion and high ethylene selectivity. The results of the gas chromatographic analysis for several different run parameters are presented. The yields from the microwave reactor were consistent with results predicted by SPYRO® in a theoretical reactor operating under similar conditions of temperature and residence time. These initial results indicate there may be advantages of microwave heat transfer compared to conventional combustion furnaces. This may allow novel reactor designs that result in more efficient production of ethylene.

Summary

The overall objective of this work was to explore microwave heating as a method to significantly decrease the energy consumed during the cracking of hydrocarbons to form ethylene. This first step was to determine feasibility and explore different laboratory scale process configurations for producing ethylene. Ceralink Inc. (company specializing in commercializing energy efficient manufacturing technologies), and Technip (a world leader in ethylene production systems and project management) collaborated to investigate the hydrocarbon-microwave interaction mechanisms through microwave experiments on the cracking of ethane and comparing the laboratory results to conventional modeling. Ceralink designed, built and performed laboratory heating studies using a single mode microwave reactor operating at 2.45 GHz. Technip participated in the test program and performed modeling work using SPYRO® to predict ethane conversion rates and product yields.

This work demonstrated that a microwave heating system could be used to produce ethylene from ethane. A series of experiments were conducted which varied microwave power, system pressure, and flow rates to explore ethane conversion and product selectivity. A gas chromatograph was used to identify and quantify the product species, which included hydrogen, methane, acetylene, and ethylene. Initial results demonstrated this heat transfer method to be feasible, and indicated that ethylene yield and selectivity were similar to results modeled with SPYRO®. This suggests that microwaves neither enhance nor inhibit the pyrolysis chemistry.

Background

Ethylene and other olefins are produced from hydrocarbon materials and dilution steam fed into furnaces that require large amounts of fuel. The gaseous fuel is burned to produce the temperatures required to drive the cracking reactions, usually in excess of 850 °C. In the cracking furnace, the heat is transferred from the combustion products to the reactor coils via radiant and convective heat transfer. Since this conventional heat transfer approach requires a substantial temperature driving force, less than half the energy released from the combustion of the fuel is actually transferred directly to the furnace coils that contain the hydrocarbons. The balance of the fired energy is recovered for feed preheating and steam production.

In conventional heating, all heat must be transferred through the outer surface of the material to the interior. Microwave heating offers an important advantage of being able to place energy directly into the work piece. Microwave heating of materials occurs through internal friction when dielectric loss mechanisms respond in a microwave field. There are many loss mechanisms that can be activated in real materials related to crystal structure, defects, bonding, surface, grain boundaries, etc. In practical terms, microwave energy can be used to directly and rapidly heat materials. There are certain frequencies that are allowed for industrial use, referred to as ISM (industrial, scientific and medical). The two most commonly used frequencies in industry are 915 MHz and 2.45 GHz.

Procedure

The laboratory scale microwave cracking system is shown in Figure 2. A single mode microwave system operating at 2.45 GHz with 2 kW available power was used. A surfaguide microwave applicator[1] specially designed for plasma was used to intensify the microwave field where the quartz tube passed through. The system also had a directional coupler for measuring the forward and reverse microwave power, along with a 3 stub tuner for impedance matching, and a variable attenuator for fine power adjustment.

This system used ethane as the hydrocarbon feedstock, and argon as the carrier gas. The flow rate for each gas was independently controlled using Brooks mass flow controllers (MFC). The gases passed through a static mixer prior to entering the quartz reaction tube. The flow rate of the MFCs were controlled and recorded using LabVIEW software, a computer graphical user interface for instrumentation. LabVIEW was also used to monitor and record inlet gas temperature, outlet gas temperature, pressure inside the system along with the microwave power and cooling water temperature. As a safety feature, an emergency stop button is located in the LabVIEW interface, which immediately shuts off the hydrocarbon gas flow and microwave energy. A Lexan safety shield is located in front of the hot zone of the quartz tube, which also incorporated microwave shielding. Microwave power was applied between 100 and 800 watts.

The gas chromatograph (GC), model HP 5890A, was fitted with a thermal conductivity detector (TCD) and used argon as carrier gas. The GC was used for analysis of the hydrocarbon products during the laboratory testing and provided the composition of the cracked gas for the following components: H_2, CH_4, C_2H_2, C_2H_4 and C_2H_6. Any species heavier than ethane in the effluent would not be detected. The GC was located next to the microwave cracking system so that samples could be taken directly from the exhaust line of the reactor tube and introduced into the GC through a gas sampling valve, minimizing contaminants. It was calibrated using a gas standard (volume %) containing 10% hydrogen, 10% methane, 30% ethylene, and 50 % ethane.

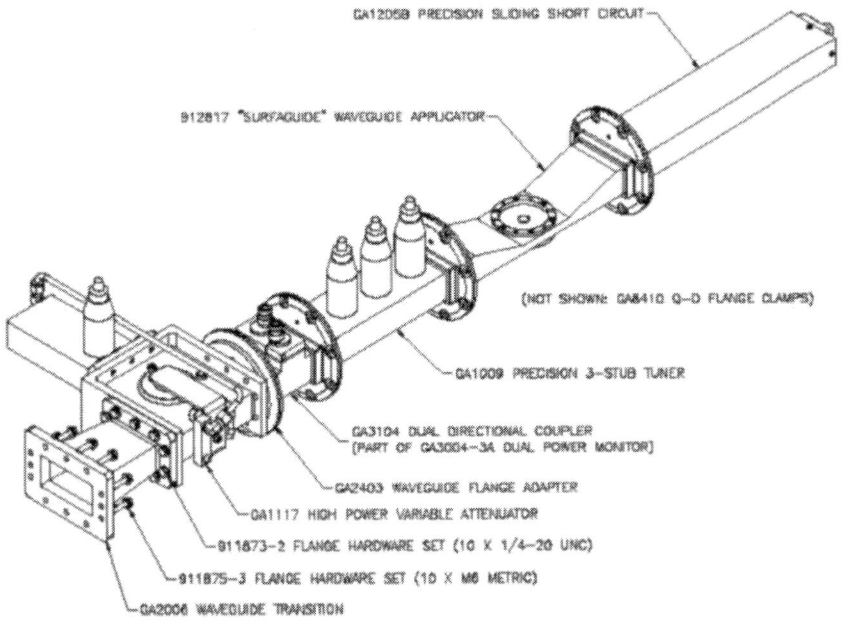

Figure 1. Schematic of the microwave single mode system from Gerling Applied Engineering

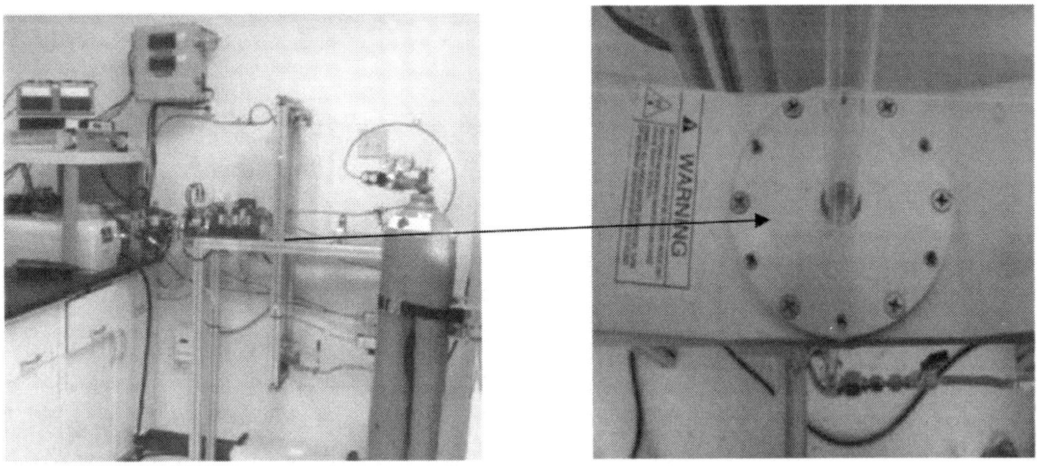

Figure 2. A) Digital image of the microwave system in Ceralink's lab. B) Close-up of where the quartz tube passes through the microwave surfaguide.

Results and Discussion

The microwave reactor at Ceralink was operated with three degrees of freedom:

- Flow rate of argon (via mass flow controller)

- Flow rate of ethane (via mass flow controller)
- Temperature (via electrical microwave input)

In order to obtain a stable microwave plasma, the reactor was operated under vacuum. The absolute pressure at the reaction zone outlet was a function of the inlet flow rate and ranged from 15 – 55 Torr. When the flow rate was adjusted higher, there was a corresponding increase in the reactor pressure, as the vacuum pump was set at all times to operate at full vacuum. These two effects nearly cancel so that the residence times were similar for most of the runs.

The initial experiments employed a microwave-generated plasma arc inside the reactor. The plasma arc was difficult to control and resulted in over-cracking of the feed. It was concluded that studying the control the plasma arc was going to take too long for the purposes of the project. Similarly, experiments at ambient pressure resulted in residence times that were too long and the ethane over-cracked to form hydrogen and carbon. Therefore, the majority of the experiments, and all the data presented here, were performed under vacuum with microwave heat input into the reactor in the absence of a plasma arc. It is likely, however, that a microwave reactor could be designed to operate at elevated pressures and using plasma arc if so desired.

It is important to note that for the conditions of the experiments, microwave exposure alone is not sufficient to crack the ethane feed. A microwave-generated plasma was induced in the reactor zone at the start-up of operation to light off the cracking reactions. After a thin layer of carbon formed on the surface of the quartz tube, the plasma was extinguished and continued microwave exposure was sufficient to continue the cracking reactions. The quartz tube was used as the reactor vessel, which is microwave transparent (little to no heating). Carbon, especially a thin carbon coating is known to be a good absorber of microwave energy. It has been reported previously that amorphous carbon has heated by microwave to >1250 °C in less than 1 minute[2].

In this case, it is theorized that the carbon layer is heated by the microwaves and this heat allowed the cracking reactions to proceed without the need to sustain a plasma phase in the reactor. (In an industrial application, the dilution steam would be heated by the microwave.) The optical pyrometer indicated a temperature reading in excess of 700 °C within 5 seconds of applying the microwave power (200 W). The temperature profile of the gas was not known precisely. A contact thermocouple placed an inch below and an inch above the microwave hot zone indicated low temperatures (< 150 °C); meaning the process gas underwent a rapid heating across the microwave zone and subsequent cooling from heat dissipation. An optical pyrometer aimed at the heated section of the quartz tube indicated a single temperature in the microwave zone (> 700 °C). This temperature reading was used as a guide rather than an absolute temperature of the reaction zone. It is believed that the gas reached temperatures well above the pyrometer measurement, perhaps 1000 °C or higher for the instances of high conversion of ethane. Evidence for this was the occasional (and accidental) deformation and melting of the quartz tube during ethane cracking at pyrometer readings of only 850°C.

The gas chromatograph (GC) provided the composition of the cracked gas for the following components: H_2, CH_4, C_2H_2, C_2H_4 and C_2H_6. Any species heavier than ethane in the effluent were not detected. The atomic molar hydrogen/carbon ratio of the reactor effluent was in the range of 3.0 – 3.1 for ethane conversions below 50%. This is good confirmation that the analytical measurements of the effluent were accurate because the effluent H/C ratio was the same as the feed ethane. At higher conversions of ethane, the associated H/C ratio of the effluent increased to the range of 3.2 – 3.3 mole/mole. This indicates that components heavier than ethane were in the reactor effluent but not detected by the GC. For example, if benzene was formed at high conversions of ethane, then the quantity of H_2 in the effluent would increase while the carbon-containing benzene would not be detected; this would lead to a H/C ratio greater than 3.0 of the effluent mixture as measured by the GC.

The experimental results were compared to simulations performed with Technip's proprietary SPYRO® program, the industry standard for hydrocarbon pyrolysis modeling. SPYRO® is recognized for its ability to simulate industrial operations. The ISO version of SPYRO® was developed for the purposes of simulating laboratory conditions like those at Ceralink. The ISO version of SPYRO® uses its kinetic model to predict yields as a function of residence time, pressure and temperature only – the reactor dimensions are not required. The model is quite stable and predicts yields under conditions such as those above, namely: argon rather than dilution steam, vacuum pressure, and very high temperatures.

Since the actual time-temperature profile in the reactor was not known, an approximation of the residence time was taken to be 0.001 seconds. This is very short by industrial standards and is a result of the very low absolute pressure in the reactor and very short reaction length (< 1 cm). Based on this residence time assumption, and the ratio of argon to ethane in the feed (1 mole/mole), SPYRO® was run at various temperatures to illustrate the relationship between temperature and ethane conversion, as illustrated in Figure 3. The "S" shape curve shows the increase in conversion as temperature increases and a gradual tailing off of conversion at temperatures in excess of 1100 °C.

The yields of the measured components, as a function of ethane conversion, are provided in Figure 4a and 4b. The yields are presented in terms of ethane conversion because the actual residence time and temperature were not known to a high degree of certainty. The yield of ethylene increases with increasing conversion and reaches a maximum of 64 wt% yield at 95% ethane conversion; thereafter the ethylene yield drops off. The yields of hydrogen, methane and acetylene show similar trends up to 95% at which point methane and acetylene yields increase significantly with the loss of ethylene yield.

It is important to note that the yields presented from SPYRO® simulations have been re-normalized to include only those components which were measured by the GC. For example, at 70% conversion of ethane, SPYRO® predicts about 94% yield of hydrogen (H_2) through ethane (C_2H_6), the balance being predominantly C_3H_6 along with other higher hydrocarbons which were not detected by the GC. So to be consistent with the lab data, the SPYRO® yields of H_2 through C_2H_6 presented here have been renormalized to 100 %.

The experimental data are presented in the following figures. The agreement between the lab and SPYRO® is very good for hydrogen (Figure 5) and acetylene Figure 6. SPYRO® under-predicts methane (Figure 7) and ethylene (Figure 8) to some degree, but the data follows the shape of the SPYRO® curves nicely. It is expected that more detailed GC analyses would result in an even closer match to SPYRO®. Considering the limited time available for the experimental program, the quality of the data seems to be quite high.

In summary, the experimental program established that microwave energy can be used to supply the endothermic heat required for the pyrolysis of hydrocarbons to olefins. The experimental results were in-line with simulations using the SPYRO® program suggesting that microwaves neither enhance nor inhibit the pyrolysis chemistry. Microwave generation requires electrical input which is inherently less efficient compared to energy generated from combustion of fuel gas as is currently employed in industry. However, conventional heat transfer necessitates many parallel reactor tubes because of limitations on heat flux because of the resistances to heat transfer and the temperature gradient from the hot combustion gas to the cooler process gas. There may be advantages of microwave heat transfer compared to conventional heat transfer which relax current constraints and thereby allow novel reactor designs that result in more efficient production of ethylene. Those design exercises were beyond the scope of this project. In addition to industrial reactor design, it would be very interesting if further experimental work could be performed at elevated pressures and with the use of dilution steam to aid the microwave heat transfer mechanism.

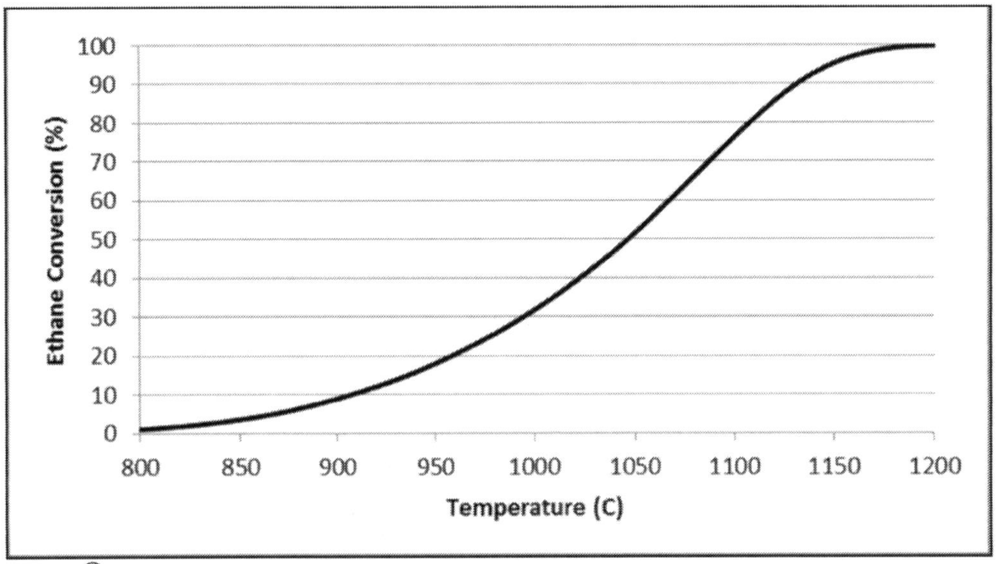

Figure 3. SPYRO® Predicted Ethane Conversion at 42 Torr and 0.001 Seconds Residence Time

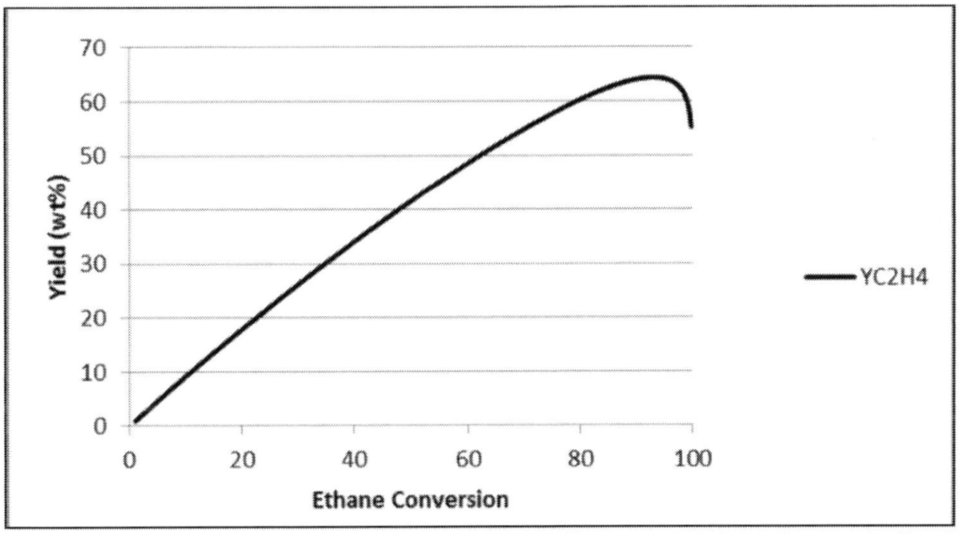

Figure 4A. SPYRO® Predicted Yield of Ethylene at 42 Torr and 0.001 Seconds Residence Time

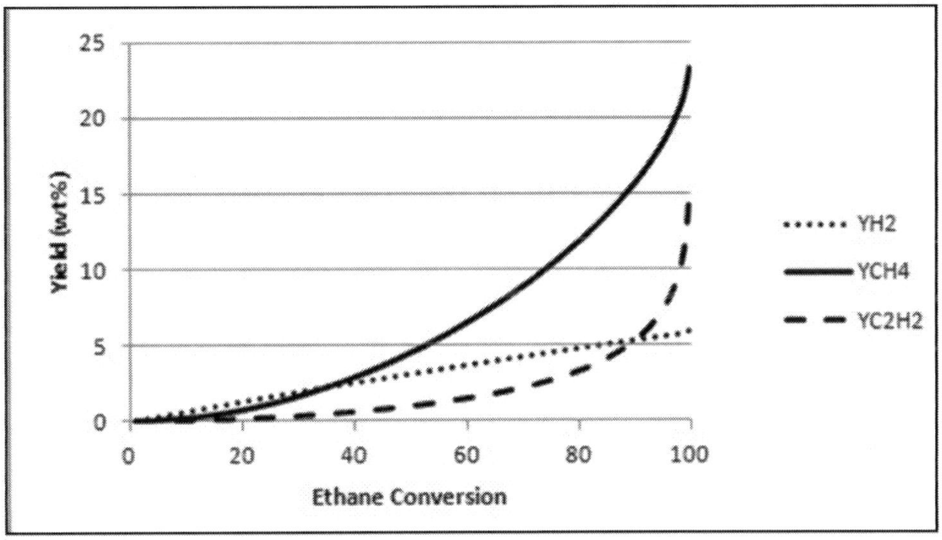

Figure 4B. SPYRO® Predicted Yields for Hydrogen, Methane and Acetylene at 42 Torr and 0.001 Seconds Residence Time

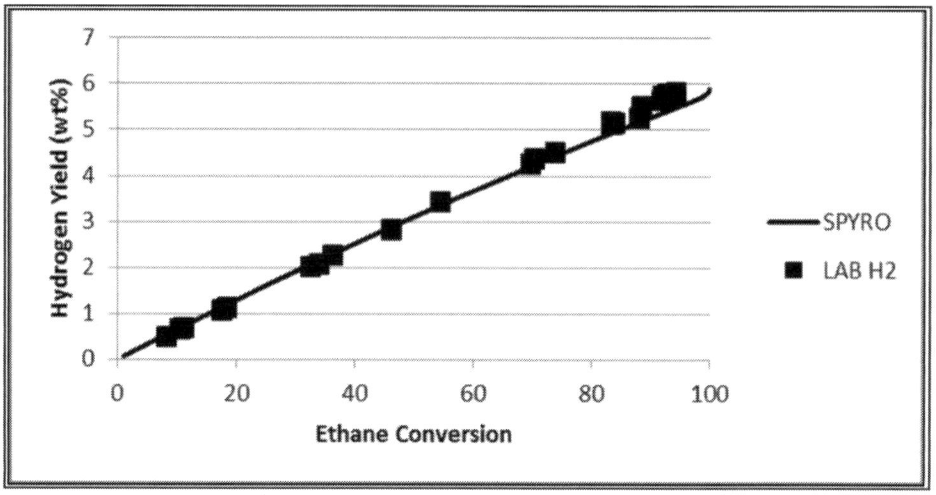

Figure 5. SPYRO® Predicted and Measured Yields Hydrogen (H_2)

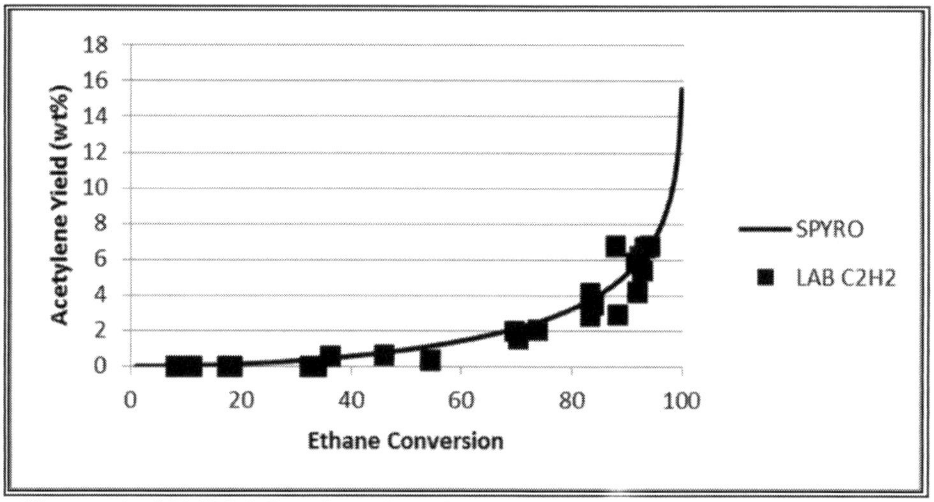

Figure 6. SPYRO® Predicted and Measured Yields Acetylene (C_2H_2)

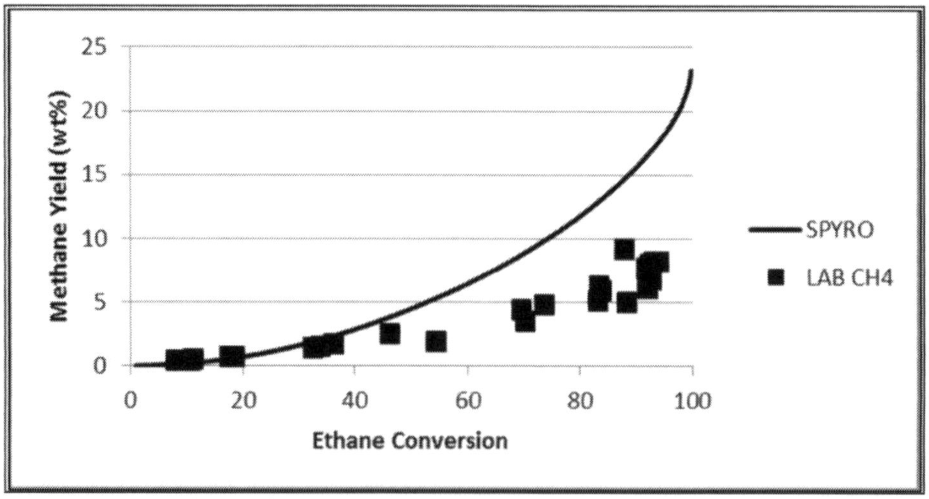

Figure 7. . SPYRO® Predicted and Measured Yields Methane (CH_4)

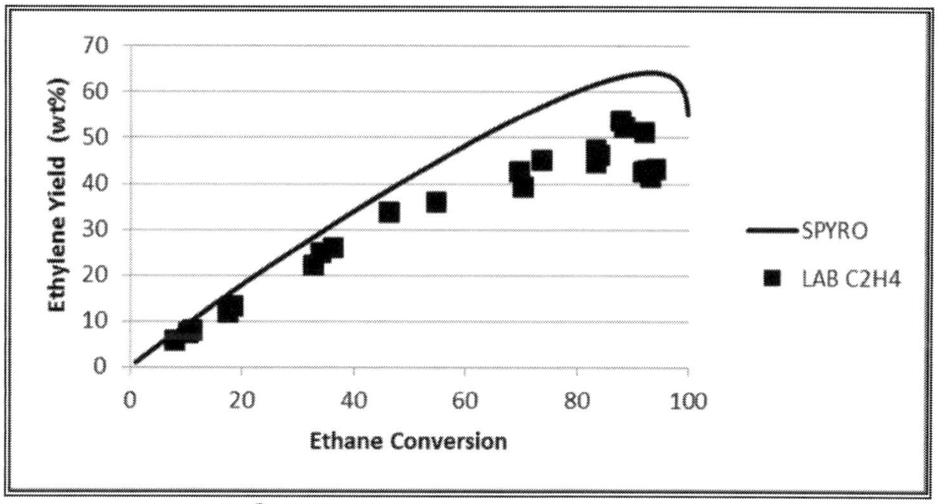

Figure 8. SPYRO® Predicted and Measured Yields Ethylene (C_2H_4)

Conclusions

Microwave cracking of ethane demonstrated good product conversion and ethylene selectivity, with a short residence time (~0.001 sec). The system was operated below atmospheric pressures, in the range of 15 – 55 torr, with argon as a carrier gas. The measured products included hydrogen, methane, acetylene, and ethylene. The results followed similar trends to those predicted by the modeling software SPYRO®, with the exception that the microwave appeared to produce slightly lower amounts of ethylene and methane, although enhanced analytical analysis should reduce the difference. Continued testing will be required to verify these results and quantify the energy consumption of microwave vs. conventional. The microwave cracking process is an attractive option due to the possibility of selectively heating the reaction volume rather than the reactor walls, which may allow novel reactor designs that result in more efficient production of ethylene.

Acknowledgments

This material is based upon work supported by the Department of Energy under Award Number DE-EE0003469.

References

1. Moisan, M., et al., *A Waveguide-based launcher to sustain long plasma columns through the propogation of an electromagnetic surface wave.* IEEE Transactions on Plasma Science, 1984. **PS-12**(3): p. 203-214.
2. Walkiewicz, J.W., G. Kazonich, and S.L. McGill, *Microwave heating characteristics of selected minerals and compounds.* Minerals and Metallurgical Processing, 1988: p. 39-42.

() Dynamic Scheduling for Optimal Decoking Operation of Cracking Furnace System

Qiang Xu, Chuanyu Zhao and Kuyen Li, Dan F. Smith Department of Chemical Engineering, Lamar University, Beaumont, TX

Thermal cracking process employs multiple cracking furnaces in parallel to convert various hydrocarbon feedstocks to multiple smaller-molecular hydrocarbon products. The operation is a typical semi-continuous dynamic process, where cracking furnaces have to be periodically shut down for hot maintenance (e.g., decoking). Given the data of multiple feed characteristics, different furnace performances, various sale income of products and manufacturing costs, as well as specified operational constraints, an optimal decoking policy for the entire cracking system has to be determined to achieve the best economic performance of an ethylene cracking furnace system. In practice, because of the feed supply and maintenance uncertainties, the furnace decoking activities are better to be conducted in a dynamic and reactive way, by which the entire furnace system operations can be timely reconciliated once the impact of an uncertainty has triggered the projection of necessary operational change. Thus, the feeds from the new delivery and the leftover inventories can be timely, feasibly, and optimally allocated to available cracking furnaces to obtain the maximum average net profit for the coming operating time period.

Facing this challenge, this paper develops a novel reactive scheduling model to help generating dynamic decoking strategy for the entire furnace system. It can optimize multiple-feed assignments, running length of each batch operation, and decoking sequence for every furnace in the system based on the new feed deliveries, the leftover feeds, and current furnace operating conditions. It also simultaneously addresses major scheduling issues for real cracking furnace operations, such as non-simultaneous decoking, secondary ethane cracking, and seamless rescheduling. The developed model is solved by an effective solving strategy. The efficacy of the study and its significant economic potential are demonstrated by comprehensive case studies

AIChE Paper Number 51E

Process of Cracking a Hydrocarbon Feedstock with a heavy tail

Johan van der Eijk

Technology Manager Ethylene

Technip

Prepared for Presentation at the 2011 Spring National Meeting

Houston, Texas, April 1-5, 2012

Process of Cracking a Hydrocarbon Feedstock with a heavy tail

Johan van der Eijk
Technology Manager Ethylene
Technip

Abstract

Cracking furnaces are the heart of the ethylene plant. These furnaces convert hydrocarbon feedstock into ethylene, propylene and other cracked gas products by thermal cracking in the presence of steam. Typical examples of hydrocarbon feeds are ethane, propane, butanes, naphtha's, kerosenes and heavy atmospheric and vacuum gas oils.

It could also be economical to crack low quality feedstock with a heavy tail. This is a feedstock with a fraction vaporises at a much higher temperature compared to rest of the feedstock. Sometimes the fraction is even not vaporisable. Examples of such components are tar like material, solid particles and the residuum of high boiling fractions.

If a feedstock with a heavy tail is cracked in a cracking furnace, the heavy tail will cause fouling in the convection section, radiant section and the transfer line exchangers. This fouling will result in short on stream time and result in an uneconomical operation.

However, the low quality feedstock could be cracked without fouling issues if the heavy tail is separated out at a suitable location in the convection section. Consequently, the heavy tail bypasses the remaining non fouling part of the cracking furnace. Special attention has to be given to effectively control the separation of the heavy tail, because the operating parameters often have a limited flexibility.

This paper presents a cracking furnace design which has the possibility to crack a hydrocarbon feedstock with a heavy tail. The design has the flexibility to effectively control the temperature of the separation device without penalties on the furnace efficiency. The temperature of the separation device is controlled by regulating the heat exchange capacity of the feed preheater upstream the separation device.

Introduction

Steam cracking is a process for the thermal conversion of hydrocarbon feedstock into ethylene, propylene and other cracked gas products by thermal cracking in the presence of steam. Various types of feedstock are used for steam cracking, such as ethane, propane, butanes, straight-run refinery fractions (naphtha, kerosene, atmospheric gas oil, vacuum gas oil) and hydro treated feedstock (hvgo and hydrowax). All these different feedstock types have in common that the feedstock is a distilled product. This means that the feedstock is clean and that no heavy ends (in the form of residue, tar, solid particles or very high boiling fractions) are present.

Nowadays, an increasing interest is found among ethylene producers to process natural gas condensates. These are basically unrefined hydrocarbons that are obtained as by-product from natural gas production. As these condensates have a wide boiling range, they are not directly suitable as motor fuel. This implies that their price level is lower than that of straight-run or hydro treated feedstock and explains their popularity as petrochemical feedstock.

The condensates available on the market originate from a wide variety of sources, and so vary in their qualities and price levels. Some of them are directly suitable as cracker feedstock, but others contain heavy ends or impurities, which have adverse effects on the run length of the cracking furnace. The heavy tail and/or impurities may cause increased fouling in the following furnace parts:

- in the convection section, due to formation of deposits,
- in the radiant section, due to increased catalytic coking activity caused by the presence of metals,
- in the transfer line exchangers, due to increased tar formation.

The innovation developed by Technip concerns a process for the on-line removal of the heavy tail inside the cracking furnace, which makes it feasible to crack low quality and low cost feedstock without expensive off-site pre-treatment steps.

Description of the process cracking feedstock with a heavy tail

In the convection section of a cracking furnace heat is recovered from the hot flue gases leaving the radiant section where the actual steam cracking takes place. The recovered heat is used for high-pressure steam generation and for preheating the feedstock and process steam. In case of liquid feedstock the vaporization takes place in the convection section.

The innovation concerns a process for removal of the heavy tail from the furnace convection section, by introduction of a vapour/liquid separation step at a temperature such that the mixture of feed and process steam is not yet fully vaporized and that the heavy tail is still present in the liquid phase. As the amount of heavy tail is very small, a high vaporization degree (in the order of 97 – 99.5 %) is required as operating point for the separator.

Having a correct operating temperature is crucial. A too low temperature implies a loss of valuable feedstock, whereas a too high temperature will cause a slip of heavy tail to the downstream furnace sections with subsequent run length problems. Part of the innovation was to develop a scheme for the control of the separation temperature that does not interfere with the normal furnace control parameters, such as capacity, severity, dilution steam ratio, etc. To this purpose an extra degree of freedom was created by having a controllable duty of the feed preheater upstream of the separator.

A usual way to control the duty of a heat exchanger is to apply bypass control: by bypassing part of the stream to be heated the heat pick-up can be reduced. However, a bypass control scheme for the feed preheater is not the most suitable option for a cracking furnace, as a very high vaporization degree in the feed preheater may result in subsequent fouling problems. Also, without further precautions a lower duty of the feed preheater will result that not all heat of the flue gases is recovered. As a consequence the stack temperature will increase and overall furnace efficiency will drop.

In order to overcome the problems linked to a conventional bypass control scheme an innovative design concept was developed, allowing the duty of the feed preheater to be controlled while maintaining full flow through it and without loss of furnace efficiency. To this purpose the feed preheater is placed between two economizer banks, heating the boiler feed water used for generation of high-pressure steam (HP steam generation is the common way of heat recovery in a cracking furnace). Usually a single economizer bank is applied for the furnace HP steam system. However, in the new concept the boiler feed water is routed under split flow control over the two economizers: as a result the flue gas temperature at the inlet of the feed preheater can be varied, and so its duty.

Although the flue gas temperatures around the feed preheater are variable now, the furnace stack temperature can be kept constant by installing sufficient surface area in the upper economizer coil: in this way the flue gas temperature will pinch and will always be close to the boiler feed water temperature. The innovation has been patented and is called the "SANDWICHED FEED PREHEATER" design. Figure 1 gives a presentation of this design.

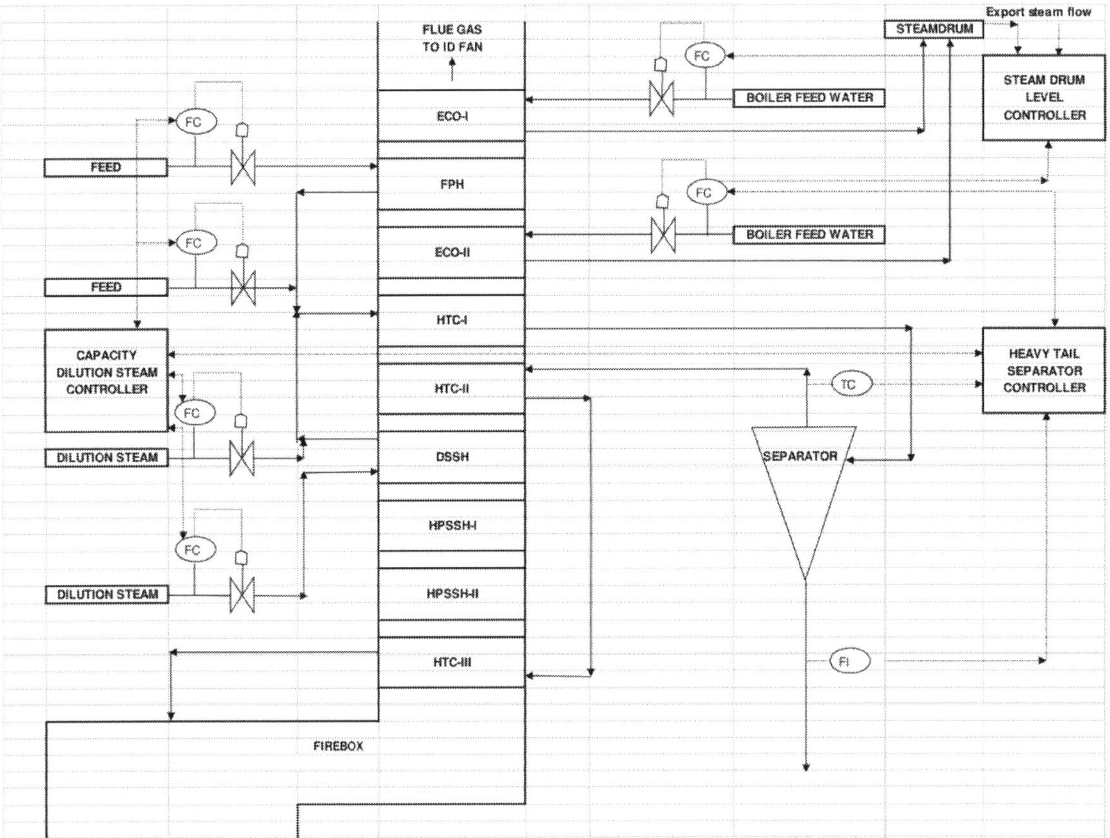

FIGURE 1: Cracking furnace with separation facilities in the convection section

Figure 1 shows that feed is routed to the feed preheater (FPH) coil. The heat pick-up of this FPH can be controlled by regulating the flow through economizer-II (ECO-II). The outlet of the FPH is mixed with superheated dilution steam and routed to the high temperature coil-I (HTC-I). The scheme presented in figure 1 also shows the possibility to introduce a feed downstream the FPH. The stream from HTC-I is routed to a separator which separates the heated feed into a vaporous fraction and a liquid fraction. The vaporous fraction is further heated in HTC-II and HTC-III and routed to the radiant section.

The inlet temperature to the separator and the liquid flow from the separator are both measured and provide feedback about the separation. This feedback will be used to regulate the flow to the ECO-II in order to control the separation fraction on the set point as defined by the operator. The rest of the BFW flow required by the process will be routed to ECO-I.

Example of the application

A natural gas condensate is routed to a convection section as shown in figure 1. The boiler feed water flow through the lower bottom economizer (ECO-II) is varied as a percentage of the total flow of boiler feed water through both economizers. The effect on the separation temperature and separation efficiency as function of the flow through the ECO-II is shown in figure 2.

Figure 2 shows that with this flow scheme a separation temperature of approximately 240°C can reached by controlling the flow through ECO-II to a value of approximately 13% of the total boiler feed water flow rate. This results in a liquid separation degree of approximately 0.45 wt%. By increasing the flow rate through ECO-II to a value of 40%, the heat pick up of the ECO-II is increased. As a result, the heat pick up of the feed pre heater located above the ECO-II is decreased. As a further consequence, the separation temperature is reduced till approximately 216°C and the liquid flow to be separated increases to approximately 1.35%.

FIGURE 2: TEMPERATURE & SEPARATED LIQUID FRACTION VERSUS BFW FLOW PERCENTAGE THROUGH BOTTOM ECONOMIZER

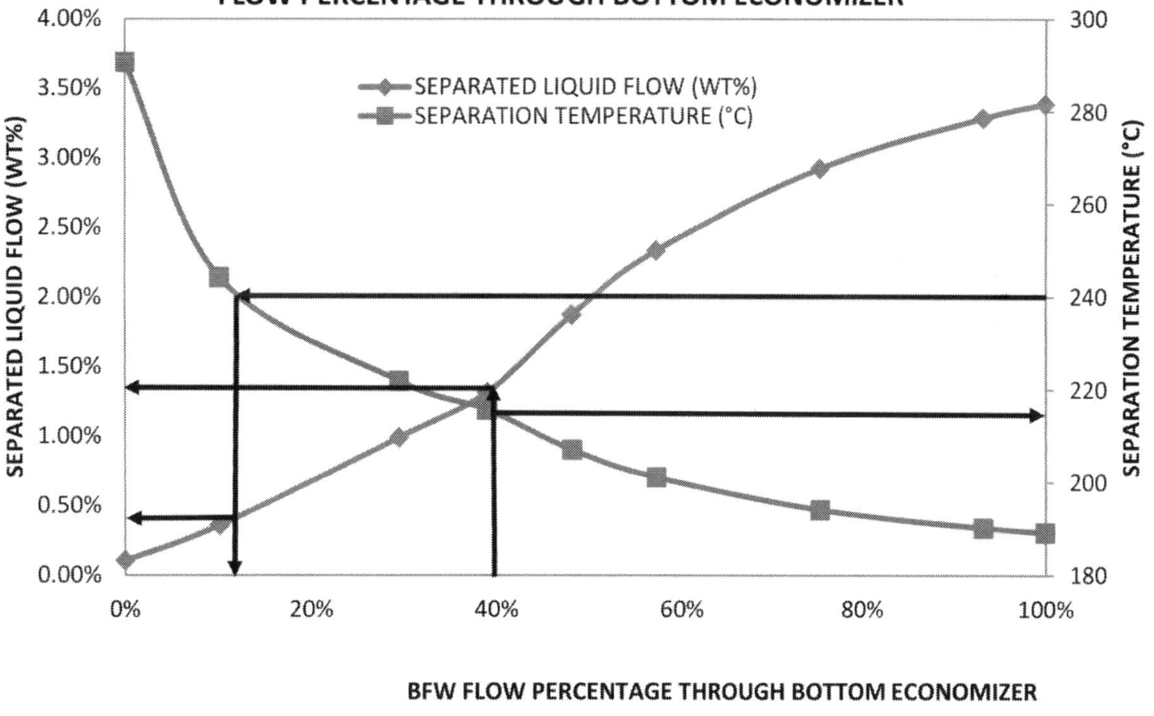

Conclusions

This paper demonstrates that a heavy tail can be separated in the convection section of a cracking furnace. The heat pick-up of the feed preheater and thereby the separation temperature of the feedstock can be controlled by regulating the heat exchange capacity of the ECO-II. Thereby, the liquid fraction to be separated can be controlled and adjusted as desired. This design allows controlled and efficient removal of the heavy tail of a feedstock from the part of the feedstock to be cracked.

Paper 51F

Bio-Ethylene Production: Alternatives for Green Chemicals and Polymers

Kevin M. Van Geem
Laboratory for Chemical Technology, Ghent University
Ghent, Belgium

Jinto Manjaly Anthonykutty
VTT Technical Research Centre of Finland
Espoo, Finland

Steven P. Pyl
Laboratory for Chemical Technology, Ghent University
Ghent, Belgium

Ali Harlin
VTT Technical Research Centre of Finland
Espoo, Finland

Guy B. Marin
Laboratory for Chemical Technology (LCT), Ghent University
Ghent, Belgium

Prepared for presentation at the 2012 AIChE Spring Meeting
Houston, TX, April 1-5, 2012

Production of bioethylene: alternatives for green chemicals and polymers

Prof. Kevin Van Geem
Laboratorium voor Chemische Technologie
Ugent – Ghent University
Technologiepark918
9052 Zwijnaarde
tel: 00 (32) 478573874 - 00 (32) 92645597
fax: 00 (32) 92645824

ABSTRACT:

There is an increasing trend of using bio-polyethylene and bio-polypropylene in Europe for making consumer goods. However there is currently very limited production capacity available for producing these base chemicals that are used in the polymerization processes. This contribution will give an overview of the presently available routes for the production of bio-ethylene and bio-propylene (bio-ethanol to olefins, methanol to olefins, hydrodeoxygenation of biomass followed by steam cracking, fast pyrolysis of biomass) and discuss advantages and disadvantages. The talk will be completed with results obtained from different pilot plant studies starting from tall-oil and waste fats and greases. In a first step these feedstocks are catalytically converted, in a second step they are cracked towards olefins. The total light olefin yield (ethylene and propylene) that has been obtained is in all cases higher than with naphtha and is above 50 wt% depending on process conditions, pretreatment and the biomass origin.

Introduction

Increasing oil prices, and the growing awareness of the detrimental environmental effects related to the use of fossil resources have focused research attention towards alternative routes and feedstocks for the production of light olefins and aromatics[1-3]. These large-volume chemicals such as ethylene, propylene, 1,3-butadiene and benzene are the building blocks for most polymers and the starting materials for many additives, solvents, etc.

Currently, steam cracking of fossil-based feedstocks is responsible for the production of these chemical building blocks, producing over 14010^6 tons of ethylene per year with a growth rate of 5.3%[4]. Large capacity crackers are operated all over the world, making the use of biomass-derived feedstocks to complement or even replace fossil feedstocks in existing units an interesting option. Several types of low cost biomass resources are, after effective upgrading, suitable renewable feeds for conventional steam crackers.

Crude tall oil is a viscous liquid obtained as a by-product of the Kraft process for wood pulp manufacture when pulping mainly coniferous trees[5]. It can be fractionated into distilled tall oil (DTO) and tall oil fatty acids (TOFA). These fractions mainly contain long chain fatty acids. DTO also contains significant amounts of rosin acids, i.e. a mixture of organic acids such as abietic acid. Catalytic hydrodeoxygenation (HDO) of both DTO as well as TOFA removes the oxygen in these acids in the form of H_2O, CO and CO_2, producing highly paraffinic hydrocarbon liquids, i.e. HDO-TOFA and HDO-DTO respectively[6].

Similarly, catalytic hydrodeoxygenation of triglyceride based biomass (TGB), such as algae oils or low cost waste greases like poultry fat and yellow grease, also produces paraffinic liquids (HDO-TGB) that are attractive feedstocks for conventional steam crackers.

In order to properly assess the potential of such deoxygenated bio-oils, steam cracking of HDO-TGB and HDO-TOFA was studied in the LCT pilot plant. The dedicated on-line analysis section includes a GC×GC-FID/(TOF-MS) and enables quantitative and qualitative on-line analyses of the entire reactor effluent with high level of detail[7]. The detailed characterization of feedstocks and pilot plant products for varying process conditions allows validation of the single-event microkinetic (SEMK) model, originally developed for steam cracking of fossil feedstocks[8, 9].

Experimental

TGB hydrodeoxygenation

A low value waste grease blend (46 wt% poultry fat, 18 wt% yellow grease 18 wt% brown grease, 9 wt% floatation grease, and 9 wt% grease from prepared foods) was acquired from Tyson Foods (Springdale AR, USA). Preliminary pretreatment involved filtration and washing to remove contaminants such as animal solids, phosphorus, and solubilized alkali/alkaline earth metals. The pretreated greases were subjected to catalytalic hydrodeoxygenation (HDO) in a two stage HDO pilot plant (Syntroleum, Geismar LA, USA). Catalysts were activated according to the Syntroleum sulfiding procedure before introduction of the feed[10, 11]. Bio-Synfining™ catalysts, i.e. commercially available hydrotreating catalysts, and the reactor operating conditions are described in more detail by Abhari et al. [11].

TOFA and DTO hydrodeoxygenation

Commercially available tall oil fractions (Stora Enso Kraft pulping facilities, Sweden) from Norwegian spruce pulping were used as raw materials. SYLVAT® 2 is a tall oil fatty acid (TOFA) with high fatty acid content (96 wt%) and a low content of rosin acids (1.8 wt%) and unsaponifiables (2 wt%). The employed SYLVATAL® 25/30S distilled tall oil (DTO) contained 27 wt% rosin acids, 70 wt% fatty acids and 3 wt% unsaponifiables. In both fractions the main component is oleic acid (C18:1). The main rosin acid in these fractions is abietic acid. Hydrodeoxygenation (HDO) of TOFA and DTO was performed in a fixed-bed reactor system at SINTEF Materials and Chemistry (Trondheim, Norway) aiming to produce 150 liters of product for further processing. The oils were processed as received on a continuous basis for 23 days without any pre-treatment/upgrading. The commercial HT-catalyst (NiMo) was presulfided prior to the experiment and the reactions were conducted in a temperature range of 320-360^0 C at 50.5 bara hydrogen pressure. Bio-oils were fed to the reactor at a constant rate (WHSV = 2 h^{-1}). Additionally, HDO experiments were also performed at VTT Technical Research Center Finland (Espoo, Finland)[6] in a bench-scale reactor in order to contrast the pilot scale and lab scale studies. The same HT-catalyst (NiMo) was used in VTT studies and the presulfidation was carried out by using H_2S/H_2 mixture for 5 hours at 400°C (H_2S/H_2 = 5 vol %). The experiments were conducted in a temperature range of 325-450 ^{o} C at a hydrogen pressure of 50 bar. The liquid feed was fed in to the reactor with three different feed rates (WHSV= 1, 2 and 0.3 h^{-1}). After a 6-hour stabilization period, liquid sample was collected and fractionated in to organic and aqueous phases.

The organic phase was further analyzed by GC-MS, GC×GC and CHNS-O elemental analyzer[6].

Steam Cracking Pilot Plant

The pilot plant at the Laboratory for Chemical Technology (Ghent University, Belgium), shown in Figure 1, consists of 3 sections: the feed section, the furnace/reactor section and the analysis section[12, 7]. The furnace, built of silica/alumina brick (Li23), is 4 m long, 0.7 m wide and 2.6 m high. It is fired by means of 90 premixed gas burners, mounted with automatic fire checks and arranged on the side walls in such a way as to provide a uniform distribution of heat. The furnace is divided into 7 separate cells that can be heated independently so that any type of temperature profile can be set. Inside the furnace a tubular reactor is mounted, in which the feedstock is evaporated, mixed with steam and subsequently cracked, at temperatures ranging from 600 to 900°C. The cracking coil, made of Incoloy 800HT, is 12.8 m long and has an internal diameter of 9 mm. These dimensions are chosen to achieve turbulent flow conditions in the coil with reasonable feed flow rates. The reactor outlet pressure is controlled by a computer regulated restriction valve. Twenty thermocouples and five pressure transducers are mounted along the coil to measure the temperature and pressure of the reacting gas.

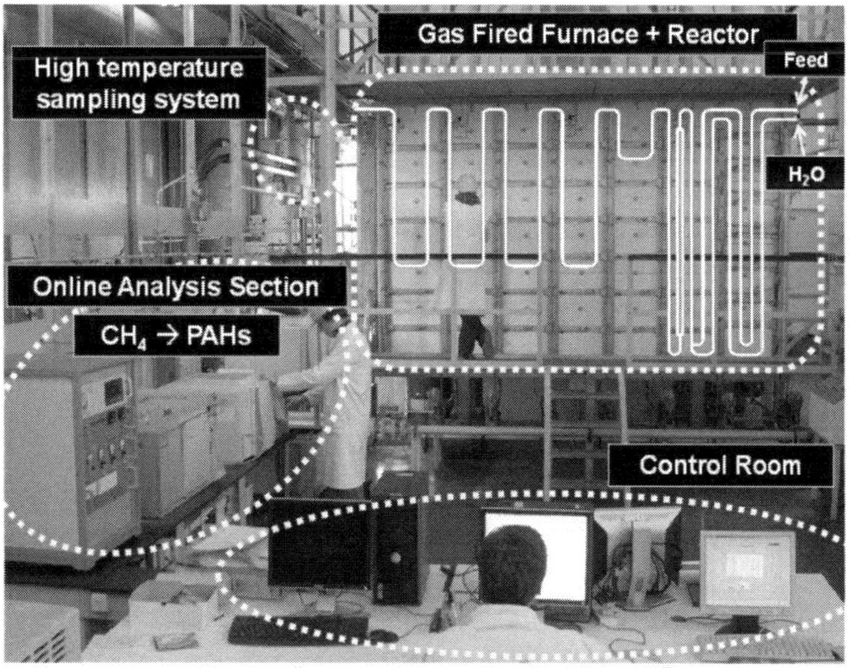

Figure 1: Steam cracking pilot plant[7]

The analysis section of the pilot plant enables on-line qualification and quantification of the entire product stream, i.e. a wide boiling mixture containing H_2, CO, CO_2 and hydrocarbons ranging from methane up to polyaromatic hydrocarbons[9, 7, 3]. The complexity of the effluents calls for several gas chromatographs: a permanent gas analyzer (PGA), a refinery gas analyzer (RGA), a detailed hydrocarbon analyzer (DHA) and a GC×GC-FID/TOF-MS. The pilot plant effluent is sampled on-line, i.e. during pilot plant operation, and at high temperature (400°C-500°C) using a heated valve-based sampling system and uniformly heated transfer lines. A fixed amount of N_2 is continuously added to the effluent, which acts as an internal standard, and permits the determination of absolute flow rates of all effluent components which allows to assess the effect of process conditions on the product distribution.

Single-event Microkinetic Model

Thermal cracking of organic components mainly proceeds through a free radical mechanism[13]. Three important reaction families can be distinguished: (i) bond scission and the reverse radical recombination; (ii) hydrogen abstraction; and (iii) radical addition and the reverse β-scission.

The microkinetic model is divided into two parts: the *μ network*, i.e. an extensive database of sub-mechanisms for the primary decompositions for large species (typically C_{6+}), and the *β network*, i.e. the core of the kinetic model that contains the interactions between smaller (typically C_{5-}) radicals and molecules [14, 15, 8]. A comprehensive group additive framework, i.e. a consistent extension of Benson's group additivity concept to transition state theory, allows automatic calculation of all kinetic parameters using a set of reference reactions and group additive values[16-20]. The SEMK model was originally developed to model steam cracking of fossil-based feedstocks, e.g. light hydrocarbons (ethane, propane, butane), naphtha and gas oils. The model therefore includes a wide range of components, such as paraffins, olefins, mono-, di- and tri-cyclic naphthenes, mono-, di- and tri- cyclic aromatics, and naphtheno-aromatics.

Coupling the microkinetic model to an appropriate reactor model and a solver for the resulting set of DAE's, allows simulation of industrial, pilot scale and bench scale reactors for a wide range of feedstocks[21]. However, as the number of species, and consequently equations and simulation time, increases exponentially with the complexity and average carbon number of the feedstock[22], *a priori* application of the quasi steady state approximation for so-called μ radicals ensures that the number of species and equations remains manageable [14, 15, 8].

Results and discussion

Feedstock Analyses

The detailed composition of the studied feedstocks was determined using GC×GC-FID/TOF-MS[7]. In Figure 2 the group-type compositions of the hydrodexygenated oils are presented as well as the composition of a typical petroleum-derived naphtha, i.e. currently the main steam cracker feedstock certainly in Europe [23]. Compared to this naphtha, the HDO-TGB contains high amounts of n-paraffins in a significantly higher carbon range, i.e. C14-C24 for the HDO-TGB versus C3-C13 for the naphtha. Only small amounts of naphthenes and no more than traces of aromatics were detected in the HDO-TGB. Also the HDO-TOFA and HDO-DTO feeds are highly n-paraffinic mixtures (C14-C24). The rosin acids present in the untreated TOFA and DTO fractions result in significant amounts of tricyclic naphthenes, such as norabietane (C19), and aromatics, such as norabietatriene (C19). In both fractions also some fatty acids methyl esters (FAME) were measured.

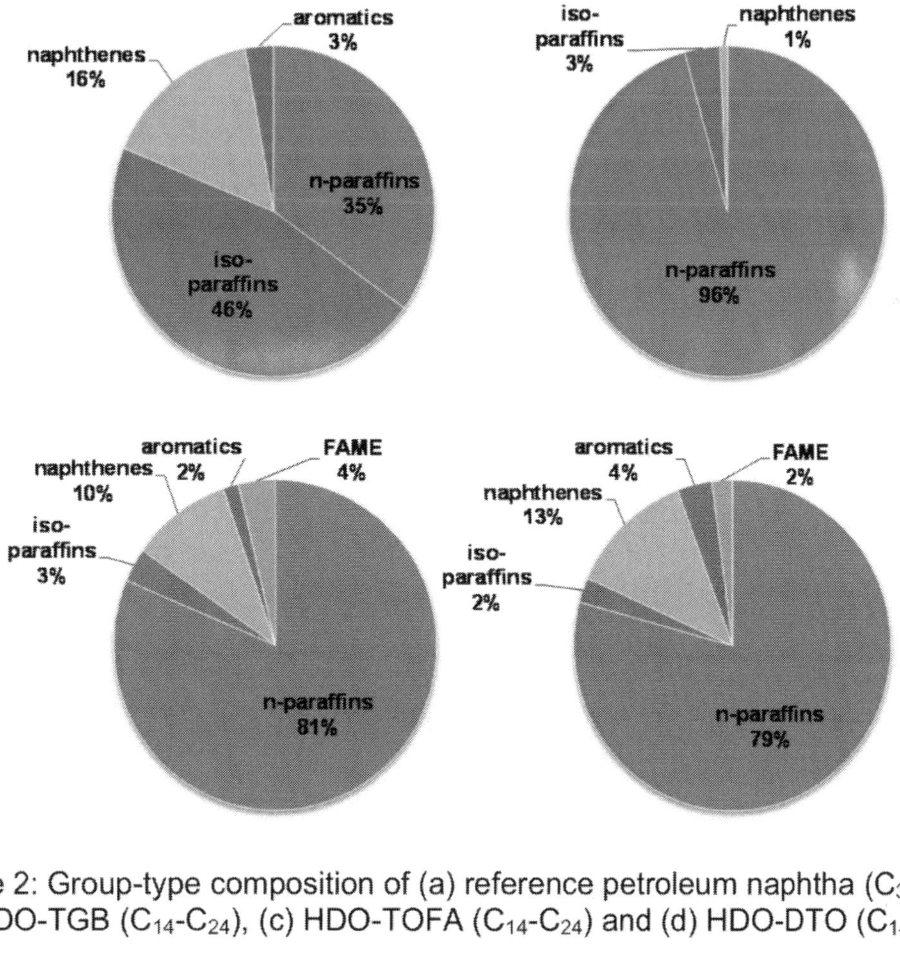

Figure 2: Group-type composition of (a) reference petroleum naphtha (C_3-C_{13}), (b) HDO-TGB (C_{14}-C_{24}), (c) HDO-TOFA (C_{14}-C_{24}) and (d) HDO-DTO (C_{14}-C_{24})

Pilot plant steam cracking

Table 1 gives an overview of the measured product yields for the various studied feedstocks at a coil outlet temperature (COT) of 820 °C and 850 °C respectively.

Table 1: Effect of feedstock composition and coil outlet temperature on product yields [δ = 0.45 kg/kg; τ = 0.3 s, COP = 1.7 bar]

Feed	**Naphtha**		**HDO-TGB**		**HDO-TOFA**	
COT	**820 °C**	**850 °C**	**820 °C**	**850 °C**	**820 °C**	**850 °C**
Methane	12.7	15.6	9.81	12.7	10.4	11.8
CO	0.02	0.03	0.02	0.04	0.13	0.10
CO_2	< 0.01	< 0.01	< 0.01	0.01	0.10	0.03
Ethylene	25.9	29.9	36.0	37.5	35.4	36.9
Propylene	17.8	16.5	19.5	18.1	17.5	15.3
1-Butene	2.37	1.44	4.18	1.63	2.20	1.05
1.3-Butadiene	4.72	5.33	7.45	7.56	4.51	4.92
Benzene	4.48	6.05	4.16	7.02	4.45	6.75
Toluene	2.14	2.47	1.29	1.86	1.42	3.33
Fuel oil	0.60	1.50	0.21	0.88	11.3	2.44

When cracking HDO-TOFA or HDO-TGB, ethylene yields are significantly higher compared to naphtha cracking for these conditions. This can be explained by the n-paraffinic nature of the renewable feeds. However, naphtha cracking results in relatively high methane and propylene yields due to the higher amount of iso-paraffins in the feed.

Cracking HDO-TOFA results in the highest amounts of CO and CO_2 which is in line with the feedstock analysis: the HDO-TOFA feed contains an important amount of esters. At a COT of 820°C quite a lot of unconverted feed was observed resulting in a large amount of so-called fuel oil, i.e. the heavy and low value fraction of the product stream.

Microkinetic Modeling

The combination of detailed feedstock analyses and pilot plant data allows validation of the SEMK model discussed above. This kinetic model was originally developed to model steam cracking of fossil-based feedstocks such as ethane, naphtha and gas oils. Table 2 compares the measured and simulated yields of some important products.

Table 2: Comparison of experimental and simulated product yields [δ = 0.45 kg/kg; τ = 0.3 s, COP = 1.7 bar, COT = 820 °C]

Feed	**Naphtha**		**HDO-TGB**		**HDO-TOFA**	
	Exp.	**Sim.**	**Exp.**	**Sim.**	**Exp.**	**Sim.**
Methane	12.7	12.4	9.81	9.48	10.4	9.42
Ethylene	25.9	25.7	36.0	35.5	35.4	33.6
Propylene	17.8	17.9	19.5	20.4	17.5	19.5
1-Butene	2.37	2.81	4.18	4.34	2.20	3.73
1.3-Butadiene	4.72	5.03	7.45	6.51	4.51	6.25
Benzene	4.48	3.32	4.16	5.35	4.45	6.54
Toluene	2.14	1.56	1.29	1.27	1.42	1.83

For the naphtha, the model performance is quite exceptional. Not surprisingly, since the SEMK model was optimized based on an extensive set of pilot plant data, mainly comprising experiments with gaseous and naphtha feedstocks. Nevertheless, the model performance is more than adequate for both HDO-TGB and HDO-TOFA. This, however, after inclusion of retro-ene decompositions for long-chain olefins. Via these concerted path reactions linear α-olefins, i.e. the primary decomposition product of paraffins, are transformed into propylene and a smaller linear olefin. Including these decomposition pathways in the microkinetic model proved to be necessary to more accurately model the measured product distribution of these highly paraffinic feeds.

As discussed above, the HDO-TOFA also contains a small amount of FAME. Since the SEMK model currently does not consider any esters, these components were represented by n-paraffins. This can explain the somewhat larger discrepancies between measured and simulated yields for this feedstock. However, including the decomposition of esters to the SEMK model should result in more accurate predictions.

Conclusions

Catalytic hydrodeoxygenation of tall oils as well as waste greases such as poultry fat and yellow grease produces paraffinic hydrocarbon liquids. These liquids are attractive renewable feedstocks that can be used in conventional steam cracking units.

Pilot plant steam cracking of such hydrodeoxygentaed bio-oils reavealed that, compared to a typical petroleum-derived naphtha, high amounts of light olefins, e.g. up to 40 wt% of ethylene, can be produced.

Acknowledgement

The authors acknowledge Stora Enso for supporting this research; Jari Räsänen and Tapani Penttinen for their vision on wood based olefins; and Dynamic Fuels, a joint-venture of Tyson Foods Inc. and Syntroleum Corporation, for providing the HDO-TOFA. The authors also acknowledge the financial support from the Long Term Structural Methusalem Funding by the Flemish Government – grant number BOF09/01M00409.

References

1. Ren T, Patel MK, Blok K. Steam cracking and methane to olefins: Energy use, CO2 emissions and production costs. *Energy*. 2008;33:817-833.
2. Carlson TR, Cheng Y-T, Jae J, Huber GW. Production of green aromatics and olefins by catalytic fast pyrolysis of wood sawdust. *Energy & Environmental Science*. 2011;4:145-161.
3. Pyl SP, Schietekat CM, Reyniers M-F, Abhari R, Marin GB, Van Geem KM. Biomass to olefins: Cracking of renewable naphtha. *Chem. Eng. J.* 2011;176-177:178-187.
4. Zimmermann H, Walzl R, *Ethylene*, in *Ullmann's Encyclopedia of Industrial Chemistry*. 2009, John Wiley & Sons, Inc.: New York.
5. Norlin L-H, *Tall Oils*, in *Ullmann's Encyclopedia of Industrial Chemistry*. 2005, John Wiley & Sons, Inc.: New York.
6. Anthonykutty JM, Van Geem KM, Pyl SP, Kaila R, Räsänen J, Penttinen T, Laitinen A, Krause O, Harlin A. *Upgrading of fatty acid containing rosin acids in to high value hydrocarbons via catalytic hydrodeoxygenation*. in *ECCE 8*. 2011. Berlin, Germany.
7. Van Geem KM, Pyl SP, Reyniers M-F, Vercammen J, Beens J, Marin GB. On-line analysis of complex hydrocarbon mixtures using comprehensive two-dimensional gas chromatography. *J. Chromatogr. A*. 2010;1217:6623-6633.
8. Van Geem KM, Reyniers MF, Marin GB. Challenges of modeling steam cracking of heavy feedstocks. *OGST - Rev. IFP*. 2008;63:
9. Pyl SP, Van Geem KM, Reyniers M-F, Marin GB. Molecular reconstruction of complex hydrocarbon mixtures: An application of principal component analysis. *AIChE J.* 2010;56:3174-3188.
10. Abhari R, Tomlinson HL, Havlik PZ, Jannasch NL, *Process for co-producing jet fuel and LPG from renewable sources*. 2008: USA.
11. Abhari R, Havlik PZ, *Hydrodeoxygenation Process*. 2009: USA.
12. Van Damme PS, Froment GF. Putting computers to work - Thermal cracking computer control in pilot plants. *Chemical Engineering Progress*. 1982;78:77-82.

13. Rice FO, Herzfeld KF. The Thermal Decomposition of Organic Compounds from the Standpoint of Free Radicals. VI. The Mechanism of Some Chain Reactions. *J. Am. Chem. Soc.* 1934;56:284-289.
14. Clymans PJ, Froment GF. Computer generation of the reaction paths and rate equations in the thermal cracking of normal and branched paraffins. *Computers & Chemical Engineering*. 1984;8:137-142.
15. Hillewaert LP, Dierickx JL, Froment GF. Computer-Generation of Reaction Schemes and Rate-Equations for Thermal-Cracking. *AIChE J.* 1988;34:17-24.
16. Willems PA, Froment GF. Kinectic modeling of the thermal cracking of hydrocarbons - 1. Calculation of frequency factors. *Ind. Eng. Chem. Res.* 1988;27:1959-1966.
17. Willems PA, Froment GF. Kinectic modeling of the thermal cracking of hydrocarbons - 2. Calculation of activation energies. *Ind. Eng. Chem. Res.* 1988;27:1966-1971.
18. Sabbe MK, Reyniers MF, Van Speybroeck V, Waroquier M, Marin GB. Carbon-centered radical addition and beta-scission reactions: Modeling of activation energies and pre-exponential factors. *ChemPhysChem*. 2008;9:124-140.
19. Sabbe MK, Reyniers MF, Waroquier M, Marin GB. Hydrogen Radical Additions to Unsaturated Hydrocarbons and the Reverse beta-Scission Reactions: Modeling of Activation Energies and Pre-Exponential Factors. *ChemPhysChem*. 2010;11:195-210.
20. Sabbe MK, Vandeputte AG, Reyniers MF, Waroquier M, Marin GB. Modeling the influence of resonance stabilization on the kinetics of hydrogen abstractions. *PCCP*. 2010;12:1278-1298.
21. Froment GF. Kinetics and reactor design in the thermal cracking for olefins production. *Chem. Eng. Sci.* 1992;47:2163-2177.
22. Broadbelt LJ, Stark SM, Klein MT. Computer-generated Pyrolysis Modeling – on-the-fly Generation of Species, Reactions, and Rates. *Ind. Eng. Chem. Res.* 1994;33:790-799.
23. Nizamoff AJ, *Renewable Liquids as Steam Cracker Feedstocks - PERP09/10S12*, in *Chemsystems PERP Program*. 2010.

AIChE Paper Number 96a

AMINE TREATING: A REVIEW OF THE TECHNOLOGY

Richard Ackman

Engineering Technology Leader

Dow Oil & Gas

Linda Pirtle

Senior Technical Services Engineer

Dow Oil & Gas

Prepared for Presentation at the 2012 Spring National Meeting

April 3, 2012

Houston, TX

Amine Treating: A review of the Technology

Richard Ackman

Engineering Technology Leader

Dow Oil & Gas

Linda Pirtle

Senior Technical Services Engineer

Dow Oil & Gas

Removal of H_2S and CO_2 from both gas streams and liquid hydrocarbon streams is important in many industrial processes. Solutions of aqueous amines have been used for more than 70 years to remove these unwanted acidic components in refinery streams, chemical processing and gas purification. Knowledge of amine treating application has grown from both practical experience and basic scientific research. An understanding of this accumulated technical knowledge is needed to design a competitive and economical gas treating plant. This paper gives a brief history of amine treating and describes the fundamentals of gas treating technology.

An understanding of amine chemistry and the physical properties related to gas treating is important for solvent selection and to take advantage of improvements in gas treating technology. New formulated solvents and tower internals technology offer opportunities for smaller and more energy efficient plants. However, contaminants and unwanted side reactions can limit plant efficiency and operability, negating these advantages. Use of simulation tools must be done correctly to take advantage of the process modeling and the latest process technology.

Background

Hydrogen sulfide (H_2S) is a toxic, flammable, slightly acidic gas that is found in natural gas and crude oil. It is also formed in certain chemical processes and hydrocarbon refining. It is corrosive to carbon steel pipes and burners, is poisonous to many catalysts and is a source of sulfur pollution. Carbon dioxide does not present the same toxicity issues as H_2S but it is usually an undesirable component in natural gas and many chemical processes. CO_2 reduces the heating value of gas, dilutes sulfur plant feed streams, will form solids in cryogenic applications and can combine with H_2S to form carbonyl sulfide (COS).

Removal of H_2S and CO_2 from gas streams has been accomplished with a number of different technologies over the years. The technology chosen for a particular removal application depended on what was commercially available and process economics, including the ever-increasing cost of waste disposal. Non-regenerable absorption technologies have been used to remove small amounts of H_2S and CO_2 from process streams. Examples of non regenerable technologies include scrubbing with solutions of caustic, potassium permanganate or triazine or adsorption onto iron oxide.

Regenerative technologies have typically been used to remove larger amounts of H_2S and CO_2. Economical regenerative technologies include scrubbing processes using carbonate salt solutions, amino acids, transition metal redox solutions of iron or vanadium chelants or ethanolamine solutions. Ethanolamines have been found to be more useful across a wider range of treating conditions and treated gas requirements than other technologies. Many ethanolamines form thermally unstable carbamates, carbonates, bicarbonates and bisulfides and have a low solution vapor pressure. They can be used in aqueous solution with low losses and have reaction energies that allow absorption and desorption of H_2S and CO_2 at economical process conditions.

The initial work on ethanolamines for acid gas removal was done in the 1930s using aqueous solutions of triethanolamine (TEA) and diethanolamine (DEA). Research in the 1950s identified several amines that could be useful for commercial applications in gas treating but the development of these amines in industrial processes did not take place until the 1960s and 1970s. Development of hindered amines for highly selective treating, removal of H_2S in preference to CO_2, took place in the late 1970s and 1980s; work with blended amines and formulated amine solutions also started in the 1980s. Formulated amines are generally based on methyldiethanolamine (MDEA) solutions with additives that modify the performance of MDEA. These formulated amines offer significant performance improvements over single, or blended generic amines and have a significant position in the gas treating market today.

Chemistry

The structure and chemical formula for the predominant generic amines use in gas treating are shown in Table 1. These amines are generally denoted as primary, secondary or tertiary, depending on the number of hydrogen substituents bonded to the active nitrogen. Alkalinity of ethanol amines generally decreases from primary to tertiary amine with acid gas solution energies following the same trend. (Table 2)

Table 1

Predominant Amines Used in Gas Treating Applications

Primary Amines	Formulas	Structures
MEA Monoethanolamine	$H_2NCH_2CH_2OH$	
DGA Diglycolamine	$H_2NCH_2CH_2OCH_2CH_2OH$	
AMP 2amino 2 methyl 1 propanol	$HOCH_2C[CH_3]_2NH_2$	
Secondary Amines		
DEA Diethanolamine	$HN[CH_2CH_2OH]_2$	
DIPA Diisopropanolamine	$[CH_3CH(OH)CH_2]_2NH$	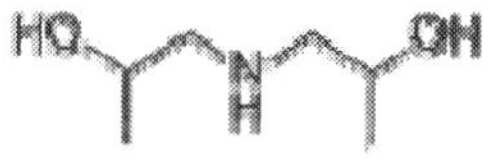
MMEA Methylethanolamine	$HNCH_3CH_2CH_2OH$	
Tertiary Amines		
MDEA Methyldiethanolamine	$CH_3N[CH_2CH_2OH]_2$	
TEA Triethanolamine	$N[CH_2CH_2OH]_3$	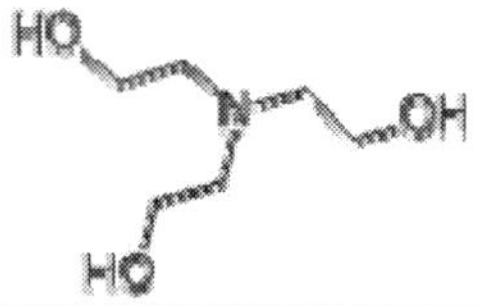

Table 2

Approximate Heat of Solution for Selected Amines

Amine	CO_2 (Btu/lb)	H_2S (Btu/lb)
MEA	825	615
DGA	820	675
DEA	700	510
MDEA	575	520
TEA	465	430

Amines react with H_2S according to the equation R-1. The slightly basic amine accepts a proton from the mildly acidic H_2S after the H_2S is absorbed into the aqueous solution. This is a very fast proton transfer reaction.

R-1: $HN[CH_2CH_2OH]_2 + H_2S <-> H_2^+N[CH_2CH_2OH]_2 + HS^- + Heat$

Primary and secondary amines react with CO_2 according to the equation R-2. CO_2 is acidic in aqueous solution, but does not have a proton to donate when reacting with amines. Primary and secondary amines have labile protons on the nitrogen that are available to react with CO_2 and form amine carbamates.

R-2: $HN[CH_2CH_2OH]_2 + CO_2 <-> CO_2^-N[CH_2CH_2OH]_2 + H^+ + Heat$

Some of the CO_2 will react with water to form carbonic acid (R-3); this is a slow reaction relative to the formation of amine carbamates and much slower than the reaction of H_2S with an amine. Tertiary amines such as MDEA do not have a labile proton so cannot react directly with CO_2. A proton donor is needed for the MDEA/CO_2 reaction to take place. In an aqueous mixture, CO_2 will react with water (proton donor) to form carbonic acid, which will then react with a tertiary amine (R-4). Since this is a slow reaction relative to the H_2S reaction, MDEA can be used to selectively remove H_2S in gas containing both acidic components. MDEA has been mixed, or formulated, with other amines and additives to enhance MDEA/CO_2 reactions kinetics. These additives, or reaction accelerators, increase the rate of CO_2 reaction with MDEA making formulated MDEA products competitive in removing CO_2 from gas streams. See R-5 (activator facilitates CO_2 reaction with MDEA and water)

R-3: $CO_2 + H_2O <-> HCO_3^- + H^+$ (slow)

R-4: $HCO_3^- + H^+ + CH_3N[CH_2CH_2OH]_2$ <-> $H^+CH_3N[CH_2CH_2OH]_2 + HCO_3^-$ + Heat

R-5: $CO_2 + CH_3N[CH_2CH_2OH]_2 + H_2O$ <-Activator-> HCO_3^- + <-> $H^+CH_3N[CH_2CH_2OH]_2$ + Heat

Another group of amines called sterically hindered amines has been developed to enhance H_2S selective removal. (Figure 1) These amines have bulky substituents that are one or two carbons removed from the central nitrogen which hinder the approach of CO_2 to the nitrogen, thus slowing the reaction between the nitrogen and CO_2. The heat of solution for H_2S is typically higher than with MDEA allowing lower concentrations of H_2S to be achieved for similar operating conditions but requiring more energy to remove H_2S from the solution.

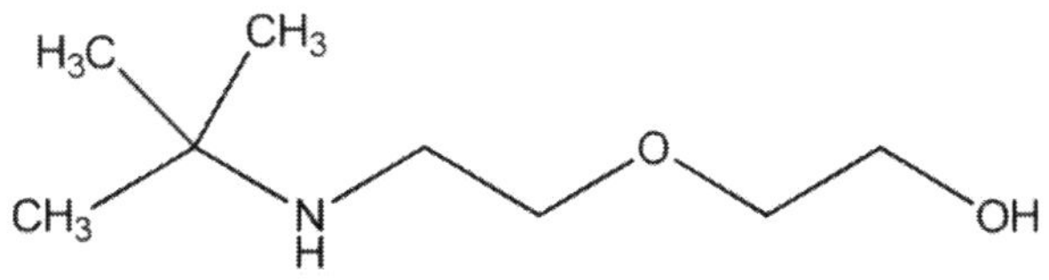

Sterically Hindered Amine
Figure 1

The different amines described have historically been used to remove H_2S and CO_2 from a number of different gas streams. These streams include coke oven vent gas, natural gas from gas fields, co-produced gas from crude oil production, various refinery streams, synthesis gas from a number of processes and purification of gases in chemical and iron production plants.

Each of these streams has characteristic components some of which are reactive to varying degree with the amines, with CO_2 or in basic environment (pH>8). These reactions can have a significant impact on the performance of the solvent and equipment in the plant. Oxygen reacts with amines to form organic acids (formic & acetic) and other amines, usually referred to as amine degradation products. Carbon monoxide reacts to form formic acid. Aldehydes undergo base-catalyzed reactions and can form a polymer residue in the system.

Organic acids (formates, acetates, oxalates, etc.) will react with amines to form thermally stable salts resulting in loss of solution scrubbing capacity. Some of these thermally stable amine salts can be recovered by addition of a stronger base to the solution; however this will increase the ionic strength of the solution and can eventually result in precipitation of salts on heat transfer surfaces as well as cause corrosion if not remediated.

The reaction of carbon dioxide with some primary and secondary aliphatic alkanolamines does not stop with the formation of an amine carbamate. Mechanisms have been identified that show that carbamates react further with amines to form diamines and other by-products, some of which have been identified as good iron chelants. These chelants can make an amine solution corrosive, particularly as the amine and CO_2 concentrations increase. The nature of some solutions to become corrosive has been identified; field experience has led to the understanding that amine concentrations should be limited. Table 3 shows the normal maximum concentration of amines commonly used for gas treating and the maximum acid gas concentration typically used for design. Tertiary amines and some of the proprietary formulated gas treating products based on MDEA do not form amine carbamates that react to form the corrosive CO_2/amine by-products. MDEA and the current generation of formulated MDEA treating products have demonstrated significant improvements in chemical stability with CO_2 vs. earlier generations of amines.

Table 3
Typical Amine Operating Capacity Comparison

Amine	Operating Concentration (wt%)	Operating Concentration (mol %)	Loading Range (high - low)	Relative Capacity* MEA = 1
MEA	18%	6.1%	0.35	1.00
DGA	50%	14.6%	0.45	3.09
DEA	28%	6.3%	0.48	1.41
DIPA	45%	10.0%	0.485	2.28
MDEA	50%	13.12%	0.49	3.02
Formulated MDEA	50%	13.88%	0.485	3.16**

* Based on molarity and differences in typical loading values
** Removal of CO2 to cryogenic specification

Solvent selection and process design

The basis of solvent selection, amine plant design and process layout are the feed gas conditions, the amount of acid gas to be removed, and treated gas specifications. The residual CO_2 (and/or H_2S) in the treated gas that is achievable is described by the characteristic solution VLE (vapor-liquid equilibrium) (Figure 2). Low temperatures and high pressure favor removal to lower concentrations. Higher temperatures and lower pressures result in higher acid gas concentrations and the need for stronger solvents (higher heat of solution).

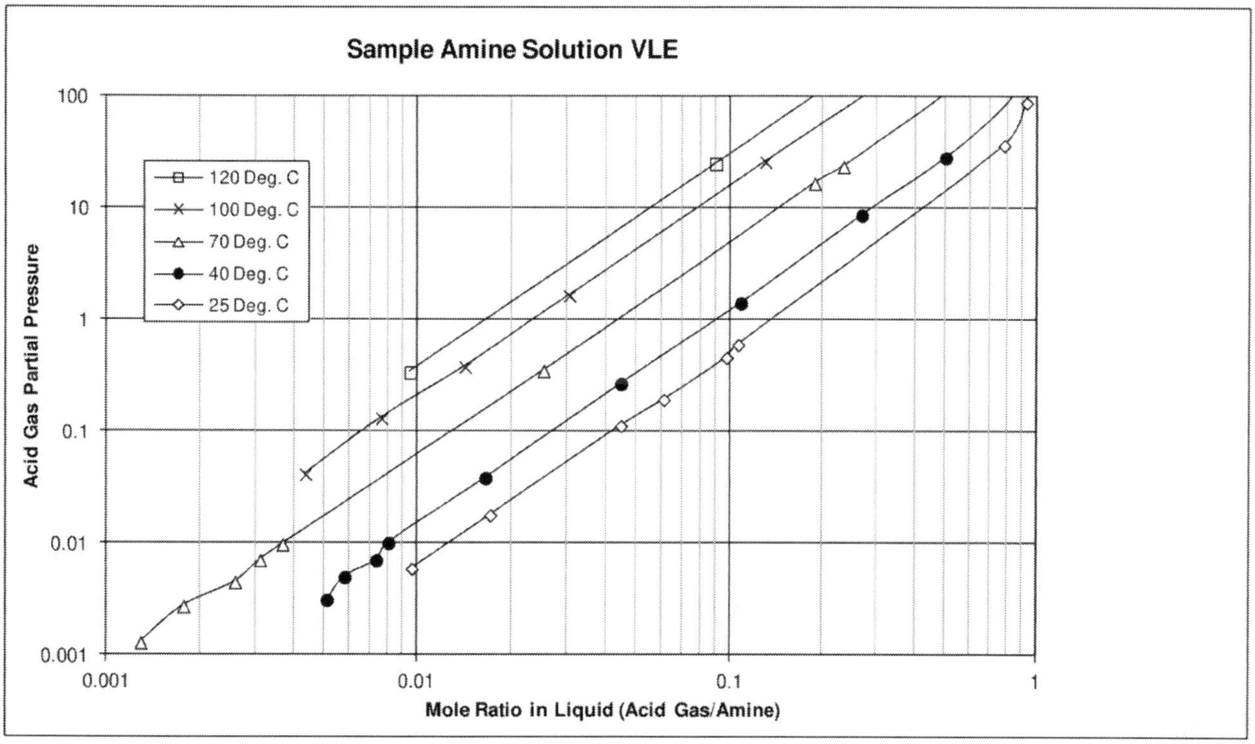

Figure 2

Stronger solvents, such as primary amines and certain formulated amines, may be needed to achieve low treated gas specifications when temperatures are high and/or pressures are low. A drawback to using a stronger solvent for acid gas absorption is a high heat of solution. The high heat of solution may cause a large temperature bulge in the absorber that can result in reaching the VLE limit; once the VLE limit is reached, absorption of acid gas stops. A higher solvent flow rate is then needed to control the temperature in the absorber. High heats of solution also require more energy to regenerate the solvent. Some formulated amine solvents have been designed to improve energy efficiency when compared to single amine solvents. The heat of solution for these products has a stronger function of acid gas concentration vs. generic amines so are more energy efficient. The circulation rate and the energy needed for regeneration for a given treating application are typically much lower with a formulated MDEA solvent than with a single solvent generic amine.

Processes

Acid gases are removed from a gas stream by contacting the gas with an amine solution in an absorber tower. Figure 3 shows a typical amine plant. Plants treating high pressure gas will have co-absorbed gas components in the rich (acid-gas loaded) solvent exiting the bottom of the absorber. Solubility of the gas components follows Henry's Law behavior. The majority of the co-absorbed components are separated from the amine solution in an intermediate pressure flash. The flashed gas is either recycled back to the high pressure absorber or treated in a low pressure absorber and used elsewhere in the process.

The rich solvent from the flash drum is heated by cross exchanging with hot lean solvent from the regenerator, then fed to the top of the regeneration tower. The acid gases are removed from the rich amine solution by steam stripping in this tower. The regenerated solution is recycled back to the absorber tower to treat more gas. Stripped lean solvent exits the regenerator at the bottom of the tower and is returned to the absorber through the cross exchanger, a cooler and filters. The acid gas exiting the regenerator top is cooled; the condensed water is recovered and returned to the system.

Absorber

Amine systems are designed around the performance of the absorber tower. The absorber tower provides intimate contact between the solvent and gas to absorb the acid gas. Absorber towers are staged counter current contactors.

Early designers used bubble cap trays or Pall rings for tower internals. As tower internals technology advanced, new internals were adapted for amine absorber use. Cross flow valve trays seem to be the most predominant tray in use for current designs. Valve trays are less expensive than bubble cap trays and provide a wider operating range than sieve trays which have limited turndown. Tray design for amine systems is described adequately by Kohl & Nielson and Kister.

Tower internals design equations consistently predicting tower flood points below real values. An empirical value called the "system factor" was developed to adjust design equations to accurately predict flooding limits. They are based on experience and are somewhat vague in their connection to particular amine solvents and to the gas being treated. System factors adjust the system capacity to allow for additional separation time needed between the gas and liquid for different solvents and gas with varying compositions. A value of 1.0 would indicate that no adjustment to the design is needed for gas/liquid separation. Generally more time for separation is needed as the amine molecule increases in size and the hydrocarbon molecule increases in size. Aromatics in particular have been identified as a cause for increased froth in amine units. Amines are considered heavy foamers and typically use tray spacing of 24 inches. Literature values for system factors vary from 0.8 - 0.73 and seem to be based on experience with DEA. Experience with MDEA and formulated solvents indicates that lean gas and syngas streams can be effectively designed with a system factor of 0.8 while gas containing aromatics (benzene, toluene, ethyl benzene, xylene) should use a value of 0.68 to 0.7.

High capacity trays have been used in amine service. These trays usually have a higher velocity gas flow vs. standard cross flow trays. The higher velocity necessitates increased tray spacing to allow for gas-liquid separation. The tower diameter is reduced but the overall height is increased usually by about 25% (24" vs. 30" tray spacing).

Packing is used in absorber towers on occasion. Trayed absorbers are sometimes retrofitted with packing to increase gas processing capacity. Packing (2" particle size) has more flow capacity than trays for the same diameter tower. Valve trays (2.5" weirs) with 24" tray spacing have approximately the same mass transfer efficiency as 2" packing with the same overall tower height. In addition to higher capacity, packing has a lower pressure drop than trays for

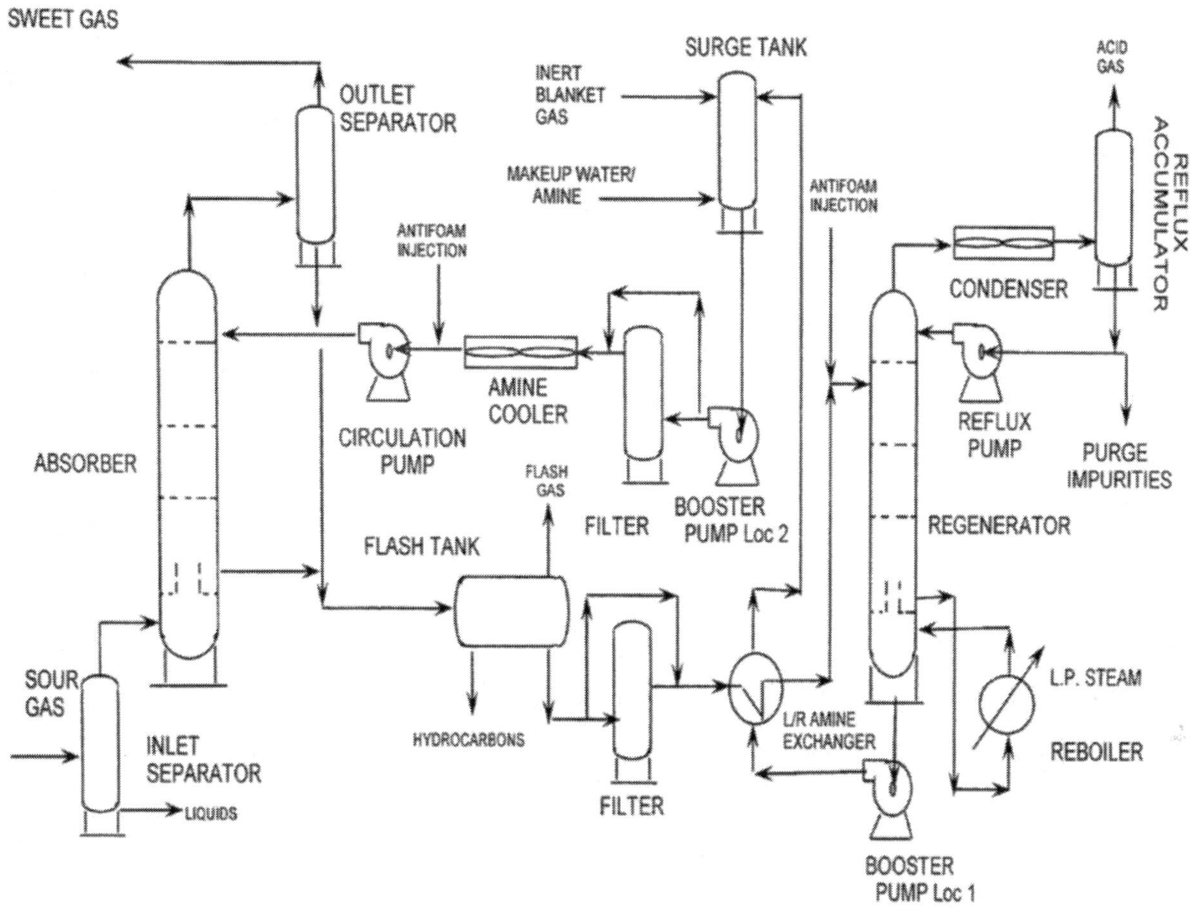

Typical Amine Treating Plant
Figure 3

the same scrubbing efficiency. One drawback of packing is that particulate contamination can foul packing, resulting in maldistribution and loss of efficiency. Gas rate turndown is nearly 100% in a packed tower. The liquid turndown however, is limited by the lower operating limit of the liquid distributor. This combination results in a very high liquid rate relative to gas rate.

Structured packing is rarely used in land-based amine systems, but is virtually standard for systems in swaying service such as offshore platforms and FPSOs (Floating Production Storage and Offloading). The corrugated metal packing limits lateral mixing and maldistribution caused by vessel sway. To compensate for any potential maldistribution, amine flow is generally increased by some factor relative to the amount of movement that will be encountered. Structured packing is reported as having higher capacity than random packing for the same diameter tower, however the cost of structured packing is also reported as being significantly higher than either random packing or trays.

Mention should be made about selective treating and packing vs. trays. The reaction of H_2S with amines is a simple proton transfer and is very fast. This reaction takes place in the liquid film just after absorption into the liquid phase. The CO_2 hydrolysis reaction takes place in the bulk liquid phase, so CO_2 must first be absorbed and travel through liquid film to the bulk phase. Packing holds up less liquid than trays and therefore has a lower concentration gradient and a lower mass transfer driving force for CO_2 when compared to trays. This will result in less CO_2 removal in a packed tower for the same H_2S removal.

The initial absorber design calculations begin by fixing the lean solvent temperature. Temperature constraints are based on the vapor-liquid equilibrium of the system and the need to keep the lowest temperature in the absorber above the hydrocarbon dew point of the feed gas after all the acid gas has been removed. Environmental constraints such as air temperature or lack of other cooling may limit the temperature. This may force the use of solvents that are less energy efficient but which result in low CO_2 and H_2S treated gas concentration. The lean flow is then varied to evaluate tower performance based on:

- achieving treated gas specification
- staying below the maximum rich loading, and
- achieving a tower temperature profile that will allow for variations (gas flow, acid gas content) in operations while meeting the required specifications (Figure 4).

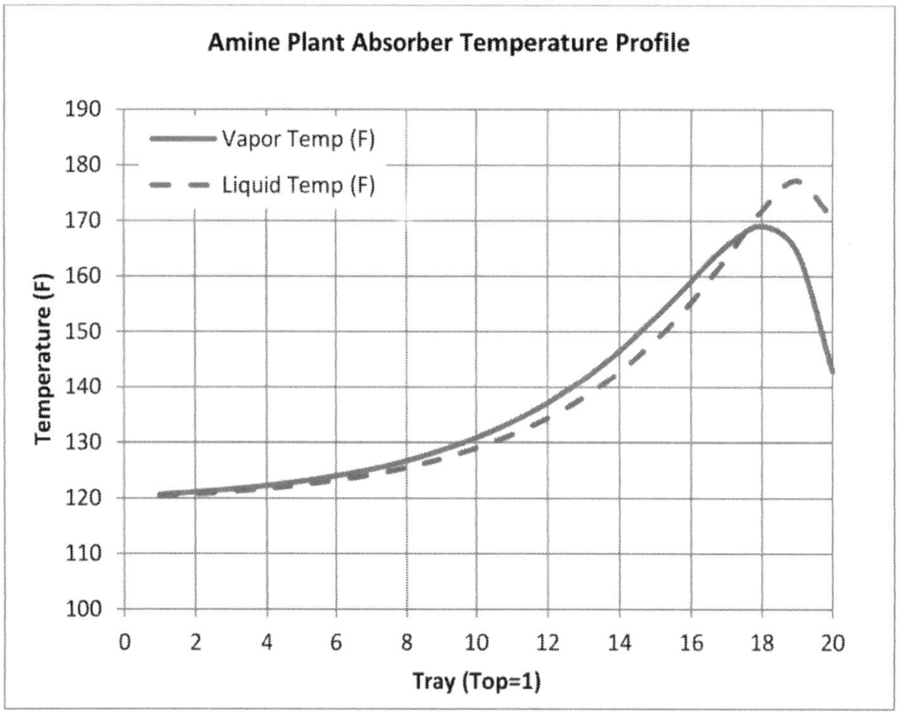

Figure 4

Different solvents may be evaluated at the same conditions to minimize plant size and cost. Evaluation of the application using specialty solvents is usually done by solvent suppliers who have experience with their proprietary models and associated physical property packages.

Tower staging is usually evaluated against the solvent circulation rate. The optimum number of trays is determined by balancing the circulation rate against the tower height and the related costs. Lower amine circulation rates are preferred but may be limited by a tower VLE pinch or maximum loading constraints. A 20 tray absorber tower is considered typical and is normally 5 to 6 theoretical stages. Absorbers removing CO_2 to less than 100 ppmv will normally be built with 24 to 30 trays depending on the processing conditions. Two ppmv CO_2 has been achieved in amine systems with specialty solvents.

Rich Amine Flash Drum

Absorber towers operate at delivered gas pressure which can vary from 5 to 1200 psig. As the pressure increases acid gas removal becomes easier, but the amount of co-absorbed gas increases. The co-absorbed gas is recovered in the flash tank where the pressure is reduced, typically to 60 - 120 psig. If the flash gas has sufficient hydrocarbon content, it can be used as fuel gas. If the flash gas contains mostly inerts, it can be vented or flared depending on permit allowances. If there is no possibility of entraining heavy hydrocarbons (C4+) in the rich amine, the flash tank can be designed as a two-phase separator. For most systems (especially refineries), the vessel is designed as a three-phase separator with the ability to skim or separate liquid hydrocarbons from the aqueous amine solution. A 20 - 30 minute residence time is recommended to affect the liquid separation. A mesh coalescing pad is optional. Figure 5 is one example of a three-phase separator.

THE GRAPHIC ON THIS PAGE WAS NOT AVAILABLE ON THE E-MEDIA VERSION OF THIS PAPER.

Figure 5
Example of a Rich Amine Flash Vessel with Three Phase Separation

Lean/Rich Heat Exchanger
Flashed rich amine flows from the flash tank to the cross exchanger where it is heated prior to regeneration. Some amount of flashing will take place in the cross exchanger. The flashing is controlled by locating the flash tank level control valve downstream of the heat exchanger to maintain back pressure and by limiting the rich exit temperature. The hot side exit temperature is usually at the same temperature as the regenerator overhead temperature unless this results in excessive flashing inside the heat exchanger.

Regenerator
Solvent regeneration is accomplished with steam stripping. Hot rich solvent is fed to the top of a tower and cascaded to the bottom while being countercurrently contacted with steam

rising from the bottom of the tower. Steam is generated in a reboiler by heating solvent with steam, hot oil or direct fire. The most common reboiler configurations are single pass heaters or kettles. True thermosiphons are not as common in gas treating as the other configurations because of turn down limitations.

The energy needed for regeneration must be enough to heat the solvent to the reboiler temperature, heat and vaporize the reflux water and reverse the acid gas/amine reaction. Enough steam must be generated to remove acid gas to the lean loading needed to achieve the treated gas specification. The steam is both an energy source and a carrier gas in the regenerator.

A typical amine plant design has 20 stripping trays. Where higher lean loadings are sufficient or regeneration energy is not a factor, fewer regenerator trays have been used. Older amine systems designed with MEA or DIPA have as few as 13 trays. Twenty trays (or the packed equivalent) have been found to be the most cost efficient. If the number of trays is reduced to less than 18, more steam is usually required to compensate for loss of staging. Increasing the staging will increase stripping efficiency but the rate of improvement decreases steadily with more trays, which is typical as the pinch point is approached. The justification for more than 18 trays seems to be the capacity to absorb variations in the rich feed without changing the stripper reboiler rate.

The operating pressure of the regenerator is usually only as high as needed for downstream processing requirements. CO_2 only systems historically have vented the overhead to the atmosphere and will operate from 8 to 10 psig at the reflux accumulator. Lower operating pressures are possible but at lower pressures the regenerator temperature can fall below the minimum effective value needed to reverse the heat of solution.

The maximum operating pressure is limited by the thermal stability of the amine. The maximum bulk temperature in the reboiler is reported to be 260 - 270 ^{o}F depending on the amine. The heating media temperature is usually calculated by limiting the heat flux into the reboiler. Various references suggest limits of flux. Normal reboiler temperatures vary from 240 – 260°F.

Filtration

Feed gas conditioning will remove entrained particulates and entrained liquids and improve operational stability. Gas liquid separators are typically used to coalesce entrained liquids to a form where they can be separated from the gas. Feed gas separation equipment usually consists of a slug catcher followed by a high efficiency coalescing separator. Current designs for amine systems include this capability.

Amine absorbers are very effective scrubbers. Particulates and entrained liquids are removed from the gas and accumulate in the amine solution. Iron sulfide and iron carbonate formed from acid gases reacting with carbon steel equipment will erode from vessel surfaces and form particulates in solution. Make-up water can also be a source of contaminants. Particulate and carbon filtration are used to control these contaminants. Normally a 10% to 20% slip stream of lean solvent is filtered through a particulate, carbon bed, and a second particulate filter in series.

Process Configurations:
The majority of amine plants have a flow scheme show in Figure 6. This is a low capital cost plant with high flexibility for applications. Variations of the flow scheme have been used when the treated gas spec and processing conditions enable an economic advantage.

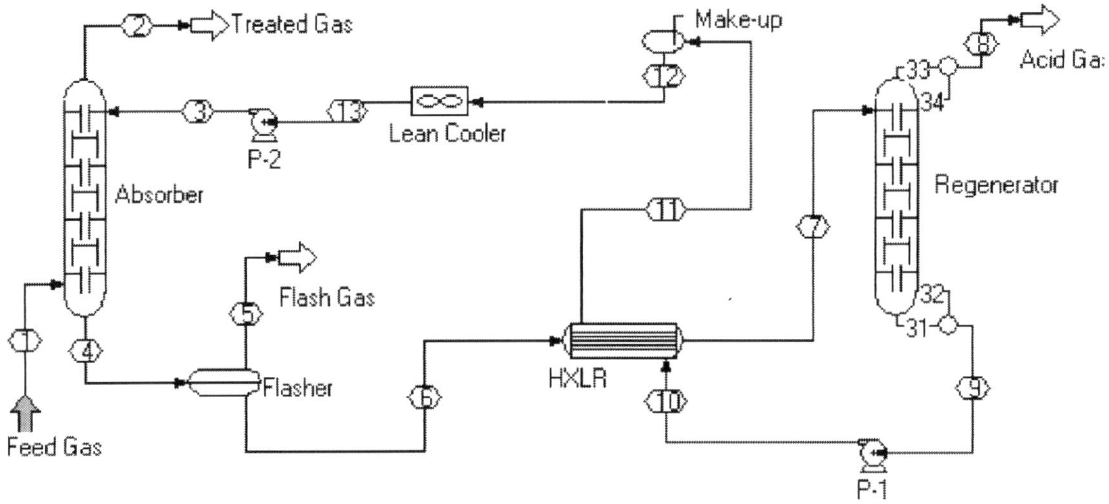

Typical Amine Plant

Figure 6

Flash regeneration

Gas with a CO_2 specification of more than one atmosphere partial pressure can be treated with a solvent that has been flash regenerated. (Figure 7) This configuration is more often used when the feed gas has high CO_2 partial pressure and the rich solvent can be loaded to a high level. Examples are co-produced gas from enhanced oil recovery facilities (70 to 90% CO_2 gas at >300 psig) or natural gas with 10 % or more CO_2 treated at 1000 psig or higher pressure. The solvent circulation rate is typically 3 to 6 times higher than a standard amine plant but the regeneration energy is usually zero. Absorbers for these facilities usually have more staging than standard amine plants (30 or more trays) to get the maximum utility from the solvent.

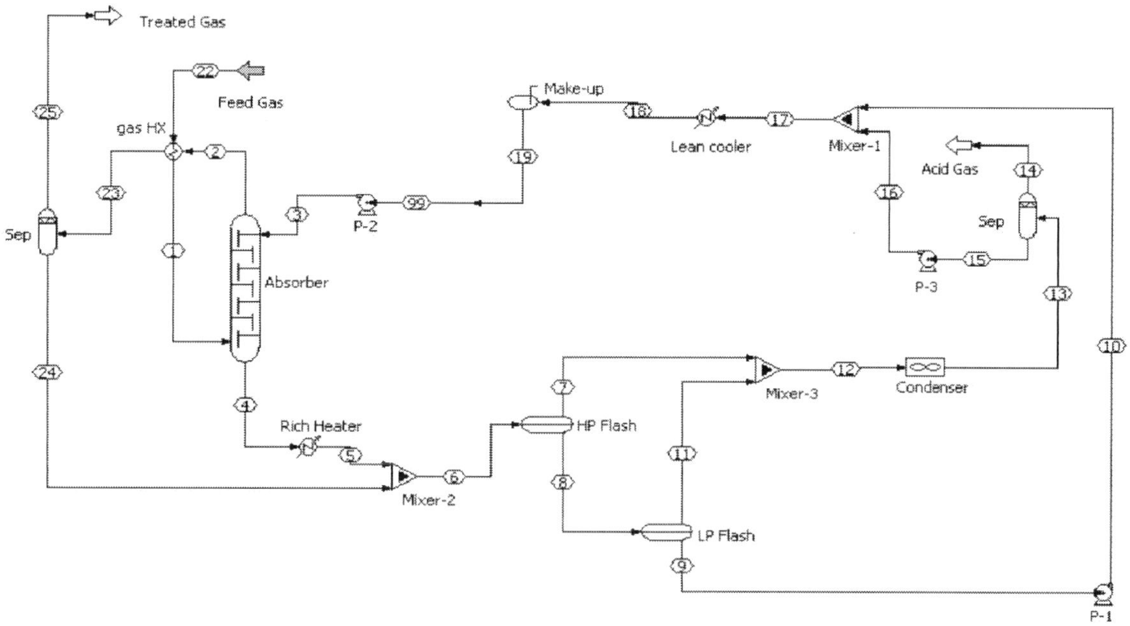

Figure 7 Flash Regeneration Flow Scheme

Split flow

Applications that need a low treated gas specification (<50 ppm CO_2; <4 ppm H_2S) require solvent regeneration by stripping. Gas with a high acid gas partial pressure may be more economically treated by employing flash regeneration, or partial regeneration, to remove some of the CO_2/H_2S and using a stripped solvent to remove the remaining acid gas. These processes can be combined into one plant with two basic configurations. Figure 8 shows a split flow process where partial regeneration is accomplished in the upper part of a regenerator tower, then the partially regenerated amine is used for bulk removal of acid gas in the bottom of the absorber tower. Figure 9 shows a flow scheme where partial regeneration is accomplished with a low pressure flash, then part of that stream is regenerated by steam stripping. The overhead from the stripper is used to heat the low pressure flash to improve CO_2 removal. The economics of the processes depend on the feed gas partial pressure and the properties of the solvent. The higher the feed gas CO_2 content, the more favorable the flash regeneration partial regeneration process becomes. The driving force for these processes is reducing the energy to remove the CO_2.

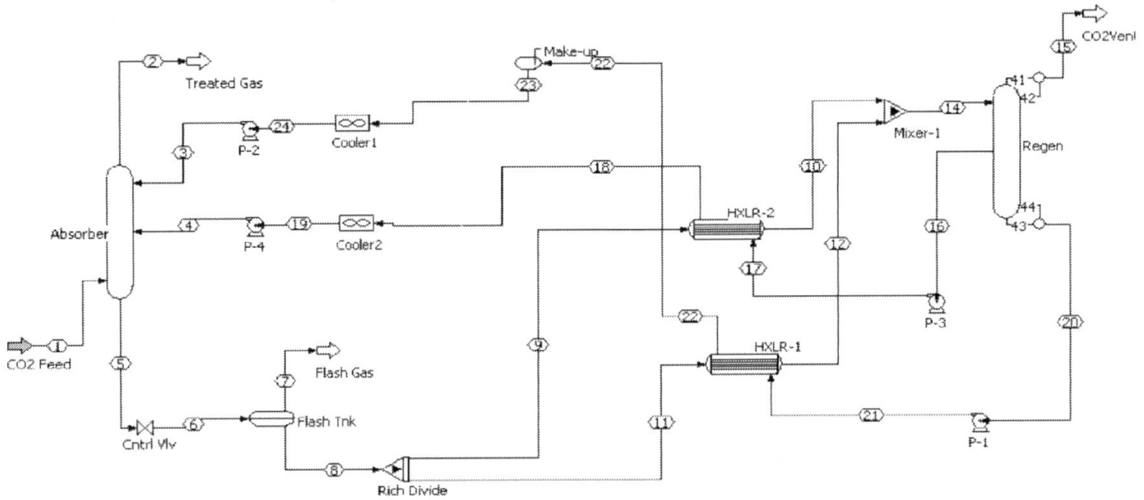

Split Flow Process; Regenerator with Side Draw
Figure 8

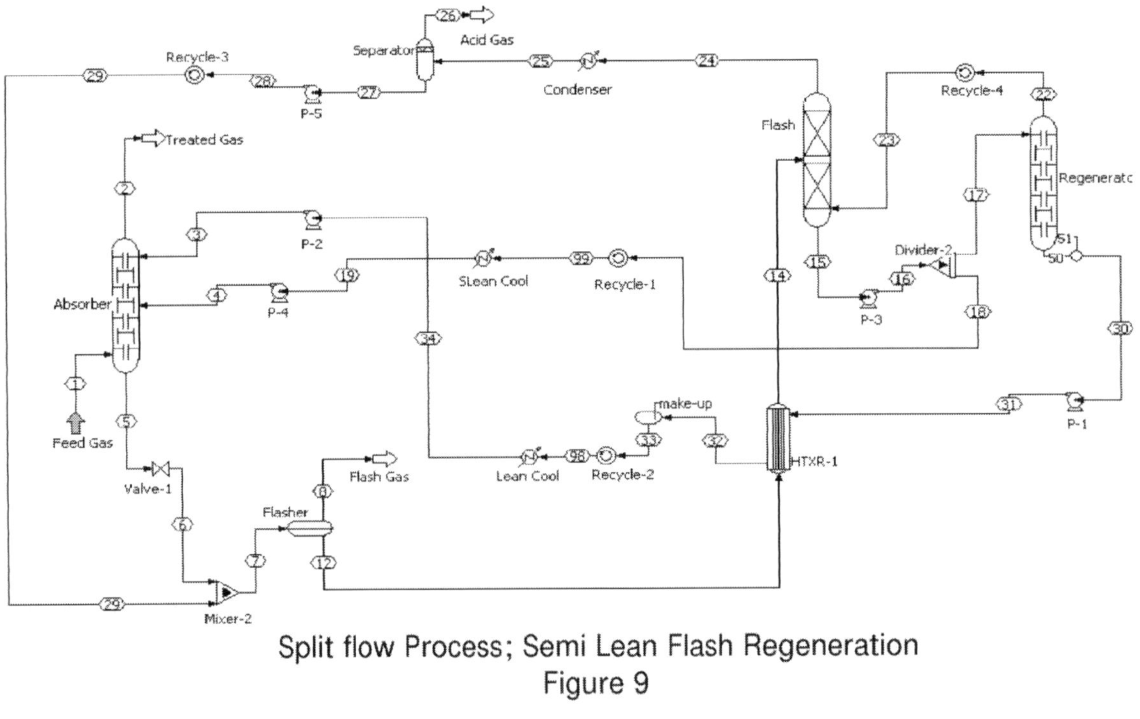

Split flow Process; Semi Lean Flash Regeneration
Figure 9

Process variations

There have been many process variations proposed that offer advantages in some treating applications which are not economical in others. Increasing the complexity of a process increases capital cost and operating complexity which is sometimes difficult to quantify in the early stages of design. Some variations include: pretreating gas with co-current contactors, absorbers with intercoolers or pump arounds, heated flash regeneration, multi-pressure stripper configurations for high pressure regeneration, regenerator pump arounds, multiple feeds to regenerators.

Process simulation

Amine process models are more complicated than typical equilibrium based distillation tools. Several companies provide rate-based process simulators specifically designed to model amine gas treating. Some of the providers of specialty formulated amine solvent have proprietary simulators designed around their solvents which give better performance predictions for their solvents. Assessing relative performance of these models is a complex task due to the variations in tower internals, differing engineering techniques, veracity of the physical property data used in the model, and varying thermodynamic properties which can be used.

Summary

This paper is a brief review of gas treating technology. What initially looks like a simple, straight forward process becomes more complex when one considers; variations in process configuration, materials of construction and corrosion, fluid dynamics and environmental issues. Recent studies of amine technology in carbon capture applications have resulted in the proposal of complex process flow schemes that are reported to operate more efficiently than current operating plants. Patent applications indicate that improvements in chemical technology may be forthcoming. The attractiveness of natural gas as an energy source promises to drive th continued development of gas treating technology.

Bibliography

Bottoms, RR, "Removing Acidic Gases by Reaction with Organic Bases", Chemical and Metallurgical Engineering, August 1931 Vol. 38 No.8, p 465

Dingman, JC, Allen, Moore, "Minimize corrosion in MEA Sweetening Units", LRGCC Conference Proceedings – 1966

GPSA Engineering Data Book, Eleventh Edition (1998), Gas Processors Suppliers Association

Hatcher, N, Keller, A, "Uncertainties In Modeling Acid Gas VLE In Amine & Sour Water Systems", LRGCC Conference Proceedings – 2005

Kister, HZ, Distillation Design, 1992 McGraw-Hill p 259-295

Kohl, A, Nielsen, R, Gas Purification Fifth Edition, 1997 Gulf Publishing, p 40-99

Lawal, Ayanduntan O.; Idem, Raphael O, "Kinetics of the Oxidative Degradation of CO_2 Loaded and Concentrated Aqueous MEA-MDEA Blends during CO_2 Absorption from Flue Gas Streams" Industrial & Engineering Chemistry Research (2006), 45(8), 2601-2607.

Lepaumier, Helene; Picq, Dominique; Carrette, Pierre-Louis., "New Amines for CO_2 Capture. II. Oxidative Degradation Mechanisms". Industrial & Engineering Chemistry Research (2009), 48(20), 9068-9075.

Kennard, ML and Melsen, A, "Mechanisms and Kinetics of Diethanolamine Degradation 1985, Industrial & Engineering Chemistry Fundamentals", vol 24 issue 2, p 129-140

Mathews, J, Arendell, S, Johnson, J "A practical Approach to Amine system Optimization", LRGCC Conference Proceedings – 2010

Reed, RM, Updegraff, NC, "Removal of Hydrogen sulfide from Industrial Gases", Industrial and Engineering Chemistry, November 1950 p2269

Thomas, JC, Bradley, A, "Operation Experience with Accelerated MDEA Solvents in CO_2 Removal Applications", LRGCC Conference Proceedings – 2008

Vickery, DJ, Campbell, SW, Weiland, RH, "Gas Treating with Promoted Amines", LRGCC Conference Proceedings – 1988

(96b) Adsorbents Used in Petrochemical Production: Safe Handling Recommendations

David Radtke, UOP, LLC, Houston, TX

The petrochemical industry has grown to rely heavily upon adsorbents to remove contaminants from hydrocarbon streams. Molecular sieves, activated aluminas, metal oxides and other adsorbents are very successful in removing contaminants down to extremely low levels protecting not only equipment, but improving the quality of the chemical products produced, and improving yield and effective life of increasingly sensitive polymerization and other process catalysts. Past presenters have discussed adsorbents in relation to the contaminant removed in a particular application. This presentation will take that a step further to increase basic understanding of the properties and nature of these materials and what this means to proper handling of both fresh and spent adsorbents. In some cases, fresh adsorbent needs to be protected from the moisture in the air. Frequently, spent adsorbents need to be prepared properly before unloading, and then handled carefully. The precautions taken will vary depending upon the class of materials. This presentation will cover some basic recommendations for the handling of key adsorbent materials used in petrochemical plants.

AIChE Paper Number

Comprehensive Quench Water And Steam Generator Management For Optimized Unit Performance

Michel Didden

Senior Tech Support, Boiler Product Applications

GE Water & Process Technologies

Richard Krichten

Group Manager, Boiler Product Applications

GE Water & Process Technologies

Patrick Lucas

Global CPI Product & Technical Leader

GE Water & Process Technologies

Prepared for Presentation at the 2012 Spring Meeting

Houston, TX (April 1-5, 2012)

Comprehensive Quench Water And Steam Generator Management For Optimized Unit Performance

Michel Didden

Senior Tech Support, Boiler Product Applications

GE Water & Process Technologies

Richard Krichten

Group Manager, Boiler Product Applications

GE Water & Process Technologies

Patrick Lucas

Global CPI Product & Technical Leader

GE Water & Process Technologies

Abstract: In any ethylene plant, the dilution steam generator system remains the most critical component, regardless of feedstock being processed. The recovery of water and the production of good quality dilution steam can be challenging processes. The quench water and dilution steam generation systems are closely linked and adjusting specific operational procedures and chemical programs can help optimize overall unit performance. This paper will review:

- fundamental goals for all sections of the dilution steam generation system
- potential problems such as corrosion, emulsification, fouling, and salt entrainment that can reduce steam production and affect plant efficiency and profitability
- problem identification and what to look at for a good diagnosis
- operational and chemical solutions

1.) Process Overview

By dilution steam generation system, we mean an almost closed loop including the quench water tower, oil/water separator, dissolved oil extraction unit, coalescer, process water stripper, dilution steam generator and pyrolysis furnace at the back end. The main purpose of the quench water system is to cool down the cracked gas coming from the pyrolysis furnace to between ~95 - 105°F (~35 - 40°C) prior to the cracked gas compression section. As a result of cooling the cracked gas, virtually all of the dilution steam from the furnaces is condensed in the quench system. Small amounts of hydrocarbon and tar are also condensed into the quench water (~175°F / 80°C). After the quench water tower, the bottoms, a combination of steam condensate and hydrocarbon, flows to a water/oil separator where hydrocarbons are removed.

Once separated, the hot quench water is used for various low level heating services (e.g., propylene splitter reboiler) throughout the plant, while the hydrocarbon (raw py-gas) is routed back as reflux to the oil quench tower, upstream of the quench water tower. The quench water is further cooled in quench water/cooling water exchangers and returned to the top and middle sections of the quench water tower.

A portion of the recirculating quench water is continuously sent to the dilution steam generator to produce dilution steam for the cracking furnace. In most plants, the quench water blowdown will then go through a coalescer or a dissolved oil extraction (DOX) system to further remove small amounts of emulsified, insoluble hydrocarbons. The water is then stripped in a process water stripper to remove dissolved gases and soluble hydrocarbons, such as benzene, styrene, and toluene. The hydrocarbons are refluxed back to the quench water tower and the stripped process water condensate (bottoms) becomes boiler feedwater for the dilution steam generator.

The dilution steam generator system consists of a steam separating drum and a number of heat exchangers (steam generators) with the boiler feedwater on the tube side, and either hot quench oil and/or medium pressure steam on the shell side. The dilution steam generator generally has an operating pressure of 75 - 110 PSIG (5 – 7.5 bars). All of the steam produced in the dilution steam generator is sent directly back to the cracking pyrolysis furnaces. The dilution steam generator blowdown is cooled and sent to the waste water treatment plant.

In gas-based steam cracking (ethane, propane, butane), a closed, re-circulating quench water column is located directly downstream of the cracking furnaces and transfer line exchangers.

In liquid feedstock steam cracking (naphthas, hydrocarbon condensates, and atmospheric gas oils), an oil quench tower (primary fractionator) is located directly downstream of the transfer line exchangers. The fractionator cools the cracked gases to between 350 - 430°F (~175 – 220°C) and removes heavy quench oil and pyrolysis fuel oil from the cracked gas.

The cracked gases and steam leaving the oil quench tower can be further cooled by two means:

1. The overhead from the oil quench tower is sent to the closed, recirculating quench water tower. In the quench water tower, most of the steam is condensed along with a large quantity of raw pyrolysis gasoline. The gasoline is separated from the steam condensate in the quench water/oil separator and is continuously drawn off. Some of the gasoline is used as reflux and the rest is sent to fuel recovery. The quench water is processed similarly to the quench water blowdown discussed above.
2. A Fin-fan condensing system cools the gases and condenses all of the raw pyrolysis gasoline and steam. The fin-fan outlet is further cooled in cooling water service exchangers. An accumulator downstream of the cooling water exchanger separates the hydrocarbon from the steam condensate. Some of the hydrocarbon is sent back to the oil quench tower as reflux, while the rest goes to fuel recovery. The net condensed steam is reused as dilution steam generator feedwater.

2.) Areas of concern

Throughout the dilution steam generation system (refer to Figures 1 and 2 below), most ethylene plants experience corrosion and fouling in multiple locations, emulsification in the quench water oil/water separator, as well as foaming and steam quality issues.

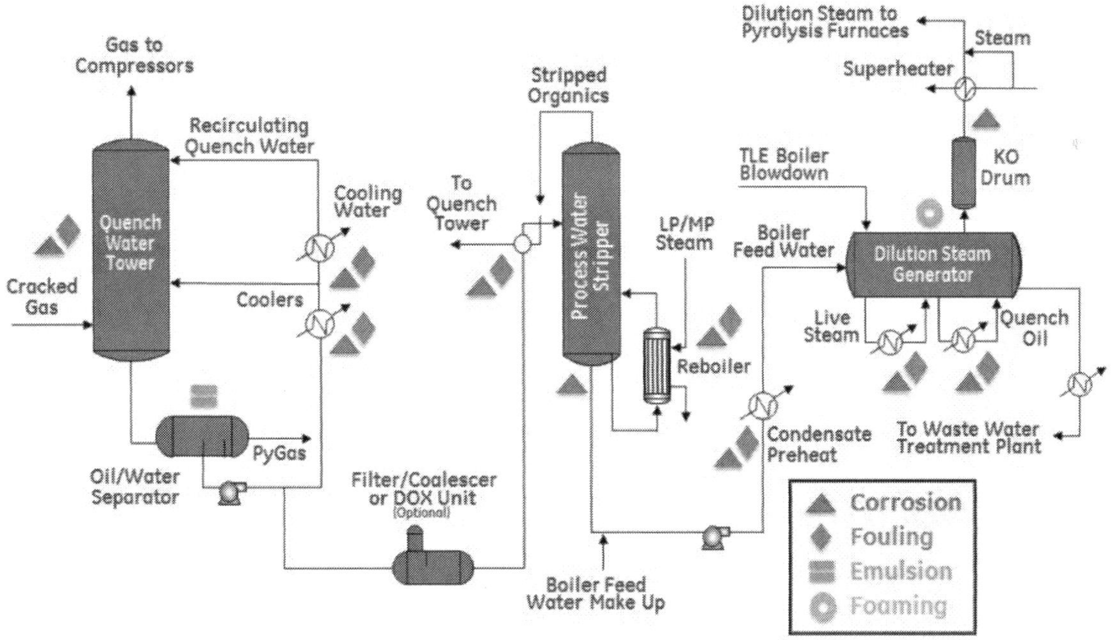

Figure 1: Dilution Steam Generation System Problem locations

Corrosion

Corrosion may occur in the quench water tower, but is mainly found in the process water stripper and dilution steam generator itself. Acids in the cracked gas can condense in the quench water causing low pH conditions. Typical acidic species formed during the cracking operation include organic acids (formic, acetic, propionic, etc,) and carbonic acid. Low pH steam generator feedwater will result in general corrosion in this system.

Dissolved oxygen (>20ppb) in the quench water can also contribute to pitting corrosion, which can result in tube failures at the process water stripper and the dilution steam generator.

Additionally, underdeposit corrosion, precipitated by fouling, is of major concern and will be discussed in more detail in subsequent sections of this paper.

Emulsification of water and hydrocarbon

The degree of organic fouling that occurs downstream of the quench water tower is directly related to how effectively quench water emulsification (water/hydrocarbon & grease) is controlled in the water/oil separator.

Factors that have a significant impact on emulsion stability in the quench water separator are residence time, and pH, as well as possible contaminants from recycle streams. The residence time for the water and hydrocarbon to separate in the separator drum is usually less than 10 minutes. Rarely is this enough settling time for good phase separation efficiency. Additionally, high pH (>8.5) can also promote or aggravate an emulsion problem.

Fouling

Fouling can be categorized as either inorganic, initiated primarily by corrosion products, or organic, resulting from entrained hydrocarbon, including tar and asphaltenes.

Excessive hydrocarbon in the recirculating water quench loop can cause fouling in the quench water column (especially packed columns) and coolers; in the process water stripper tower and reboiler; and in the dilution steam generator preheater, reboilers, and separation drum.

Generally speaking, organic fouling in the dilution steam generation system is not of great concern, as long as the level of organic components in the feed water is below 10 - 20 ppm. However, changes in feed quality can significantly impact the acid/base balance, leading to pH swings, which has a serious impact on water/hydrocarbon separation. In any case, organic olefin compounds will easily polymerize, and therefore organic fouling may occur even at low levels of contamination.

The organic hydrocarbons will polymerize or heat degrade, forming insulating deposits on heat transfer equipment surfaces throughout the dilution steam generation system. When the fouling becomes severe enough to significantly limit heat transfer, the steam generator (reboiler) tubes must be mechanically cleaned.

Fouling of the circuit heat exchanger results in very significant economic losses, and as a result, the production of dilution steam decreases and it becomes necessary to take more expensive steam from other sources. Generally speaking, every 1 mt/h of steam that is not internally produced, results in an additional cost of approximately $30,000/year. In many cases, the high fouling potential often necessitates the plant to maintain a high blowdown in the dilution steam generator.

Fouling also results in underdeposit corrosion of generator tubes. The corrosion is typically very aggressive, with localized, gouging-type attack.

Foaming

The presence of organic materials in the dilution steam generator, as well as possible high levels of dissolved salts can result in foaming problems, increasing the carry-over with the steam. If caustic is added to control feed water pH, it may result in sodium carry-over to the furnaces, thereby contributing to coke formation and reduced coil life.

3.) Overall Treatment Program Goals

Typically, producers expect to achieve 8-10 blowdown cycles at the dilution steam generator, before fouling diminishes heat transfer efficiency enough to require a shutdown for cleaning. Therefore, the primary goal of any treatment program is to help plant operators achieve the longest run possible, while keeping the overall system reliable and efficient. Meeting this goal requires maintaining a balance of many complex, interrelated process and operational variables. These include adequate pH control, efficient water / hydrocarbon separation, maximum hydrocarbon removal at the process water stripper, minimized fouling and corrosion throughout, as well as optimum water and dilution steam quality.

As previously mentioned, proper pH control in the quench water tower, process water stripper, and dilution steam drum will help minimize corrosion and emulsion stabilization throughout the system,

Likewise, optimizing water and hydrocarbon separation at the quench water/oil separator, the DOX unit, and the coalescer helps assure a high quality process condensate stream.

Ultimately, careful monitoring and adequate control of pH, fouling, corrosion, foaming and emulsification, water balance, sodium levels, and amine loading can help meet and maintain dilution steam quality goals, as well as overall system reliability.

4.) Problem Diagnosis and Treatment

Corrosion Control

Minimizing corrosion throughout the dilution steam generator system requires a thorough understanding of overall system design, feedstock source and composition, upstream treatment, makeup of recycle and blowdown streams, fouling potential and initiators, as well as the operational and treatment options available to monitor and control pH.

A successful corrosion control program will include an in-depth evaluation of existing program performance; chemical treatment and application technology designed specifically for each system; and ongoing equipment inspection and monitoring to maintain overall system reliability. Equipment inspection and monitoring of corrosion activity can identify the most critical areas to treat. Monitoring pH, total and soluble iron levels, as well as sodium, ammonia, and amine concentrations helps assure optimum treatment performance,

Monitoring the pH of upstream units, such as the oil quench tower, compressor section, and transfer line exchangers (TLE), as well as the recycle and blowdown streams will reveal locations where chemical treatment would be appropriate. Other factors to take into account include pH and ammonia levels in the flare drum condensate; organic acid concentration and pH of water from the gasoline hydrogenation unit (depentanizer overhead); as well as possible caustic entrainment in hydrocarbon leaving the spent caustic settler.

An appropriate chemical treatment program should be tailored to meet the specific needs of each system. Consideration needs to be given to the use of a non-sodium versus caustic-based program, or possibly a coordinated phosphate program, which might provide adequate protection with improved cost performance.

Some plants feed caustic soda or ammonia to neutralize acids. However, overfeeding may result in undesirable levels of ammonia in the finished ethylene product, and caustic soda, if not applied properly, can contaminate the dilution steam with sodium, contributing to coking in the pyrolysis furnaces.

An alternative to the use of caustic or ammonia, are organic neutralizers with that can provide protection against acidic corrosion in both liquid and vapor phases and condense with the initial aqueous condensate. Depending on the salt properties of these neutralizers, operators can expect uniform pH control with reduced risk of caustic embrittlement; minimized sodium carry-over; improved pH buffering. These organic neutralizers, with some volatility, can also help prevent acidic condensate formation in the steam piping.

One proprietary technology from GE Water & Process Technologies is a water-soluble, process-side blend of neutralizing amines designed to reduce corrosion by inhibiting dissolved acidic compounds (such as H_2S, CO_2S, sulphoxy and carboxylic acids). Its neutralization salts migrate to the water phase, and will not harm downstream catalysts. A convenient feedpoint for this product is into the recirculating quench water loop. Target pH is typically 6.5 - 7.5. An organic neutralizer can also be used to protect the process water stripper and control pH in the dilution steam generator.

If corrosion problems occur in the process water stripper reboiler, a non-volatile water soluble filming inhibitor can be applied in conjunction with the neutralizer. Additionally, if the oxygen content of the process water stripper feed or bottoms and dilution steam generator water exceeds 20 ppb, the use of an oxygen scavenger is recommended.

Congruent pH/Phosphate (PO_4) program

Another option to prevent corrosion in the dilution steam generator system is to use GE's congruent pH/PO_4 program. Dilution steam generator boiler water has little buffering ability. As previously stated, low pH conditions may develop if the feedwater has been in contact with acidic, water-soluble hydrocarbons. Underdeposit, alkaline corrosion can occur if "free caustic" ((unbuffered moles of OH (hydroxide), in excess of moles of Na_2HPO_4 (sodium hydrogenphosphate)) which may have been added for neutralization purposes, is present in the boiler water. Therefore, the recommended boiler water treatment program for the dilution steam generator is based on a controlled pH/phosphate mole ratio, or congruent control (Figure 3). However, successful operation of dilution steam generators has also been achieved using phosphate chemistry for pH buffering, without coordination (controlling strict Na/PO_4 mole ratio). The dilution steam generator's blowdown rate (cycles of concentration) is generally not controlled based on blowdown water chemistry. It is controlled by system water balance. Typical operating cycles are 8 – 10 (i.e. the blowdown flow is >11% of the feed water flow).

Congruent pH/PO4 control in the dilution steam generator is achieved by maintaining boiler water phosphate concentrations and pH within specified boundaries (refer to Figure3). This is normally accomplished via adjustments in the feedrate of phosphate, caustic, and / or neutralizing amine. The congruent pH/PO_4 control assures that no "free caustic" will develop by maintaining correct pH to PO_4 control ranges and as a result the risk for caustic concentration under deposits and subsequent gouging is minimized. The HPO_4^{2-} acts as a buffer to neutralize the caustic, should caustic find a concentration spot, for instance, under a deposit. Because dilution steam generation systems vary in design and water chemistry, exact control parameters should only be set after the system has been carefully studied. An example of a typical control range could be 4 – 10 ppm PO_4 and a corresponding pH range of 9.0 – 9.8.

Basis of congruent phosphate buffer control

The use of congruent (coordinated) phosphate control minimizes the corrosion of carbon steel by providing elevated boiler water pH with "captive" or phosphate-balanced alkalinity, and sufficient chemical buffering capacity for low-level system contaminants. A phosphate buffer is used to prevent under-deposit caustic gouging that could occur in the presence of free/unbuffered caustic (NaOH). The basis of this chemistry is:

$$NaOH + Na_2HPO_4 = Na_3PO_4 + H_2O$$

The risk of "free caustic" in the boiler is further reduced with the introduction of congruent phosphate treatment, where the Na/PO_4 mole ratio is controlled below 2.8.

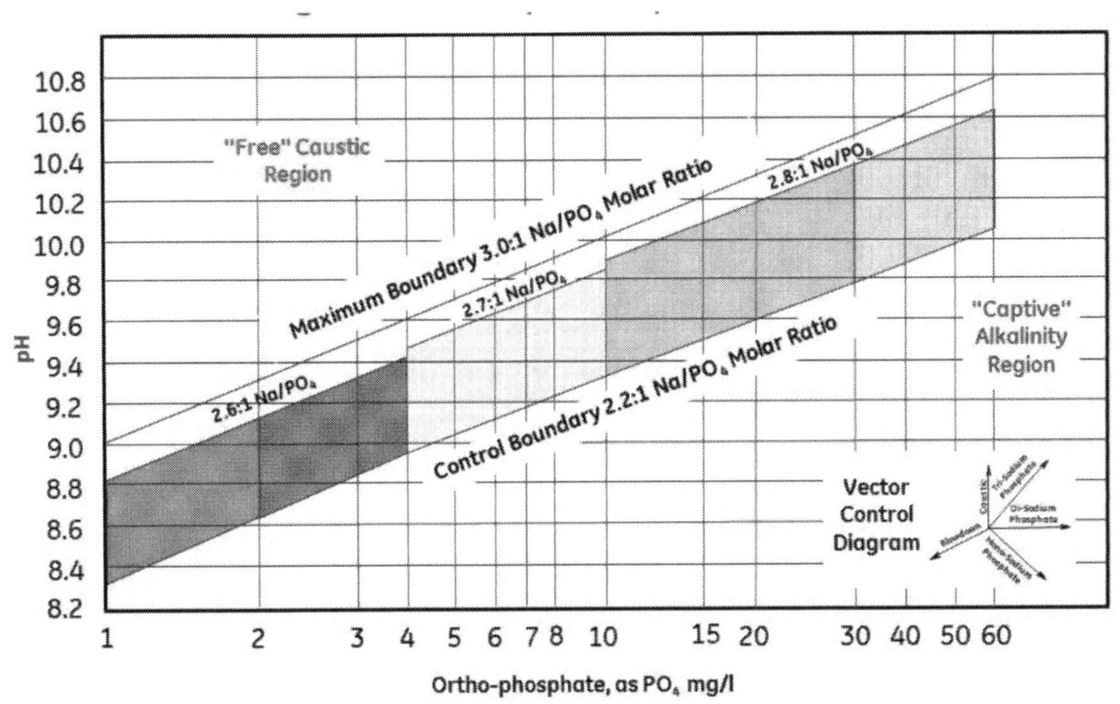

Figure 2: Congruent Phosphate/pH Control Chart

Polymeric dispersants are also included in this treatment approach to disperse iron oxides entering the dilution steam generator, minimizing the risk for deposit formation and making them available for removal through the generator blowdowns. The combination of dispersants to prevent deposits, and phosphate buffer to prevent free NaOH protects the system.

Emulsification control

If untreated quench water samples contain significant levels of emulsified oil (>50 ppm hydrocarbon & grease), an emulsion breaker should be applied to clean up the water. Proper application of the emulsion breaker at inlet to the water/oil separator, together with a well-designed and maintained coalescer, can result in acceptable hydrocarbon & grease content at the coalescer outlet.

An appropriate water-soluble emulsion breaker will help resolve water-continuous (reverse) emulsions in petrochemical aqueous systems. The technology coalesces hydrocarbon, which separates faster from the water phase and improves water clarity. This would help to:

- Minimize emulsion in quench water systems
- Enhance operation of DOX separation units
- Reduce hydrocarbon and total organic carbon (TOC) concentrations downstream in the process water stripper.

Fouling control

Fouling in the process water stripper and dilution steam generator reboiler can be controlled by feeding an organic and inorganic dispersant or a polymerization inhibitor.

The dispersant technology prevents the agglomeration of organics, allowing them to flash overhead with the dilution steam or if heavy, leave via the generator blowdown water.

Dissolved oxygen analyses can be used to verify dissolved oxygen levels and can be addressed with a water soluble oxygen scavenger. Removing oxygen also reduces fouling caused by oxygen initiated free radical mechanisms.

A polymerization inhibitor should be used if styrene in the process water stripper and dilution steam generator reboilers is forming polystyrene deposits on the tubes.

Foaming control

In general, the most appropriate way to minimize a foaming problem is to reduce the amount of organic materials carried from the feed water, and limit the total dissolved salts into the blowdown of the dilution steam generator to a maximum of 700ppm (equivalent to 1000microS/cm2).

The root cause solution to this problem is as follows:

- Maintain pH control of the water (quench loop - circuit up to the stripper bottom)
- Proper water/oil separator operation
- Effective process water stripper operation
- Control Total Dissolved salts (TDS) and blowdown flow of the steam generator

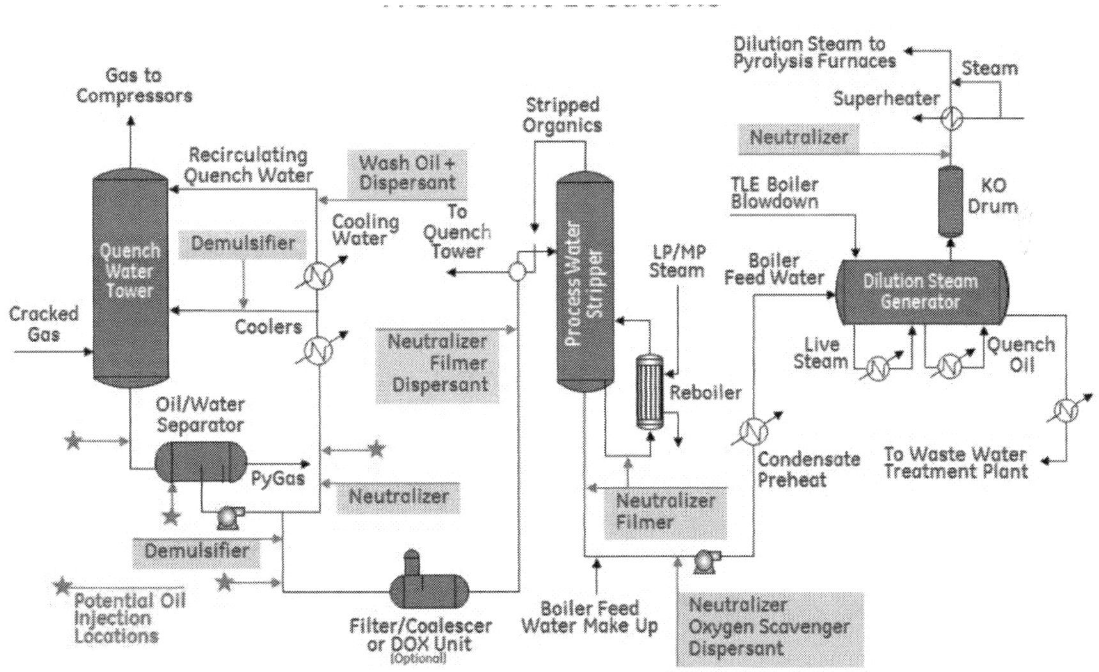

Figure 3: Dilution Steam Generation System Treatment Locations

5.) Best Practices Check List

pH Control

Understand the impact of upstream treatment, and recycle or blowdown streams that can cause pH excursions:

- Monitor the pH in the primary fractionator/stripper overhead, compressor section, and transfer line exchanger (TLE) blowdown, where a neutralizer can possibly be fed,
- Check amines and pH from boiler feed water treatment,
- Evaluate possible caustic soda entrainment with hydrocarbon leaving the spent caustic settler of the caustic scrubber,
- Measure pH, and ammonia in the flare drum condensate,
- Check pH in the water drain from gasoline hydrogenation unit (depentanizer overhead), source of high organic acid concentration.

Select a chemical program based on system needs:

- Non-sodium versus caustic based program,
- Use of phosphate coordinated program to maintain pH buffer

Efficient water / hydrocarbon separation

- Define water quality goal, e.g. DOX outlet turbidity of <100 ntu,
- Adequate hydrocarbon removal from quench water/oil separator, DOX unit, and coalescer,
- Optimize quench water/oil separator interface level to maximize water residence time
- Select and apply a proper demulsifier (water-soluble breaker), if needed. An effective emulsion breaker can be identified by jar testing at different pH levels.
- Improve the settling time by the addition of hydrocarbon (DeC_4 bottom, py-gas stripper bottom, $C9^+$) to reduce specific gravity from ~ 1 to <0.9, leading to a faster separation of hydrocarbon and water and minimize deposition of hydrocarbon in the quench water tower and separator,
- Optimize the pH control range (6.5 to 7.5) for better emulsion resolution and corrosion control,
- Understand impact of recycle streams that contain surface active components, solids, chemicals,
- Understand the impact of furnace sulfiding program on emulsion stability.

Proper operation of the process water stripper

Evaluate stripper design efficiency:

- Look at and optimize the steam input (live steam and/or reboiler) to improve hydrocarbon removal efficiency
- Increase the stripping steam to enhance VOC (volatile organic compounds) removal

Analyze feed and bottom samples for water quality:

- Hydrocarbon level ((pH, oil & grease, TOC (total organic carbon),turbidity, appearance))
- Styrene content

Minimize organic and inorganic deposition

Understand the impact of upstream treatment, and recycle or blowdown streams that can contribute to fouling:

- Hydrocarbon from spent caustic settler, compressor wash oil,
- Green oil, corrosion products, TLE blowdown (phosphate based products).

Improve the water quality upstream by minimizing organic and inorganic deposits: respectively styrene, hydrocarbon carryover, polymers, corrosion products, and inorganic salts.

Manage system water balance

Controlling and adjusting the blowdown flow.

- A high blowdown rate (low cycles)
  - Increases waste water treatment load,
  - Increases requirement for neutralizer and chemical costs,
  - Decreases concentration of contaminants in the dilution steam generator,
  - Generates loss of energy via the blowdown.

- A low blowdown rate (high cycles)
  - Increases the concentration of the contaminants in the dilution steam generator, causing an increase in fouling and foaming, resulting in neutralizer/caustic loading with the dilution steam into the furnace.

The cycles of concentration can be optimised with the blowdown via the process water stripper bottom and by applying an intermittent blowdown procedure.

- Measurement of cycles: Typical dilution steam generator systems target 8 to 10 cycles
- Run water analysis such as conductivity, total dissolved salts (TDS), flow measurement
- Control liquid level, which can impact foaming, salt entrainment and conductivity
- Optimize the blowdown to help circulate dispersant (if used)

Dilution steam quality standards (sodium levels, pH, amine and styrene loading)

Dilution steam quality is a good indicator of the dilution steam generation system efficiency. Indicators and/or parameters to look for are the following:

- Low conductivity, indicative of good steam quality,
- pH impacts the partitioning of organic acids/amines/ammonia into the dilution steam generator. The higher the pH, the more organic acids in the liquid phase and the more neutralizing amines in the vapor phase.
- Dilution steam pH>8,5 is required to avoid corrosion of the dilution steam line,
  A volatile amine is required for protection of the steam line (however this could contribute to the formation of ammonia in the cracking furnace)

Steam line corrosion is an issue when condensation occurs

- Limit sodium carryover in dilution steam to <250 ppb (0 ppb for millisecond furnaces)
- Excessive OH from "Free caustic" increases the potential of foaming in dilution steam drum
- Styrene carryover can cause polymerization in the demister pad and steam line

6.) Possible causes for poor performance of dilution steam generator

- Preheater/dilution steam generator exchangers are fouled (tube and / or shell side)
- Return valve/line is restricted, increasing drum pressure
- Dilution steam generator boiler feed water/steam flow meter is wrong
- Excessive blowdown
- Superheater plugging is causing back pressure on the drum
- Steam traps are not operating properly, preventing steam flow to the generator or insufficient condensate removal
- Poor removal of non-condensables from the steam chest
- Heat flux to reboiler is too high, with a high percentage evaporation rate with respect to the circulation flow (% steam produced over the circulation flow). This is a current issue on horizontal U tube reboilers, with water on the shell side. Reboilers of this design can experience a departure from nucleate boiling, formation of a water / steam interface, fouling at the interface, and then underdeposit corrosion.

Case History (1)

Quench water separation with emulsion breaker and pH/PO_4 corrosion control in dilution steam generator at BASF Antwerp, Belgium

The quench water and dilution steam from a steam-cracker flows though different sections of the process, like furnaces, oil quench and water quench towers, quench water/hydrocarbon separation drums, several heat exchangers, process water stripper and dilution steam generators, The quality of the quench water has an impact on each section. The control parameters are pH, iron concentration & TOC/hydrocarbon in water content.

The quench water pH is dependent on the amount of absorbed organic acid, sulfur components and NH_3 levels, but also the type and amount of pH controlling additive. The quench water pH is critical for corrosion control in the quench water, process water stripper and dilution steam generation system and also for the phase separation in the quench water/hydrocarbon separation drum.

The levels of hydrocarbon contamination in the quench water are dependent on the phase separation, which is pH dependent, composition of the gasoline (density, end point, naphthalenes, etc.) and affected by the presence of surface active component, e.g. organic acids. Severe contaminated quench water leads to fouling in the different process water sections. With no additional neutralizer injection to the quench water loop in order to control the pH at the desired level, an emulsion breaker injection was applied to achieve an adequate process water quality, lowering the hydrocarbon levels and fouling potential, and secured with turbidity measurements and regular TOC/hydrocarbon in water analysis.

Downstream fouling and corrosion protection in the dilution steam system is achieved by a caustic injection to secure the pH in combination with a GE W&PT boiler treatment program. The pH/PO4 program has been applied successfully since mid '90's in order to avoid caustic corrosion which was observed during the start-up up of the cracker. Besides corrosion control, polymer fouling was minimized by applying a dispersant technology, which has proven itself as very effective in terms of heat exchanger performance and run-length.

Case History (2)

Corrosion control program helps keep DSG reboilers in service for over 6.5 years
IRPC, Thailand

Leaked tube sheet

IRPC Thailand had been experiencing severe corrosion in their dilution steam generator reboilers since initial start-up in 1997. It is a naphtha cracker with 9 DSG reboilers which use circulating quench oil as the heating medium. Reboiler tube leaks began occurring every 4 to 6 months, due to the corrosion and an unusual reboiler design.

In 1999, GE implemented a customized anti-corrosion program consisting of a neutralizing amine, a coordinated pH/Phosphate, an oxygen scavenger, and caustic provided by IRPC. With the new treatment program, the reboilers continued to fail, but the frequency and number of leaking tubes dramatically decreased.

Tube Sheet (GE Treated)

During a turnaround in 2000, all of the tube bundles were replaced with a new tube side design that increased the process water flow to the reboilers. For approximately the next 6.5 years, no reboiler tube leaks occurred. In 2006, unrelated to program performance, GE lost this business to another chemical supplier.

After only 2 years of treatment by the other chemical supplier, the DSG reboilers started to leak again. The GE anti-corrosion program was restarted in 2008, and as of this writing in January 2012, the program has been successful, with no tube leaks.

Case History (3)

Phase Separation Improvements Increase Plant Flexibility and Profitability

A North American ethylene unit, designed to crack gas feeds, began to crack heavier feeds in the late 1990's in order to expand the plant's feed stock flexibility and increase profitability. Excellent corrosion control was maintained in the dilution steam generators using a neutralizing amine for pH adjustment. However, poor separation of the hydrocarbons and water in the quench water separator resulted in high organic levels in the process water stripper condensate. The poor water quality caused significant organic fouling of the downstream piping, steam drums, and generators reducing process reliability and increasing maintenance costs.

To improve the separation efficiency in the separator and DOX unit, GE recommended a program feeding an emulsion breaker into the quench water tower recycle, in conjunction with the injection of a process oil stream into the quench water separator. The emulsion breaker destabilized the emulsion in the quench water tower. The process oil injection aided the separation of the quench water and condensed oils by solubilizing heavy hydrocarbons and floating hydrocarbons out of the water. The addition of process oil decreased the density of the hydrocarbon phase to <0.90 which improved the driving force for the separation. The result was a significantly cleaner process condensate with a turbidity of <50 ntu from the DOX unit. A dispersant was also applied to prevent deposition of hydrocarbons downstream of the DOX unit.

These improvements made a significant impact on generator run length. Cleaning costs for the generators were reduced by 85 %. Run lengths were increased from weeks to over one year. As part of GE's continuous improvement program, operating and chemical guidelines were developed for each different feed slate the plant wanted to run. The plant was able to crack up to 50 % more heavier feeds in their feed slate enabling them to improve profitability.

With the recent move back to more economically favorable gas cracking, the plant has continued to utilize the phase separation improvement program. Generator run lengths have increased to more than two years between cleanings. Since the plant commissioning, corrosion control using the non-volatile neutralizing amine has resulted in no tube failures in the generators.

7.) Conclusion

Controlling corrosion and fouling are critical aspects of the operation of the quench water and dilution steam generator system. Diagnosing the problem and identifying the root cause is key to a successful chemical treatment program. Designing an effective treatment program that complements operational procedures and helps optimize overall unit performance is not an easy task. It requires a solid understanding of the process, proven analytical capabilities, and extensive experience. Within this paper, GE W&PT has shared some of its best practices which can help operators to diagnosis problems and better understand proposed solutions.

AIChE Paper Number 96d

Root Cause of Corunna Debutanizer Tower Operational Upsets

Feridoun Fahiminia

Research Scientist

NOVA Chemicals Research and Technology Centre, Calgary, AB, Canada

André Bernard

Process Improvement Engineer

NOVA Chemicals (Canada) Ltd

Patricio Herrera (Retired)

Senior Research Scientist

NOVA Chemicals Research and Technology Centre, Calgary, AB, Canada

Morgan Stante

Research Technologist

NOVA Chemicals Research and Technology Centre, Calgary, AB, Canada

Prepared for Presentation at the 2012 Spring National Meeting

Houston, TX, April 1-5, 2012

Root Cause of Corunna Debutanizer Tower Operational Upsets

Feridoun Fahiminia

Research Scientist

NOVA Chemicals Research and Technology Centre, Calgary, AB, Canada

André Bernard

Process Improvement Engineer

NOVA Chemicals (Canada) Ltd

Patricio Herrera (Retired)

Senior Scientist

NOVA Chemicals Research and Technology Centre, Calgary, AB, Canada

Morgan Stante

Research Technologist

NOVA Chemicals Research and Technology Centre, Calgary, AB, Canada

Abstract: The NOVA Chemicals Debutanizer Tower at the Corunna site experienced operational upsets. Gamma scans revealed that the Tower had flooded, and upon opening the tower it was found that the bottom trays fouled. An investigation was initiated to identify the root causes of the fouling as well as to aid in optimizing conditions under which fouling is minimized. Fouling deposit samples were collected at different times and characterized employing various analytical tests. Laboratory fouling experiments were designed and carried out to test fouling propensity of the tower bottom streams. In addition, experiments were run under different conditions to see how process variables such as temperature and residence time impact the debutanizer bottom fouling rates. A range of temperature and residence time conditions were identified under which fouling might be minimized. Also, plant data was gathered, and the tower was simulated to understand process conditions and tower performance.

Introduction

NOVA Chemicals Corunna site is an olefins plant with a front-end crude unit and back end aromatic unit. The plant was originally designed as a naphtha cracker but has since developed feed flexibility through expansions and revamps. The latest revamp was executed in 2005 and pushed the plant capability in processing NGL feed by an additional 25%. High NGL diets caused a significant change in heater effluent composition, which is also observed by a rate reduction in the C4 area.
A general decline in performance on the debutanizer was first noticed in August of 2008. The performance kept deteriorating until February 2009, where it became a limitation to plant throughput. A gamma scan of the tower was then conducted, and it indicated that the tower had flooded from tray 48 to tray 6. The unit was shut down and the debutanizer tower was opened up; extensive fouling was seen in the lower trays of the tower. This observation was contradictory to previous inspections, which reported a fairly clean tower.
Although for many years Debutanizer towers have not experienced significant fouling, in recent years this trend has changed, i.e. more frequent operational upsets due to fouling have been observed. Arnst (2006) stressed that aggressive optimization of the antifoulant treatment programs, extension of plant run length, increased capacity, and higher cracking severity conditions are potential causes of more frequent fouling.

Debutanizer Description

The Debutanizer is a 97 ft tower with a diameter of 7 ft and a total of 49 high capacity trays. Low pressure steam supplies the bottom reboilers and cooling water provides cooling to the overhead condenser. As illustrated on Figure 1, the tower is fed at tray 29 from the Depropanizer bottom. The overhead mix C4 product is collected from the reflux drum and sent to the C4 reactors after treatment in a caustic tower. C5+ material

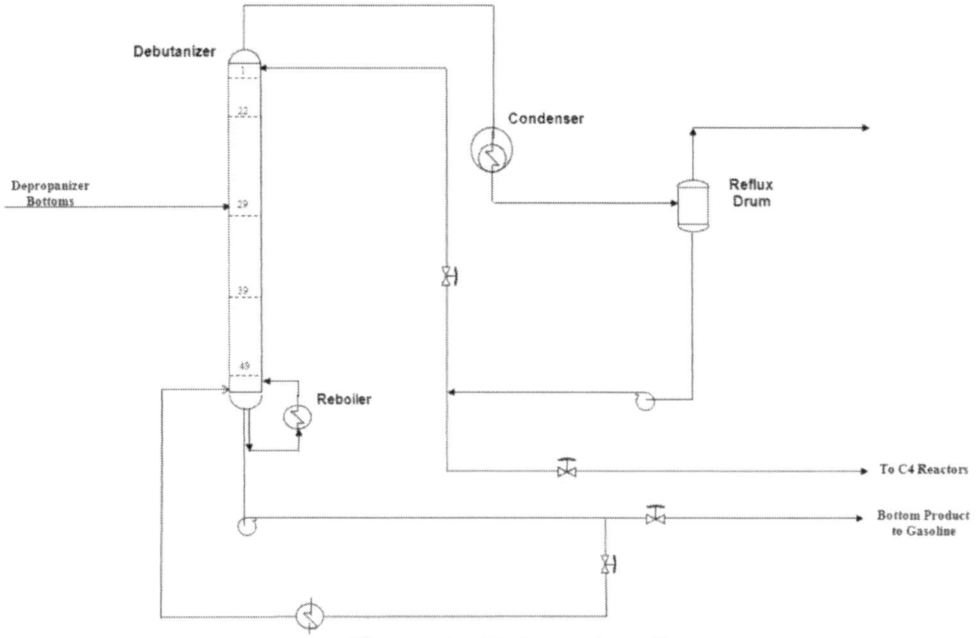

Figure 1 – Debutanizer Tower

collected from the bottom is mixed with gasoline extracted in the hot end of the plant and sent to storage for further processing in the aromatics unit. A slip stream from the bottom of the tower is recycled back to the tower sump after cooling against cooling water. This recycle stream was not part of the original design but was added in an attempt to reduce reaction of cyclopentadiene to di-cyclopentadiene in the tower sump.

Debutanizer Inspection Findings

Inspection of the tower revealed that Tray 48 was heavily fouled (Figure 2) - not the downcomers but the deck of the tray itself. Tray 47 was less fouled, and tray 46 even less so. All the trays above that were clean. In September of 2009, a mini turnaround took place in Corunna. When the debutanizer tower was opened up again, fouled trays were observed in similar locations as seen in February. Fouling deposit samples were taken for later characterization tests.

Interestingly enough, the bottom tray (tray 49), the tower sump and the reboiler were found relatively clean.

Trouble Shooting

Considering that this fouling occurrence with the tower was unexpected and unusual given previous tower inspection reports, something with the tower or process conditions had to be different. The first step in the trouble shooting exercise was therefore to focus on identifying all the key changes since the 2005 turnaround. Table 1 summarizes all the key changes with their expected consequences. An understanding of the chemistry was also deemed necessary as it would identify the reactants and conditions responsible for

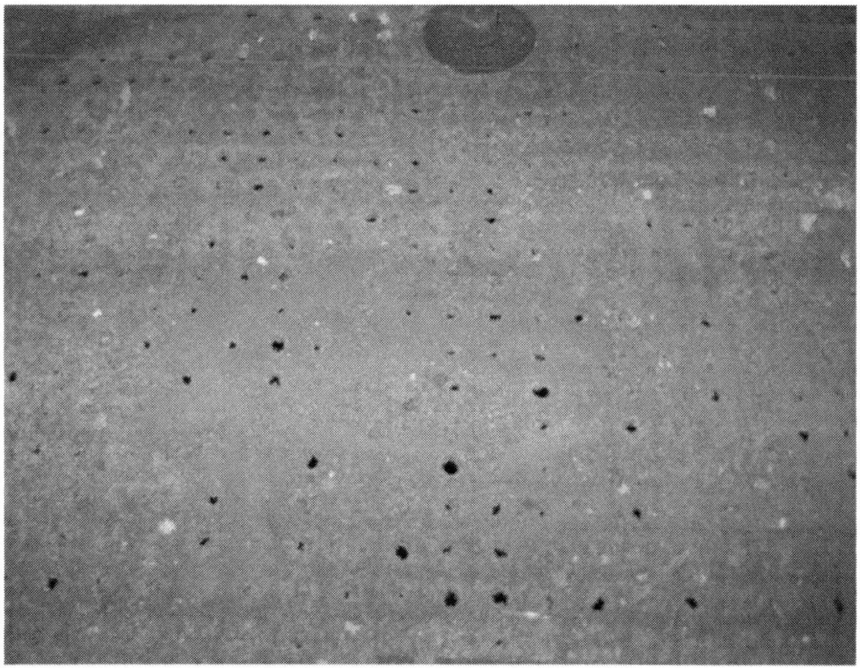

Figure 2 - Picture of Fouled Tray in Debutanizer Tower from February 2009

Table 1 - Key Debutanizer Tower & Process Changes

Changes	**Expected Consequences**
Plant Feedstock - Higher NGL Diet	Feed Composition Changes Lower Feed Rates in the C4 Area
Decline to Elimination of External C4 Processing	Lower Feed Rates in the C4 Area
Revamp of the Cooling Water System	More condensing of gasoline components in the Charge Gas Compression train leading to lower feed rate in the C4 area and compositional changes.
Higher Bottom Temperature in the tower bottom	Higher temperature increase the rate of fouling
Tray weir height reduction from 4" to 3"	Reduce liquid hold up on the tray (shorter residence time) Reduce tray pressure drop. Reduce downcomer backup, entrainment rate and weeping tendency.

the accumulation of material on the trays. In order to address the tower operational upsets, the following tasks were initiated:

- Deposit Characterization - characterization of deposits collected in February and September 2009 from the Debutanizer.
- Fouling Experiments - fouling experiments employing the tower bottom liquid samples to study the effect of process variables.
- Gathering Historical Data and Process Simulation - Gathering of plant data and process simulation to further understand process conditions and tower operation. This would also help connecting the dots between the fouling conditions in the tower and laboratory results.

Fouling Characterization And Experimental Apparatus

a) Deposit Characterization

The following analytical techniques were employed to characterize either fouling deposits or liquids:

- Thermogravimetric analysis (TGA) coupled with Fourier Transformed Infrared Spectroscopy (FT-IR) was used to investigate the loss in weight of the sample as it was oxidized in a controlled environment. The coupled FT-IR test provided basic information about the type of gases evolving during the oxidation process.

- Energy Dispersive X-Ray Spectroscopy (EDX) was used for semi-quantitative elemental analysis of the sample's outer layer (1-5 microns).
- Fourier Transformed Infrared Spectroscopy (FT-IR) was used to determine the nature of organic matter in the sample. FT-IR is a very powerful tool for identifying types of chemical bonds (functional groups) in a sample.
- Scanning Electron Microscopy (SEM) was used to study the surface morphology of deposits at magnifications up to 100,000X.
- Bulk Elemental Analysis was performed to measure carbon, hydrogen, nitrogen, oxygen and sulphur contents in the samples.
- Pyrolysis-GC/MS was performed to identify organic components and polymer monomers in the deposits.
- Proton Induced X-Ray Emission (PIXE) analysis was used for ppm level detection of inorganic components from sodium through uranium.
- Soxhlet Extraction was employed to test the solubility of fouling deposits in various solvents under continuous reflux conditions.
- Nuclear Magnetic Resonance (NMR) was used to identify key molecular structures in the samples.
- Simulated Distillation was used to estimate boiling ranges of various cuts as well as determine corresponding carbon ranges.

b) Fouling Batch Experiments

The bench scale experimental set-up using Fouling Batch Mixed Reactors (FBMRs) is shown in Figure 3. Each reactor consists of a 600 mL (2.5 in ID, 8 in height) stainless steel, removable, pressure vessel, a magnetic stirrer with 1/4 hp variable speed stirrer motor. Rupture disk and pressure safety valves were installed to protect the reactors against over pressurizing.

Heating was provided by an 800 W Aluminum block heater and a temperature controller, controlling the bulk temperature of the test sample at the required level. A check valve was installed in the inlet gas line to prevent back flow of the reactor during the experiments. In each reactor, two K-type thermocouples were used to measure the temperature, one located in between the reactor wall and block heater, and the other located inside of reactor at a depth 1.5 cm from the bottom of the reactor to directly measure the liquid temperature. The latter served as the input signal for the temperature control.

For each experimental run, 300 mL of test sample was prepared accurately and loaded into the reactor, followed by nitrogen pressurization up to 500 psig for the leak test. The reactors were left under pressure overnight at room temperature to detect any possible leaks. After ensuring that the reactors were leak free, they were depressurized and then heated to the pre-determined temperatures. The stirring rate of 20 rpm was used throughout the experiments. During each experiment, 60 mL liquid samples were taken from each reactor at various time intervals. Gum content of these samples along with those taken before and after the fouling test was measured using Existent Gum Measurement unit. In some of the experimental runs, withdrawn samples were filtered using 0.45 μm pore-size filter papers to detect formation of any particles.

C) Existent Gum Measurements

Existent gum measurement was a key experimental tool in the fouling tests on the debutanizer bottoms. The existent gum unit was used to measure the progress of the fouling in the FBMR unit. The procedure to measure existent gums involves measuring out a known amount of liquid and evaporating the liquid under controlled conditions of heat and nitrogen flow. The residue remaining in the pre-weighed beaker is recorded and reported in milligrams per 100 ml.

Figure 3 - Fouling Batch Mixed Reactors (FBMRs) Employed for Fouling Tests

Results And Discussion

Characterization of Deposits from Debutanizer

Digital pictures of the foulant samples are shown in Figure 4 below. The deposits were collected from the debutanizer after opening the tower in February and September 2009. The February sample was made up of many small particles varying in color from brown or black to orange and off-white. The particles themselves varied greatly in size and shape. The sample received in September appeared very similar to the sample received in February. It was composed of particles in an array of shapes and sizes. The color of the sample varied again from off-white or an orange color to dark brown and black. TGA analysis of February sample revealed that it was 94% organic and 6% inorganic, whereas the September sample was 78% organic and 22% inorganic. The largest components of the TGA ash of both February and September samples were iron, oxygen, and carbon.

Fouling Deposit Collected in February 2009

Fouling Deposit Collected in September 2009

Figure 4 - Pictures of Debutanizer Deposits as Received

The FTIR showed presence of olefinic hydrocarbons and oxygenated compounds in both February and September samples. The pyrolysis and headspace GC/MS analyses revealed that cyclopentadiene, cyclohexadiene, and methyl-cyclopentadiene were the major components identified in both February and September samples with minor traces of styrene, indene, toluene, and naphthalene derivatives. Also, Soxhlet Extraction experiments showed that both February and September samples were hardly soluble in typical organic solvents like toluene and dichloromethane indicating high cross-linking structure within the samples.

Impact of Temperature And Residence Time On Gum Formation

Fouling Batch Mixed Reactors (FBMRs) were run using debutanizer liquid samples at three different temperatures of 90, 100, and 110°C. More detailed information on operating conditions of FBMR reactors are presented in Table 2. Gum content of the samples, withdrawn from the FBMR reactors during batch fouling experiments, was measured using Existent Gum unit. Figure 5 shows a photograph of the gum (formed at 110°C after 24 hours of fouling experiments) left after the existent gum test. The formed gum has a yellowish color with a boiling range above 180°C. The gum measurement results for fouling experiments ran at temperatures of 90, 100, and 110°C are presented in Figure 6. It is seen that at each temperature, as the time goes by, gum content (in terms of mg/100 ml) increases and approaches an asymptote. There are no significant differences in the amount of gum formed at temperatures of 90 and 100°C. On the

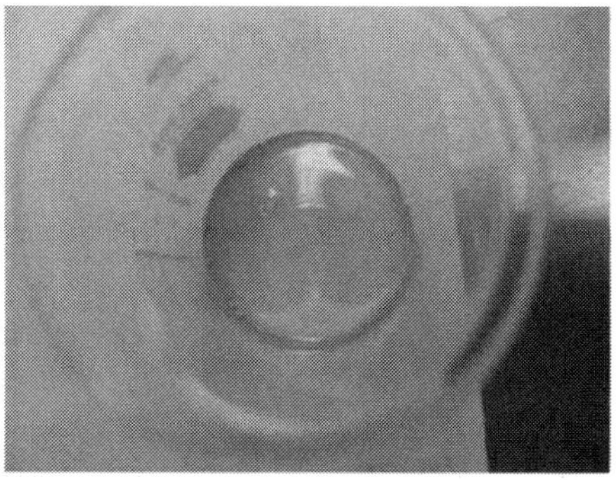

Figure 5 - Gum Content of Debutanizer Bottom Liquid Samples (After 24 Hours of Fouling Experiment at 110 °C)

Table 2 - Operating Conditions of Fouling Batch Experiments

Run	**Temperature** (°C)	**Pressure** (psig)	**Purging Media**	**Sample Volume** (mL)	**Run Duration** (hr)	**Stirring Rate** (RPM)
1	90	38	N_2	300	216	20
2	100	45	N_2	300	216	20
3	110	70	N_2	300	240	20

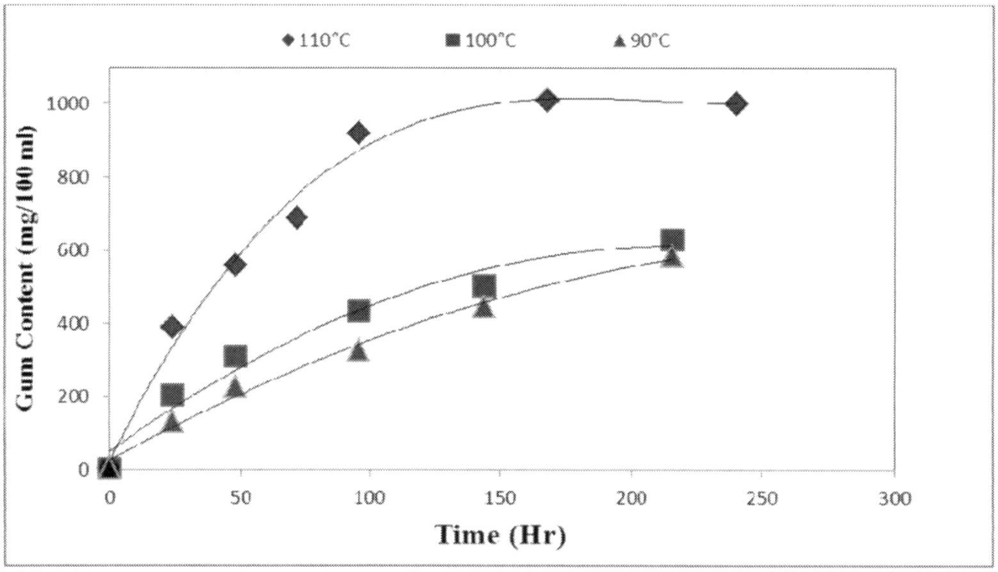

Figure 6 – Impact of Temperature on Gum Formation

other hand, when the temperature is raised to 110°C, an increase of approximately 100% is observed. This reveals that the threshold temperature limit is 100°C, and the temperature in both debutanizer tower and the re-boiler should not exceed this value. If the temperature exceeds the threshold limit, more gums will form, which will eventually be precipitated as fouling deposits, especially in the areas where the liquid is stagnant.
Another set of experiments were conducted to investigate the time required for fouling deposit formation under different temperature conditions. Liquid samples were withdrawn at different time intervals and filtered using 0.45 μm pore-size filter papers to

detect presence of particles. For instance, in one experiment that continued for about 24 hours at 110°C, the particle concentration was found to be 25 ppm. Figure 7 shows the filter paper and the trapped particles generated from this test. The results of particle formation onset times are presented in Table 2. It reveals that at temperatures of 90 and 100°C, the onset times are 216 and 168 hours, respectively. However, at 110°C the onset time is 24 hours which is substantially shorter than those at 90 and 100°C. This behavior is similar to what was observed for gum formation, indicating that the same approach, for example, lowering temperature, may be used to prevent fouling deposit formation.

The aforementioned results show that in order to reduce fouling deposit formation in the debutanizer tower and re-boiler, two approaches should be considered. The first is lowering the temperature below its threshold limit of 100°C. The second approach is reducing the fluid residence time in the tower and re-boiler. The fouling experimental results shown in Table 3 are a good guidance for this approach if the real residence times in the tower and re-boiler are known or can be estimated. These mitigation techniques might be implemented simultaneously or sequentially.

Table 3 - Fouling Deposit Onset Time at Different Temperatures

Temperature (°C)	90	100	110
Fouling Deposit Detection Time (hr)	216	168	24

Figure 7 - Fouling Deposit Particles (25 ppm) Formed During 24 Hours of Fouling Test Conducted at 110 °C

Other characterization tests such as elemental analysis, simulated distillation, and pyrolysis GC/MS were conducted on the gums. The elemental analysis results are presented in Table 4. The C:H ratio is 0.78, which is close to that of dicyclopentadiene

Table 4 - Elemental Analysis Results for the Debutanizer Gums

Sample	C (wt%)	H (wt%)	N (wt%)	O (wt%)	S (wt%)	C:H
Debutanizer Gums	88.73	9.48	<0.3	2.07	<0.3	0.78

(0.83) or its derivatives. The pyrolysis GC/MS leads to the same finding, i.e. dicyclopentadien and derivatives make up the main portion of the sample.

Data Gathering & Process Simulation Evaluation

Operating data has been gathered to compare the facility's 2005 pre and post revamp operation. A sample equivalent to 1 year of data based on daily averages was collected from the plant data historian for both periods. 2007 data has been selected for the post revamp operation to eliminate noise from transition and plant optimization conducted shortly after the revamp commissioning phase. The main variables collected include the tower feed rate, bottom and overhead products as well as the control tray temperature. As illustrated on Figure 8, the 2007 tower feed and overhead product rates distinguish themselves from the pre revamp operation by a tighter operating range. The larger operating range observed on the 2005 data could be explained by external C4 processing, which is not processed on a full-time basis. The operating range becomes narrower after 300 days' time on Figure 8. This narrower range could be related by an interruption in external C4 processing. As shown on Figure 8, the narrower operating window observed on the overhead product rate support the above suspicions. The bottom feed rate suggests an increase in bottom product for the post revamp conditions. The total tower feed seems to be in the same order of magnitude for the pre and post revamp data. As a general conclusion, the pre and post revamp feed rates are more or less similar. However, the C4 to C5+ split seems different with a higher C5+/feed ratio on the post revamp conditions. It is evident from Figure 8d that the control tray temperature runs warmer in the post revamp conditions by as much as 10°C at time.
This information alone is not sufficient for evaluating tower conditions that may lead to fouling. Process simulation of the tower for the two operating modes will help identifying other variables that may relate the process changes described on Table 1 to the fouling conditions evaluated in the laboratory.
Two sets of daily plant data reflecting the pre and the post revamp conditions were selected as inputs to the simulation. Tower feed composition was estimated from cracking heater yields input into a general material balance model. Adjustments on the feed composition were based on material balance reconciliation. The model was then validated by comparing key simulated outputs to plant values as described on Table 4. The pre revamp data provides a relatively good accuracy in predicting the reboiler duty. However, a 10% deviation is observed on the reflux prediction. On the other hand, the post revamp simulation provides a better match on the reflux but shows a significant discrepancy on the reboiler. Given that this evaluation was done in the post revamp period, access to post revamp data and sensitivity analyzes was easier for validating

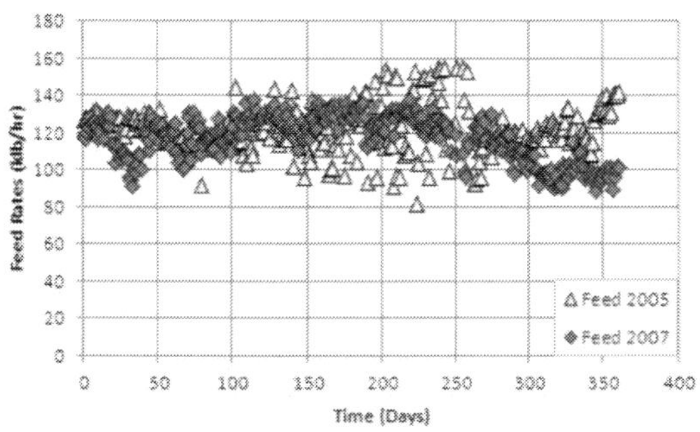

Figure 8a: Tower Feed

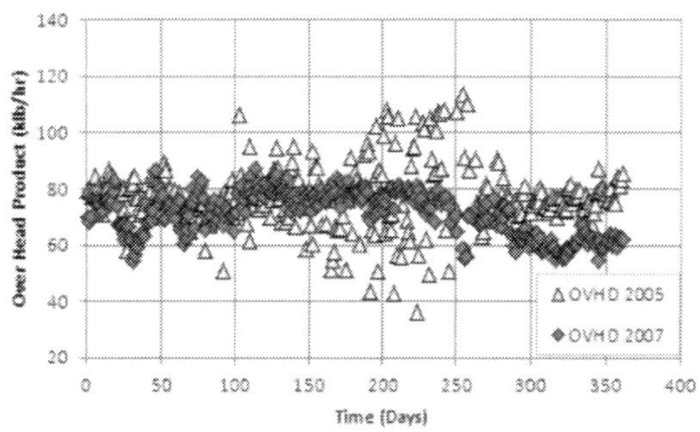

Figure 8b: Tower Overhead Product

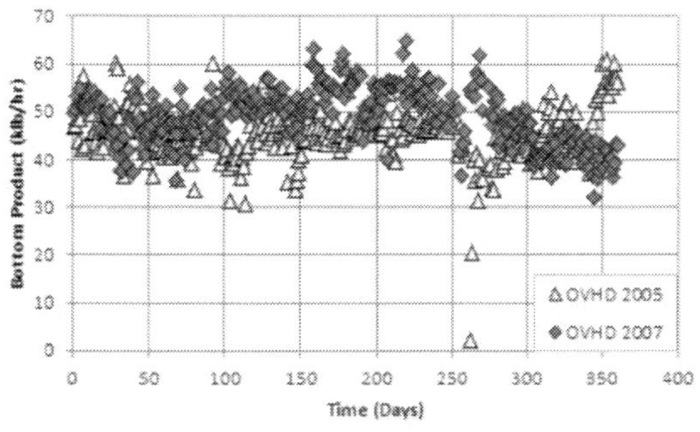

Figure 8c: Tower Bottom Product

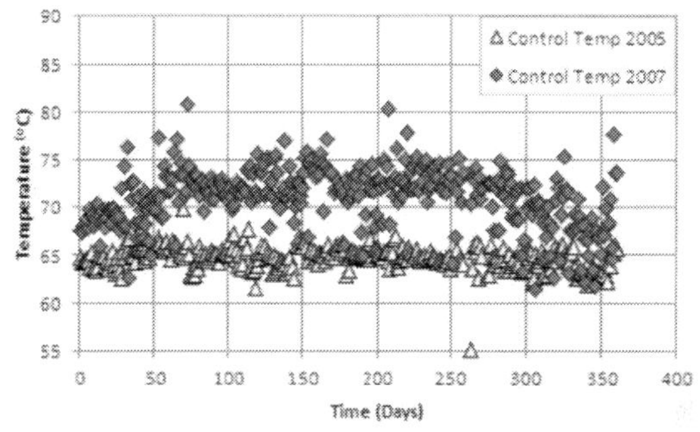

Figure 8d: Control Tray Temperature

Figure 8: Pre and Post Revamp Plant Data

data. The Debutanizer simulation has been tuned against post revamp data and in most occasions a match within 4% was obtained on the reflux. A larger difference was observed on the reboiler duty, with a deviation varying between 10 to 15% at the most. Note that the simulated reboiler duty is compared to the steam flow converted in duty by assuming saturated steam condensing to saturated liquids. The tower has two reboilers, one in operation and one on standby. The reboilers are switched over based on loss of performance caused by fouling over time. Therefore, depending on where the reboilers are in their cycle, it is expected that they will use more or less steam and that steam and condensate conditions will deviate from the above assumptions. In all cases the predicted reboiler duty was always under estimated when compared to plant data. Another cause for this deviation in heat duty is suspected to come from the water cooler shown on Figure 1, which cools a slip stream from the tower bottom that is recycled back to the tower sump. The tower sump contains a baffle that separates the reboiler from the sump side. However, a significant hole exists in the baffle, which causes the level on both sides of the sump to level and also causes heat exchange between the two sides. This external cooling is compensated by additional reboil and defeats the purpose of this system, which was intended to keep the sump material cool to prevent reaction of cyclopentadiene.

Given the nature of this analysis, which relies on gathering historical data, it is impossible to further tune the model by performing detailed laboratory analysis of each stream and validating plant instrumentation. It is understood that the simulation model does not provide the typical level of accuracy desired. However, its accuracy is considered acceptable for a comparison between pre and post revamp conditions.

Table 5: Simulated vs. Plant Data

Case	Feed (klb/hr)	Reflux Rate Actual (klb/hr)	Reflux Rate Sim (klb/hr)	Reflux Rate Difference (%)	Reboiler Duty Actual (klb/hr)	Reboiler Duty Sim (klb/hr)	Reboiler Duty Difference (%)
Pre Revamp	126	86.9	77.7	10	24.6	25.6	4
Post Revamp	136	80.3	77.1	4	27.1	21.3	21
Post Revamp (2nd set)	126	87.3	84.5	3.2	27.1	24.04	11

The feed composition estimation for the pre and post revamp conditions reveals a slight difference in component distribution as shown on Figure 9. Consistent with the increase in bottom flowrates, the post revamp feed composition increased by over 5% in C6+ material. The other interesting information is the C5/C6+ ratio between both periods. The pre revamp period has a C5/C6+ ratio of 1.3 compared to 0.86 for the post revamp. The higher tower bottom C6+ content is consistent with the post revamp higher temperature reported on Figure 8d. The simulated temperature profile is illustrated on Figure 10 and shows a warmer profile in the stripping section for the post revamp conditions. The difference varies between 2 and 10 °C throughout the stripping section. The tray control temperature is approximately 14 °C warmer for the post revamp conditions, which is consistent with the plant data illustrated on Figure 8d. The temperature in the rectifying section is similar between the two sets of conditions.

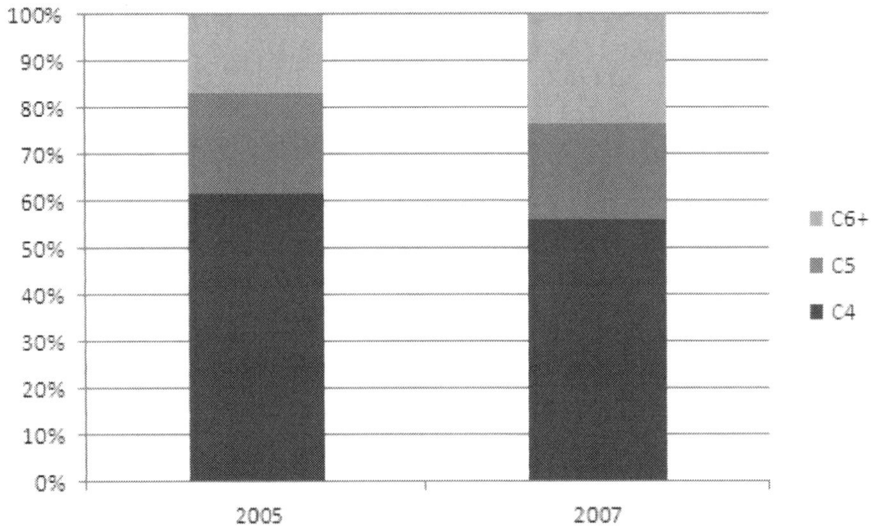

Figure 9: Debutanizer Feed Composition Distribution

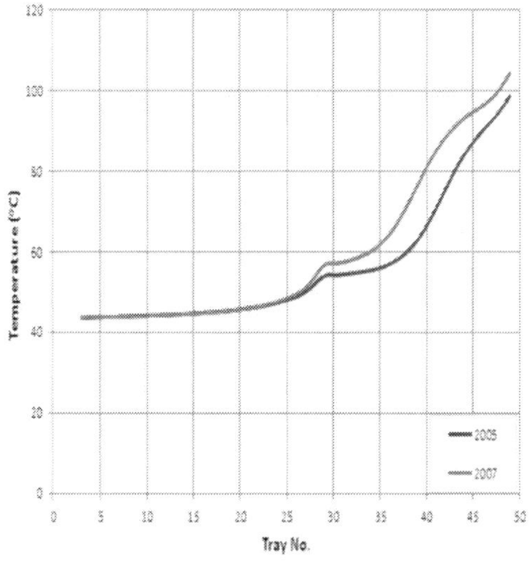

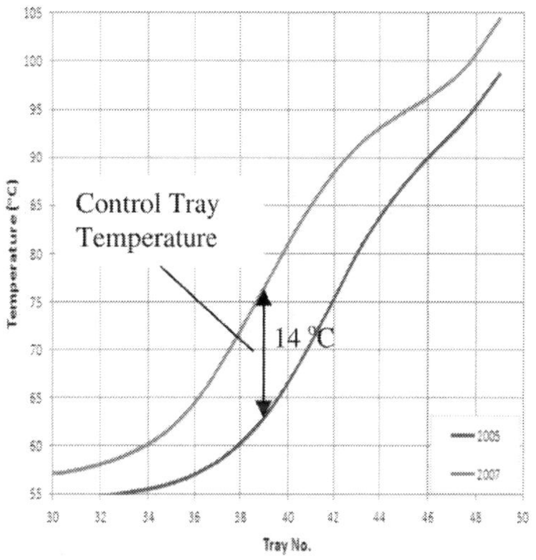

Figure 10a: Temperature Profile Tower

Figure 10b: Temperature Profile Stripping Section

The tower composition profile is illustrated on Figure 11a, where components are grouped into C4's, C5's and C6+. The solid line represents the pre revamp data and the dash line the post revamp conditions. The post revamp conditions exhibit a lower C4 composition in the stripping section, which again is consistent with the warmer temperature observed in that section. C5 and C6+ are higher in the post revamp condition for most of the stripping section. The shape of the profile is similar to the pre-revamp conditions down to tray 45, where a larger drop in C5 is observed on the post revamp conditions leading to a higher C6+ and lower C5 composition on trays 45 to 49. Figure 11b illustrates the composition profile of 1,3-butadiene (BD) and cyclopentadiene (CPD), two known reactants in this system. The post revamp conditions show a lower BD profile but higher concentration of CPD are observed between trays 25 to 49 with a small peak on tray 48. Given the lower composition profile at post revamp conditions, BD can be ruled out as the potential fouling precursor. The CPD composition however, is found 2 to 5% higher between trays 35 to 48, with the largest difference at tray 40. The difference for trays 46 to 48 (fouled trays) vary between 2.25 to 2.9. Interestingly enough, the CPD composition is only 1.2% higher than the pre revamp conditions at tray 49.

Tray Loadings

The vapor and liquid loads have been estimated from the internal traffic and properties obtained from the simulation. A correspondence between the bottom product rate and steam flow to the reboiler can be establish from the simulation data to correlate the vapor and liquid load. By doing so, estimation of the tray load can be plotted from plant data as illustrated on Figure 12. The red dash horizontal lines represent the tray vapor loads limits between weeping and jet flood. The blue dash lines represent the tray liquid load limits between unseal downcomers and downcomer choke. As demonstrated by Figure 12 and unlike initial suspicions, the tray load did not decrease with the post

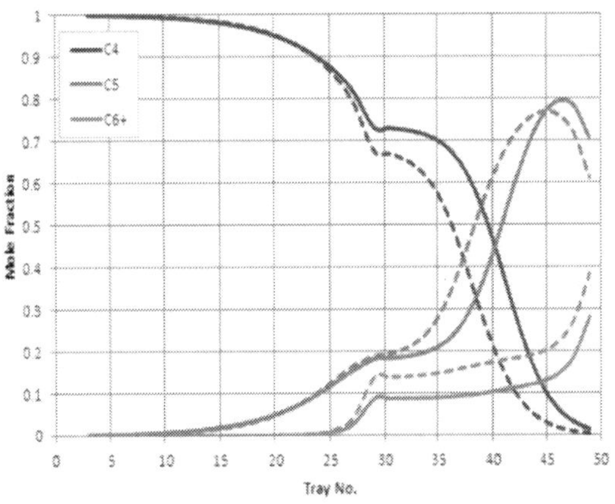

Figure 11a: Composition Profile
Pre Revamp: Solid Line
Post Revamp: Dash Line

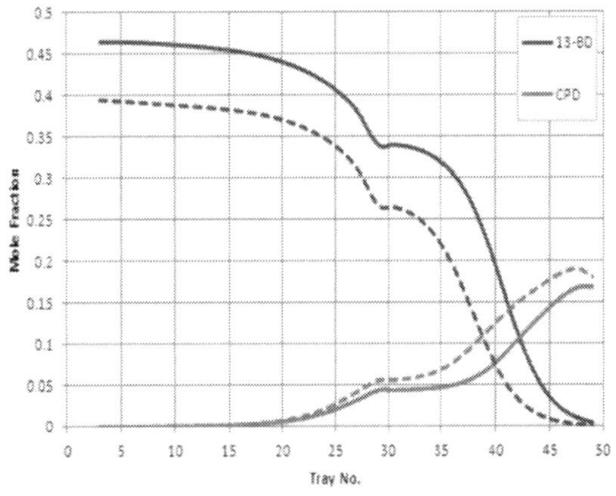

Figure 11b: BD & CPD Profile
Pre Revamp: Solid Line
Post Revamp: Dash Line

revamp conditions but did in fact increase. Accumulation of polymer on the tray deck was suspected to plug the holes with the trays operating below the weep point. Based on the loads illustrated on Figure 12, those speculations can be ruled out.

Residence Time

Residence time is an important factor to understand when correlating results from Fouling Batch Mixed Reactors (FBMR) to actual plant operation. From a flow through standpoint, the liquid residence time on a tray is in the order of magnitude of seconds. However, because of vapor/liquid equilibrium and depending on their volatility; components will cycle internally in the tower and could stay significantly longer then the hydraulic residence time. Estimating the residence time of reactants in a distillation tower is not a straight forward exercise. The vapor/liquid environment along with internal recycle from tray to tray makes it even more difficult. Tracer injection and residence time distribution (RTD) is a method that can be used experimentally for determining the average residence time of components within a system. There are however a number of difficulties that make this method not practical for this evaluation. Building on the concept and using a dynamic simulation of the Debutanizer makes the tracer injection and RTD method accessible. Furthermore, it allows defining a tracer that has the desire characteristic from a volatility standpoint. A pseudo component having the same boiling point, molecular weight and specific gravity as CPD was configured in the dynamic model and used as tracer for estimating the average residence time of CPD. Table 5 compares the residence times evaluated for the pre and post revamp conditions. Unlike expectation, the post revamp conditions exhibit a lower residence time then the pre revamp. The difference is however small within a range of approximately 1 minute. It is interesting to note that the tray residence time is higher in the bottom trays. Which would be expected as CPD recycle to the higher trays is caused by its presence in the vapor phase. Therefore the CPD internal recirculation is higher in the bottom trays given the composition and temperature profile. Based on the above observations it can be concluded that the tray residence time was not a contributing factor for tray fouling in the post revamp period.

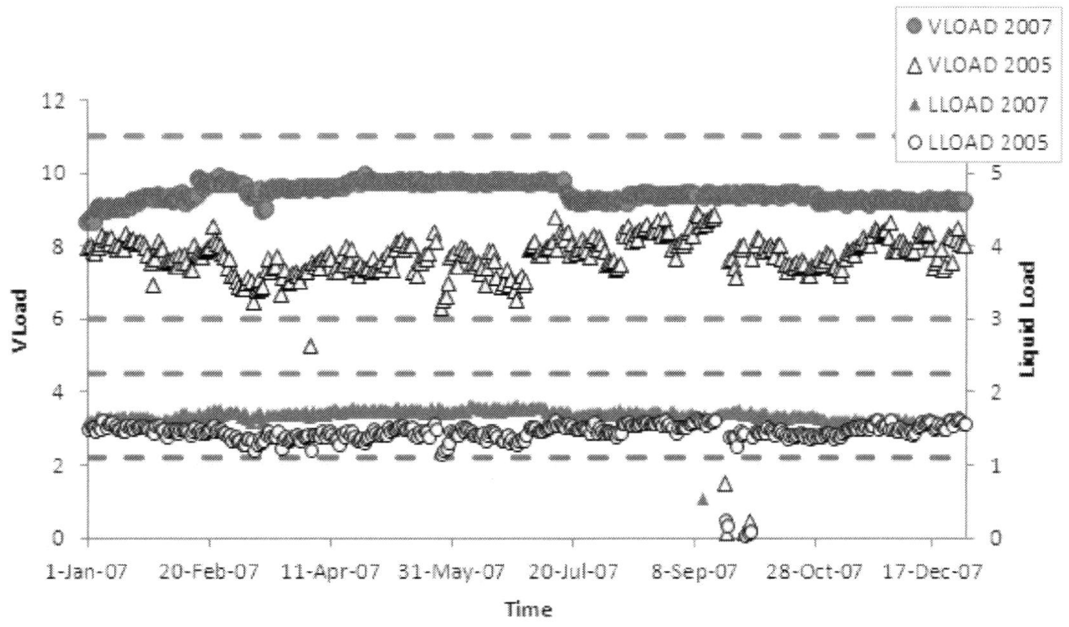

Figure 12 - Bottom Trays Loads

Connecting the Dots

Review of the key process changes stated on Table 1 with the findings from the laboratory work, plant data and process simulation suggest that higher bottom temperature along with higher CPD concentration are primarily responsible for the fouling excursion encountered.
Review of process data demonstrates that even though the plant feedstock transitioned to lighter feed, the Debutanizer throughput remained more or less the same. However, the change in feed diet did indeed cause compositional changes on the Debutanizer. Even though lighter then pre revamp; the feed diet for the period evaluated was at the bottom end of the NGL feed spectrum. It is anticipated that the Debutanizer throughput will reduce significantly at the higher NGL feed slate.
Suspicions that tray weeping would have cause polymers to plug the holes have been ruled out by comparing the tray loadings against the expected weep point. The post revamp loadings were also found higher than those evaluated for the pre revamp

Table 6 - CPD Average Residence Time Estimate

Tray	2005 Average Residence Time (min)	2007 Average Residence Time (min)
40	39.8	40.0
42	41.5	41.2
43	43.2	42.5
45	45.0	43.8
46	46.8	45.4
48	49.7	48.1
Tower	53.4	52.1

conditions. However, tower slumps during plant upsets or shut down may cause any chunk of solids or polymers to block the holes. Unless already plugged, the jet flow of vapor is believed able to keep the hole free from any material during normal operation of the tower.

The reduction in tray weir height from 4" to 3" is not a contributing factor. This reduction would in fact reduce the weep point and should have a favorable impact in reducing tray liquid residence time as well.

Review of the Debutanizer operation concluded that the tower bottom recycle to the sump with a cooling water exchanger was providing little if no benefits to the operation. In fact, it is adding unnecessary load to the reboiler. Recommendations were issued to operation for terminating this practice.

Tray and tower residence time estimates are significantly lower than those observed in the FBMR reactors. However, one needs to realize that the accumulation of polymers on the trays is the result of years of operation. A rough estimate of gum formation has been done using the data from Figure 6 at 100 °C, the estimated tower residence time and the tower bottom rate. This estimation is based on a simple homogeneous continuous system where the outlet rate of material is equivalent to its rate of formation. In other words, none of the material accumulates in the system. A rate of 6.5 lb/hr was estimated, which is equivalent to approximately 57 klb of solids in one year of operation. Obviously, not all of this material is contained on the tray and a fairly high percentage is flushed out with the tower bottom liquid. Furthermore, the use of anti-foulant reduces the rate formation.

Conclusions

The characterization test results reveal that: the fouling deposits from the debutanizer tower were mostly organic; iron and oxygen were the main elements in the inorganic portion of the fouling deposits; cyclopentadiene was the pre-dominant fouling precursor in the debutanizer deposits and gums; the fouling deposit samples were highly cross linked and insoluble in typical organic solvents like toluene and dichloromethane.

Results of fouling experiments indicate that: fouling in the debutanizer tower is temperature dependent; liquid residence time at high temperature regions play a significant role in fouling formation; no significant differences were seen in gum formation between 90 and 100°C; but at 110°C the gum formation was increased by approximately 100%, and the gums formed are largely composed of cyclopentadiene polymers.

Review of plant data and process simulation results concluded that the warmer tower bottom temperature and compositional changes were the major contributors to the fouling excursion observed. After these findings, some adjustments were made to the fouling inhibitor injection program.

It is important to note that each debutanizer tower is unique in terms of operating conditions and feed compositional profile. This means that no exact prescriptions can be offered to mitigate fouling in all debutanizer towers. Therefore, it is recommended that debutanizer towers be studied separately and fouling mitigation strategies appropriate for the tower conditions be developed accordingly.

In order to get a broader picture of the performance of any debutanizer tower, it is recommended to employ a systematic approach where a combination of various activities such as conducting key analytical tests, performing fouling experiments, gathering historical data, and simulating the tower is undertaken.

Acknowledgements

NOVA Chemicals' permission to publish this work is highly appreciated.

References

Arnst T.C., 2006, Optimize Until It Hurts, 18st Ethylene Producers' Conference, AIChE 2006 Spring Meeting.

AIChE Paper Number 96e

Dissolved Oxygen Content Measurement in Aqueous and Hydrocarbon Liquids

Feridoun Fahiminia

Research Scientist

NOVA Chemicals Research and Technology Centre, Calgary, AB, Canada

Ted Salari

Technical Team Leader

NOVA Chemicals Corporation, Corunna Manufacturing Site, ON, Canada

Steve Schmidt

Senior Technical Coordinator

NOVA Chemicals Corporation, Corunna Manufacturing Site, ON, Canada

Prepared for Presentation at the 2012 Spring National Meeting
Houston, TX, April 1-5, 2012

Dissolved Oxygen Content Measurement in Aqueous and Hydrocarbon Liquids

Feridoun Fahiminia

Research Scientist

NOVA Chemicals Research and Technology Centre, Calgary, AB, Canada

Ted Salari

Technical Team Leader

NOVA Chemicals Corporation, Corunna Manufacturing Site, ON, Canada

Steve Schmidt

Senior Technical Coordinator

NOVA Chemicals Corporation, Corunna Manufacturing Site, ON, Canada

Abstract: Dissolved oxygen in aqueous and organic liquids affects a variety of process operations. Even at trace levels, it can lead to fouling in processing units such as separators, heat exchangers, pumps, compressors, and reactors, causing unplanned shutdowns. It can also promote corrosion in processing units and storage tanks. This study was initiated to develop a proper liquid sampling procedure which minimizes air contamination during sampling and to measure the dissolved oxygen content of different liquids employing GC-MSD method and a direct measurement using an oxygen analyzer sold under the trademark Orbisphere by Hach Lange. After developing an appropriate experimental setup using the aforementioned methods, the dissolved oxygen content of various liquids such as toluene, normal heptane, 1-hexene, heavy reformate, and caustic aqueous solutions were measured. The results are in reasonable agreement with reported literature under similar conditions. Selected samples were purged with air to establish oxygen saturation levels, or purged with nitrogen to evaluate the use of this gas to reduce dissolved oxygen concentrations.

Introduction

1 - Autoxidation Fouling

Oxygen presence in liquid hydrocarbons affects a variety of process operations, leading to unplanned shutdowns. Its presence, even at trace levels in hydrocarbons, can lead to fouling. It can promote corrosion either in storage tanks (containing organics, water, or a mixture of both), or processing units.
Although rigorous exclusion of oxygen, or addition of anti-oxidants may eliminate such fouling, in many industrial situations fouling cannot be prevented. The chemical mechanisms for such fouling are very complex, nevertheless the generation of fouling precursors can be described by the following simplified equation (Li and Watkinson, 2007):

Hydrocarbons + O_2 ⟶ Hydroperoxides (soluble)

Hydroperoxides (soluble) ⟶ Soluble polymeric peroxides (soluble gum)

Soluble polymeric peroxides (soluble gum) ⟶ Insoluble (gum) polymeric peroxides

Gum will precipitate from bulk solution when its solubility limit is reached, leading to fouling of surfaces, such as in heat exchangers, distillation columns, etc. The limit of gum solubility depends on temperature, and the nature of the gum and the other species in the mixture. Fouling of heat exchangers occurs when soluble gum is converted to insoluble material on the hot surface, or when precipitated gum particles adhere to the heat transfer surface. Hence, this simplified scheme can apply to fouling in different types of process units.
The rate of gum formation depends on the composition and structure of reacting species, the dissolved oxygen content, temperature, and the presence of catalysts. Dissolved oxygen content is one of the variables, which has attracted more attention during recent decades. Li and Watkinson (2007) examined the effects of dissolved O_2 on peroxides formation in blended naphtha oxidation at a constant temperature (120°C) with air-N_2 mixtures, covering the range of 11% to 100% air. Their findings are presented in Figure 1. It shows that a decrease of the O_2 partial pressure generally resulted in a decreased peroxide level and a prolonged induction time, with the exception of the data at 11% air.
One mitigation technique would be minimizing the dissolved oxygen content of the processing streams. But an important prerequisite is determining how to measure dissolved oxygen concentration. Several researchers have tried to develop measurement methods applicable either to batch samples collected from plant site at analytical laboratories, or to flowing streams at the plant site. For the laboratory measurements of batch samples, collecting hydrocarbon liquid samples without air contamination is the main challenge. However, for on-line measurements, fluctuations in process variables such as temperature, pressure, and the presence of particles may interfere with the measurement.

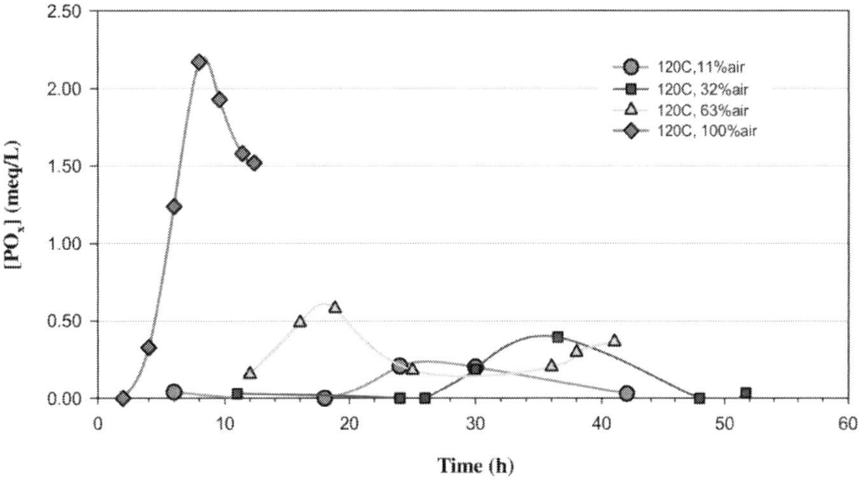

Figure 1 – Effect of Oxygen Presence on Peroxides Formation in Blended Naphtha Oxidation at 120°C

This work was initiated to address oxygen-induced fouling problems causing operational upsets. Three important tasks were identified to proceed with this project:

- Develop dissolved oxygen measurement setups appropriate for batch samples or flowing streams at plant sites.
- Apply developed methods to simple or complicated samples, and then compare the findings with literature data.
- Develop sampling methods without air contamination as well as facilitating sample processing.

2 - Dissolved Oxygen Detection Techniques

There are a variety of measurement techniques for determining the solubility of dissolved oxygen in liquid hydrocarbons. These techniques have used polarographic and electrochemical procedures, in addition to sophisticated gas chromatographic procedures. Gas solubilities can be determined physically; for instance, a volume change can be measured by removing the dissolved gases from the liquid matrix (Taylor and Wallace, 1967). Another technique of measuring oxygen content in a real time fashion is by using an electrolytic-probe contained in a room-temperature jet fuel.

Petrocelli and Lichtenfels (1959) were the earliest known authors to adopt gas chromatography to measure dissolved oxygen in petroleum fractions. Chromatographic techniques for measuring dissolved oxygen are common, but the associated precision has been historically low (error greater than 3%) compared to physical measurements. Taylor (1967), Braun and Hausler (1976), Asomaning and Watkinson (1995), Crittenden and Khater (1987), and Wilson and Watkinson (1994) adopted ASTM D2779 and ASTM D3287 methods to estimate dissolved oxygen concentration using the partial pressure of oxygen in the vapour phase of other systems (No measurement of the dissolved oxygen was made).

3 - Principles Behind Orbisphere Analyzer Performance

Orbisphere analyzers are designed for on-line measurement of various gases, including oxygen, in aqueous streams. A picture of the analyzer sensor head is illustrated in Figure 2, showing important parts of the measurement sensor. The analyzer uses electrochemistry principles to detect dissolved oxygen in liquids. The oxygen sensor has a centre electrode (the cathode) and a counter electrode (anode) immersed in an electrolytic solution. The electrodes and the electrolyte are separated from the sample by the gas permeable membrane. Gas penetrates through the membrane into the cell and dissolves into the electrolyte. It then undergoes a reaction at the cathode, causing a measurable electric current to flow. This current will be proportional to the amount of gas entering the cell. The amount of gas entering the cell is proportional to the partial pressure of the gas in the fluid, which can then be converted into a concentration value. Orbisphere analyzers can be installed in-line for process control, or in flow chamber for spot measurements. In storage tanks, the sensors can be used for measuring oxygen concentration either in headspace gases or in the liquid phase.

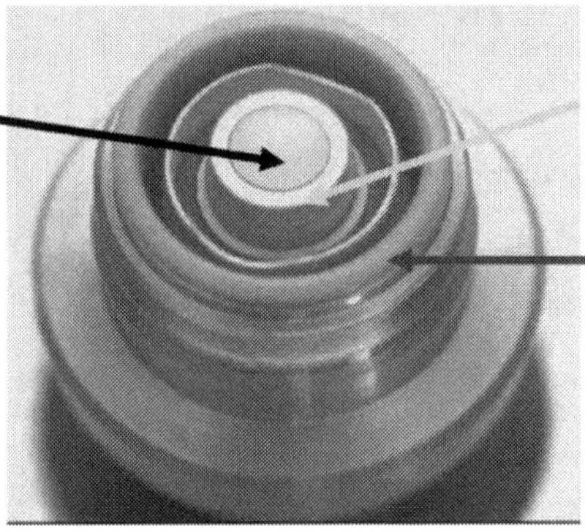

Figure 2 - A Picture of Orbisphere Analyzer Sensor Design

4 - Variables Affecting Orbisphere Analyzer Measurements

A variety of parameters impact Orbisphere analyzer performance, among which the important ones are temperature, pressure, ΔP (pressure drop across the experimental setup including the analyzer), membrane type, purging media, flow chamber orientation, sensor conditioning, flow rate, solution nature, and presence of particles.

Temperature affects the solubility of oxygen in liquid hydrocarbons. In general, as the temperature increases, the oxygen solubility decreases and vice versa. Therefore, from the plant operational point of view, it means that maintaining the liquid streams at higher temperatures lowers the oxygen content, which is desirable. On the other hand, higher temperatures promote chemical reaction fouling, which is not desirable. Hence, it is likely to have an optimal temperature at which both oxygen content and chemical reaction fouling balance each other, leading to fouling mitigation. For the Orbisphere instrument, a temperature compensating curve is built into the analyzer software with a defined mathematical relationship. This implies that the Orbisphere measurements will include the temperature changes into the final outcome. However, it is best to calibrate the unit at temperatures close to the unit operating temperature.

Pressure and pressure drop across any experimental setup, integrated with the oxygen Orbisphere analyzer, are important as well. A higher pressure drop, ΔP, will release part of the dissolved oxygen, interfering with the measurement. In order to overcome this problem, it is recommended to reduce the pressure drop ΔP as much as possible. Membrane type is another variable affecting Orbisphere analyzer performance, especially when using complicated hydrocarbons that may contain corrosive components. They may change the physical properties of the membrane influencing oxygen measurements indirectly.

Depending on the dissolved oxygen content of the liquid, purging media impacts Orbisphere results. It is always recommended to use a nitrogen purge to evacuate any air that may have been trapped inside the experimental setup before making any measurements. In addition, the oxygen sensor should be conditioned with nitrogen.

Flow chamber orientation affects analyzer accuracy and may cause operational problems as well. Figure 3 illustrates a proper orientation of a flow chamber in which the flow pattern of the test fluid minimizes particle adhesion to the sensor membrane, thereby increasing membrane lifetime and reducing operational problems.

Flow regime, solution nature, and presence of particles impact the Orbisphere performance as well. Over the course of this study, several experiments were performed to identify more important variables and take them into account in designing a new experimental setup.

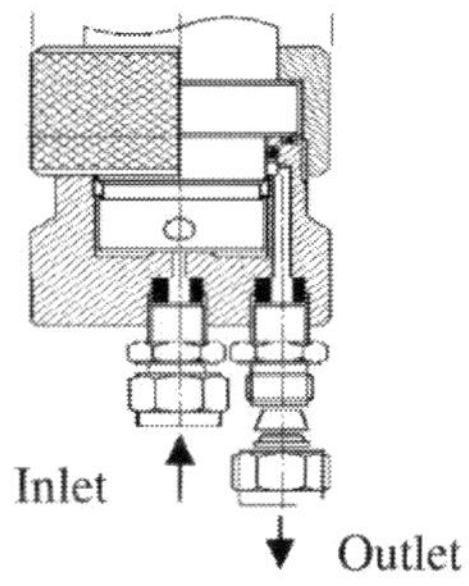

Figure 3 - Orbispher Analyzer Flow Chamber

Experimental Apparatus And Methods

1- Measurement of Dissolved Oxygen Employing GC-MSD

As mentioned earlier, Petrocelli and Lichtenfels (1959) were the earliest known authors to adopt gas chromatography to measure dissolved oxygen in petroleum fractions. After that, several researchers have tried to improve the associated precision. In the present study, a Hewlett Packard 6890/5973 Gas Chromatograph-Mass Selective Detector (GC-MSD) was utilized. The GC portion of the technique provided an isothermal separation (GC oven at 50°C) of the dissolved gases from the rest of the sample. A split/splitless injector was used with a split ratio of approximately 10:1 and an injector temperature of 50°C with helium carrier gas. The temperature of the injector was chosen to provide the best chromatographic peak shape for the dissolved oxygen. Ambient air was used as standard for oxygen measurement by assuming 20.9% (v/v) O_2. A 5 µl Hamilton syringe was used for the air sampling in a 0.1µl interval. Triplicates were performed in each volume.

2- Dissolved Oxygen Measurements Employing Orbisphere Analyzers

In the initial stages of this study, the general configuration, in which nitrogen is used to force liquid samples flowing over the membrane (once-through process), was employed for measuring dissolved oxygen in simple hydrocarbons. However, there were concerns associated with the membrane resistance to corrosive components in hydrocarbons, which could reduce accuracy as well as the membrane life time. New membranes have been developed recently that are more resistant to corrosive components and therefore, membrane related concerns were resolved. In order to be able to control flow rate more efficiently and run experiments when lesser amount of sample is available as well as allow more time to reach steady state conditions, the once-through setup was modified as shown in Figure 4. It consists of a flow loop in which the test liquid is continuously recirculated from a sampling cylinder through a pump, a filter, a rotameter, Orbisphere analyzer, and back to the sampling cylinder. The rotameter enabled us to control and monitor flow rate during each measurement.

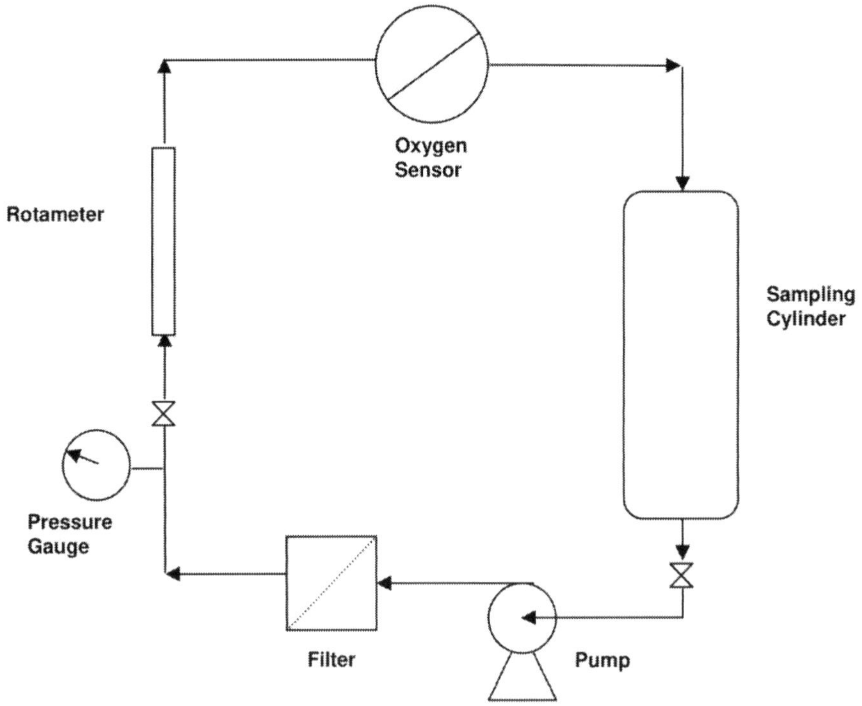

Figure 4 - Schematic of Modified Setup of Dissolved Oxygen Detection Apparatus (DODA)

3- Interpreting Orbisphere Analyzer Results

In spite of the GC-MSD method that measures dissolved oxygen contents of liquids quantitatively, the Orbisphere oxygen analyzer outputs are qualitative. To conduct an experiment employing the Orbisphere analyzer; first, a measurement is made with air (contacting the sensor with air). This yields a value called S, which represents the air solubility in ppm; second, the analyzer should be assembled in a proper experimental setup in which the test sample is forced to flow over the sensor membrane. This will generate another value called C, which is a measure of the dissolved oxygen content (ppm) in the test liquid; finally, Equation 1 can be used to

$$P = 100\frac{C}{S} \tag{1}$$

calculate the percent of saturation P. According to Equation 1, the ratio of the two measurements represents the percent of saturation. In general, C is smaller than S ($P \leq 100$). If $P > 100$, then the test fluid is supersaturated with oxygen. This procedure implies that the final outcome of each Orbisphere measurement is the percent of saturation and not the absolute value of the dissolved oxygen content. In order to find the absolute dissolved oxygen content, one needs to know the saturation concentration of the test sample. For simple hydrocarbons, such as toluene, styrene, heptane and others the saturation concentrations are available in the literature as shown in Table 1 (Hale, 2002).

Table 1: Solubility of Oxygen in Simple Hydrocarbons (Hale, 2002)

Test Liquid		**Solubility Weight (ppm)**
Paraffins	Pentane	118
	Hexane	101
	Heptane	95
	Octane	89
	Decane	68
	Dodecane	54
	Pentadecane	41
Naphthenes	Cyclohexane	76
	Methycyclohexane	83
	Cyclo-octane	54
Aromatics	Benzene	61
	Toluene	67
	Paraxylene	71
Non-organic	Water	8-10

But for more complicated solutions, no data is available. That is the reason why researchers have tried to develop other methods to generate such data for complicated hydrocarbon mixtures. For instance, one common method is expressing the saturation concentration of a liquid mixture as a function of composition and saturation concentration of each component. Although it has been employed for simple hydrocarbon mixtures, it is difficult to use it for real liquids that can sometimes comprise of more than 150 components. Therefore, it is essential to employ other techniques to establish saturation concentrations. In this study, the GC-MSD method was used to interpret the Orbisphere results.

Plant engineers are more interested in monitoring the fluctuations of the dissolved oxygen content of various streams. Therefore, from the plant operational point of view, the oxygen Orbisphere analyzer is a useful tool that may be installed in-line at the plant site for monitoring dissolved oxygen content fluctuations. Also, by interpreting the Orbisphere analyzer output using other techniques, the in-line measurements potentially can provide historical data to plant engineers who may wish to develop fouling models.

4- Importance of Sampling

Sampling is an important prerequisite for dissolved oxygen content measurements in batch samples. For instance, if the sample is contaminated with air during sampling, regardless of the measurement accuracy, the final result will be useless. One of the potential solutions to this problem would be to completely fill the sampling cylinders, but this approach has a safety concern associated with temperature variations. If the sampling cylinder is loaded without any head space and during the transportation temperature increases even in the 5 to 10°C range, the pressure will rise drastically. A high pressure increase may cause a hazard to people and is not accepted by Transport of Dangerous Goods (TDG) regulations.

To avoid this problem, it was decided to use piston cylinders instead. These cylinders are equipped with a piston that divides the cylinder into front and rear chambers. The liquid sample is charged into the one side of the piston and the other side can be filled with an inert gas like nitrogen. This configuration allows the liquid to be expanded without a significant pressure increase, in case a temperature rise occurs. In order to eliminate sample contamination with air during sampling, a step-by-step procedure was developed.

Results And Discussion

1- GC-MSD Results

Before making any real measurements with the GC-MSD method, the main effort was to test and validate its performance. This was done using heptane as the test fluid. After performing several measurements, a comparison was made between our results and the literature data. In addition to that, some samples were purged with air followed by measuring their dissolved oxygen contents. Also, some of the samples were purged with nitrogen to see whether the oxygen content drops. The results are presented in Table 2.

Table 2: GC-MSD Results of Heptane (Either Purged or Non-purged)

	Purged with N_2	Purged with Air	Lab Sample
Oxygen Content (ppm)	51	102	102
Oxygen Content (ppm) from Literature	-	-	95

Table 2 reveals that for heptane, our measurement result of 102 ppm is in good agreement with the literature data value of 95 ppm. Results of oxygen measurements using air-purged heptane illustrates that there is no difference between the oxygen content of heptane samples kept in the lab and that of air-purged samples. Also, Table 2 shows that the oxygen content of heptane decreased from 102 ppm to 51 ppm when it was purged with nitrogen. This observation is consistent with findings of previous studies. For instance, Figure 5 shows a case history of an HDS (hydrodesulphurization) unit for Virginia naphtha before and after retrofitting the unit with an oxygen stripper (Vadekar, 2003). These results suggest that several techniques can be employed to reduce dissolved oxygen content.

Further measurements were conducted for heptane samples purged with nitrogen for longer times, but the results were not consistent. Reviewing measurement procedures revealed that GC-MSD is accurate only when the sample is either saturated with oxygen, or the oxygen content is above 10% saturation level. On the other hand, when the

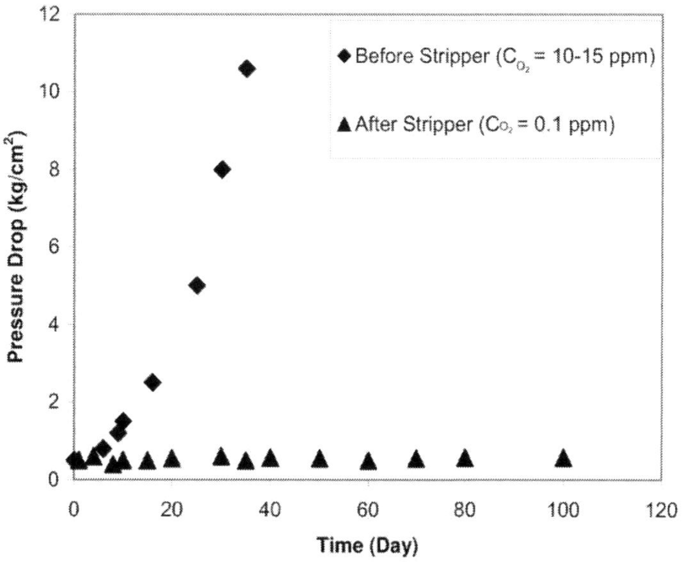

Figure 5 - Impact of Oxygen Stripping on Oxygen Induced Fouling in an DHS (Dehydrosulphurization) Unit (Vadekar, 2003)

sample has trace levels of oxygen, GC-MSD will not generate accurate results due to sample contamination during sample introduction to GC. Hence, it was decided to first employ GC-MSD to establish saturation concentration levels; second, employ Orbisphere analyzers to convert relative dissolved oxygen measurements to absolute values.

2- Orbisphere Oxygen Analyzer Results

The setup (Figure 4) in which the Orbisphere Oxygen Analyzer used is called the Dissolved Oxygen Detection Apparatus (DODA). The first set of measurements was conducted employing a simple hydrocarbon like toluene to test and validate the performance of the DODA. In the second set, more complicated samples were employed as the test samples.

Experiments were performed under the ambient conditions, and like the GC-MSD measurements, toluene was purged with either air or nitrogen followed by an oxygen measurement to see the influence of purging. The results are presented in Table 3. It is seen that our result of 62 ppm is in reasonable agreement with that of literature value of 67 ppm. In addition, there was no difference between the oxygen content of the air-purged sample and that of the toluene kept in the lab. On the other hand, purging the sample with nitrogen reduced the dissolved oxygen from 62 to 43 ppm.

Table 3: DODA Results of Toluene (Either Purged or Non-purged)

	Purged with N_2	Purged with Air	Lab Sample
Oxygen Content (ppm)	43	62	62
Oxygen Content (ppm) from Literature	-	-	67

3- Employing GC-MSD & Polarographic Techniques

As mentioned earlier, GC-MSD generates absolute values for dissolved oxygen contents, whereas Polarographic techniques generate relative values. Therefore, it is not possible to compare both techniques unless literature data is used as a reference base. In general, Polarographic techniques are appropriate for in-line measurements at plant sites where engineers are interested in variation of dissolved oxygen contents rather than absolute values. It is important to note that limited data is available in the literature and therefore, for some simple hydrocarbons whose data are not available, accurate measurement techniques should be employed. Also, process samples are complicated and comprised of several compounds. This makes the measurements challenging and time-consuming.

Further measurements were conducted employing three samples; 1-hexene (for which no data is available in the literature), heavy reformate (a mixture of several compounds), and a caustic solution (an aqueous base solution). Both GC-MSD and Polarographic techniques employed for 1-hexene and heavy reformate, whereas for the caustic solution only Polarographic method used. As mentioned earlier, GC-MSD and Orbisphere analyzer were used sequentially to complete each measurement.

For 1-hexene samples, in addition to regular measurements, air and nitrogen purge was used and simultaneously dissolved oxygen content was measured and recorded. Figures 6 and 7 show the oxygen content variation (as a function of time) for air and nitrogen purge. Figure 6 shows that the sample becomes saturated with air quickly. Figure 7 shows that when the sample is purged with nitrogen, dissolved oxygen contents drop rapidly.

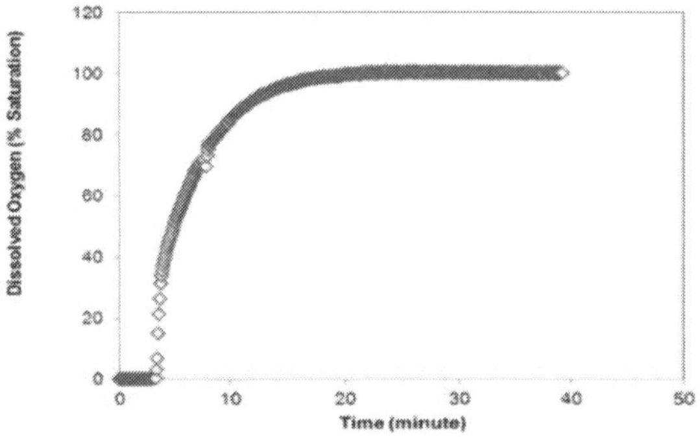

Figure 6 - Impact of Air Purging On Dissolved Oxygen Content of 1-Hexene

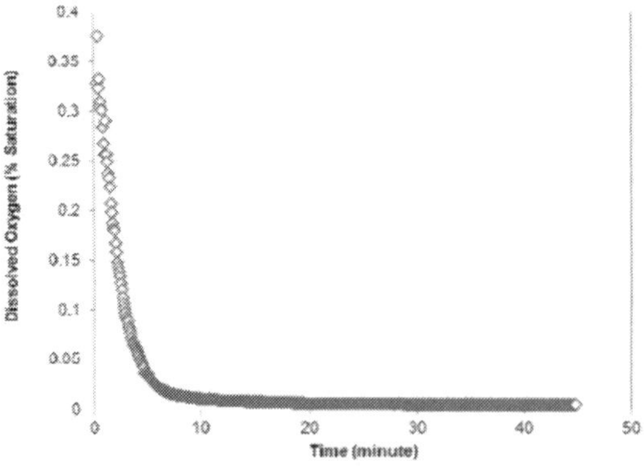

Figure 7 - Impact of Nitrogen Purging On Dissolved Oxygen Content of 1-Hexene

Measurement results for toluene, heptane, 1-hexene, heavy reformate, and caustic solution (10%) are presented in Table 4. It is seen that for toluene and heptane, saturation levels of dissolved oxygen are 62 and 102 ppm, which are in reasonable agreement with literature data, i.e. 67 ppm (for toluene) and 95 ppm (for heptane). For 1-hexene and heavy reformate, no data is available in literature, and this study generated values of 84 and 42 ppm for 1-hexene and heavy reformate, respectively. For caustic solution (10%), the solubility was measured using Orbisphere analyzer and the saturation level was found to be 8.0 ppm, which is close to 8.2 ppm reported in literature. This confirms that oxygen solubility in organic solutions is much higher than that in aqueous base compounds.

Table 4 – Dissolved Oxygen Measurement Results

Liquid	Measured Solubility (ppm)	Literature Solubility Data (ppm)
Toluene	62	67
Heptane	102	95
1- Hexene	84	NA*
Heavy Reformate	42	NA*
Caustic Solution (10%)	8.0	8.2

*Not Available

The next step of this study would be to conduct similar measurements for process streams as well as investigate the possibility of extending the methods to on-line measurements at plant sites.

Conclusions

1- It was shown that the GC-MSD results for dissolved oxygen content are accurate when oxygen concentration is high enough, i.e. close to saturation levels, whereas the polarographic method can detect trace levels of oxygen. The GC-MSD technique generates absolute values of dissolved oxygen content, but it is accurate when the oxygen concentration is high enough. On the other hand, the polarographic methods generate relative values of dissolved oxygen content, but are accurate even at trace level concentrations. Therefore, both methods should be used to measure absolute values of dissolved oxygen content in liquid hydrocarbons (simple or complicated).

2- In order to measure dissolved oxygen in batch samples, a new experimental setup (DODA), integrated with an Orbisphere analyzer, was developed and tested for simple and complicated liquid hydrocarbons. For heptane and toluene, the saturation concentrations of oxygen were found to be 102 and 62 ppm, respectively. The differences between measured values of this work and literature data are less than 8%, indicating a reasonable agreement between our results and the literature data. No data is in the literature for 1-hexene and heavy reformate, and this study generated values of 84 and 42 ppm for 1-hexene and heavy reformate, respectively. For caustic solution (10%), the solubility was measured using Orbisphere analyzer and the saturation concentration was found to be 8.0 ppm, which is close to 8.2 ppm reported in literature. Also, measurement results of dissolved oxygen in 1-hexene while simultaneously purging the sample with nitrogen indicated a quick reduction of oxygen concentration suggesting that nitrogen purging is a useful technique for reducing dissolved oxygen in liquid hydrocarbons.

3- It is critical to note that proper sampling is the most important prerequisite of any oxygen measurement. In order to minimize air contamination during sampling as well as meet TDG regulations, it was decided to employ piston cylinders for sampling.
4- In addition, well-defined procedures should be strictly followed during liquid sampling to eliminate air contamination. The polarographic technique is the most practical method that can be employed for on-line oxygen measurement at plant sites.

Acknowledgements

NOVA Chemicals permission to publish this work is highly appreciated.

References

1. Li, Y-H. & Watkinson, A. P.; "Foulant Formation in Blended Naphtha Under Oxygenated Conditions", Final Report, 2007.
2. Vadekar, M.; "Oxygen Contamination of Hydrocarbon Feedstocks", PTQ (Winter Issue), January 2003, pp. 87-93.
3. Hale J.M., "Oxygen Measurements in Non-Aqueous Process Fluids", Orbisphere Technical Bulletin, Unpublished Study, 2002.
4. Wilson, D.I., Watkinson, A.P., 1995, Model Experiments of Autoxidation Fouling Parts A & B, Trans. I. Chem. Eng. Vol.73, pp. 59-76.
5. Asomaning, S., Watkinson A.P., Heat Exchanger Fouling by Olefin-Kerosene Mixtures, *Can. J. Chem. Eng.,* Vol. 70, 1992, pp.444-451.
6. Taylor, W. F. & Wallace, T. J.; "Kinetics of Deposit Formation from Hydrocarbon Fuels at high Temperatures – General Features of the Process", Ind. Eng. Chem. Proc. Res. Dev., 6, 1967, pp. 258-262.
7. Crittenden, B. D. & Khater, E.M.H.; "Fouling from Vapourizing kerosine", Journal of Heat Transfer (Trans, ASME), Vol. 109, 1987, pp.583-589.
8. Braun, R.; Hausler, R. H.; "Contribution to Understanding Fouling Phenomena in the Petroleum Industry", Paper 76-CSME/CSCHE-23, 16th National Heat Transfer Conference, Aug 8-11,1976, S. Louis. MO.
9. Petrocelli, J. A.; Lichtenfels, D. H. "Determination of Dissolved Gases in Petroleum Fractions by Gas Chromatography", Anal. Chem., 31, 1959, pp. 2017-2022.

96f

NATURALLY OCCURRING RADIOACTIVE MATERIAL (NORM): DETECTION & PROTECTION IN THE PETROCHEMICAL INDUSTRY

Justin Harris
Radiation Safety Services Manager
Quantum Technical Services, LLC

Prepared for Presentation at the 2012 Spring National Meeting
Houston, TX., April 1 – 5, 2012

NATURALLY OCCURRING RADIOACTIVE MATERIAL (NORM): DETECTION & PROTECTION IN THE PETROCHEMICAL INDUSTRY

Justin Harris
Radiation Safety Services Manager
Quantum Technical Services, LLC

Abstract: Naturally Occurring Radioactive Material (NORM) is very prevalent in all aspects of oil and gas production. Depending on geological formations, the rate of contamination from the rate of production also varies. Radon progeny contamination of natural gas production facilities contribute to possible exposure to long lived isotopes plating out on internal surfaces of equipment. NORM contamination is fairly common with facilities that remove ethane and propane from natural gas. This paper will discuss the instrumentation needed to effectively monitor for NORM, overview of when decontamination will be needed, the government regulations and the NORM training requirements for personnel working in a facility containing NORM. Having a NORM plan in place during turnaround activities will dramatically reduce cost, as well as downtime, and also help avoid unnecessary delays.

Naturally Occurring Radioactive Material (NORM): Detection and Protection in the Petrochemical Industry

Survey Instruments

There are various types of survey instruments that can be used when dealing with radioactive material. It is imperative that a survey instrument be used that is suitable for the energies and types of radiation being emitted. In the petrochemical industry, possible exposure to alpha, beta and gamma types of radiation are very likely.

External Screening Tools

During routine operation of a gas processing facility, small traces of Radon decay products can sometimes be detected on outer surfaces of equipment. The majority of the time these external gamma readings are low (μR/hr), but, they sometimes can be concentrated at levels that would warrant precautionary measures be taken to control general public exposure. During the decay of Radon gas (alpha emitter) a very short lived isotope is produced (Bismuth -214). Bismuth-214 is primarily a beta emitter but carries a fairly energetic gamma peak, which is often detected on the outer surfaces of process equipment. External surveys of process equipment during normal operations are typically used as a screening tool to identify areas of possible internal contamination. The instrument of choice for these measurements would be a handheld survey instrument with a scintillation detector. This type of survey meter should be capable of reading radiation exposure rates at the μR/hr level. It is important to note that a lack of an external reading does not necessarily mean a lack of internal contamination. Therefore, it is important to make an internal assessment of contamination levels once the equipment is opened for maintenance. Areas such as pump inlets, pump discharge, control valves, compressors and reflux pumps are common areas where elevated readings may be found. Areas that tend to have a temperature, pressure and/or direction change, tend to yield the highest readings.

Internal Contamination Assessment

During a plant downtime or shutdown, equipment is purged to remove any hydrocarbons or other hazardous material from the lines in order to be opened safely. During this time, some Radon gas will also be removed. However, the radon gas is constantly decaying and depositing solid particles such as Lead-210 (Pb-210) and Polonium-210 (Po-210) on the internal surfaces of equipment. Pb-210 has a half-life of ~22.3 years. Po-210 has a half-life of 138.38 days. These alpha and beta isotopes are the major concern when opening up and performing maintenance on equipment. The concern with alpha and beta particles is ingestion or inhalation.

Once the equipment has been purged, the focus should then be on using a survey instrument that is capable of detecting alpha and beta particles. The most commonly used detector for measuring alpha and beta surface contamination is the G-M "pancake" detector. The pancake detector is capable of detecting alpha, beta and gamma radiation. Survey meters that are coupled with pancake detectors typically read out in Counts Per Minute (CPM). The pancake detector will need to be placed near the internal surfaces of the equipment, but avoiding actual contact, to eliminate cross contamination. The limits for surface contamination release criteria are generally in Disintegrations Per Minute (DPM), so make a conversion based on the efficiency of the detector being used. (Any reputable licensed company can perform an efficiency on any detector at the time of calibration using standardized NIST traceable sources.) The meter reading that is taken at the surface is called the "fixed" reading. At this point, take another reading which will be considered a "removable" contamination test. This would simulate how much radiation would be removed during normal working conditions on or around the surface of the equipment. The amount of removable radioactive material per 100 cm2 of surface area should be determined by wiping that area. With a dry filter or soft absorbent paper, apply moderate pressure, and assess the amount of radioactive material on the wipe with an appropriate instrument of known efficiency. All surveys and efficiency determination shall be made with the detector's active surface no greater than one centimeter from the surface being surveyed, wipe being analyzed, or source being used. A scaler must be used when evaluating wipe samples and count times must be sufficient to detect 10% of the applicable limit with 95% confidence that the activity would be detected. Typical surface contamination limits for Pb-210 and Po-210 are:

- Fixed – 5,000 dpm/100cm^2
- Removable – 1,000 dpm/100 cm^2

These limits are for equipment to be released as "unrestricted." Maintenance can still be performed on the equipment as long as the appropriate Personal Protective Equipment (PPE) is used.

NORM Decontamination

NORM contaminated equipment in the petrochemical industry has been a health and environmental concern for a very long time. Usually during normal maintenance, there is no reason to decontaminate equipment, as long as it stays at the facility and will be put back into service. However, the workers performing the maintenance on that particular piece of equipment must be protected. If the contaminated equipment is going to be shipped off-site, it must either be below the regulatory limit or shipped to a company that is specifically licensed to work on NORM equipment.

Any company performing NORM decontamination services must be specifically licensed to perform such services. If the licensee is covered by a particular state's radiation control program, reciprocity must be requested before performing work in another state. How do you determine if the activity is considered to be "decontamination"? There has been a lot of confusion over routine vs. non-routine maintenance over the past 20 or so years. Let's clarify. Any maintenance that provides a different pathway for exposure than is found in daily operations and that increases the potential for additional exposure is NOT considered routine. Each case by case scenario should be evaluated to ensure compliance with state and federal regulations.

Regulations

Several states have adopted specific regulations for the control of NORM. States that do not choose to have specific NORM regulations are still required to comply with the general requirements found in their ***"Standards for Protection Against Ionizing Radiation"***. All applicable exposure and contamination limits found in these sections must be followed even though a particular state may not have NORM specific regulations. The following states in Table 1 below have specific NORM regulations:

Table 1

Arkansas	Georgia	Louisiana
Maine	Mississippi	New Jersey
New Mexico	New York	Ohio
Oregon	South Carolina	Texas
	West Virginia	

It is recommended that anyone dealing with propylene, ethane and propane processing streams conduct baseline and periodic surveys to monitor for the presence of NORM. It is important to note that the survey results can vary from time to time due to a change in the feedstock.

Training

All facilities having identified the presence of NORM, shall train all personnel according to their job task. Basic NORM training is usually given to individuals that only work around NORM equipment but do not have the responsibility of performing the monitoring. Some topics covered in Basic NORM Training might include: Introduction to NORM, Health Effects, Radiation and Contamination, Methods of Controlling Dose, Personnel Monitoring and State and Federal Regulations.

If an individual is required to physically perform the radiation monitoring or testing, a more in depth training program would be required. This training would include all of the items listed above, as well as, thorough training on specific instrumentation and survey documentation. Before performing in-house monitoring, it is imperative to choose the correct survey instruments for the particular need. Contact a radiation safety consultant to gain knowledge or consulting for a certain particular need(s).

Proactive NORM Program

It is very important to have a proactive NORM program in place to eliminate problems which could be encountered during routine or non-routine maintenance. A written program should be in place. This program would outline all equipment maintenance procedures for in-house, as well as, procedures for shipping to an outside facility.

Some things to consider when planning a plant turnaround/shutdown:

- Will potentially contaminated equipment need to be shipped off-site? If yes, it must be decontaminated. Decontamination vendors require several days of set-up, as well as notification to state agencies.
- Will your in-house employees be working on the contaminated equipment? If yes, monitoring is required and surveys have to be performed to ensure no contamination is present. Proper PPE must be addressed according to the task being performed.
- Will entry occur inside vessels that are potentially contaminated? If yes, constant monitoring of individuals will be required to ensure no one is leaving a restricted area with NORM contamination on their person.
- Will vessel internals such as trays be replaced in NORM suspect equipment during the turnaround? If yes, what is the procedure for the disposal of the scrap metal?

References

1. 25 TAC 289.259 *"Licensing of Naturally Occurring Radioactive Material"*

2. Canadian Association of Petroleum Producers "*Guide – Naturally Occurring Radioactive Material"* June 2000 pp. 6-9

3. Tenorm.com "*States with TENORM regulations*", Table 17

4. Wilson, William F. (1994) "NORM – *A guide to Naturally Occurring Radioactive Material*", pp. 171-174.

AIChE Paper Number 97a

Piping Life Extension Through Assured Mechanical Integrity of CUI Prone Piping at Dow Hydrocarbons Plants

Doug Walker

Reliability Associate

The Dow Chemical Company

Prepared for Presentation at the 2012 Spring National Meeting

Houston, TX, April 1- 5, 2012

Piping Life Extension Through Assured Mechanical Integrity of CUI Prone Piping at Dow Hydrocarbons Plants

Doug Walker

Reliability Associate

The Dow Chemical Company

Abstract:

The purpose of Dow's Global Mechanical Integrity Safety Standard (GMISS) is to reasonably ensure that equipment used to process, store or handle chemicals are designed, constructed, installed and maintained to help minimize the risk of chemical releases. This standard ensures compliance with applicable legal requirements and incorporates industry consensus, commonly called "Recognized and Generally Accepted Good Engineering Practices" (RAGAGEP).

Piping systems are covered by Dow's GMISS. However, in 2008 a Dow hydrocarbon plant experienced a sudden and complete line failure which was caused by Corrosion Under Insulation (CUI) as defined in Dow's GMISS under API 571 damage mechanism 46. API 571 damage mechanism 46 covers corrosion resulting from water trapped under insulation or fireproofing on carbon steel, low alloy steels, 300 Series Stainless Steel and duplex stainless steels.

This paper summarizes the approach taken by Dow's Hydrocarbons (HC) Business to reasonably ensure the mechanical integrity of CUI prone piping including:

- Organizational Requirements
- Work Processes
- Registration
- Inspection
- Repair
- QA/QC

Introduction

The purpose of Dow's Global Mechanical Integrity Safety Standard is to reasonably ensure that equipment used to process, store or handle chemicals are designed, constructed, installed and maintained to help minimize the risk of release of chemicals. This includes equipment such as piping, pressure vessels, storage tanks, relief devices and expansion joints.

Dow's GMISS ensures compliance with applicable legal requirements and incorporates industry consensus, commonly called "Recognized and Generally Accepted Good Engineering Practices" (RAGAGEP).

On Mar. 13, 2008 a DOW HC plant experienced a sudden and complete line failure on an 8 inch regeneration line (figure 1) caused by CUI (corrosion under insulation.) Dow immediately took steps to address the issue.

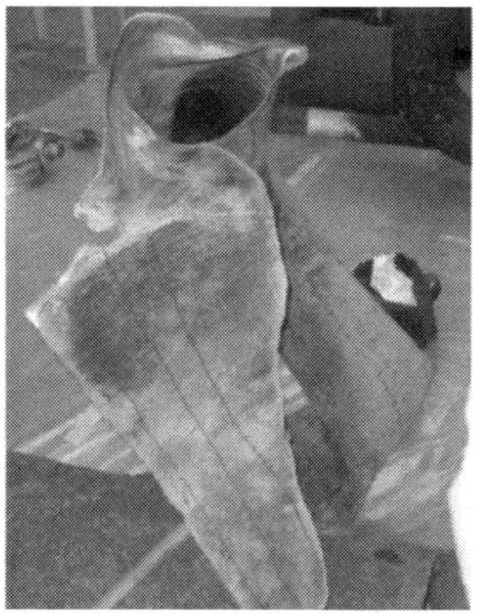

Figure 1 – Failed Pipe

The Call to Action

This incident triggered a call to action for Dow HC plants to immediately define, identify and inspect those piping systems deemed to be most critical as related to CUI susceptibility and consequence. For Dow HC these "critical" piping systems were defined as those lines associated with the Cracked Gas Dryer Regeneration System. These regeneration systems by design are in cyclical operation; ambient temperate until mole sieve regeneration is needed at which time temperatures may reach 230 ^{o}C.

By the end of 2008 Dow's HC plants worldwide had reviewed previous inspection data on their "critical" piping systems to ensure mechanical integrity. In some cases, insulation was removed and a visual inspection was completed (figures 2, 3, 4, 5). All deficiencies found were resolved immediately.

These initial inspections and the overall inspection process were of course very high profile in Dow and highly scrutinized by Dow. This was Dow HCs' opportunity to analyze the entire work process and use the lessons learned to ensure the effectiveness of their inspection program.

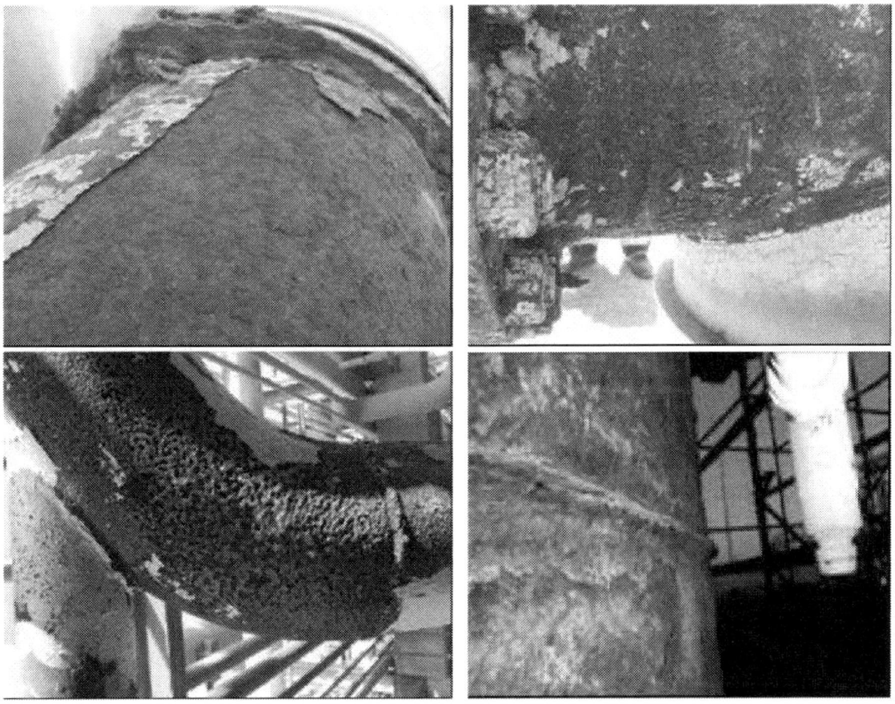

Figures 2,3,4,5 – CUI Examples

CUI Piping Mechanical Integrity Process

Organizational Requirements

In 2008 Dow's HC Business created an internal "Olefins PCE Steering Team" consisting of the HC Business Reliability & Maintenance Director, HC Process Safety Director, HC Reliability Associates and a Dow Global Inspection Associate (chart 1). This team was tasked with designing a process to ensure the mechanical integrity of Dow's HC Facilities worldwide including defining all work processes, organizational requirements and creating an environment to stimulate "smart" inspection techniques in order to satisfy the vision of "Elimination of CUI Susceptibility".

The effectiveness of the process includes the following conditions: standards must be defined and followed; specifications must be available and followed; craft requirements must be communicated and validated; quality must be verified.

THE GRAPHIC ON THIS PAGE WAS NOT AVAILABLE ON THE E-MEDIA VERSION OF THIS PAPER.

Chart 1 - Organization

Each Dow HC plant shall have a Local Plant Assessment Team typically led by the maintenance engineer and consisting of a process knowledgeable person, qualified inspector, maintenance representative, (and metallurgist as needed). This team will meet as needed to review recent inspection information and to plan future inspections. This team has the authority to accept inspection reports, request more data, plan any required mitigation or require immediate action.

The Olefins PCE Team led by the HC reliability associate is made up of the Local Plant Assessment Team leaders, qualified inspectors and HC maintenance managers. This team is tasked with ensuring that all operating discipline, best practices, and inspection experiences are shared, followed, and updated as well as providing training on the new work processes.

Taskforce Teams were created as sub-sets of the Olefins PCE team:

- Assure Mechanical Integrity – Instrumental in leveraging and tracking of key performance indicators (costs, inspections planned/completed, deficiencies found/resolved)
- Inspection – Identify and implement new "smart" inspection techniques
- Repair – Define best repair practices and ensure quality of repairs
- New Coating and Insulation Technology – Seek technologies not currently used in the business. (insulating coatings, thermal spray aluminum...)

Registration Definition

Within Dow the definition of "registration" required clarification as follows: To ensure that all governmental and Dow requirements are met, each piece of equipment shall have a uniform method of field identification, all applicable governmental registration requirements shall be met, all equipment design and maintenance activities (including inspection and repair) are properly documented in a mechanical integrity equipment file, piping circuits containing hazardous chemicals shall be defined for inspection purposes and mechanical integrity equipment files shall be established and maintained. Based on this definition and the findings of the regeneration piping inspections, all steps of the registration and inspection processes were re-evaluated.

Registration Requirements

The Dow HC Business redefined its registration requirements for CUI prone piping. The Olefins PCE Steering Team developed operating discipline to:

1. Prioritize piping circuits as Severe, Serious and Other
   - CUI flagging Severe - Cyclic service (> 10 cycles/year and $\Delta T > 100$ C)
   - CUI flagging Serious - Intermittent service (< 2 cycles/year, transition zones, T< dew point)
   - CUI flagging Other – Remaining CUI Circuits
2. Record the required registration and circuit information such as Class, Chemical, and Operating Conditions etc. This included requirements for identifying circuits on Piping and Instrument Diagrams (P&IDs), for completing isometric drawings with potentially susceptible areas identified and for documenting inspection types, inspection intervals and date of last/next inspection.

Inspection Requirements

Further clarification was also needed in the area of inspections to ensure mechanical integrity. Inspections shall be based on failure mechanisms; installed insulation and coatings shall be reviewed to verify the appropriate use of standards (spot checks may be conducted to verify the status of coating systems), and deficiencies in the insulation or coatings require further review by the Local Plant Assessment Team to develop an action plan. Furthermore, the inspection strategy must be based on Dow's GMISS and API 570 code (table 1 & 2).

Dow will conduct owner visual inspections (completed by trained operators or contractors) at 2.5 year intervals. These inspections must be scheduled in the approved maintenance management system for traceability, shall be reviewed by the Local Plant Assessment Team and the data shall be stored in the Mechanical Integrity Equipment File. External inspections and Non-Destructive Examination (NDE) will be based on API 570 Chapter 6 and shall be completed by API 570 certified inspectors.

Table 1 – Dow GMISS Inspection Timing (ref API 570 Table 6-1)

Class	Thickness Measurement	Visual External
1	5 Years	5 Years
2	10 Years	5 Years
3	10 Years	10 Years

Table 2 – Dow GMISS Required Insulation Removal (ref API 570 Table 6-2)

Class	Approximate Amount of Follow-up Examination with NDE or Insulation Removal at Areas with Damaged Insulation	Approximate Amount of CUI Inspection by NDE at Suspect Areas on Piping Systems within "Susceptible Temperature Ranges as indicated in API 574"
1	75 %	50 %
2	50 %	33 %
3	25 %	10 %

Risk driven external inspections and NDE shall be the strategy to ensure optimal inspection efficiency. As such, pre-inspections shall be required to assess insulation condition, identify potential damaged areas and develop the inspection plan. Each pre-inspection starts with challenging the need for insulation (personnel protection may be satisfied by using insulating coatings or wire mesh cages). Inspection plans shall be marked up on isometric drawings, detailing all suspect and damaged areas. Cost estimates are completed to determine the amount of scaffolding and insulation removal as it may be more cost effective to strip an entire circuit instead of stripping separate sections. Once insulation is removed the paint condition is assessed and scored based on Dow's Paint Condition Level codes. Any evidence of wall thickness reduction is quantified, documented and evaluated to determine if immediate action is required (figures 6 & 7).

Inspection closure requires several validation steps by the Local Plant Assessment Team. Inspection results must be evaluated to determine if the amount of insulation removal and number of inspection points meet Dow's GMISS requirements. All follow-up actions including deficiency resolution and repairs to coatings and insulation must be documented; assigned to an individual; and must include due dates for completion. Finally, subsequent inspection dates must be scheduled and all results must be included in the Mechanical Integrity Equipment File. It is at this point that the inspection can be closed.

Figures 6,7 – CUI Examples

Repair Requirements for Coatings and Insulation

Coating applicators are responsible for proper surface preparation and shall document all QA/QC inspections including pre-job, pre-blast, post-blast, coating application and final acceptance. Applicators shall use only abrasive blasting for all surface preparation including field welds (figure 8). (Power tools may be used in some cases if the surface area is less than 25 cm^2 but hand tool cleaning per SSPC SP-3 is strongly disapproved.)

Figure 8 – Field weld CUI

Prior to coating application, all process conditions (temperature range, cyclic or continuous service) must be agreed on by the coating subject matter expert (SME) and the process representative. The process conditions will be used help to select the appropriate Dow painting repair specification.

Insulation will always be challenged, as permanent removal is the most effective method to eliminate CUI. However, when required for process reasons, the HC Business has defined specific insulation systems for hot (above 27 °C) and cold (below 27 °C) services. Cyclic services are evaluated by the insulation SME for the appropriate system (figures 9 & 10).

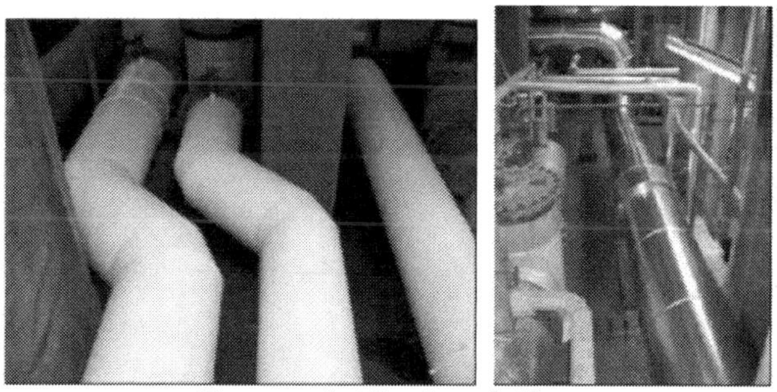

Figures 9,10 - Insulation

For the Dow HC Business the involvement of SMEs and effective QA/QC programs is required, as only with these in place can we reasonably ensure effective coating and insulation projects. The Dow HC Business will therefore field validate each coating project with independent "CIP Level 1" or "FROSIO White Level 1" certified owner's representatives. Training was developed for the Dow insulation standards and specifications and all insulation projects are validated with trained insulation inspectors. Contractor programs are evaluated, all work and QA/QC activities are spot checked before, during and after the job and final documentation for all painting and insulation work shall reside in the Mechanical Integrity Equipment File.

Ensuring Mechanical Integrity

Finally, the process must be sustainable; in other words it must ensure mechanical integrity. In order to accomplish this, training programs were developed for business leaders and local teams. Roles and job duties were defined, parameters for critical activities were developed and organizational requirements were spelled out. But was that enough? No! The last step was to validate the process at each Dow HC plant. Show me the data! Site visits were made to each Dow HC plant worldwide to assess the closed inspections (mechanical integrity equipment files) and progress towards completion of all CUI inspections. Inspection plans were reviewed to ensure codes and standards were being followed, QA/QC documentation was reviewed to ensure all requirements were met, and follow-up documentation was validated to ensure appropriate Fitness-For-Service calculations were available.

Conclusion

In response to a sudden and complete piping failure in 2008, the Dow HC Business used a focused global multi-year initiative to inspect and verify the mechanical integrity of CUI prone hazardous chemical containing piping systems. In the past the process was driven by the inspection discipline. Now, a robust, multi-disciplined process is in place to ensure the continued long term success of Dow's GMISS relating to CUI prone piping. The need for insulation is always challenged, and insulation is always removed where possible thereby helping to eliminate CUI. The quality of inspection and repair has been drastically improved by closing the loop from inspection to repair and back to inspection and by implementing the required organizational changes. We can conclude that these improvements will result in lower Long Term Cost of Ownership (LTCO) at higher reliability levels.

AICHE PAPER NUMBER: 97b

EQUIPMENT LIFE EXTENSION USING THERMAL SPRAY ALUMINUM TO MITIGATE CUI

Chris O'Rear

Maintenance Engineer

The Dow Chemical Company

Prepared for Presentation at the 2012 Spring National Meeting

Houston, TX April 1-5, 2012

EQUIPMENT LIFE EXTENSION USING THERMAL SPRAY ALUMINUM TO MITIGATE CUI

Chris O'Rear

Maintenance Engineer

The Dow Chemical Company

Abstract: Corrosion under insulation (CUI) of carbon and low alloy steels is one of many failure mechanisms that decrease reliability and equipment life while increasing the cost of ownership. Carbon and low alloy steels are most susceptible to CUI when operated in or cycled through temperatures ranging from 10°F (–12°C) to 350°F (175°C). Because CUI is "hidden", the inspection costs to locate this failure mechanism are high when compared to external corrosion of non-insulated piping and equipment. The key to managing CUI is to break the CUI cycle by applying surface coatings that protect the substrate from water and oxygen, the catalyst of corrosion.

Traditional surface coatings such as epoxies and acrylics have a service life of 5-8 yrs before the bond begins to breakdown and corrosion begins. Surface imperfections during the initial application of traditional coatings and/or damage of the coating during installation of the insulation result in an immediate location for CUI to begin. Reduction of CUI requires a surface coating that has a service life of > 20 years and is tolerant to surface imperfections.

This paper will focus on Thermal Spray Aluminum (TSA). TSA is a surface coating that has been proven to have a long service life and is tolerant to surface imperfections due to its anodic properties.

Problem Statement

Corrosion under insulation (CUI) of carbon and low alloy steels is one of many failure mechanisms that decrease reliability and equipment life while increasing the cost of ownership. Carbon and low alloy steels are most susceptible to CUI when operated in or cycled through temperatures ranging from 10°F (–12°C) to 350°F (175°C). Because CUI is "hidden", the inspection costs to locate this failure mechanism are high when compared to external corrosion of non-insulated piping and equipment. The key to managing CUI is to break the corrosion cycle.

What is Corrosion?

Corrosion or the corrosion cycle is a natural process that reduces the binding energy in metals. The corrosion cycle involves a metal atom (M) that loses one or more electrons (e^-) and leaves the bulk metal (Equation 1). This is called oxidation. The lost electrons are conducted through the bulk metal to an area where they combine with another metal ion (H^+) (Equation 2).

$$M \rightarrow M^+ + e^-$$

Equation 1 – Metal Atom Loses an Electron

$$H^+ + e^- \rightarrow H$$

Equation 2 – Metal Ion Gains an Election

The corrosion cycle consist of 4 elements: an anode, a cathode, an electrolyte, and an electron path (Figure 1). An anode is a negatively charged site where electrons are lost. The cathode is a positively charged site where electrons are transferred to. An electrolyte is a conductive solution that is in contact with an anode and cathode. An electron path is a conductive path through which electrons flow from an anode to a cathode.

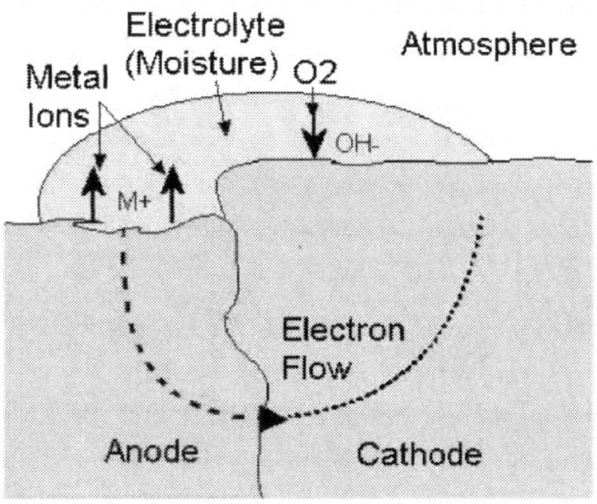

Figure 1 – Corrosion Cell

On a microscopic level, a carbon steel surface consists of numerous individual areas divided by grain boundaries in the metal. The individual areas have potential differences (Figure 2) that make some areas cathodes and other areas anodes. When an electrolyte is introduced the corrosion cycle begins. Because electrons are constantly moving from one carbon steel area to another carbon steel area, the potential differences are constantly changing and the corrosion cycle continues. The corrosion cycle will continue until one of the 4 corrosion cycle elements is removed. If unprotected, the carbon steel will be the first thing to be removed from the corrosion cycle.

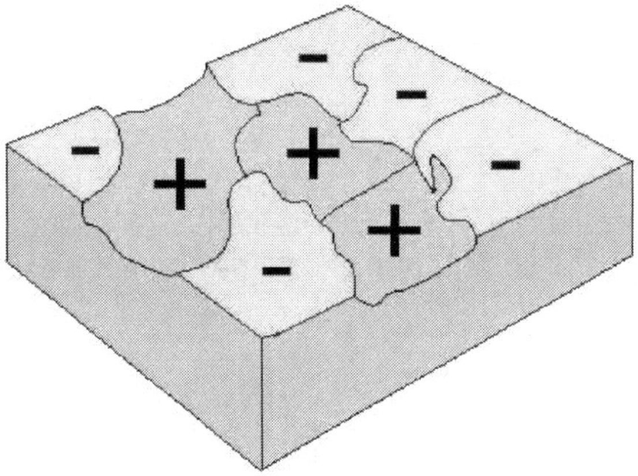

Figure 2 – Carbon Steel Surface with Grain Boundary Potential Differences

How to Break the Corrosion Cycle

In order to break the corrosion cycle, one or more of the elements must be removed. Traditionally, the corrosion cycle is broken by removing the electrolyte from the cycle. This is accomplished by using barrier coatings such acrylics, enamels, and epoxies. These coatings if applied with zero imperfections provide excellent protection against corrosion, however due to imperfections caused by improper surface preparation, mixing errors, application flaws, and mechanical damage the electrolyte is allowed to enter the cycle leading to corrosion.

An alternative to barrier coatings is an anodic coating such as galvanizing or thermal spray aluminum (TSA). Thermal spray aluminum breaks the corrosion cycle by removing the electrolyte from the cycle like barrier coatings, and provides anodic protection and passivation of surface imperfections. When a carbon steel surface is coated with TSA, the carbon steel's potential energy is positive in relation to the TSA coating (Figure 3).

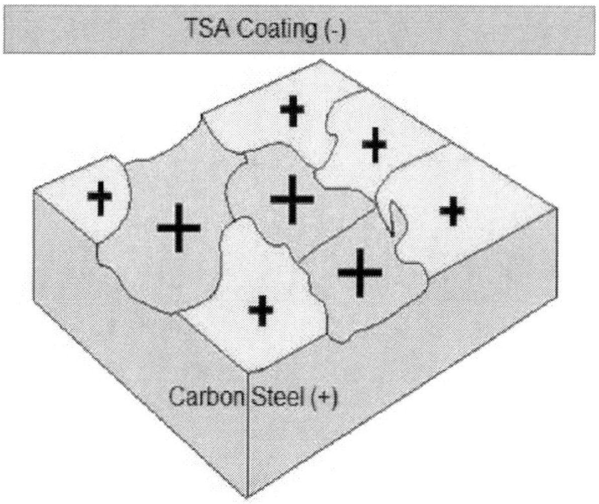

Figure 3 – Carbon Steel Surface with TSA Coating

This changes the elements of the corrosion cycle such that the carbon steel is the cathode and the aluminum is the anode. Additionally, the carbon steel substrate is protected by passivation. Passivation is the formation of a thin ion plating of corrosion product, passive film, on a metal's surface that prevents oxidation (Figure 4). With TSA, the passive film is aluminum oxide which adheres to the carbon steel. Aluminum oxide is not water soluble; therefore, it is resistant to weather. Unlike barrier coatings that allow corrosion at imperfections, the passive film from TSA provides a barrier at imperfections that prevents corrosion.

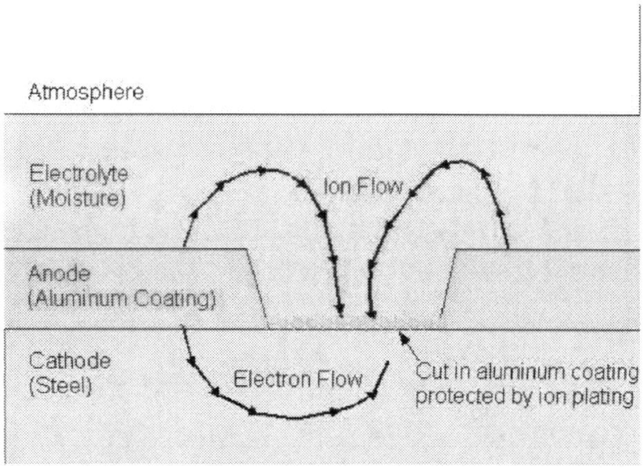

Figure 4 – Passivation Schematic

How is TSA Applied?

Thermal Spray Aluminum is applied not by welding, but by melting the aluminum with a heat source and accelerating the melted aluminum towards a substrate (Figure 5).

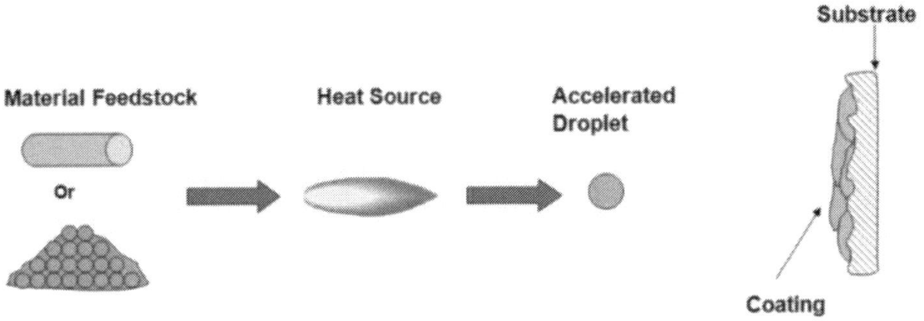

Figure 5 – Basic Application Flow of TSA

Upon cooling, the accelerated aluminum droplets become mechanically bonded to the substrate and other aluminum droplets. This application results in a bond strength of > 1200 psi. The typically bond strength of barrier coatings is 600-700 psi. Because of the high bond strength of a TSA application and the increased ductility of aluminum compared to barrier coatings, which are brittle, a TSA application is more forgiving of rough handling during transport or installation. After a shop application where a barrier coating was applied, it is a must that the coating be touched-up after installation. This leads to increased cost and additional time needed in the work schedule.

Surface Preparation

Because poor substrate preparation can lead to low or no bonding strength and decreased service life, the TSA application must start with quality substrate preparation. Prior to abrasive blasting, weld spatter and slag must be removed, and sharp edges rounded. The substrate must also be tested for chlorine levels, which must be below 10 $\mu g/cm^2$. If chlorine levels are greater than 10 $\mu g/cm^2$, then the surface must be washed and retested. Next, the substrate must be abrasive blasted to a "White Metal Blast" with a sharp angular profile ranging from 0.003 in to 0.005 in (75 µm to 125 µm).

Before applying TSA, all blast media, dust, and loose materials must be removed from the substrate. Additionally, the substrate must dry and free of rust bloom. To avoid the formation of rust, it is recommended that TSA be applied within 6 hours of blasting. When surface preparation is complete, TSA can be applied. TSA thickness is 0.01 in to 0.018 in (250 µm to 450 µm) applied in multiple layers within one working step to prevent oxidation between layers. To ensure uniform coverage, the layers should be overlapping and in multiple directions.

Flame Spray Process

The TSA application depicted in Figure 5 is accomplished by two common methods. The first method, called Flame Spray, uses a tool called a wire flame spray gun (Figure 6).

Figure 6 - Wire Flame Spray Gun (Sulzer-Metco Series 16E)

The wire flame spray gun applies TSA by continually feeding an aluminum wire through the gun and into a flame to melt the aluminum. A compressed air stream at the tip of the flame accelerates the melted aluminum forward (Figure 7).

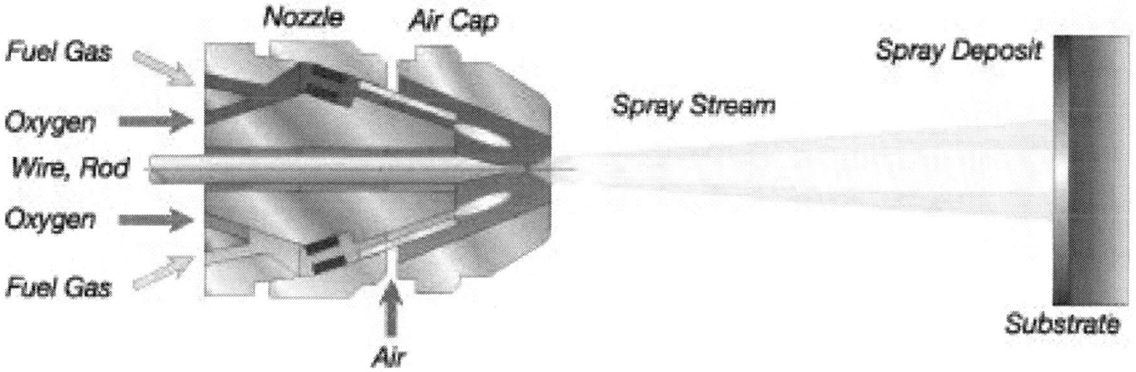

Figure 7 – Wire Flame Spray Gun Operation

The wire flame spray gun applies thermal spray aluminum at a rate of up to 20 lbs/ hr. It is used for touch-ups, small applications, field applications where a potential ignition source needs to be concentrated, shop applications, and applications with complex geometries.

Electric Arc Spray Process

The second method for application of thermal spray aluminum is called Electric Arc Spray. Electric Arc Spray uses a tool called an ARC gun. The ARC gun applies TSA by continually feeding two aluminum wires that are energized with a 480V source. As the energized wires come together in the head of the gun, the conducting heat from the electrical charge melts the aluminum wires. The melted aluminum is accelerated forward by compressed air (Figure 8).

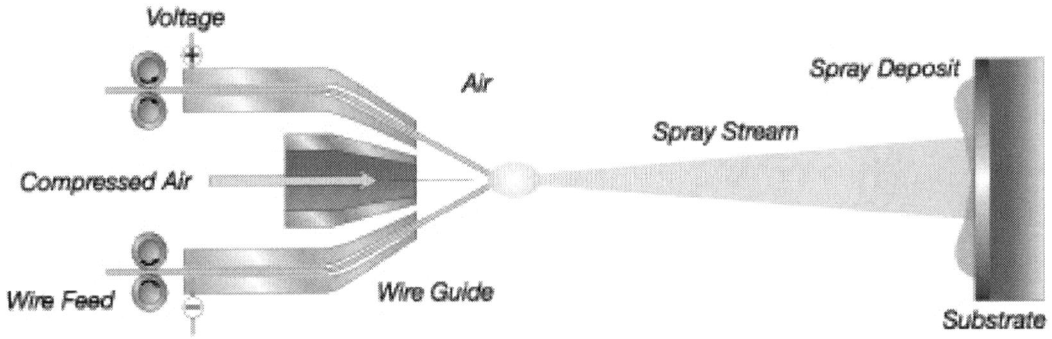

Figure 8 – Electric Arc Spray Gun Operation

The electric arc spray gun applies thermal spray aluminum at a rate of up to 55 lbs/ hr. It is used for large jobs, field applications where a potential ignition source can be managed, and shop applications.

TSA USE AT THE DOW CHEMICAL LOUISIANA OPERATIONS SITE OLEFINS UNITS

It was first applied in 2010 to the header boxes on a series of retubed fin-fan exchangers. It has also been applied via a shop application following the shop repair of a drum, a separator, and a reboiler.

Picture 1 – Dow Fin-Fan Header Box TSA Shop Application

Picture 2 – Dow Separator TSA Shop Application

Picture 3 – Dow Drum TSA Shop Application

Because TSA is a longer term solution for breaking the CUI cycle and preventing costly damage due to corrosion, it was realized that the field application of TSA was required. There were apprehensions concerning field application because the TSA application introduces a potential ignition source into an operating facility. To begin, the two processes for applying TSA in the field, flame spray and electric arc spray were evaluated. Following previously established Dow EH&S guidelines for performing field work that has the potential to generate a source of ignition in areas where flammable or combustible materials may be present, a procedure was developed for applying TSA in an operating facility. The first field application of TSA using the procedure was the coating of the feed and dilution steam piping on an ethylene cracking furnace in mid-2011. The second application was to a column in late 2011.

Picture 4 – Dow Column CUI

Picture 5 – Dow Column CUI Remediated with TSA in the Field

RECOMMENDATION FOR BEST PRACTICE

TSA is a superior coating for breaking the CUI cycle thus reducing the long term cost of ownership of carbon steel equipment. It is recommended for use on insulated carbon steel equipment that operates in the temperature range of 10°F (–12°C) to 350°F (175°C), insulated carbon steel equipment that operates in a cyclical temperature scheme, and non-insulated carbon steel equipment that is in an area were atmospheric corrosion is accelerated such as near cooling water towers or marine environments.

References

1. Suzler Metco Ltd (2011). Thermal Spray – Combustion Processes; Combustion Wire. *Sulzer Metco Products & Services*. Retrieved November 2011, http://www.sulzermetco.com/en/desktopdefault.aspx/tabid-4021/7681_read-17567/

2. Suzler Metco Ltd (2011). Thermal Spray – Arc Processes; Electric Arc Wire Spraying. *Sulzer Metco Products & Services*. Retrieved November 2011, http://www.sulzermetco.com/en/desktopdefault.aspx/tabid-4021/7684_read-17372/

3. Weber, James K.(2009), "B&H Distributors TSA and Corrosion", B&H Distributors

Acknowledgements

Contributions received from Duane Turner (The Dow Chemical Company Plaquemine, LA Coatings Inspector), Kevin Adams (The Dow Chemical Company Global Hydrocarbons PCE/ Inspections/ NDT SME), and Doug Walker (The Dow Chemical Company Global Olefins Reliability Associate).

AIChE Paper Number 97c

CRACKING FURNACE GASKET SELECTION – A KEY FACTOR IN FURNACE LIFE EXTENSION

Kathryn Worden
Applications Engineer
Flexitallic, L.P.

Mark Karrs
Advisor
Becht Engineering Co., Inc.

David Sankey
Furnace Specialist
BASF North America

Prepared for Presentation at the 2012 Spring National Meeting
Houston, TX, April 1 - 5, 2012

CRACKING FURNACE GASKET SELECTION – A KEY FACTOR IN FURNACE LIFE EXTENSION

Kathryn Worden

Applications Engineer

Flexitallic, L.P.

Mark Karrs

Advisor

Becht Engineering Co., Inc.

David Sankey

Furnace Specialist

BASF North America

Abstract: Flanges on ethylene cracking furnaces are necessary for proper maintenance, and offer specific challenges in terms of leakage avoidance and safety assurance. On cracking furnaces, any gasket failure will require a cold shutdown for repair. As both radiant coil and refractory service life are strongly impacted by hot-cold cycling, avoidance of gasket failure events can be key to maintenance minimization and furnace life extension.

The steam cracking process normally takes hydrocarbon feed from roughly 120°F to above 1,600°F. The thermal expansion of the heating surface requires provisions to absorb the resulting movement which may result in significant stress on joints. In addition, the furnace operation is cyclic in terms of thermal cycling as well as oxidation/reduction cycling due to the intermittent decoking operation.

The subject paper reviews the common flange types and locations on a cracking furnace and front end piping and outlines important considerations for gasket specification/selection. Suggested gasket solutions for each location are presented, which consider the fluid, the operating conditions, code requirements and safety related issues.

Key Factors to Consider

Process Conditions

Temperature and pressure are obvious factors that will impact gasket selection as well as choice of joint configuration, flange type and rating. However, other process fluid and exposure factors should also be considered in the gasket selection procedure.

Temperature Cycling

Any furnace related gasket service will be subjected to hot/cold cycling, with the number of cycles a function of the requirements for shutdown and restart. In the case of a cracking furnace, it should be an operations goal to minimize the number of unnecessary shutdown events, not only from the perspective of lost production but also because of the detrimental impact on radiant coil creep.

Oxidation/Reduction Cycling

A cracking furnace run length and associated decoking cycle will expose certain gaskets to alternating reducing atmosphere (cracking reaction produces a H_2 rich hydrocarbon-steam cracked gas mixture) followed by an oxidizing atmosphere (steam air decoking of the radiant coil). This may result in gasket degradation if the gasket material is not resistant to oxidation, e.g. graphite based.

Fluid Contaminants

Certain components that may be present in the fluid being handled may impact flange, piping, and gasket integrity. Compounds which may form strong acids or bases as a result of reaction or phase change, such as sulfur, chloride, or sodium should be considered during material selection.

Pyrophoric Fluid

Leakage of any fluid which is above its auto ignition temperature will result in spontaneous ignition upon contact with air, irrespective of any ignition source or spark. Many of the flange joints near the radiant coil of a cracking furnace are handling mixtures which are above their auto ignition temperatures. Thus any leakage resulting from joint failure will be a serious safety issue, a fugitive emission, as well as a maintenance issue. Auto ignition temperatures of common cracking furnace feed and product molecules are listed in Table 1.

Table 1 *(1)*

Auto Ignition Temperature	**°F**	**°C**
Hydrogen	932	500
Methane	1,076	580
Ethane	959	515
Ethylene	914	490
Propane	878	470
Propylene	856	458
n-Butane	761	405
Naphtha	1,022	550
Gas Oil	637	336

Personnel Exposure

All flanged joints are normally accessible for maintenance purposes, but certain flanges are located in areas of frequent personnel traffic. This factor should be considered when selecting suitable gaskets.

Fire Exposure

As fires in and around the cracking furnaces may be anticipated, the proper gasket selection for fuel piping should take into account fire resistance. NFPA 54 National Fuel Gas Code outlines requirements for gasket performance that include fire exposure. However, this code normally applies only to fuel gas piping inside occupied buildings. In the absence of code requirements, application of fire-safe ratings such as API or PVRC FITT to gasket selection on cracking furnaces becomes a matter of corporate or plant safety policy.

Code Requirements

Hydrocarbon piping and furnace coils are usually designed to ASME B31.3 piping code or the equivalent European or Japanese codes, and gaskets in those lines should conform to the applicable code. Modern cracking furnace design will typically incorporate a 1,200 to 1,800 psig (80-120 bar) steam system integrated into the waste heat recovery. The bolted flange joints and gaskets must comply with the applicable ASME and boiler code requirements.

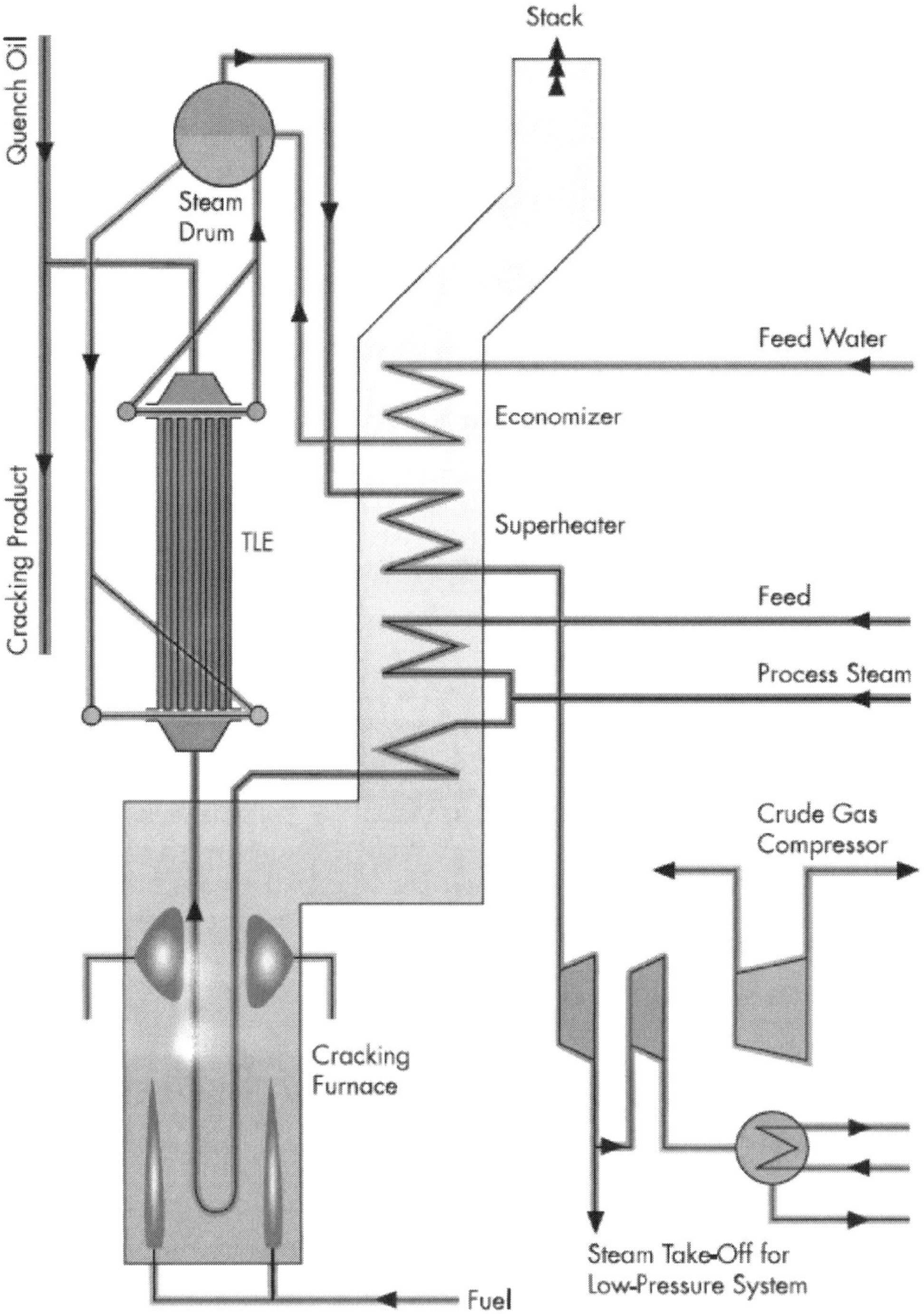

Ethylene Cracking Furnace Schematic *(2)*

Gasket Selection

Although a gasket may be a very small component of a cracking furnace, proper gasket selection and installation means the difference between continuous operation or joint failure leading to downtime and lost productivity while possibly compromising the safety of plant personnel. The gasket is only one component of a bolted joint that includes the flanges, fasteners, installation procedure, tools for assembly, and a human factor. Proper gasket selection is the first step to eliminating leaks and eventual unplanned outages due to joint failure. A severe incident can start with one gasket leak and can ultimately cost a plant operator millions in lost production and rehabilitation costs.

Flanged joints on cracking furnaces can be grouped into three categories with similar requirements: process, steam and fuel streams. Process streams are those handling fluids associated with the cracking feed or effluent, including dilution steam, as this "dirty steam" stream normally has organic and inorganic impurities far in excess of the streams associated with steam generation. Similar considerations apply to gasket selection for the process, steam and fuel streams of a cracking furnace.

PROCESS: Hydrocarbon Feed Inlet

For liquid crackers, this stream is normally a virgin naphtha or gas oil with significant organic sulfur. The temperature is normally low (< 120°F) and the pressure elevated to account for the furnace pressure drop but not to a point where a RF Class 150 flange cannot meet the requirements. Hot/Cold cycling is negligible and oxidation/reduction cycling is not present, as the decoking air stream is normally introduced further downstream to reduce the pressure required. The fluid is a liquid far below the auto ignition temperature, and the flange location is normally at the highest point on the furnace far away from fire exposure or any regular personnel traffic.

Recommended Gasket

The recommended gasket selection in this case could be based mainly on economics. A spiral wound gasket (SWG) with an outer ring will provide blow out resistance and enough resiliency to respond to joint relaxation. Thermal cycling is negligible and requires no additional recovery; therefore a 304SS winding wire is standard for this application with carbon steel flanges. For the filler material, graphite is an excellent sealing material, compatible with hydrocarbons and suitable for the low temperatures. "Inner rings for flexible graphite-filled spiral-wound gaskets shall be furnished unless the purchaser specifies otherwise." *(3)* (Inner rings for large diameters and high pressures are still mandatory for all filler materials.) Inner rings provide multiple advantages including significant benefits to sealing as well as eliminating the problem of buckling. In general, inner ring metallurgy is chosen to match the winding wire.

Less expensive filler materials are available (chlorite graphite type or similar), but they can easily be mistaken for graphite. Most plants decide that the *minimal cost advantage* is not worth the risk of accidentally installing a material with lower chemical and temperature resistance.

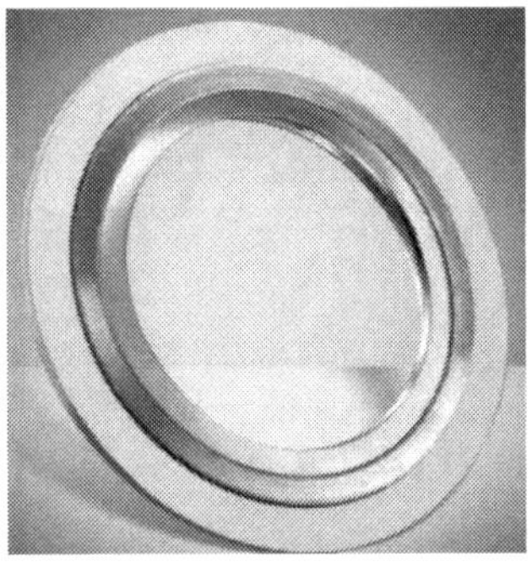

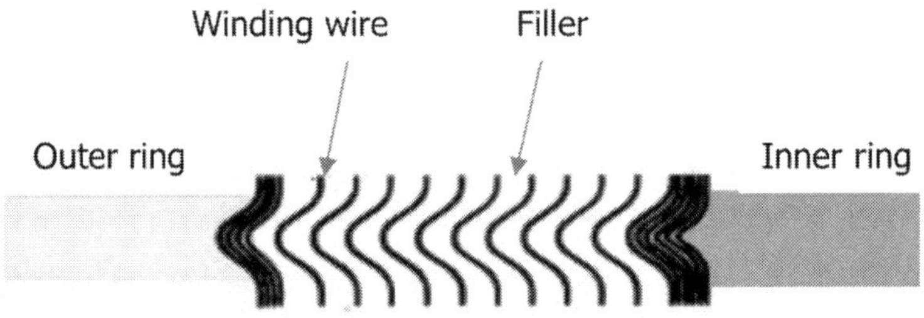

Spiral wound gasket with inner and outer ring

For gas crackers, the Hydrocarbon Feed stream will be a low molecular weight alkane, typically ethane or propane which may have trace sulfur added to inhibit the coking rate. The pressure and temperature will be in the same range as for a liquid feed. The vapor is still far below the auto ignition temperature and the gasket selection would follow the same criteria.

PROCESS: Dilution Steam Inlet

At this point, Dilution Steam is mixed with the partially vaporized liquid feed or the preheated vapor feed. Hot/Cold cycling will apply but operating and ambient temperatures will remain below 250°F. Two phase flow at these locations may result in some level of vibration/mechanical cycling. Oxidation/Reduction cycling will likely be present as the decoking air stream is normally introduced to the dilution steam flow prior to the mix point.

The fluid is still below the auto ignition temperature and the flange location is normally at an intermediate point on the convection section away from fire exposure or any regular personnel traffic.

Recommended Gasket

The typical gasket choice is a graphite filled spiral wound (conveniently the same as mentioned above). One significant advantage to standardizing gasket material is inventory consolidation and reducing risk of the wrong gasket being put in service – a common mistake when many different types are in the warehouse.

PROCESS: External Crossovers

The fully preheated mix is transferred from the convection coils to the radiant section coils via external crossovers which are normally constructed from 300 series stainless alloy. The normal process temperature will be in the range of 1,100 to 1,250°F and the mix will consist of steam and hydrocarbon components far above the auto ignition temperature.

Thermal cycling between ambient and process temperature will occur at every cold shutdown of the furnace, and oxidation/reduction cycling will occur at the end of every furnace run and associated decoking cycle. Sulfur will be present in dilute concentration.

The crossover flanges are usually located at the arch level, accessible from platforms such that movement of the crossover piping can be monitored and instrumentation on the crossovers can be serviced. Instrumentation such as thermowells and pressure transmitters may also have flanges associated with them. Crossover movement and mechanical stress on joints can be significant, as the flexibility in the crossover piping usually must absorb the movement of the radiant inlet associated with the thermal growth of the radiant coil tubes.

Recommended Gasket

All these factors combine to make joint sealing and gasket selection especially challenging for any flanges required in external crossover piping. Leakage from crossover flanges will smoke immediately and pose a hazard to personnel as well as other critical equipment at the arch level, e.g. TLE's.

Gasket selection should be mindful of the elevated temperature, thermal cycling, as well as risk and tolerance for leakage. The risk of fire, product loss, process shutdown, and personnel safety should drive gasket selection. The best choice is a spiral wound gasket with an inorganic filler material with good sealing properties, along with an extremely resilient winding wire. A common example of such winding wire is Heat Treated Inconel X-750. The best example of such a filler material available today is one using specially formulated combinations of different types of exfoliated vermiculite. *(4)* Going forward, this filler material will be referred to as 'exfoliated vermiculite'.

Vermiculite is an inorganic phyllosilicate mineral with extreme temperature ($1800^{\circ}F^{+}$) and chemical resistance. Exfoliated vermiculite is fire safe, maintains its integrity and sealability at extreme temperatures, and will not oxidize at high temperatures due to its inorganic structure.

Graphite Oxidation

The recommendation for an inorganic sealing material is to avoid the main pitfall of graphite – that of oxidation. When graphite is exposed to oxygen from the external air or internal process gases (steam, air, CO, CO2, etc), the oxygen reacts with the carbon in graphite and produces carbon dioxide. Upon discovery, the graphite is simply missing from the gasket. It seems to have *disappeared*.

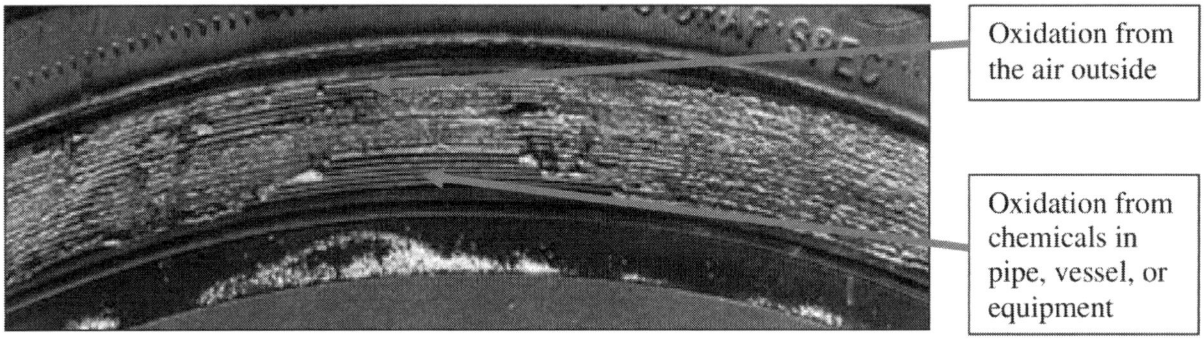

A spiral wound gasket (with an inner and outer ring) shows visible oxidation of graphite filler from outside air and internal chemicals where it is missing between the winding wire.

Oxidation means there is volume loss to the sealing element of the gasket. Too much volume loss leads to lost bolt load and resultant gasket load. End users should always inspect gaskets upon removal, and look for signs of oxidation indicated by missing graphite. It is important to catch it prior to a major joint failure, or eliminate the risk entirely by using a material that is not susceptible to oxidation but capable of producing a gas tight seal.

Note that graphite's rate of oxidation is highly dependent on the concentration of oxygen, the temperature, tightness of the joint (bolt load and resultant gasket load especially after relaxation at high temperatures), gasket type (sheet versus semi-metallic) and quality of graphite (impurities accelerate oxidation while inhibitors may decelerate it). Variation in each of these contributing factors makes oxidation very difficult to predict and furthermore, a maximum temperature limit on graphite very difficult to assign.

Not including extreme oxidizers like NO_x gas, examples of oxidized graphite in gaskets have been found below 700°F. The development of inhibited grades of graphite has improved graphite's ability to withstand oxidation but has not eliminated it. Different manufacturers employ different methods and produce different grades – each having a different resistance to oxidation. Although improvements continue to be researched and employed, the best inhibited graphite typically adds 100 degrees or less additional temperature resistance to that of its standard graphite counterpart.

End user specs and drafts of upcoming industry standards suggest maximum long term operating temperature of even the best super-inhibited graphite to be 850°F, with the poorest being limited to 575°F. If interested in a test for graphite quality, Fluid Sealing Association procedure FSA-G-604-07*(5)* can indicate graphite quality through Thermo Gravimetric Analysis.

Ineffective Alternatives

Being able to resist oxidation is not a sealing material's only requirement. If a material can be used in extreme temperatures but cannot produce a gas tight seal, there is little value in it. Prior to the availability of a sealing material made of exfoliated vermiculite, manufacturers were required to use high temperature materials like mica or ceramic. Unfortunately, they are very poor sealing materials and easily permeated. Attempts to use them as an oxygen barrier around graphite may delay the oxidation process but it will not prevent it. Oxygen permeates the material and still oxidizes the graphite.

The tendency of graphite filler to oxidize through a mica heat shield has been shown in multiple real world examples since the 1990's. The typical construction of such a spiral wound gasket has wraps of mica filler on the ID and OD with graphite between.

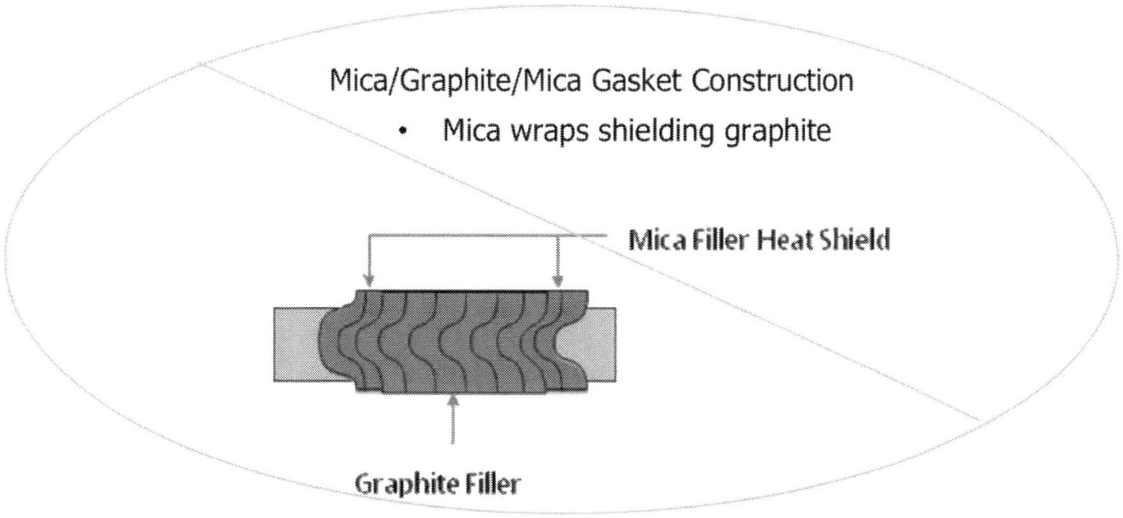

Oxidation Example 1

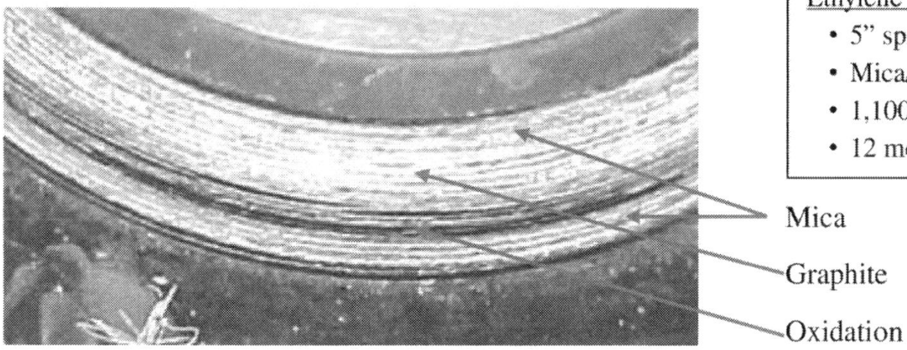

Ethylene Furnace Application

- 5" spiral with inner ring
- Mica/graphite/mica filler
- 1,100°F
- 12 months

Photo illustrates an ethylene manufacturer's furnace gasket where graphite oxidized through the mica after 12 months at 1,100°F.

Successful Application 1

This plant began using exfoliated vermiculite filled spiral wounds in 2003 to eliminate oxidation at this flange and others.

Product:	Exfoliated vermiculite SWG
Temperature:	Range 1,100°F to 1,250°F
Pressure:	20 psig
Since:	February 2003

Oxidation Example 2

Beginning in late 2005, a Japanese ethylene producer began researching alternative sealing materials to asbestos following the country's legislative ban of the material. Statistical analysis of mica/graphite/mica spiral wounds showed significant leakage after relatively short periods of service, as shown in the excerpt of test results in Table 2. Upon investigation of the gaskets in leaky joints, the graphite was found to be all or mostly "disappeared".

Conditions of sealing included:

- Normal Operation: Hydrocarbon gas, Steam; 1,110°F; Maximum pressure 51 psig
- At Decoking: Air, Steam; 1,020°F; Maximum pressure 18 psig

Table 2. Ethylene producer tests mica/graphite/mica performance

Gasket Type:	Period of Operation	
Spiral wound, mica/graphite/mica	3.7 months	5.8 months
Qty ON TEST at start of operation	22	25
Qty LEAKING at end of operation	6	15

Successful Application 2

All gaskets were eventually replaced with exfoliated vermiculite filled spiral wounds to eliminate the oxidation related leaks and were found to be "superior in sealing quality and long endurance".

Product:	Exfoliated vermiculite SWG
Temperature:	Up to 1,110°F
Pressure:	51 psig maximum
Since:	November 2005

Exfoliated Vermiculite

In the case of vermiculite (and graphite for that matter), the exfoliation process of the natural flake is key to generating good sealability and low porosity. Unlike mica, vermiculite can be exfoliated. Unlike graphite, vermiculite will not oxidize. Exfoliated vermiculite has been used as a sealing material with tremendous success since 1997 and is now in tens of thousands of joints replacing poor sealing materials or those that suffer from oxidation.

Vermiculite Flake

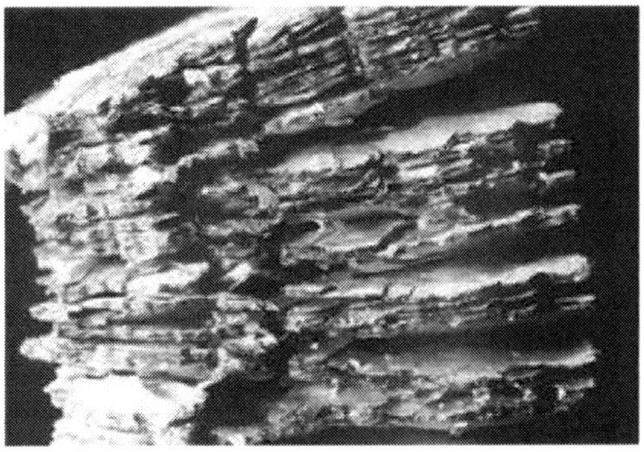

Exfoliated Vermiculite Flake

Heat Treated Inconel X-750 Winding Wire

Extreme thermal cycling causes differential thermal expansion and contraction of components in a bolted joint. It is difficult for the gasket to compensate for fluctuating load conditions, and sometimes extra resiliency must be built into the joint. A special heat treated (HT) Inconel X-750 winding wire has been developed in conjunction with the nuclear industry to ensure that joint integrity is maintained during thermal cycles. Heat treating must be specified in order for the precipitation hardening process to be performed by the manufacturer. *(6)*

Testing from a nuclear OEM & a nuclear power producer shows that heat treated Inconel X-750 has twice the usable recovery over 316L Stainless Steel (refer to Table 3). Usable recovery is different from overall recovery. A gasket's ability to spring-back/recover while still compressed in a bolted joint, or its usable recovery, is more significant than how much it recovers after disassembly. The HT Inconel X-750 material has been used in cycling applications since the early '80s with remarkable success.

Table 3. Test results showing usable recovery of different winding wire in SWGs.

Winding Material	316L SS	Heat Treated Inconel X-750
Initial Thickness	0.178″	0.179″
Compressed Thickness	0.122″	0.121″
Total Springback	0.011″	0.013″
Springback to Leakage (2500 psig test press.)	0.0038″	0.0078″

HT Inc X-750 has twice the usable recovery over 316LSS.

Full Scale Test Results (Averaged) on Reactor Coolant Pump Gaskets (40-5/8 x 42 x .175)

Successful Application 3

An ethylene manufacturer discovered vermiculite's resistance to oxidation and the resiliency of heat treated Inconel X-750 after a test application and has been using them both in spiral wound gaskets in their external crossovers since September 2002.

Product:	Exfoliated vermiculite SWG, HT Inc X-750 winding
Temperature:	1,300°F maximum
Pressure:	30 psig
Since:	September 2002

PROCESS: Radiant Outlet / TLE Inlet

The cracked gas must pass from the radiant coil to the quench in a manner so as to minimize residence time at temperature, and thus avoid secondary reactions that can degrade olefin yields. At the same time, flanges are desirable at the radiant coil/TLE transition for cleaning, inspection and other maintenance related purposes, such as coil repair or replacement. The TLE will normally have a removable "inlet cone" that is flanged at both ends.

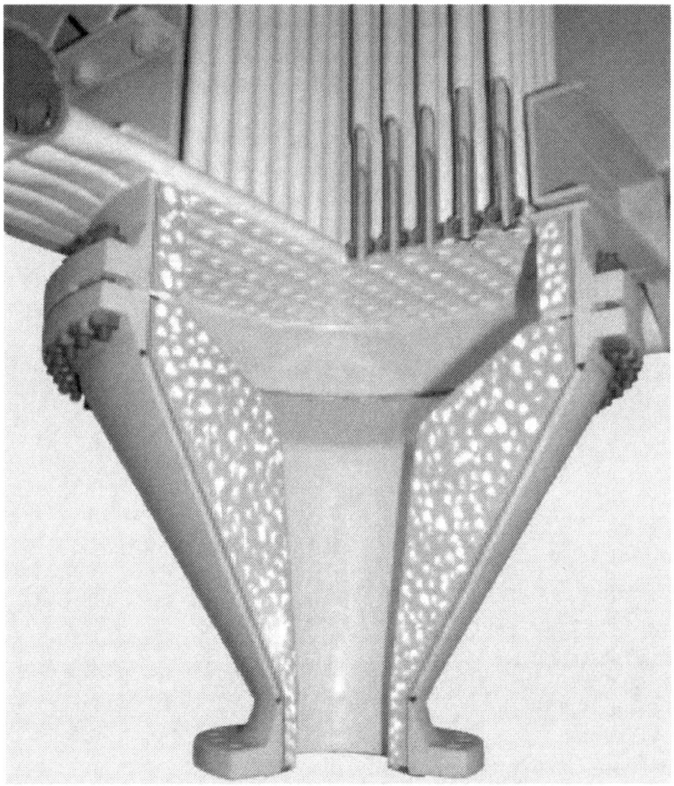

The inlet flange of the cone is normally directly coupled to the radiant coil outlet while the upper "body flange", also known as the "girth flange", is bolted to the inlet channel. Both flanges are under considerable mechanical stress, as they are supporting the dead weight of some portion of the radiant coil.

While the pressure at the coil outlet is kept as low as possible, normally 10 – 20 psig, the coil outlet temperature required for the cracking reaction will be in the range of 1500 – 1650°F necessitating the use of bulky flanges.

Internal refractory lining can be used at both the inlet flange and the body flange to reduce the operating temperature of the joint significantly below the flowing process temperature.

Transfer Line Exchanger Inlet Cone (2)

The cracked gas will be high in steam and H_2 and will thus normally be a reducing atmosphere. However, the intermittent decoking will expose the radiant outlet flange to a steam air mix and high temperature oxidizing atmosphere that may also see erosive coke fines. Also consider that even though the internal media may be reducing, exposure to the outside air can still cause oxidation. Sulfur will be present in both modes, predominantly in the form of H_2S.

Thermal cycling will be over an extreme range (ambient to 1650°F), and radiant outlet and associated flanges on the TLE inlet cone and body flange can also be large diameter joints, further complicating the gasket sealing requirements.

Recommended Gasket

These are the most difficult sealing applications on a cracking furnace. Critical joints require special focus on not only the gasket and process conditions but also on installation. Even the best gasket will still leak if not installed with an evenly distributed, adequate level of load. The following items are critical along with the gasket: bolt tightening sequence, uniform and adequate bolt stretch, fastener condition, and using tools with high level of accuracy. For some guidance in this area refer to *ASME PCC-1-2010: Guidelines for Pressure Boundary Bolted Flange Joint Assembly*.

Proper gaskets in this service would use exfoliated vermiculite, due to the extreme temperature requirements. A spiral wound gasket with heat treated Inconel X-750 winding wire is an ideal choice for extreme thermal cycling.

On flanges where available bolt load is limited, a kammprofile (serrated solid metal with a facing material) can be a useful alternative. The minimum required seating stress is less than half that of a spiral wound gasket. Therefore, it allows for sealing on non-standard flanges where bolting is inadequate to seal or maintain a seal with a spiral wound gasket. Kammprofiles do not respond as well as heat treated Inconel X-750 in cycling but because they require less seating stress than a spiral, many have found success with them in cycling applications especially with a higher initial load applied at assembly.

The kammprofile's solid metal core also eliminates some common pitfalls of a SWG. They do not buckle. They do not spring apart. Handleability is greatly improved especially on large diameters where spirals tend to come apart from flexing, poor handling, or when trying to install in flanges with poor accessibility. Kammprofiles do not require a compression stop and essentially cannot be crushed. They are advantageous on narrow flange widths where an inner ring cannot be accommodated with a spiral. (Kammprofiles inherently do not require an inner ring.) They are also less sensitive to bolt up inconsistencies.

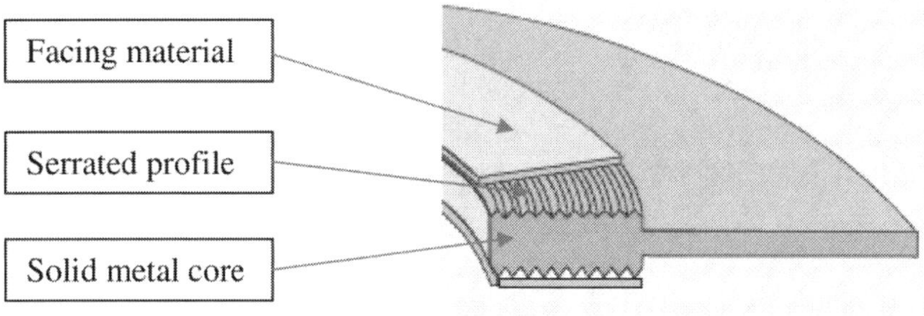

Cross-section of kammprofile style gasket

Successful Application 4

Beginning in November 2005, BASF Port Arthur converted all TLE inlet and outlet cone gaskets to kammprofiles faced with exfoliated vermiculite after a successful in-service test.

Product:	Exfoliated vermiculite kammprofile
Temperature:	1,500°F
Pressure:	65 psig
Since:	November 2005

Successful Application 5

Beginning September 2002, this ethylene producer converted all inlet flanges to the primary TLEs to exfoliated vermiculite filled spirals with heat treated Inconel X-750 winding wire. The prior gaskets using graphite filler oxidized after 6 months in service, causing flash fires, downtime, and lost productivity costing "millions".

Product:	Exfoliated vermiculite SWG, HT Inc X-750 winding
Temperature:	Approximately 1,650°F
Pressure:	30 psig
Since:	September 2002

Successful Application 6

One particular ethylene producer was plagued with weekly fires due to graphite oxidation and extreme thermal variation across the flange face because the TLE was without refractory lining. After several iterations of gasket metallurgy and filler materials including ceramic and mica, the final construction was a carrier ring of an exotic alloy with HT Inconel X-750 winding wire and exfoliated vermiculite filler for the spiral wound. This design is employed on 10 furnaces.

Product:	Exfoliated vermiculite SWG, HT Inc X-750 winding
Temperature:	1545°F process; 1,650°F decoke

Pressure: 65 psig

Since: 2003

PROCESS: Effluent Transfer Line Piping

The cracked effluent will pass to the main transfer line and quench sections of the plant at low pressure via large diameter piping. The temperature cycling of this effluent piping between the End-of-Run (EOR), Start-of-Run (SOR), decoking, and ambient startup will be very significant. The thermal expansion and associated movement of this large piping will result in significant stresses. Layout, stress and support design for the effluent piping involves some of the most difficult and sophisticated engineering in an ethylene plant.

The effluent piping will see oxidation/reduction cycling because of the decoking operation and sulfur will be present in the effluent, mainly in the form of H_2S. The fluid will be above the auto ignition temperature at any point upstream of the oil quench for liquid crackers. For gas crackers, the TLE outlet temperatures will normally be in a range below the associated auto ignition temperature.

Recommended Gasket

Because of the large diameter flanges involved, the difficult service conditions, and the extreme fouling nature of the fluid, kammprofile gaskets are favored for these applications. Condensation of heavy tars at the inside surface of effluent piping and associated joints will cause coke formation at the inside gasket sealing surface. To prevent or reduce coke formation, kammprofile inner diameters can be designed to the flange bore as long as there is enough bolt load to adequately seal the additional gasket area. Consult your gasket manufacturer for the best design.

As the transfer line will normally be a large diameter piping system spanning the entire heater row and flowing to the compressor house, use of a fire safe gasket such as graphite or vermiculite should be considered to avoid loss of containment and de-inventory of the hot section in case of an external fire. Vermiculite is inherently fire safe [(7)] due to its inorganic composition and is certified to API 607[(8)] (1,200°F exposure for 30 minutes).

FUEL: Fuel Gas Piping

Fuel to fire the cracking furnaces is normally made up of a H_2/methane mix recovered from the cracked gas during fractionation. It will normally be returned to the furnaces at ambient temperature and roughly 50 psig in the fuel header. Since NFPA 54 applies only to furnaces and boilers located inside buildings, cracking furnaces are not normally required to comply with the National Fuel Gas Code. However, NFPA 54 encourages the use of fire resistant gaskets in fuel distribution piping to address the potential risk of flange leakage feeding methane to a fire in the area containing the fuel piping.

Recommended Gasket

As the operating conditions of the fuel piping are not severe – low temperature and pressure, no thermal cycling, no oxidizing or reducing service, below the auto ignition temperature – a low cost spiral wound gasket is normally specified. If external fire exposure is judged to be a significant factor or concern, selection of a fire-safe material per API or PVRC FITT is recommend, i.e. graphite or vermiculite.

STEAM: High Temperature Steam, Boiler Feed Water Piping

Certain plants, usually those designed for very heavy feed, may superheat the Dilution Steam to a very high temperature in a separate coil, and then inject it. Not all steam in an ethylene plant is clean deaerated steam and some steam services can also be oxidizing.

Exfoliated vermiculite gaskets have replaced graphite in hundreds of steam flanges beginning at temperatures as low as 600°F and covering industries from refining to power generation to pharmaceutical. Among the dozens of documented successful applications, two for the ethylene industry are listed below.

Successful Application 7

This refining and petrochemical end user installs exfoliated vermiculite kammprofiles or spiral wounds throughout their plant in applications exceeding 750°F. In their ethylene manufacturing units, they have been using exfoliated vermiculite faced kammprofiles in their dilution steam injection flanges since early 2002.

Product:	Exfoliated vermiculite kammprofile
Temperature:	1,000 – 1,100 °F
Pressure:	25 – 60 psig
Media:	Superheated Steam
Since:	Early 2002

Successful Application 8

Since September 2002, this Ethylene manufacturer in the Gulf Coast has been using exfoliated vermiculite gasket material in their dilution steam injection flanges. They started with spiral wounds and on occasion switch to kammprofiles on larger flanges where better handleability is desired.

Product:	Exfoliated vermiculite SWG and kammprofile

Media: Superheated Steam

Since: September 2002

Modern cracking furnace designs will typically incorporate 1,500 to 1,800 psig (100 – 125 bar) steam loops in the waste heat recovery section of the furnace convection surface. Steam piping can be designed to either B31.1 or B31.3 codes, and flanges in these services are historically Ring Type Joints (RTJ). However, before deciding whether to use RTJ flanges, it's important to recognize that they may be problematic over time and alternatives are available.

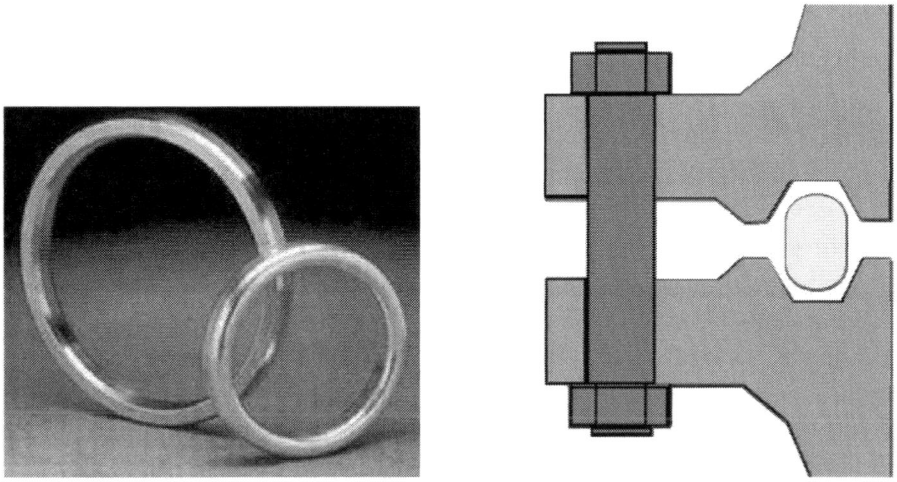

Octagonal and Oval RTJ gaskets and an Oval RTJ gasket in a flat bottom groove

RTJ gaskets are for high pressure and high temperature services but they have no recovery and are therefore not good for cyclic conditions. An extremely high level of stress is required to compress and seal a ring joint gasket which can be associated with certain risks.

When an RTJ is bolted up, a very high surface stress is generated between the RTJ gasket and the flange groove. This stress causes plastic deformation – permanent deformation of the metal. The RTJ gasket hardness increases due to the work hardening after it is compressed. The difference in initial hardness between the flange and gasket should be 15-20 Brinell. If an RTJ gasket is reused (a common mistake in the field), it eventually becomes harder than the flange causing cracking and permanent damage to the flange, typically in the groove itself. RTJ gaskets should never be reused. If this flange and gasket design is chosen, this point must be stressed.

Depending on material selection (of the piping, flange, and gasket components) and if a corrosive environment is present (Chlorides, Hydroxides, Sulphides, Hydrogen, etc), the RTJ groove might be susceptible to stress corrosion cracking or hydrogen embrittlement due to the extreme loads present. Material selection is important to preventing this.

One way to reduce the high stresses that influence corrosion is to use a spiral wound or kammprofile which are manufactured for flanges up to Class 2500 and are able to handle the pressure associated with those flange ratings. Octagonal RTJs can also be made with a serrated profile and layered with soft facing material in the area that contacts the flange to help reduce localized stresses. This design is known as a Kamm-ORJ. As opposed to graphite, vermiculite is non-conductive and can be used as the facing material if chloride-related corrosion is a concern.

Improved Assembly

Gasket selection is a key first step to eliminating downtime and extending furnace runtime and overall life. Unfortunately, even if a gasket is chosen properly, a poor installation will still cause the joint to leak. Even bolt load and sufficient gasket stress are mandatory for achieving a leak free joint. On all bolted connections, critical joints in particular, it is important to stress proper installation (uniform gasket stress by utilizing a proper tightening sequence) and adequate gasket stress. Note that achieving minimum pre-load is not always enough. High temperature joints will see extensive bolt relaxation causing a reduction in bolt stress and ultimately reduced gasket stress. Adequate preload must be applied to overcome these. As long as the gasket, bolts and flanges are not over-stressed, higher gasket stress means a tighter joint with lower emissions.

Increasing the accuracy of the tools being used will reduce the chance of leaks. Unfortunately, the most common tools in a plant are also the most inaccurate – hand and impact wrenches. On critical joints, consider upgrading to hydraulic torquing, tensioning, or (at minimum) ultrasonics to check bolt stretch. For guidance on improving assembly, refer to *ASME PCC-1-2010 Guidelines for Pressure Boundary Bolted Flange Joint Assembly.*

References:

(1) Data from The Engineering ToolBox, www.engineeringtoolbox.com

(2) Graphics courtesy of Alstom Power Energy Recovery GmbH

(3) Per Paragraph 3.2.5 of ASME Standard B16.20 - 2007 Metallic Gaskets for Pipe Flanges: Ring-Joint, Spiral-Wound, and Jacketed

(4) Developed by Flexitallic® under the trade name Thermiculite™

(5) Fluid Sealing Association STANDARD FSA-G-604-07 Oxidation Test Standard for Flexible Graphite Gasket Materials

(6) Developed in conjunction with Flexitallic®

(7) API 607 Fire Test reports are for Thermiculite™ and available upon request from Flexitallic®

(8) API Standard 607 Fire Test for Quarter-turn Valves and Valves Equipped with Nonmetallic Seats. Thermiculite™ tested to the 4th Edition.

AIChE Paper Number 97d

Inspection and Detection of Corrosion Under Insulation

Ann C. MacDonald
Inspection Team Leader
NOVA Chemicals (Canada) Ltd.

David J. Oulton
Corporate Metallurgist
NOVA Chemicals (Canada) Ltd.

Prepared for Presentation at the 2012 Spring National Meeting
Houston,TX, April,1-5, 2012

Inspection and Detection of Corrosion Under Insulation

Ann C. MacDonald
Inspection Team Leader
NOVA Chemicals (Canada) Ltd.

David J. Oulton
Corporate Metallurgist
NOVA Chemicals (Canada) Ltd

Abstract: Corrosion under insulation (CUI) is recognized by most plant owner-operators as one of the leading threats to process containment due to the potential for large and sometimes catastrophic losses. A major cost element in dealing with CUI is often associated with the inspection and detection of CUI damage particularly with respect to piping. NOVA Chemicals is no different in this regard to the rest of the chemical processing industry (CPI) and, like other members of the CPI, has embarked upon an initiative to reduce the likelihood of this insidious threat. Of special interest are non-destructive examination (NDE) methods that require the least intrusion but highest probability of detection to process piping systems which up until now, have involved wholesale removal of insulation and the construction of numerous scaffold assemblies in order to perform the simplest of inspection tasks. This paper reviews past efforts employed in the inspection of CUI and looks at the role of newer inspection technologies in achieving both cost reduction and improved reliability of detection of the problem.

Introduction

Corrosion under insulation (CUI) is recognized by most plant owner-operators as one of the leading threats to process containment due to the potential for large and sometimes catastrophic losses. A major cost element in dealing with CUI is often associated with the inspection and detection of CUI damage particularly with respect to piping. Current proactive mechanical integrity programs focus on the prevention of CUI as opposed to inspecting for it and in this respect advances in coating technologies such as thermally sprayed aluminum (TSA), improved materials selection, engineered cladding systems and proper insulation selection are all good practices at the design stage. However, there remains ageing vessels, instrumentation and miles of existing process piping where the insulation and cladding systems have experienced deterioration over numerous years of service.

It is these systems that still require inspection in order to ascertain the condition of the components. Historically, inspection for CUI has been performed by the laborious method of selecting, stripping and visually inspecting the equipment which is then followed up by remediation and re-insulation. Needless to say, the overall process is expensive and can quickly work its way through a limited maintenance budget. In the past, this approach was more reactive in that it looked at similar systems rather than being proactive in nature.

Many facilities have adopted inspection management programs that trend towards a risk-based inspection (RBI) method for selecting areas for inspection. Typical RBI programs address design and operating parameters, history, current physical condition as well as a number of other considerations such as release size and process material. Selections are then made about what equipment to inspect, how often and in what manner.

A number of companies in the chemical process industry have introduced CUI management planning initiatives to reduce the likelihood of occurrence of the problem. Of special interest to the Corunna, Ontario facility of NOVA Chemicals are non-destructive examination (NDE) methods that require the least intrusion but possess the highest probability of detection in screening process piping systems.

The principal attributes of a CUI management plan are discussed which are based on the current understanding of the problem and the reality of implementation in the field using the newer inspection technologies and inspection management databases.

CUI: The Basics

The material of construction that is most commonly damaged by CUI is carbon steel. The material when insulated is very susceptible to wet, cyclic and corrosive conditions especially in the absence of any protective coating. The type of insulation will also influence the initiation and progression of CUI to some extent but CUI has been shown to be independent of the insulating material over time. The temperature 'window of opportunity' for CUI is quite broad and has been expanded over the years based on more field case history becoming available and specific research programs on the subject. Typically, CUI can be active within a closed system at temperatures marginally above freezing to as high as 100°C with instances peaking in the 60-80°C range. However, special circumstances have been shown to extend the range to equipment operating at 200°C. A distinction should be made at this point between 'active' range and 'operating' range such that equipment operating temperatures can be well above (or below) the temperature range for CUI activity but the equipment becomes vulnerable when it cycles through the active range. This is why there is sometimes an element of surprise over newly discovered areas of CUI damage on equipment that would normally be expected to be immune from the problem.

As with the temperature, the locations at which CUI manifests itself are constantly being updated by virtue of field experience. This is solely on the basis of situations where the conditions are conducive to creating the active environment for CUI as noted above.

CUI is not limited or confined to straight runs of pipe and vessel shells. Some of the more aggressive and intrusive CUI damage that has been observed at NOVA are in areas where identification and access for inspection are more difficult. For example,

- Any kind of penetrations or transitions between insulated and uninsulated components:
  - Vents and drains
  - Vessel protrusions or appendages such as insulation supports, vacuum stiffener rings, flanges, lugs, and manways.
  - Inspection monitoring ports
- Equipment that may be continually exposed to moisture from cooling tower plumes, steam or venting and water from deluge or fire protection systems
  - Includes damage from leaking steam tracing, etc.
- Damaged insulation and coatings
- Deadlegs, specifically where there may be a wet transition
- Components in cyclic service, especially those fluctuating above and below the dew point

CUI: Detection

Historically, inspectors have focused on selecting, stripping insulation and visual inspection. In some situations, this remains the best way to get a very high confidence level about the true condition of the equipment and for a long time operating temperatures and the condition of the insulation and coating systems were the main indicators used to select inspection locations. In addition to conventional visual inspection and manual localized ultrasonic scanning there are a number of enhanced methods available today to detect the presence of CUI

- Guided wave ultrasonic testing (GWUT)
- Automated ultrasonic testing (AUT)

- Infrared imaging
- 'Flash' digital radiography
- Eddy current decay – Incotest

NOVA's Experience with Different NDE Methods

Note: The cases below describe NOVA's specific applications and experiences. NDE limitations and advantages listed relate to these specific examples and may not apply to all situations.

Guided Wave Ultrasonic Thickness Testing (GWUT):

Scope: Guided wave ultrasonic thickness testing was performed on a solvent line extending from one unit into another. The line was insulated and located in the high line therefore access was limited. GWUT was selected because it could be performed remotely and used to detect both internal and external wall loss.

Results: GWUT was successful in providing the inspection group with an idea of the condition of the line and locating areas of worst damage. Results were validated by targeted insulation removal and manual UT.

Advantages

- Remote. Do not need to scaffold entire section of the line and strip all the insulation
- Can be used around bends on piping
- Large areas can be scanned relatively quickly and results can be used to help focus inspection efforts to areas where anomalies are detected

Limitations

- Requires a certain amount of volumetric wall loss in order to register (small pits or linear indications may be difficult to detect)
- May be difficult to discriminate between internal and external damage
- Can get a 'shadow' effect at branch connections
- Results are more qualitative

Automated Ultrasonic Thickness Testing (AUT)

Scope: An automated ultrasonic thickness scan was performed on the inside surface of a vessel skirt to evaluate the extent of damage caused by corrosion under fireproofing (CUF). AUT was selected because it could be performed much faster than a manual grid and the results would be more extensive and comprehensive.

Results: Significant CUF was uncovered (Fig. 1). An engineering evaluation was performed based on the results. A repair and remediation plan was implemented (coating and re-fireproofing).

Advantages

- Uses a crawler. Did not require 100% of the surface to be scaffolded.
- Results are easy to understand and interpret. Report includes a visual map of the areas and quantifiable extent of damage.
- Much easier and quicker than a manual UT scan

Limitations

- Requires insulation to be stripped or access to the other side of the surface (inside of the skirt in this example)
- Certain limitations when equipment is operating above a certain temperature or when equipment has a radius below a certain size.

Figure 1 Example of corrosion under fireproofing. Skirt was scanned using AUT.

Digital Radiography

Scope: Digital radiography was used to perform a survey on vents and drains in a cryogenic refrigerant system.

Results: A number of drains and vents were identified with extensive CUI found at the insulation interface (Fig. 2).

Advantages

- No film or processing chemicals.
- Faster than conventional RT.
- In most cases did not require removal of insulation.

Limitations

- Still requires full access to the component
- Still requires all precautions necessary when using a radioactive source.

Figure 2 CUI detected during a vent survey.

CUI: Protection

It is clear that the avoidance of CUI lies in preplanning and design insofar as the recognition and elimination of susceptible locations and the proper protection for those configurations that are a necessary part of the equipment detail. Nevertheless, engineered protective coatings, when properly applied and maintained have a demonstrated track record of mitigating CUI. The coatings systems themselves can be high-performance inorganic zinc-hi-build epoxy and sprayed aluminum applications. A solution that is often employed but sometimes overlooked is to simply remove the 'I' from 'CUI', in other words permanently remove the insulation from the problem areas following an evaluation of the original application.

The Management of CUI

Damaged or poorly installed insulation is a common culprit for water ingress and can lead to CUI if left unaddressed. A line that has been recently insulated and inspected can become quickly prone to CUI damage if the insulation is damaged immediately following an inspection. For this reason a combination of good engineering procedures for insulation installation, frequent visual inspections and good coating systems beneath the insulation remain very important steps in the CUI management plan. When using a RBI program as the basis of such a plan the likelihood of insulation damage based on the location, for example, or the quality of the engineering standards that govern the installation of the insulation and whether these procedures are followed, should be a strong consideration in defining a 'next inspection' date.

Reactive inspections such as those triggered by findings at other facilities can provide some of the input required for a CUI management program but these should not detract from the CUI management program as a whole. Looking at all equipment where there is the potential for crevice corrosion under insulation, for example as would be found on a pipe rack, is an important exercise and should be part of the whole inspection program which includes deadlegs, vessels operating in cyclic service, etc. Building on this approach it is possible to identify three essential components to the creation of a comprehensive plan for managing CUI throughout the facility, these are;

- The Inspection Plan
- The Repair and Remediation Plan
- the Long Term Management Plan

Important considerations within the plan would be;

1) The methods of detection to be employed;
    - Invasive – removal of insulation
    - Non-invasive – in-service, on-line

2) The principal factors driving the CUI problems;
    - Mechanical damage
    - Over-exposure to water
    - Inability of insulation to dry-out
    - Styles of penetration into the insulation e.g. rings, brackets, hangers etc.

3) The insulation system by function;
    - Energy conservation
    - Personnel protection
    - Fireproofing

4) Identifying the resources and challenges, for example
   - The industry is well-served by national publications
      - API 571 Sec 4.3.3
      - API 581 App. N
      - NACE RP0198
      - HSE (UK) Guides and Technical Reports
   - "How do you estimate for a CUI plan?"
      - Defining the scope of annual activities
      - Impact on overall annual maintenance budget
      - Capital vs Major maintenance
      - Significant escalation of the budget for T/A years
      - CUI abatement strategy requires long-term financial commitment
   - Integration of CUI into existing RBI programs
      - How-to, legacy issues
      - Difficulty in predictability , statistical sampling
      - 'Scorecard' performance tracking

5) Identifying the obvious pitfalls;
   - The popular approach to using operating temperature as a screening tool for determining where to look for CUI poses numerous potential problems. Similarly, using temperature as a guide for selecting systems for inspection can oversimplify the situation and areas of aggressive CUI can easily be missed
   - Consider the operating temperature as only one of many factors in determining CUI susceptibility. Two examples follow:

      - Cyclic systems such as regeneration systems that may operate above temperatures in the CUI range and regenerate at temperatures far above the generally accepted 'upper limit' for CUI may be found to have corrosion rates far in excess of systems operating in the 'accepted' CUI range This is particularly true if the systems spend any time at ambient conditions between runs.
      - Deadlegs are locations where process fluid can accumulate and may actually operate at temperatures lower or higher than the general line operating temperature.

Therefore it can be seen from the above mentioned examples that the CUI problem cannot be generalized by temperature 'zoning' alone because of the variety of ways that CUI can present. Figure 3 shows two idealized piping systems, one hot (System 1) the other cold (System 2). As the hot system is shut down or cycles for another reason, the pipe temperature drops into the temperature range of activity for CUI and stays there until the process is restored to normal. A similar situation applies to System 2 which will warm up to the CUI zone when shut down (or in some cases steamed out). Ordinarily this would not be a problem in either case if the insulation systems are not compromised but if the piping systems are already exposed to the necessary environmental conditions which will initiate CUI then corrosion damage is assured over time.

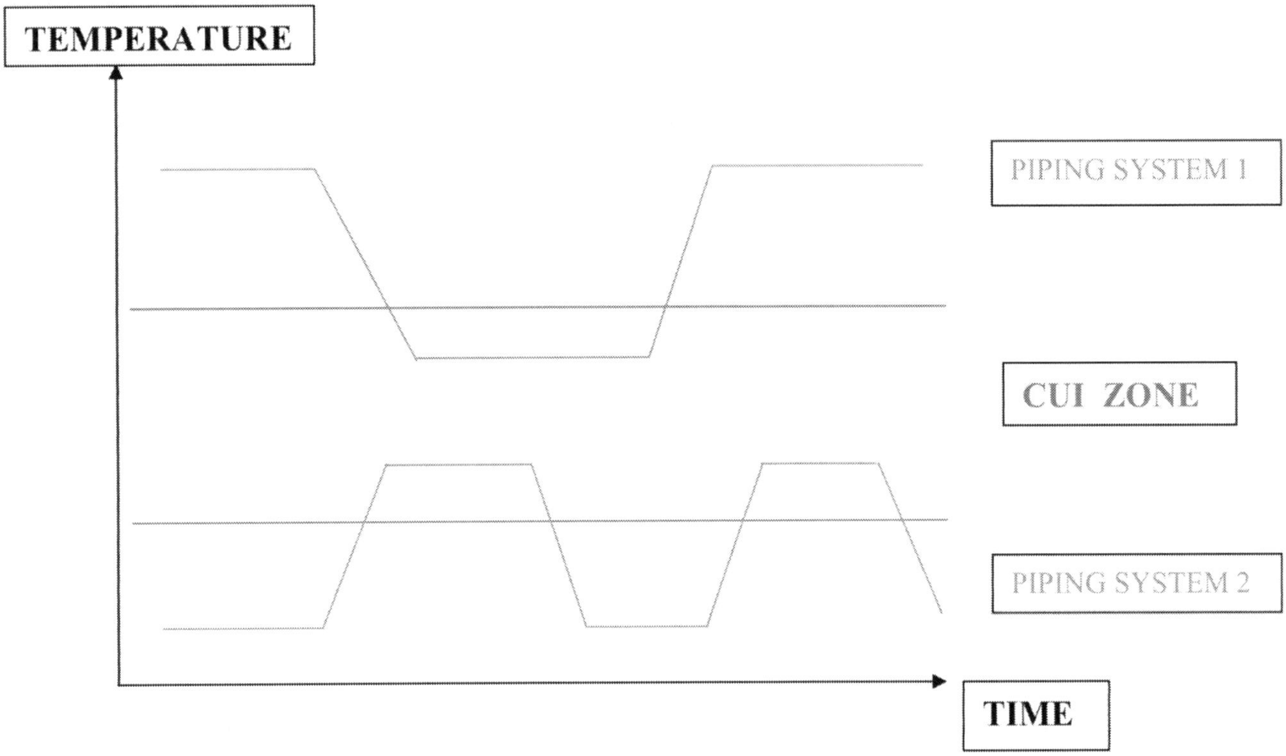

Figure 3 Cyclic aspect of CUI in piping systems

The Role of Risk Based Inspection in NOVA's CUI Management Program

The use of risk-based inspection (RBI) methods has become widespread in the oil, gas and petrochemicals industry as a process used to help determine what equipment to inspect on what frequency. As opposed to using a pure time-based approach RBI allows the inspection group to evaluate the risk of failure based on both probability and likelihood to rank the pressure retaining components in the system.

RBI programs can range from qualitative to fully quantitative. Qualitative methods use more subjective assessments of information, engineering judgment and ranking to evaluate risk. Quantitative methods require precise numerical data, modeling and costing to generate risk information. NOVA uses a semi-quantitative method of evaluation, using a combination of calculations, historical data and judgment. The API580/ 581 – *Risk Based Inspection* documents do an excellent job of outlining the recommended requirements when performing RBI and are used by NOVA as a guideline.

CUI as an external damage mode differs from most conventional internal damage modes and needs to be considered as somewhat unique when performing an evaluation using RBI. The external drivers, corrosion rates and effects can vary significantly even within one piping system and therefore NOVA has found that in many situations using two separate risk evaluations, one for internal damage modes, and a second exclusively for external damage modes has been quite effective. This has been found to be the best method to ensure that the risk of external damage does not get overshadowed by the risk of having a failure due to internal damage.

When evaluating the likelihood of failure the following are common considerations:

- historical findings on the system at this site or elsewhere
- service
- climate and sources of moisture ingress
- normal operating temperature and upset temperatures
- condition, design, age and type of insulation and coatings
- effectiveness of inspection methods

Coordinated effects of internal damage on the equipment from service-related deterioration such as corrosion or mechanical damage need also be considered. Method of inspection, whether it be visual inspection, selective or wholesale stripping of insulation, or another non-destructive method should be evaluated for effectiveness and confidence in the ability to properly portray the condition of the equipment.

The assessment of consequence of failure tends to include an evaluation of:

- the commodity (flammability, health hazard)
- amount of commodity that would be released
- the operating pressure and temperature
- the location of the equipment
- the type and size of the failure (pinhole or burst)
- environmental, legal and business impact (costs, fines and corporate image)
- current mitigation systems (fire suppression, hydrocarbon monitoring, etc.)

Some judgment may be required when evaluating a large piping system or vessel where the design or operating conditions can vary over the length. API 581 includes provisions for performing risk assessments on distinct components or areas within systems to determine risk. This may be appropriate for systems with a number of different operating envelopes.

Summary

One company's implementation of a risk-based CUI management program has proven to be an effective tool in the inspection and detection of the problem. Advanced methods of inspection demonstrated advantages in accessibility, detection and productivity over the more traditional equipment inspection practices. The limitations of each of the inspection technologies within the given application were also taken into account during their execution.

References

1. API Recommended Practice 580, "Risk-based Inspection Downstream Segment", Second Edition, November 2009
2. API Recommended Practice 581, "Risk-Based Inspection Technology", Second Edition, September 2008
3. de Bruyn, Hendrik J. ,Baecke, Antoine and McQuillan, Kevin W. "How Big is the Impact of CUI?" European Ethylene Producers Committee (EEPC) Corrosion Under Insulation (CUI) Study Group Report

AIChE PAPER NUMBER 97e

FURNACE CONVECTION SECTION CLEANING SYSTEMS - ENGINEERED SOLUTIONS FOR CRACKING FURNACES

John Ghetti
President
FINFOAM

Prepared for Presentation at the 2012 Spring National Meeting
Houston, TX, April, 1 - 5, 2012

FURNACE CONVECTION SECTION CLEANING SYSTEMS - ENGINEERED SOLUTIONS FOR CRACKING FURNACES

John Ghetti
President
FINFOAM

Abstract: In today's globally competitive petrochemical industry, manufacturers are under pressure to increase production, improve efficiency, and extend equipment life. A case study will be presented demonstrating how convection section cleaning technology improved the fuel efficiency and production throughput of an ethylene cracking furnace. An additional benefit was also recognized as this convection section was considered to be near end-of-life; however, the performance improvements noted justified extending the life of this convection section.

Introduction

A case study will be presented regarding an ethane gas-cracking furnace located at a North American Gulf Coast ethylene facility. This furnace has been in service about 20 years. It is Stone & Webster design with a "W" coil configuration and a mix of floor and wall burners. The burners are fuel gas fired.

Feed rates to this furnace had gradually decreased over time due to excess oxygen limits as measured at the crossover section between the radiant and convection sections. Even at significantly reduced feed capacity, excess oxygen levels could not reach the 2.5-3% target (Figure 1).

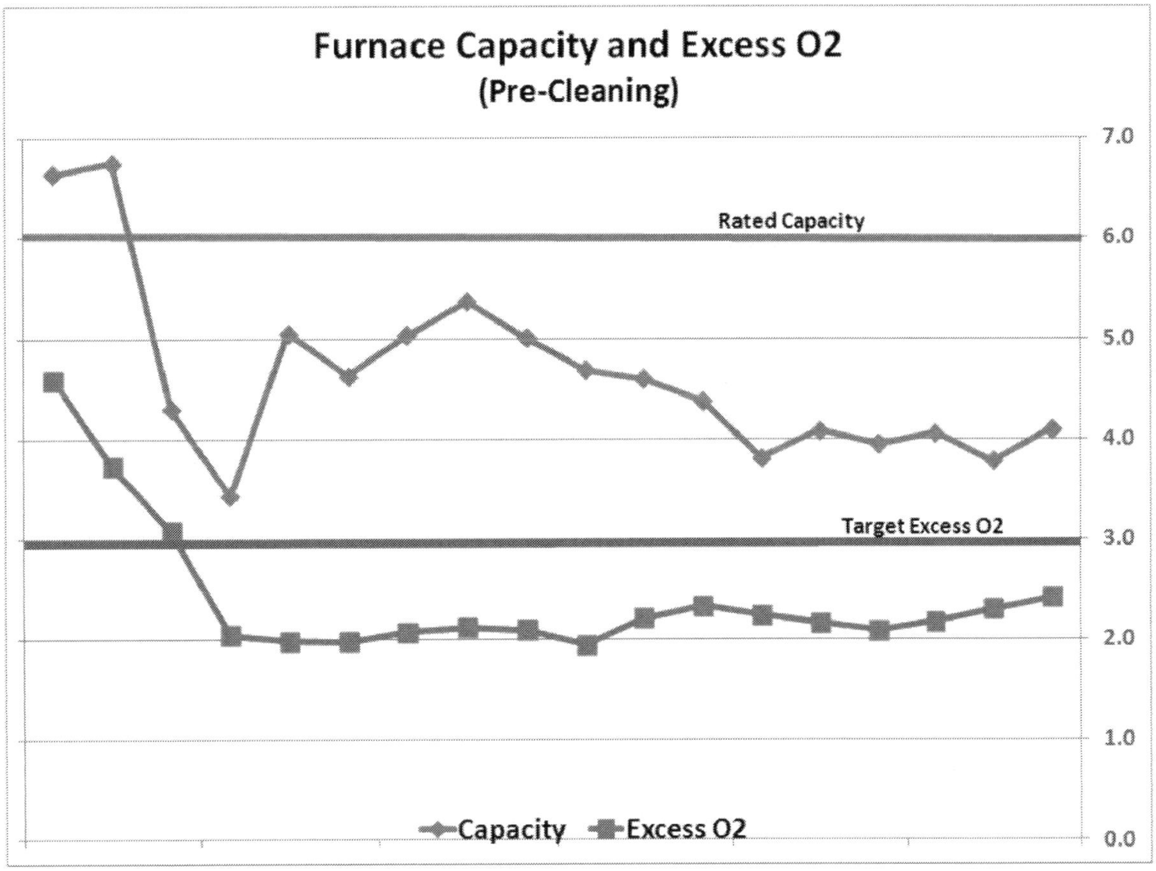

Figure 1—Furnace Capacity and Excess Oxygen Prior to Cleaning

At the same time, draft control barely maintained negative crossover pressures (Figure 2).

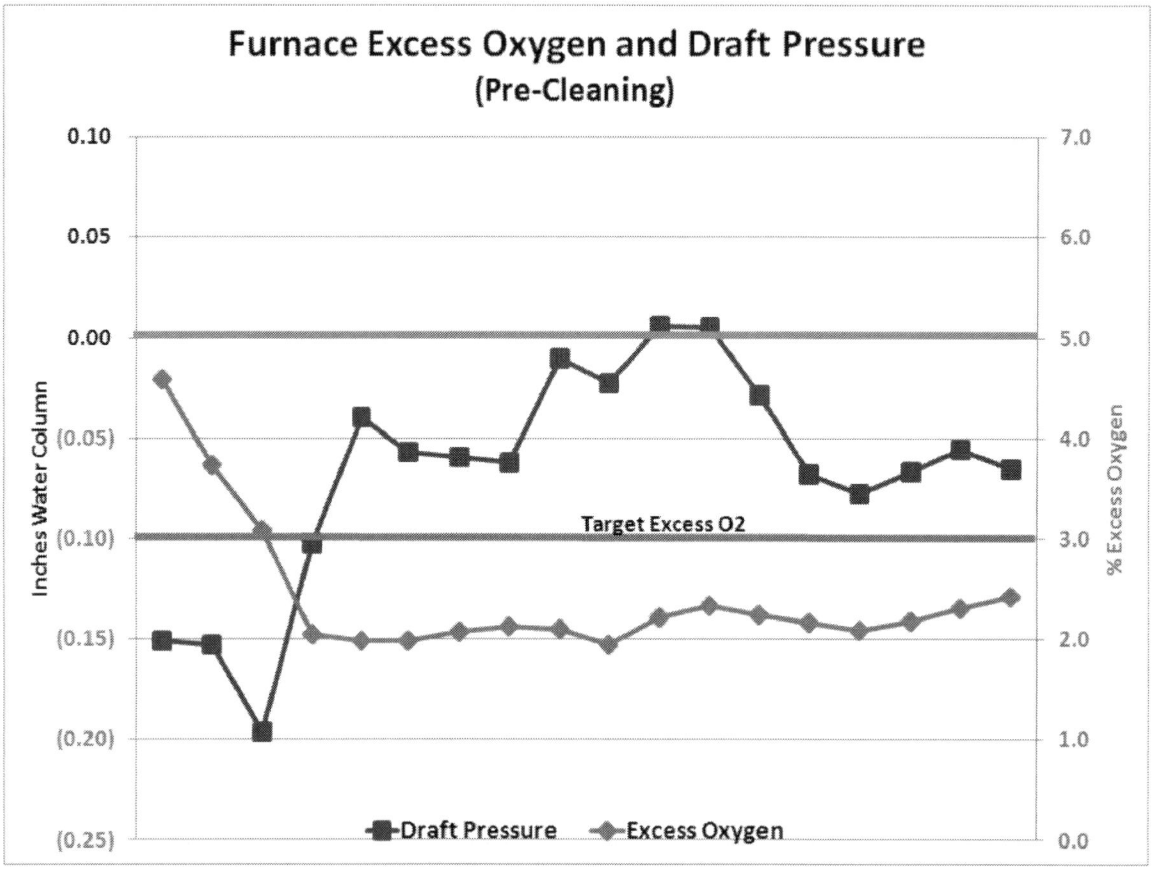

Figure 2—Furnace Excess Oxygen and Draft Pressures Prior to Cleaning

Investigation

Several steps were taken to identify the root cause:

- <u>Oxygen analysis verification</u> —independent excess oxygen measurement confirmed the accuracy of the online analyzer.
- <u>Burner air adjustments</u>—Air dampers on the floor burners and secondary air registers on the wall burners were adjusted to various open, closed, and intermediary positions. No improvements were realized. Excess oxygen levels did not increase.
- <u>Induced draft fan (IDF) review</u>—the furnace was taken down specifically to inspect, overhaul, and test the ID fan. No issues were discovered.

- Inlet IDF louver inspection—while the furnace was down for the ID fan inspection, the draft control louver was inspected and tested. Again, no issues were found.

By process of elimination, area personnel determined the flue gas flow had to be restricted in the convection section. Because of limited access and visibility, a borescope was inserted at various inspection ports to determine if anything was constraining the flue gas, such as loose refractory. This borescope inspection was performed with little success. An IR camera scan was also performed, but once again, nothing obvious was noted that could cause restricted flue gas flow.

Closer review of pictures taken of some convection section tubes, however, revealed significant build-up of debris within the fins (Figures 3, 4, and 5).

Figure 3—Finned Convection Tubes Filled with Debris #1

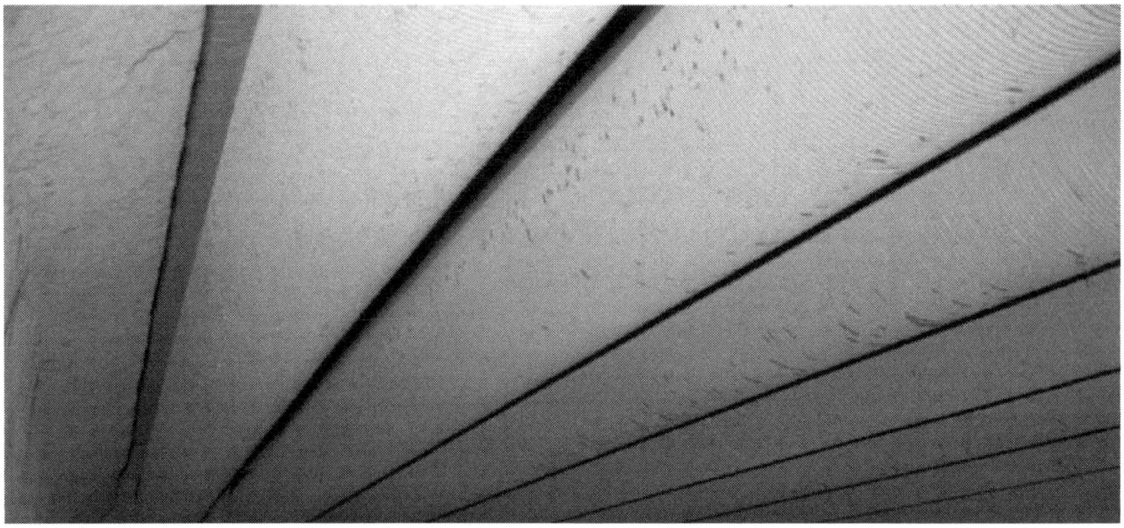

Figure 4—Finned Convection Tubes Filled with Debris #2

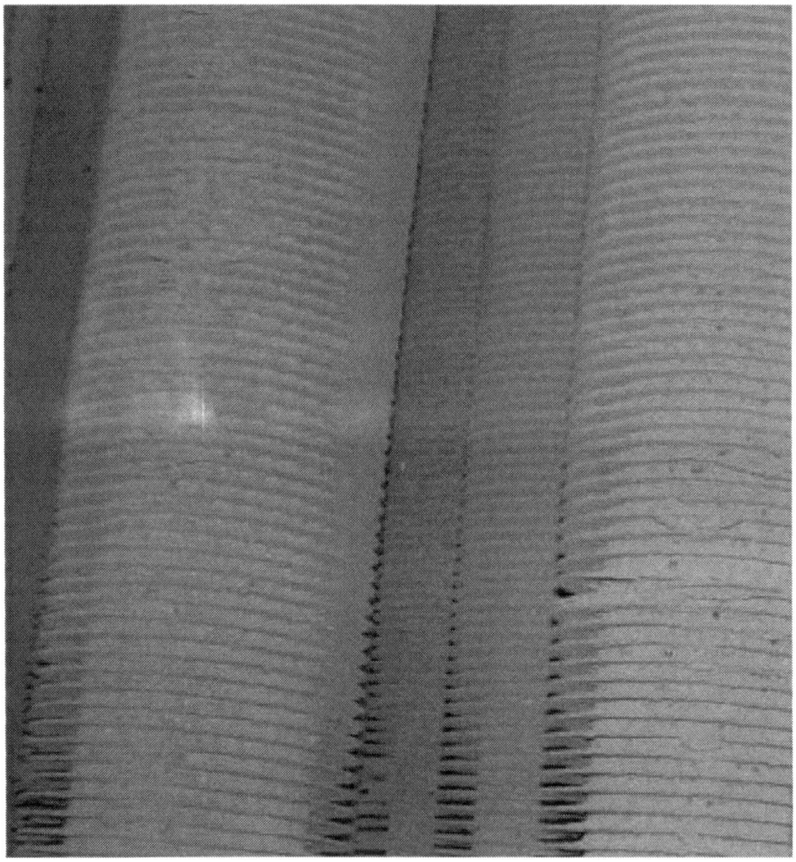

Figure 5—Finned Convection Tubes Filled with Debris #3

Because this furnace had been in service about 20 years, concerns were raised that the convection section might be at end-of-life. The area began budget planning for complete convection section replacement. In the meantime, area personnel decided to try cleaning the convection section to regain some furnace capacity until complete convection section replacement could occur.

Convection Cleaning

Several different cleaning methods were considered such as cleaning with dry ice, blasting with walnut hulls, and air blowing. However, these were dismissed mainly because of limited access to all the convection tubes. Sufficient access would have required cutting in several access doors. Consideration was then given to foam cleaning. Previous experience on other site equipment had shown this method to be effective, especially when access was limited to all the surface area.

This foam cleaning method did pose the following concerns:

- Protecting refractory from water damage
- Keeping water out of the firebox
- Collecting effluent foam and rinse water
- Subsequent waste water treatment

Each issue was addressed, and a game plan was developed between the plant personnel and the foam cleaning company.

Set-Up

The foaming company uses an environmentally safe low pressure / high volume foam process to chemically clean Pyrolysis Cracking Furnaces. FINFOAM's mobile foaming units are self contained. They simply require access to equipment, a 1 ½ inch water supply, and a frac tank for effluent foam and rinse water.

Required (facility) permits are acquired, equipment is staged, the area is barricaded, and 1 ½ inch hoses are routed from the foaming unit to a hydrant and the furnace (typically by a crew of three men with no less than 18 years combined experience.)

The furnace is then entered, and a series of tarps are positioned on the furnace walls and floors to protect the refractory and firebox from water damage.

The Chemical

The chemical employed with the cleaning of the convection section coils is a blend of anionic and non-ionic polymerized surfactants with biodegradable aliphatic having a content of approximately 25% Glycol Ether. The concentrated chemical is added to day tanks having a capacity of 750 gallons (Figure 6).

Figure 6—Foaming Equipment

Application of Foam

The readied furnace is entered by crew member while maintaining communication via radio with hole watch and foaming unit operator. Foam is applied to furnace tubes and provides 100% external surface contact of every tube (top to bottom) (Figure 7). Foam is applied until foulant(s) on the saturated tubes loosen and travel down with the foam (Figure 8). This is then followed by high volume, low pressure rinse. Rinsing continues until clear water is seen.

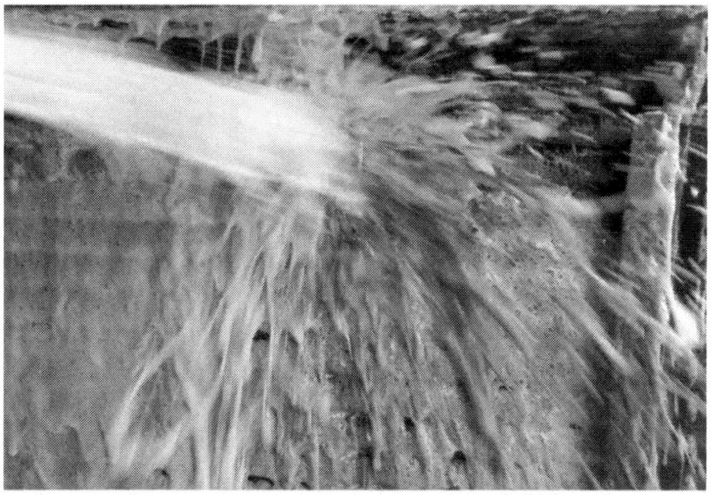

Figure 7—Application of Foam to Finned Tubes

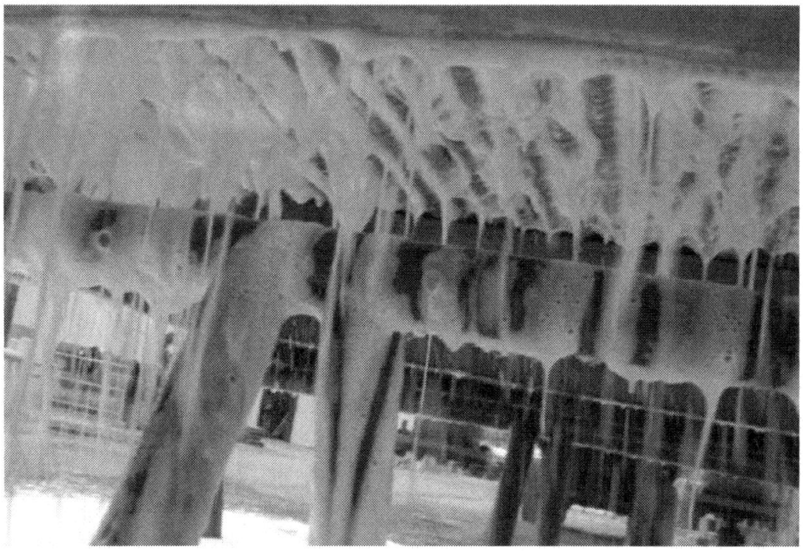

Figure 8—Foam Penetrating Through Finned Tubes

Foam Effluent Collection

At the lowest inspection port of the convection section, a 14" flex hose is connected to port then routed to frac tank (Figures 9 and 10).

Figure 9—Routing Effluent from Bottom of Convection Section

Figure 10—Into Frac Tank

Effluent Foam and rinse water is collected through this 14" flex hose, diverting effluent into frac tank (Figure 10). Collected effluent is then tested by environmental and the decision is made if any subsequent wastewater treatment is required.

Debris Removed

Debris removed is carried by the foam to the frac tank and furnace floor. Debris on the furnace floor is generally removed by a vacuum truck.

Figure 11—Example of Debris Removed by Foam Cleaning Process

Cleaned Tubes

The foam application generally takes 4 to 6 hours, which is plenty of time to easily remove scale buildup and ensure the tubes are clean (Figure 12).

Figure 12—Convection Tubes After Cleaning

Prior to returning the furnace to cracking service, a refractory dry-out step and slow heat-up are performed to remove residual water.

Results

Excess oxygen levels were immediately improved. As a result, capacity was not only regained, it exceeded target capacity (Figure 13).

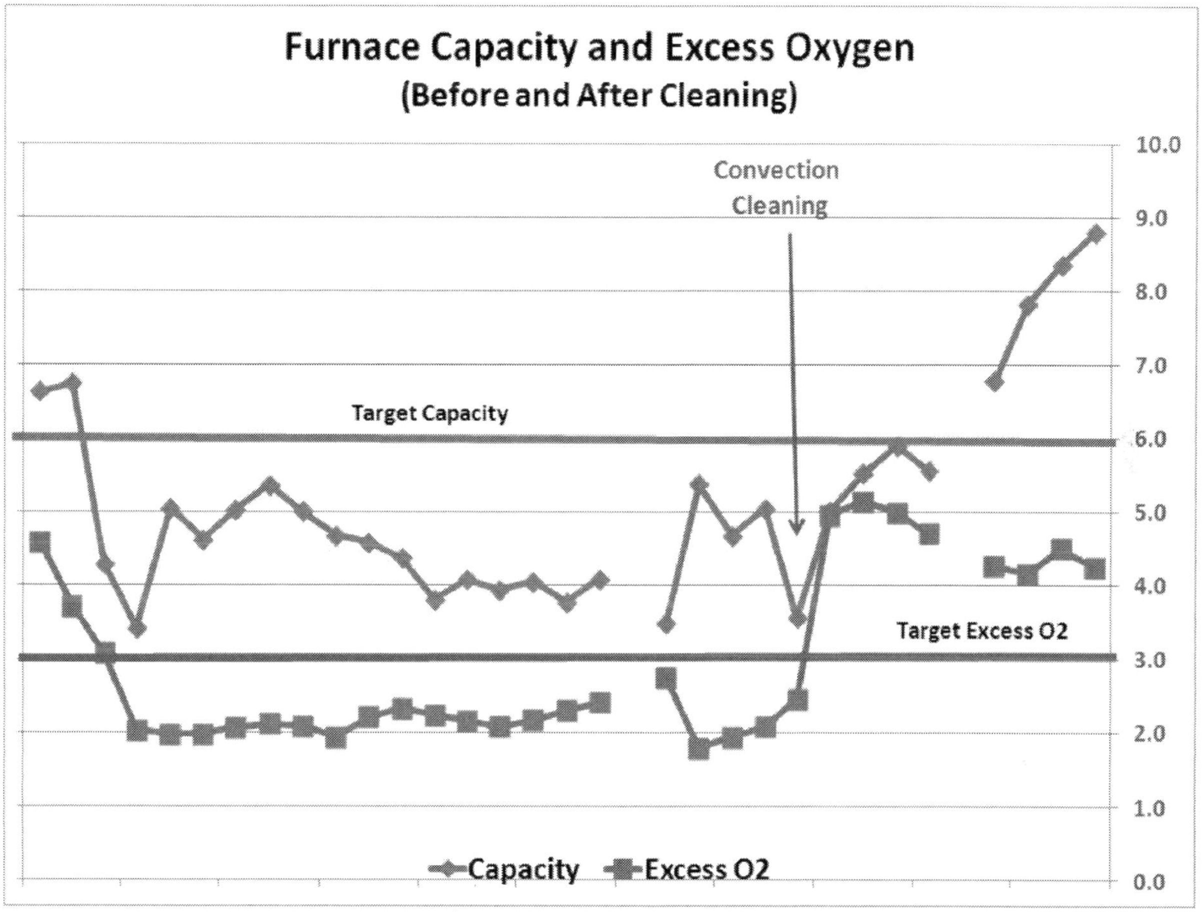

Figure 13—Furnace Capacity and Excess Oxygen Comparison Before and After Cleaning

The draft pressure control issues were also eliminated—allowing easy maintainability of negative pressures at the crossover (Figure 14).

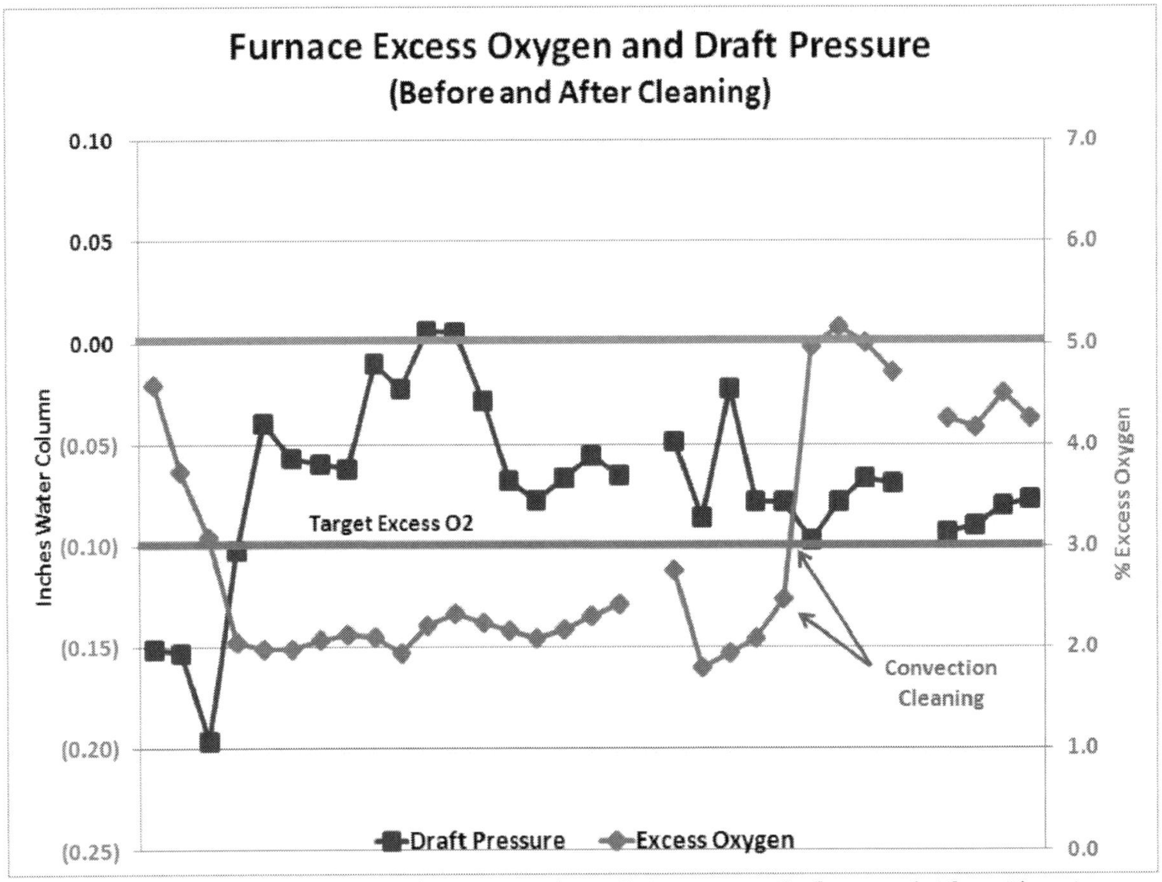

Figure 14—Furnace Excess Oxygen and Draft Pressures Before and After Cleaning

The furnace's stack temperatures were also running about 10°F less after the convection cleaning than that noted beforehand (Figure 15).

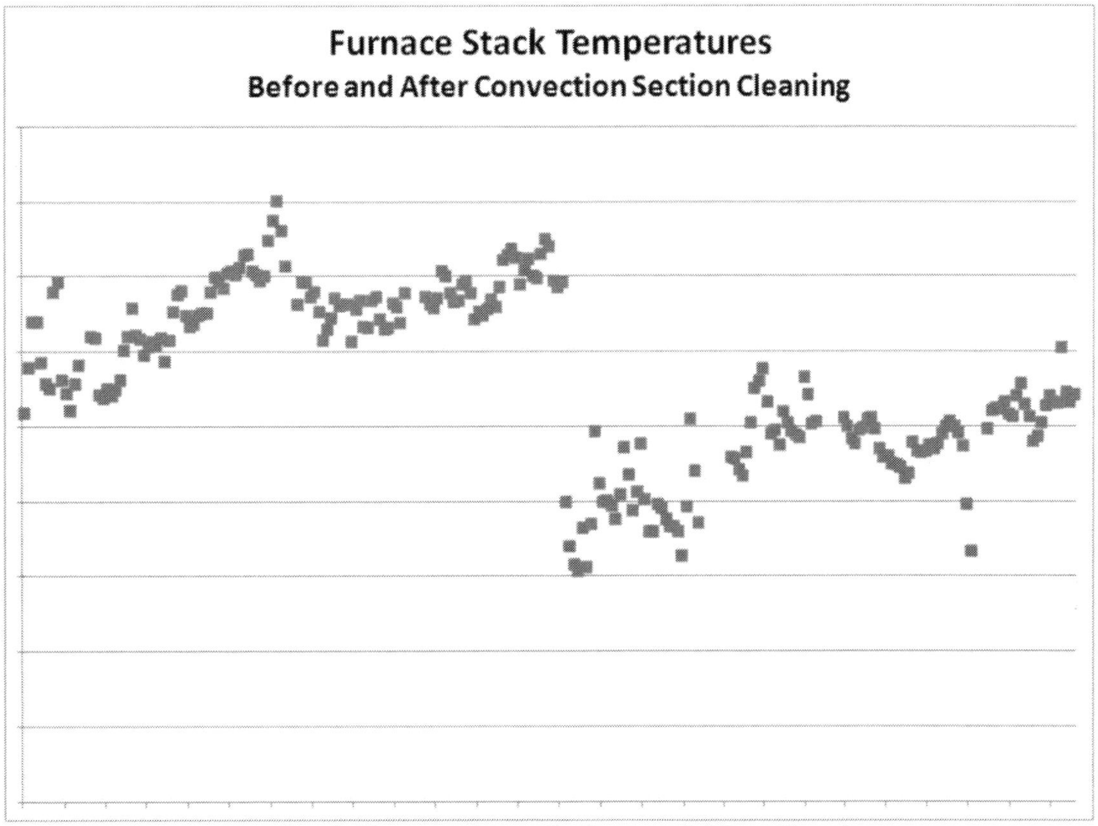

Figure 15—Stack Temperatures Before and After Cleaning

The furnace's fuel efficiency (Btu/lb ethylene) immediately increased by about 5%. Long term, the fuel efficiency improvement has sustained at 2.5%.

Conclusion

Because of the process improvements noted, the facility has postponed the immediate need for convection section replacement. Higher than target capacity has been maintained since without issue and plans are underway to clean other furnaces at the facility.

AIChE Paper Number 97f

The Application of Finite Element Analysis to the Remaining Life Assessment of Ageing Cold Boxes

David J. Oulton
Corporate Metallurgist
NOVA Chemicals (Canada) Ltd

Kyle C. Koppenhoefer
Principal
Altasim Technologies Inc, Columbus Ohio

Prepared for Presentation at the 2012 Spring National Meeting
Houston, TX, April 1-5, 2012

The Application of Finite Element Analysis to the Remaining Life Assessment of Ageing Cold Boxes

David J. Oulton
NOVA Chemicals (Canada) Ltd, Corunna, Ontario

Kyle C. Koppenhoefer
Altasim Technologies Inc, Columbus, Ohio

Abstract: Cold boxes, or more precisely, the brazed aluminum heat exchangers (BAHX) that comprise them operate unseen and do not lend themselves to the inspection methods typical of most other critical process equipment. Furthermore, BAHXs are not immune to process upset conditions which over time can threaten the structural integrity of these vital components in the form of thermal fatigue cracking. The challenge to the chemical processing industry is therefore one of attempting to establish the condition of ageing cold boxes especially with regard to quantifying the cumulative damage resulting from multiple normal and emergency upset events.

NOVA Chemicals is one of a growing number of companies that are using finite element methods to closely analyze the long-term response of BAHXs to such transient conditions and to interpret the results in the broader context of fitness-for-service and remaining life. The same methodology is also being employed to look at newer BAHX designs in order to assess their vulnerability to future thermal fatigue damage.

This paper describes a recent case history undertaken at NOVA Chemicals in which a number of older BAHXs were analyzed using finite element modeling (FEM).

Introduction

Brazed aluminum heat exchangers (BAHX) are complex fabrications and operate unseen inside cold boxes performing a vital plant function. Given the nature of their construction BAHXs do not lend themselves to the inspection methods typical of most other critical process equipment. Furthermore, brazed aluminum heat exchangers are not immune to process upset conditions which over time can threaten the structural integrity of these important process components in the form of thermal fatigue cracking. The challenge to the chemical processing industry is therefore one of attempting to establish the condition of ageing cold boxes especially with regard to quantifying the cumulative damage resulting from multiple normal and emergency upset events.

Leveraging off numerous successful previous applications of investigative finite element analysis (FEA), an earlier generation cold box at one of the NOVA Chemicals facilities was made the subject of this particular remaining life assessment.

Cold Box Construction

'Cold box' is the term given to the structure that houses one or more brazed aluminum heat exchangers. The actual function of the cold box is to provide an enclosure, or casing, that can support and insulate the internal BAHXs and their associated components such as piping and separator drums. The cold box which is the subject of this paper encloses eight exchangers and three separators and is shown in Figure 1.

Figure 1

The principal function and materials of construction of a cold box are:

- BAHXs (Cores) – aluminum alloy, pressure boundary equipment
- Drums – aluminum alloy, pressure boundary equipment
- Internal piping and valves – aluminum alloy, stainless steel, pressure boundary components
- Casing – carbon steel, support, insulation and containment structure

An internal detail common to all BAHXs is shown below in Figure 2,

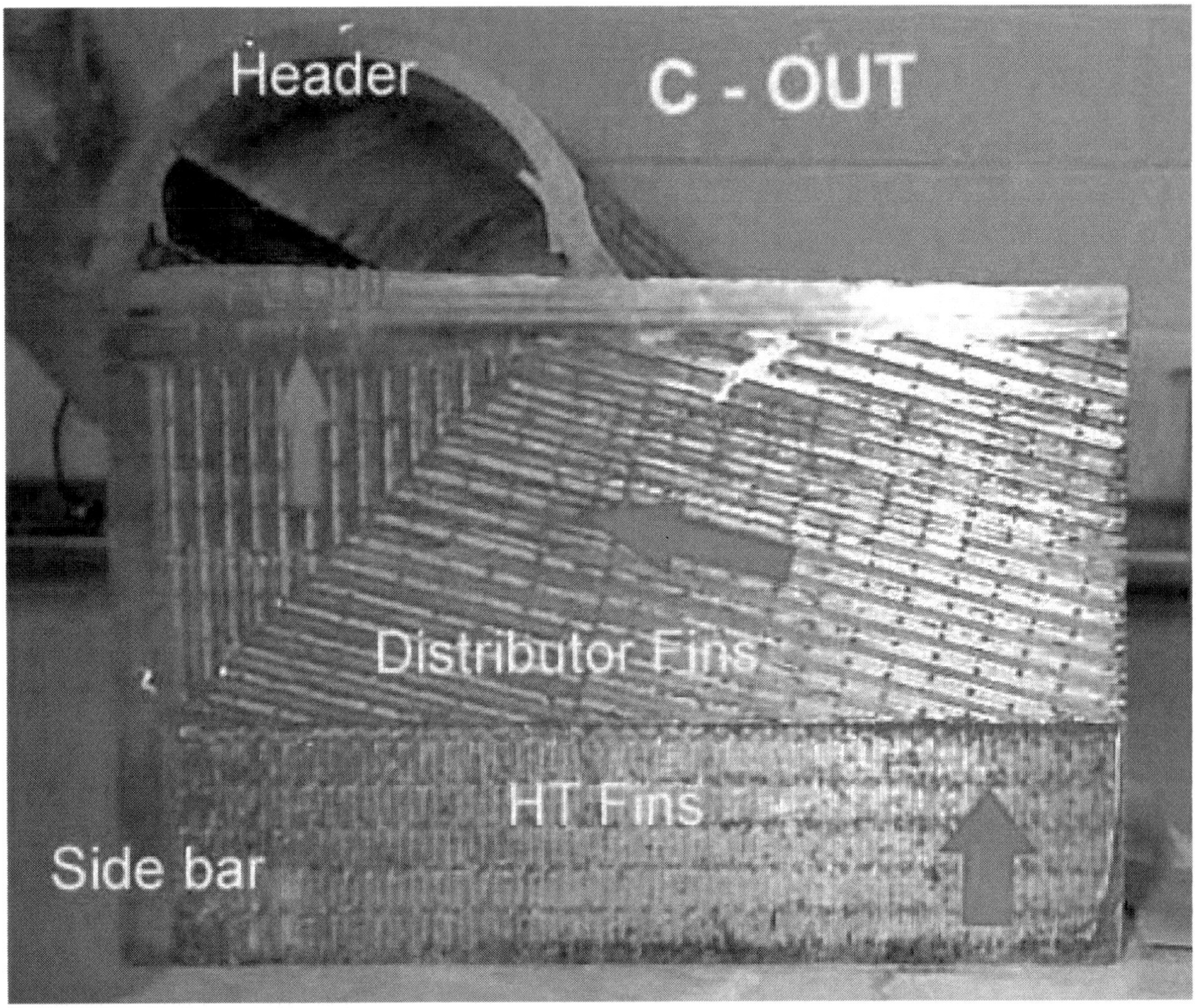

Figure 2

It is because of the internal plate and fin arrangement that BAHXs can be constructed to handle multiple process streams independent of other parts of the same assembly.

Cold Box Operation

The cold box of interest has been in service for more than thirty years and in that time two of the eight BAHXs had been replaced and a further two had received minor repairs.

The cold box primarily recovers ethylene and hydrogen from cracked feed and delivers the remaining constituents, mostly methane, as fuel gas. Typical stream operating conditions within a single BAHX can range from 2 to 460 psig and -265°F to 80°F.

Cold Box Risk-Informed Condition Assessment

A cold box falls into the general category of equipment which has either exhibited significant deterioration or damage during its operating life or whilst operating satisfactorily, there is a deficiency of information to know if damage is accumulating. In other words, each and every cold box will be subject to varying levels of uncertainty when it comes to assessing their respective condition.

This particular cold box had a risk ranking such that a medium level of risk was associated equally with each of the eight cores based on a reasonable understanding of specific damage mechanisms to cold boxes in general and a credible expectation of the consequences arising from such failures. However, as with all RBI programs some cores will have a higher risk importance than others with equal risk ranking.

The likelihood of failure can be divided into two mechanisms, chronic and acute. An example of a chronic condition would be thermal fatigue whereas an example of the latter would be mercury attack. Global experience has shown that many cold boxes have failed or at least encountered some degree of impairment within 10-25 years of operation with the majority most likely of having failed as a result of thermo-mechanical damage. Part of the problem with the failure statistics is that for a long time very little was published in the public domain in comparison to other equipment failures, although this is gradually being rectified over recent years through forums such as the EPC conferences. Risk-informed condition assessment of cold boxes is further hampered by the inspectability and subsequent inspection history of the units themselves. Cold boxes are a unique configuration and are not easily tested. In addition, the analysis of core failures is rare with respect to a published database and pressure integrity testing is not without its own risk.

The consequence of failure may be modeled under three main failure event scenarios;

External pressure envelope failure – worst case would be the potential for large volume release of high pressure (HP) streams, subsequent initial chill-down and brittle fracture of the casing and static ignition of the product from high velocity contact with the perlite insulation. Older cold boxes are often in areas of high-density equipment population so there would be a multiplication factor for collateral damage.

External pressure envelope chronic leak – An initial small leak creates an eddy effect in the surrounding abrasive perlite, the leak gets bigger as perlite continuously erodes the pressure envelope, hydrocarbon (HC) detectors subsequently alarm and the plant goes into emergency planning mode with a loss of containment incident.

Internal cross-pass leak – worst case would be if multiple passes are involved, processes become unworkable without significant recovery and recycling, planning required for next available shutdown (S/D), repair or replacement options currently unknown.

As part of any condition assessment review of a cold box, the following questions become self-evident:

- What is the most probable cause of the failures that a plant has experienced? - e.g. many BAHXs are simply repaired after experiencing internal cross-pass leaks with virtually no opportunity for a failure analysis.
- What is the likelihood of a future failure? - tracking of repair history, introduction of a new product mix, plant expansion or upgrade etc.
- What is the contingency plan in the event of a future failure? – twinning, by-passing, repair etc
- What is the long-range plan (5-10yrs) if needed? – retirement, modification, overhaul etc
- How can the condition of the cold box be monitored? – periodic pressure testing, NDE, FEA etc

Cold Box Failure Mechanisms

Cold boxes can succumb to a number of failure mechanisms in their operating lifetime chiefly:

1. Thermal shock – This is an opportunistic event affecting the cores and has a severe effect on joint integrity. It is because the cores consist of closely spaced thick and thin sections of a high thermally responsive material that temperature differentials (ΔT's) are an important consideration in the operation of a cold box. Separation of brazed joints between parting sheets and side bars as a result of thermal shock is the leading mode of failure in this type of exchanger assembly. Such damage is commonly discovered within approximately a one foot radius around headers and is usually concentrated towards one or both outside sheets of the exchanger.

2. Thermal fatigue – Produces a cumulative effect over time with a gradual loss of joint integrity.

3. Plugging with ice or debris – Results in mechanical or corrosion damage. Ice damage often shows up as symmetrical bulges visible on the outside sheets indicating the presence of similar damage inside the core at dead zones. Less common is internal freeze damage without any external local deformation. Failures due to plugging as a result of fouling are not uncommon but tend to rely on a corrosion mechanism between the fouling material and the aluminum fins and parting sheets

4. 'Slugging' - Liquid slugs in 2-phase flow produce a ballistic impact on fins and can have an erosive effect on headers in more extreme cases. The damage is restricted to inlet sections where the two-phase stream enters the core. Definitive evidence of this form of damage is the presence of extensive buckling of the fin edges under the header.

5. Mechanical restraint – affects both cores and piping by restricting thermal movement especially during the normal transient periods of starting up and shutting down. The problem is manifested by cracked braces and supports and has the potential to crack nozzle joints.

6. Metallurgical factors
   - temperature-driven sensitization effect on the Al-alloy material of construction
   - environmentally-assisted cracking in Al-alloy piping
   - internal corrosion damage during normal transient operations

In assessing the condition and remaining life of equipment in general it is often necessary to determine its tolerance to deterioration by the active damage mechanism. Published documents such as API 579 and BS 7910 have procedures to help determine fitness-for-service by taking into account the damage tolerance of the structure under investigation. However, there is a need to consider the uncertainties and limitations of the data especially with respect to general corrosion modes of deterioration and internal thermal fatigue cracking. Cold boxes may be considered to be a special case in this regard whereby internal cracking, for example, may not proceed at a constant rate but will progressively increase with time or will simply initiate at a later date in operation. Remaining life based on 'bathtub' formulae may therefore no longer apply and any planned fixed interval inspection plan would most likely be skewed towards the anticipated end of life.

Cold Boxes, Thermal Fatigue and Leak-before-Break Failure

The philosophy behind the leak-before-break criterion is that any plausible defect in a pressure boundary element can be predicted to grow through the pressure boundary in a stable manner thereby creating a detectable leak for which the consequences will be manageable. The method also requires analytical modeling of the defect to calculate critical size and response (behavior) to the chemical and thermo-mechanical environment in service.

The defects of particular interest in the case of ageing cold boxes are thermal fatigue cracks and the application of finite element modeling tools provides a quantitative understanding of the conditions that are likely to result in cracking at locations within the structure of a BAHX.

Finite Element Analysis

To this end, a series of finite element analyses were conducted to determine the effect of temperature on the fatigue susceptibility of the eight BAHXs in the cold box referred to in Figure 1. The exchangers had been initially prioritized according to ΔT variation between streams. To conduct these analyses, a solid model was constructed to represent two passes of a BAHX, as shown in Figure 3. The upper pass runs parallel to the long axis of the heat exchanger while the lower pass runs at an angle of 30° offset from the long axis. This model represents a transition region within the exchanger (e.g. Fig 1). Initial work showed offset angles of 30° and 60° produced larger thermo-mechanical stresses compared to passes at 0° and 90°. Thus, the 30° transition provides a worst-case condition to be considered in this work.

The material properties used in this model represent aluminum. Table 1 provides the material properties specified for this model. The mechanical boundary conditions shown in Figure 4 restrain the model against rigid body motion while still maintaining the freedom to expand and contract due to a homogeneous temperature change without producing thermo-mechanical strains. The thermal loading on the model is defined by specifying a convection constant on all surfaces that contact either a stream or air. For the internal surfaces, the convection constant equals 3.5 BTU/hr-ft²-°F (20 W/m^2 K). The far field temperature is specified from the process data. Figure 5 shows typical temperature traces for two streams. For the exterior surfaces, an ambient temperature of 70 °F was specified with a convection constant equal to 0.35 BTU/hr-ft²-°F (2 W/m^2 K). All other surfaces where specified as having no heat flux (i.e., insulated).

The thermal conditions specified in the model produce a temperature gradient in the BAHX that varies with time. This spatial variation in temperature, as shown in Figure 6, induces thermo-mechanical stresses and strains in the model. These stresses/strains vary with time due to the temporal variation of the stream temperatures shown in Figure 5. The Mises stresses produced from the temperature gradient are shown in Figure 7. The maximum principal strain due to a temperature differential between the two streams of 120°F is shown in Figure 8. By calculating strain over a range of temperature differentials observed in the process excursions, a relationship between temperature and strain can be developed. These strains form the basis for determining the thermo-mechanical damage produced by a temperature excursion.

To assess the damage these strains impart on the material, two criteria were developed. The first uses a principal strain criterion while the second uses a Brown-Miller criterion

$$\frac{\Delta\varepsilon_1}{2} = \frac{\sigma_f'}{E}(2N_f)^b + \varepsilon_f'(2N_f)^c$$ Principal Strain

$$\frac{\Delta\gamma_{\max}}{2} + \frac{\Delta\varepsilon_n}{2} = 1.65\frac{\sigma_f'}{E}(2N_f)^b + 1.75\varepsilon_f'(2N_f)^c \text{ -}$$ Brown Miller Combined Strain

Where,

E : Elastic Modulus

σ_f' : Fatigue Strength Coefficient

ε_f' : Fatigue Ductility Coefficient

b : Fatigue Strength Exponent

c : Fatigue Ductility Exponent

$\frac{\Delta\varepsilon_1}{2}$: Amplitude of Maximum Principal Strain

$\frac{\Delta\gamma_{max}}{2}$: Amplitude of Maximum Shear Strain

$\frac{\Delta\varepsilon_n}{2}$: Amplitude of the Strain Normal to the Maximum Shear Strain

Table 2 includes the material parameters used in this assessment. Based on these material parameters, estimates of the failure curves were calculated for the braze metal (Al-10Si) and the base aluminum (Al-3003). Figure 9 shows the relationship between cycles and strain to failure based on the principal strain criterion and Figure 10 shows this same relationship based on the Brown-Miller Criterion.

Using the relationship between stream temperature differential and strain from the finite element analyses, a curve was constructed that relates temperature differences to transients to failure (Figure 11). Based on the operating history of the BAHX units of interest, a transient count of 1000 was chosen as the cycles to failure for the unit. These results indicate that a temperature differential of 109°F between two streams could produce cracking in 1000 cycles.

Table 1 Material Properties

Material Property	English	SI
Thermal Conductivity	1460 BTU-in/hr-ft^2-°F	210 W/m-K
Specific Heat Capacity	0.215 BTU/lb-°F	0.9 J/g-°C
CTE	14.2 1/°F	25.5 1/°C
Modulus of Elasticity	9860 ksi	68 GPa
Density	0.0975 lb/in^3	2.70 g/cc

Table 2 Fatigue parameters for AL-3003 and AL-10Si (Carter, et al., 1996)

	Al-3003 (-148 °F)	Al-10Si (-148 °F)
E (MPa)	7.0E4	7.0E4
σ_f' (MPa)	230	230
ε_f'	0.46	0.02
b	-0.12	-0.12
c	-0.6	-0.6

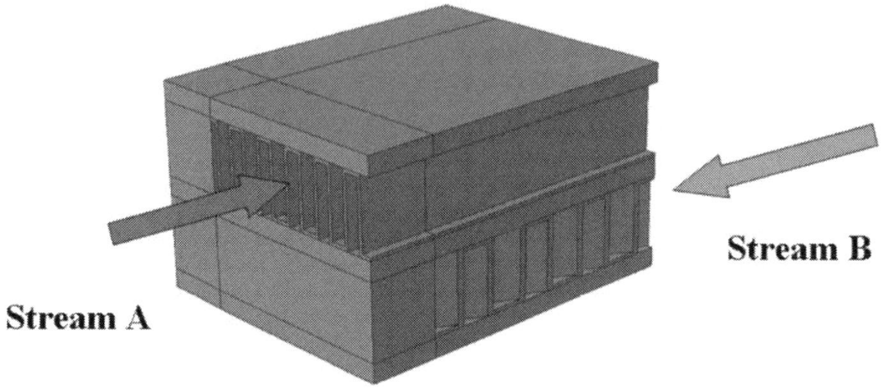

Figure 3 Solid model of two passes in a BAHX with streams flowing in different directions.

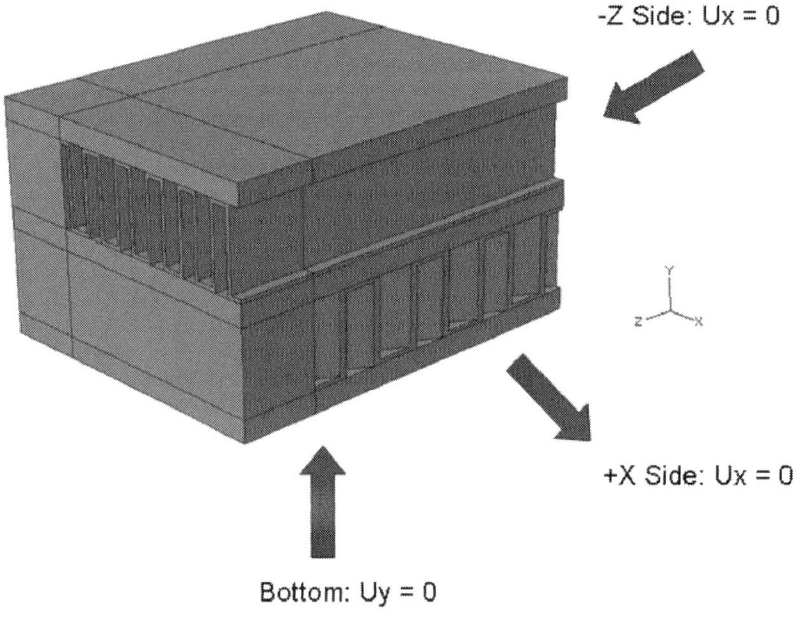

Figure 4 Boundary conditions applied to finite element model.

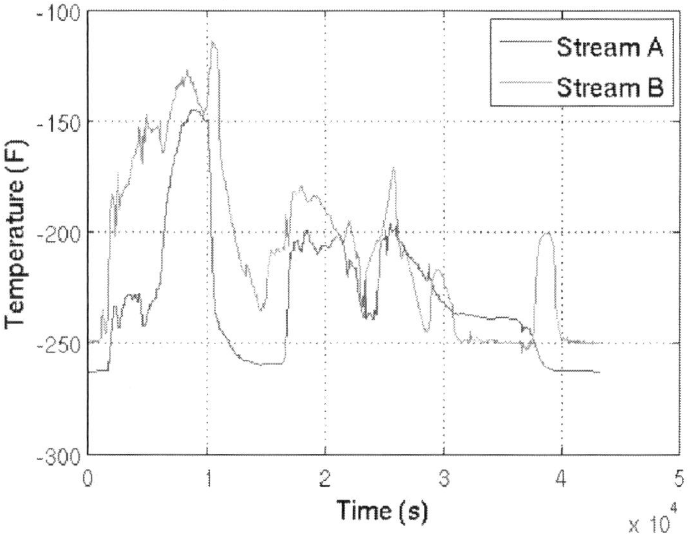

Figure 5 Typical transient of stream temperatures specified in the model.

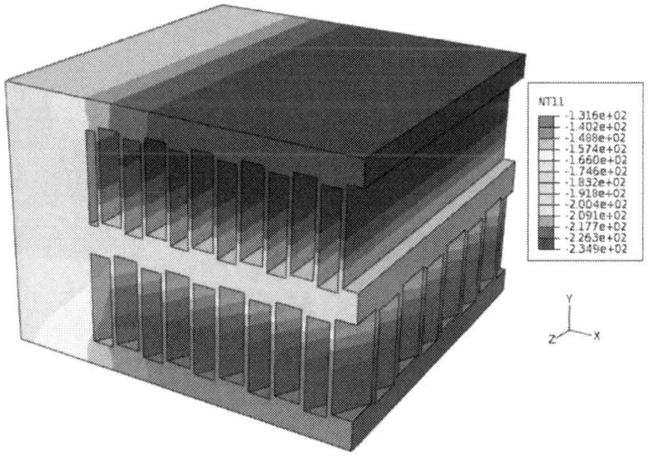

Figure 6 Temperature distribution (in °F) for a specified temperature of Stream A and Stream B.

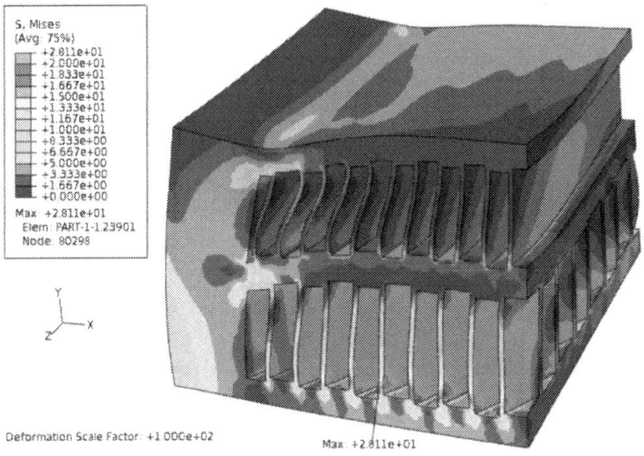

Figure 7 Thermo-mechanical stresses (in ksi) due to temperature gradient in model.

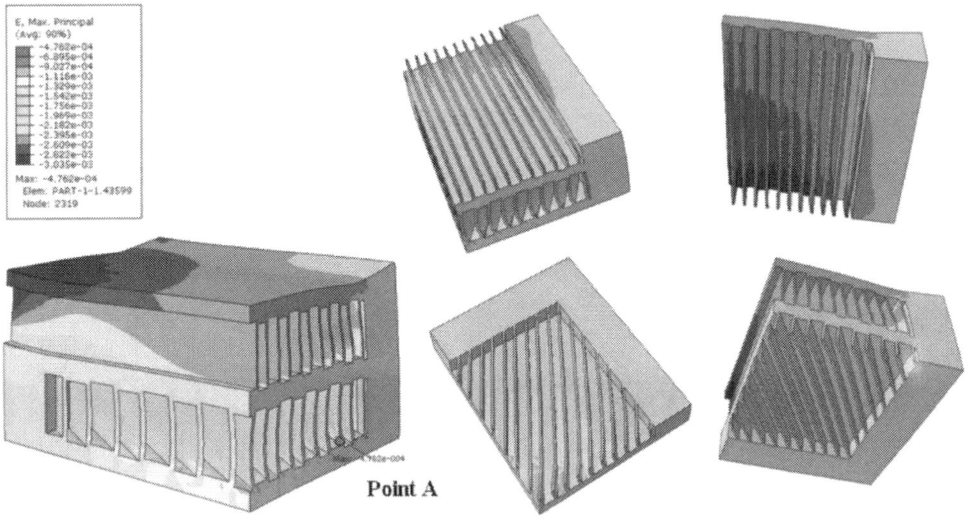

Figure 8 Maximum Principal Strain T_B-T_A=120 °F

.

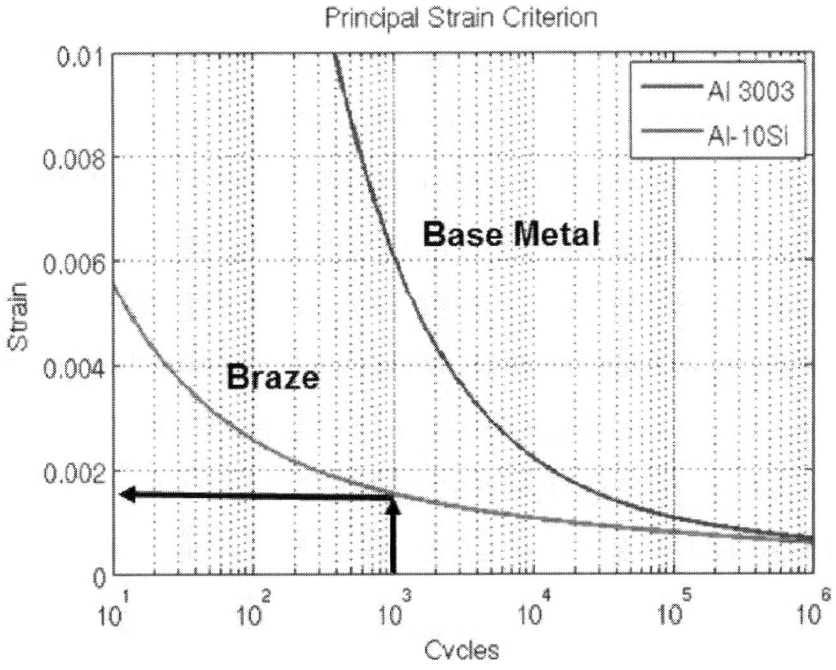

Figure 9 Strain to Failure for Base Metal and Braze using Principal Strain Criterion.

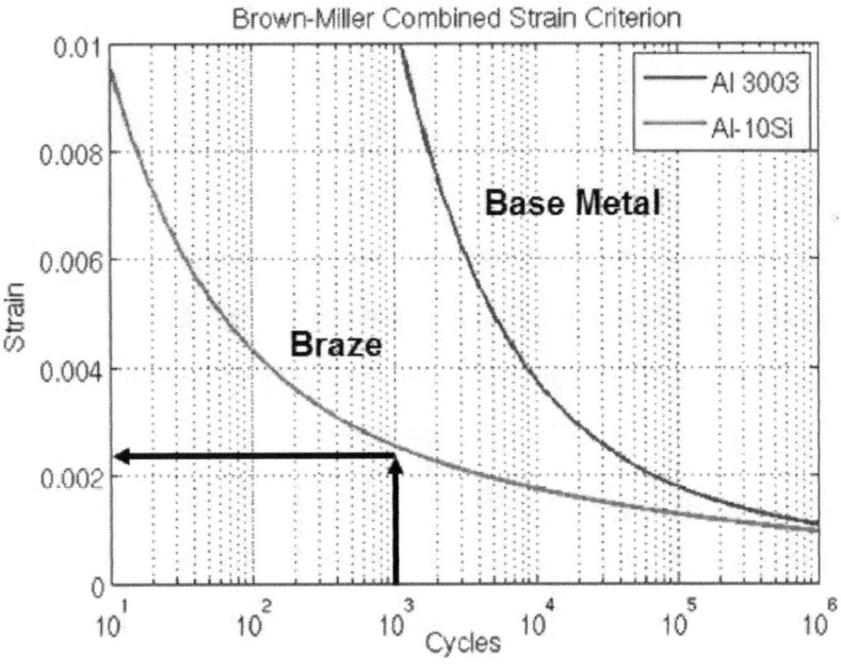

Figure 10 Strain to Failure for Base Metal and Braze using Brown-Miller Criterion.

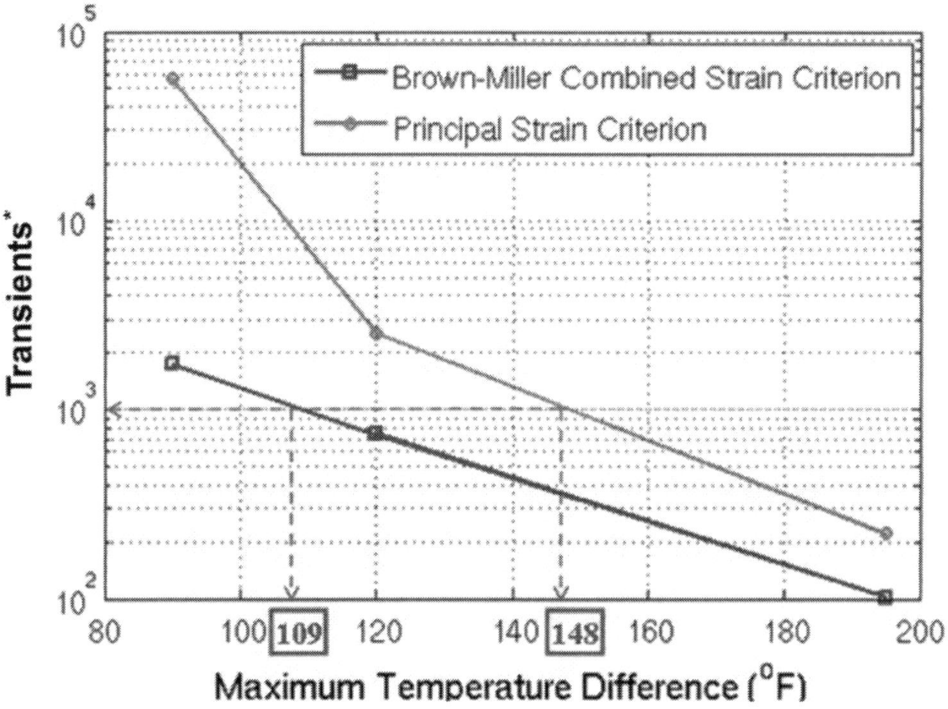

Figure 11 Effect of Temperature Difference Between Passes on Transients to Failure.

Post-Analysis Initiatives

The BAHXs that were identified as having the lowest number of cycles to failure and therefore, by inference, the highest probability of having some level of thermal fatigue damage were subjected to a pneumatic test at their respective stream design pressures during the next scheduled plant maintenance shut-down in order to verify the integrity of the design envelope. The test program was successful thereby providing an increased confidence level in the equipment and a valuable update to the RBI management program. The modeling tool employed in this assessment exercise was also applied to a later and more complex design of a BAHX in the same facility.

Summary

Finite element analysis was shown to be effective in ranking BAHXs in terms of their susceptibility to thermal fatigue damage based on exposure to normal and upset transient conditions. It was determined that the highest thermal stresses developed around distributor fins with geometries of 30-60° and that the magnitude of the temperature differentials encountered during upset conditions can significantly influence the cycle life of the exchangers.

References

1. API 579-1/ASME FFS-1, "Fitness-For-Service", (API 579 Second Edition), June 5, 2007
2. British Standard BS 7910-2005, "Guide to methods for assessing the acceptability of flaws in metallic structures"
3. Brown, M.W. and Miller, K.J., Initiation and Growth of Cracks in Biaxial Fatigue, Fatigue and Fracture of Engineering Materials and Structures, Vol. 1, No. 2, pp. 231-246, 1979.
4. Carter, P., Cater, T. J. and Viljoen, A., Failure Analysis and Life Prediction of a Large, Complex Plate Fin Heat Exchanger, Engineering Failure Analysis, Vol. 3, No. 1, pp. 29

AIChE 2012 Paper Number 127a

Analysis of Turnarounds in Olefin Plants: A 16-Year View

Claire L. Cagnolatti

Vice President, Chemical Studies

HSB Solomon Associates LLC

Richard B. Jones, Ph.D.

Director, Statistics and Risk Modeling

HSB Solomon Associates LLC

Prepared for Presentation at the 2012 AIChE Spring National Meeting

Topical 4: The 24th Ethylene Producers Conference

Houston, TX (April 4, 2012)

Analysis of Turnarounds in Olefin Plants: A 16-Year View

Claire L. Cagnolatti

Vice President, Chemical Studies

HSB Solomon Associates LLC

Richard B. Jones, Ph.D.

Director, Statistics and Risk Modeling

HSB Solomon Associates LLC

Abstract: Comparative Performance Analysis™ (CPA™) benchmarking studies of the olefin industry have been conducted by HSB Solomon Associates LLC (Solomon) since 1987. This paper will highlight the results of a 16-year analysis of trends in turnaround performance by the industry. Data will be analyzed and compared by one or more of the following variables: age of plant, feedstock type, and whether the plant uses a mini/major turnaround cycle. Statistical correlations of turnaround duration versus turnaround interval will provide insight into the tradeoffs between these two parameters. The impact of lengthening turnaround interval on cost, duration, work density, and reliability also will be examined. Other discussion topics include turnaround timing, critical path for completion, use of contractors, and work hours spent by equipment type in an olefin plant.

INTRODUCTION

To take fullest and best advantage of the now-profitable ethylene and olefins markets, manufacturers want to position themselves to run their ethylene plants, also known as olefin plants or crackers, as reliably as possible. One of the key parts of a sound reliability program is the plant turnaround, wherein critical repair, inspection, and cleaning of process equipment can prepare a plant for many years of reliable operation. However, turnaround planning and execution are increasingly complex events that must be carefully managed, even with the advent of more capable and more functional computer applications for this purpose. Moreover, the lengthening of turnaround intervals (years between major turnarounds) means it is becoming more and more difficult to retain knowledge and develop a knowledge base for turnarounds. When turnarounds occurred every 4 years or so, it was fairly likely that the turnaround manager who planned and managed the execution of the previous major turnaround was still working at the site. But with average turnaround intervals in North America exceeding 6 years and expected to approach 7 years in the future, this continuity will not be available to many plants.

HSB Solomon Associates LLC (Solomon) gathers data on ethylene plant turnarounds in the *Worldwide Olefin Plant Performance Analysis (Olefin Study)*, the most recent of which was conducted for operating year 2009. Solomon examined turnaround data for North American study participants as well as for the worldwide group including all study participants going back to 1993. Unless otherwise cited, all references and data are from Solomon's *Olefin Study*.

In each biennial study, participants provide data about their most recent turnaround, including duration, total cost, total work hours expended during the turnaround, and the interval between the two most recent turnarounds. In addition, the *Olefin Study* obtains data about the critical path for each turnaround as well as the major factor in scheduling when the turnaround will occur.

For the purpose of this paper, trends in all turnaround data were examined. In addition, a frontier analysis was performed to show the progress of two variables, such as turnaround interval and duration, with respect to each other. To study the effect of the amount of work scheduled during a turnaround, an index was created called work density, which is the number of turnaround work hours divided by the number of hours of downtime. The total number of work hours includes company, contract, straight time, overtime, supervision, management, and all work hours charged to the turnaround, which is reported by all participants in the *Olefin Study*.

Analysis Method

Solomon used simple averages (means) of turnaround data tracked over nine *Olefin Studies*, covering a span of 16 years. For the purpose of this paper, to protect the confidentiality of the study results yet still show trends, the study year 1995 was taken as the base year (100), and all other data are indexed to the 1995 data. The reason that 1995 was taken as the basis rather than 1993 is that the 1995 *Olefin Study* participation increased to above 65% of the global ethylene capacity eligible for benchmarking (fully operational, not located in US trade sanction countries, and ethylene capacity greater than 200,000 metric tons). Prior to that, the participation had been 40% or less of the world ethylene capacity. The 1995 study was also the first study to have participation from all five major ethylene-producing regions (Europe, North America, South America, Asia/Pacific, and Middle East/Africa).

These simple averages were plotted over time with each plot showing the progress of two key turnaround variables on the same chart. Trends for single variables over time had been examined in the past, but we had not examined the trends of two variables at the same time. Solomon applied this technique to olefin plant turnaround data, to see where the leading edge lies for turnaround duration with respect to interval, and work density with respect to turnaround duration and interval. Data were analyzed for the global *Olefin Study* population, and also more in-depth for North American plants, from 1993 to 2009. Further peer analysis by plant age, feedstock type, and reliability quartile, plus analysis of critical path, turnaround timing, and use of contractors was incomplete at publishing deadline but will be discussed in the presentation at the conference in April.

Analysis

Global Trends in Turnaround Duration, Work Hours, and Interval

An initial analysis using worldwide data from 1993 through 2009 for turnarounds reported by plants participating in the *Olefin Study* gave some unusual results. The trends in the number of work hours attributed to turnaround showed an alarming increase since 1993.

A global trend of turnaround interval plotted against turnaround duration (see Figure 1, next page) showed that interval grew by almost 50% from 1995 to 2009, but turnaround duration grew only about 12%. These results begged the question whether more work is being done in a shorter period of time during the turnarounds that occur after longer intervals. The assumption is that, with a longer interval between turnarounds, there are more repairs and replacements needed, which should increase

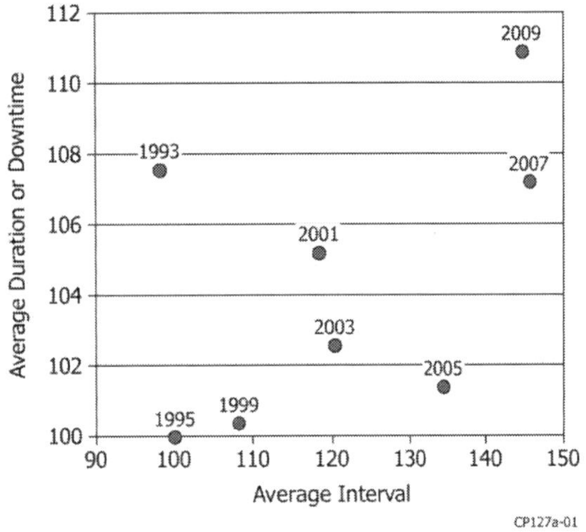

Figure 1. Turnaround Interval vs Duration Global Data Index to 1995 = 100

the scope of turnaround work and require more downtime. However, this 4-to-1 growth in interval compared to the same period growth in duration implied that additional work was being done without a commensurate increase in turnaround downtime. This was the reason for creation of turnaround work density.

Work density is total hours worked to complete the turnaround (meaning the full turnaround cycle—including accrued hours worked before, actual work hours during, and cleanup work hours after) divided by hours of plant downtime (feed-out to feed-in). The question arose whether plants were trying to do the additional work implied by a nearly 50% increase in interval while adding on average only 12% to the turnaround duration. This would be a concern for safety and efficiency reasons. The work density trend appeared to confirm the idea of more work being completed during the turnaround, with the work hours per hour of turnaround downtime nearly doubling with a >50% increase in turnaround interval, when examining global data (see Figure 2).

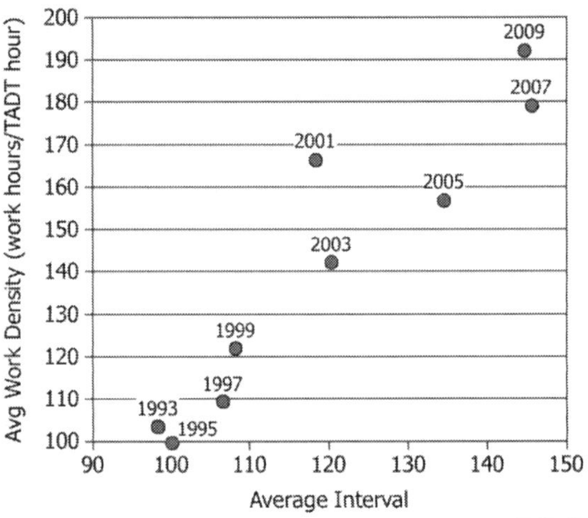

Figure 2. Turnaround Interval vs Work Density Global Data Indexed to 1995 = 100

Further analysis of the global data showed that the changes in study population caused this increase in work hours and work density from 1995 to 2009. The participation of plants in low-wage areas of the world increased significantly over this period. These plants have extremely large numbers of work hours reported for their turnarounds due to the low cost of labor, and this had skewed the results.

Work hours were not the only contributors to making the global analysis difficult, however. Other factors such as regional differences in use of contractors, differing accounting and accrual standards, large variances in plant age, and lack of continuity of data over the 16 years being analyzed also made the analysis on a worldwide basis less clear.

North American Trends in Turnaround Duration, Work Hours, and Interval

After discussions with the Operations and Maintenance subcommittees of the Ethylene Producers Committee (EPC), Solomon proceeded with a more in-depth and detailed analysis of the plants from North American region (US and Canada). This regional group has plants of similar age and more consistency of study participation than the global group. North America also has a more homogeneous wage rate across its ethylene-producing regions than does the total study population, making it more suitable for the long-term turnaround work hour analysis.

Analysis of the North American plants from 1993 to 2009 revealed that turnaround interval has grown nearly 80% over this time period while duration had increased just over 30% (see Figure 3). Again, the question arises as to whether the lower increase in duration over the same time period implies more work done in a shorter period of time to allow the duration of turnarounds to increase at a rate equal to only 38% of the growth rate in turnaround interval.

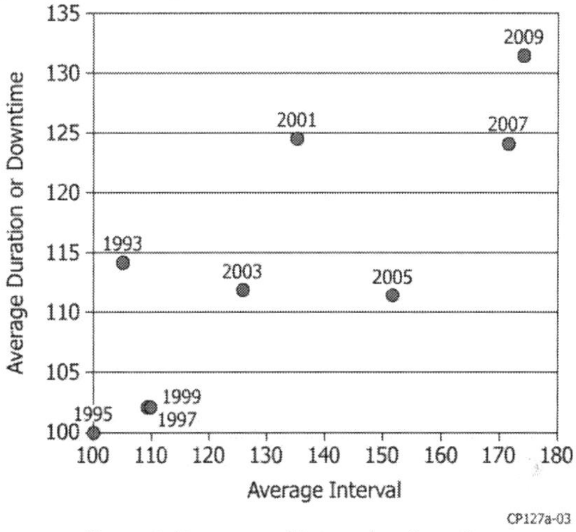

Figure 3. Turnaround Interval vs Duration
North American Data Indexed to 1995 = 100

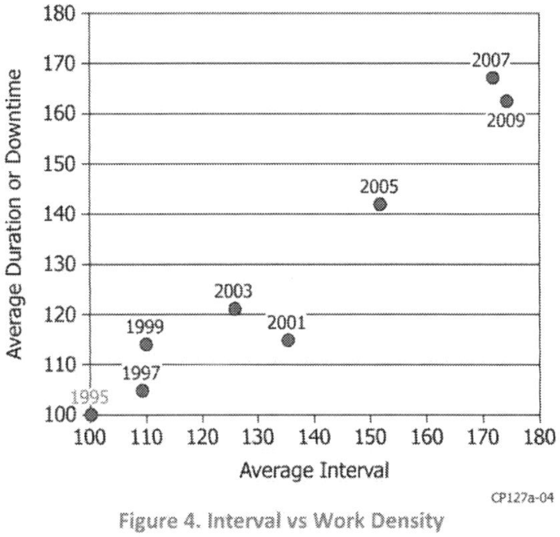

Figure 4. Interval vs Work Density
North American Data Indexed to 1995 = 100

We examined the progress of work density with turnaround interval for this same time period, illustrated in Figure 4. We saw a pattern similar to that observed with the global data. While average North American turnaround interval grew nearly 80%, North American turnaround work density increased nearly 70%—undeniably similar rates of increase. This apparently confirms the need to fit more work hours in each hour of turnaround downtime as turnaround interval grows four times faster than turnaround duration.

Other than larger work scope, there could be several reasons for the additional turnaround work hours reported by North American ethylene producers in recent years. In discussions with clients, we learned that there has been a conscious decision to have additional fire and safety watches for turnarounds. Also, we learned that many plants are locating temporary offices and workshops farther from the process unit due to fire and process hazard concerns raised after analysis of a fire and explosion during a turnaround at a Texas refinery. However, it is unlikely that these other reasons would explain the almost 65% increase in work density from 1995 to 2009.

While modern turnaround planning tools might make the planning of such busy turnarounds possible, manufacturing companies must be aware of the possible safety, efficiency, and manageability issues raised by the increased work scope and work density. Depending on equipment layout and plant design, each plant has a limit as to how many workers, supervisors, materials, vehicles (trucks, cranes, forklifts, carts) and safety personnel, can work in each plant area. The recognition of this limit will determine how much work can be safely and efficiently completed in any one area during a turnaround.

Because the purpose of a turnaround is to prepare the olefin plant for years of reliable operation, it is interesting to analyze the North American data by reliability performance quartile. The North American first quartile in reliability performance (the 25% of plants with the best reliability performance in each study) showed a 55% increase in turnaround work density for an 85% increase in turnaround interval. The fourth quartile in reliability performance (the 25% of plants with the poorest reliability performance in each study) had an incredible 90% increase in turnaround work density for a 70% increase in turnaround interval. Does the high work density introduce inefficiencies that compromise the quality or the thoroughness of the turnaround work, resulting in a negative impact on reliability, or is this observed result due to differences in plant age or design?

In general, we do not find great disparities in plant age profiles among reliability quartiles in our Olefin Studies. We normally see all plant Age of Technology (combined furnace, compressor, and plant age) groups represented in all reliability quartiles. This discounts the notion of plant age being a factor.

Difference in plant design could be a contributor—some plants may have a more favorable design with respect to reliability: more spares and bypasses, more reliable process control, better spacing of equipment to facilitate repairs, etc. However, in our single-client consulting experience, where Solomon does performance improvement consulting for single clients, we see that practices and methods are important indicators of turnaround performance. The best performers employ best practices when planning, executing, and budgeting their turnarounds, and that interval and duration are optimized based on the scope of the turnaround.

Additional analysis is in progress at the publishing deadline and will be covered in the presentation of this paper at the conference in April.

AICHE Paper Number 127b

EHS Management of Dow's Ethylene Plant Turnarounds

Jon Pucheu

Hydrocarbons Business Turnaround Leader
The Dow Chemical Company

Prepared for Presentation at the 2012
AICHE Spring Meeting
Ethylene Producers' Conference
Houston, TX April 1-5, 2012

EHS Management of Dow's Ethylene Plant Turnarounds

Jon Pucheu

Hydrocarbons Business Turnaround Leader
The Dow Chemical Company

The Dow Hydrocarbons Business has a thorough process for managing excellent EHS performance during Turnarounds. We name an EHS Focal Point for each Turnaround 2 years in advance. A Turnaround EHS Preparation and Execution plan is developed and followed. An EHS Readiness Assessment is conducted with a team of experts from the various locations around the globe 3 to 6 months in advance of execution. Follow up items are documented and corrected. Then the plan is executed with sufficient oversight.

This paper will describe the plan and assessment process and oversight details during execution.

Background Information

The Dow Hydrocarbons Business has conducted four major Turnarounds since 2010. The Recordable Injury Rate (RIR) for these four turnarounds is .07. Furthermore, environmental performance has been excellent.

The Process

The Business names an EHS Focal Point for each Turnaround 2 years in advance. This key leader drives the preparation of the EH&S Management Plan.

The EHS Management Plan is intended to cover all activities and areas related to the Turnaround. The Table of Contents is shown here:

- EH&S Vision
- Roles and responsibilities
- Indoctrinations
- Expectations
- Accountability
- Communication
- Investigation Requirements
- Contractor Safety Technicians
- Safety Technician Ratio
- Pre-Job Hazard Assessment (PJHA)
- Individual Hazard Assessment (IHA or STAC)
- At Risk Employees
- Contractor Leadership
- Hazard communication (HAZCOM)
- Tailgate Safety Meetings
- Safety Preparedness Drills and Inspections
- Stopping Work Criteria/Safety Stand Down
- Removing Workers
- Emergency Response
- Assembly and Evacuation
- Spill Reporting
- Inclement Weather
- Training and Qualifications
- Work/Rest Policy
- Near Miss Program, Observations & Critical Unsafe Behaviors
- Recognition Program
- Personal Protective Equipment (PPE)
- Safe Work Permits (SWP)
    - Line and Equipment Opening (LEO)
    - Confined Space Entry (CSE)

    - Hot Work Permits
    - Internal Combustion Engine (ICE) Requirements
    - Critical Lifts
    - Excavations
    - Electrical
    - Hydroblasting and Pressure Washing Permit
    - Radiography Permit
- Isolation of Energy Sources
- Identification Tags
- Temporary Portable Buildings
- Traffic Management
- Mobile Equipment
- Fuel
- Barricade Management
- Elevated Work
- Manual Lifting
- Scaffolding
- Steam Hazards, Plant Chemicals and MSDS
- Tools and Equipment
- Camera / Photograph Capable Equipment
- Electronic Devices
- Housekeeping
- Environmental Protection
- Waste
- Emission Control and Flaring
- Equipment Clearing
- Hydrostatic Testing
- Heat Stress
- Lead Materials
- Galvanized Materials
- Asbestos
- Fugitive Emissions
- Use of Containers
- Inert Compounds & Nitrogen Use
- Naturally Occurring Radioactive Material (NORM)
- Decontamination
- Equipment Demolition
- Flanged Connections
- SHEL and Area Reviews
- Variances
- Pyrophoric & Reactive Chemicals
- Approval
- Document History

This plan is used to communicate and structure the EHS Strategy and to serve as a guide for a positive safety culture and positive safety results during the Turnaround. The focal point assembles an EHS Team to assist in the development and implementation of the various parts of the plan.

The plan will continue to be developed, followed, and communicated as the Turnaround approaches. Approximately 3 to 6 months in advance of Turnaround execution, a team of experts from various locations around the globe perform an EHS Readiness Assessment.

The Assessment Team typically consists of:

- Maintenance EH&S Business Operations Leader
- EH&S Business Operations Leader
- Hydrocarbons Global turnaround Leader
- Process Safety representative
- EH&S TA SME's from other locations(preference to those with recently completed or pending turnarounds)

Readiness Assessment Best Practices

1. Maintenance EH&S BOL will initiate the process approximately 12 months prior to the TA start date by sending the workbook to the team.
2. TA Team will own this document, link it to the TA EH&S Plan and assemble a TA EH&S Team to begin preparing for the review. The TA Team prepares for the review by insuring that the workbook is populated with remarks and links.
3. Maintenance EH&S BOL will assemble the off-site review team and work with the TA Manager to select a date for the review. The Maintenance EH&S BOL will serve as the review Facilitator.
4. TA Team will forward the completed Pre-TA EH&S Review Workbook to the review team one month prior to the review.
5. Review is scheduled for four hours with the Facilitator and Global TA Leader traveling to the site. Remainder of Review Team to join by Live Meeting.
6. Facilitator and Global TA Leader arrive in advance and spend one day on-site reviewing procedures & preparations and conducting interviews of key personnel.
7. Local participants in the review include Production Leader, TA Manager, TA EH&S Lead and the core TA Team.
8. Written report is prepared by the Facilitator and submitted to the EH&S BOL, Global TA Leader, PL, TA Manager and TA EH&S Lead.
9. TA Team addresses action items.
10. Review facilitator to visit the TA during the first week to assess the practical implementation of the EH&S plan and provide feedback on execution.

11. TA team schedules a Post-TA EH&S highlights, plus/delta, learnings and advice for the next TA Team along with suggestions for improving the review. 30 minutes are also allocated for Q&A.

There are eight topics to be covered in the review process. They include Organization, Communications, Meet Behavior Expectations, Hazard Control, Industrial Hygiene, Environment & Waste Management, Process Safety, and Emergency Response & Security. The Plant EH&S Team populates the worksheets to prepare for the review. The details indicate questions and focus areas for the Review Team. Each of these topics will now be explained in more detail.

Organization

The Organization Plan must include the following:

1. Turnaround EHS support roles defined.
2. Lead contacts for each element of the EHS Plan established.
3. All contacts know their roles in the Turnaround.
4. Details for providing adequate CSE attendants.
5. Details for providing adequate fire watches.
6. Details for providing adequate rescue capabilities.
7. Adequate Dow Supervision to contactor ratio.
8. One Safety Advisor per 25-35 "do work" contractors.
9. Using the variance process.
10. Necessary construction permits in place.

Communications

The Communications Plan must include the following:

1. Ensuring that all persons working in the facility have received site and facility indoctrination.
2. A defined communication process for the contractor to the Turnaround Manager if an injury occurs.
3. The scheduling of all groups safety meeting including participation expectations, and details about large group safety meeting for all on day one.
4. The process to communicate first aid or environmental near misses.
5. The procedure for a safety "stand down" that will require a complete work stoppage at the facility should serious or repetitive issues arise.
6. The process to stop work during bad weather such as lightning.

Meet Behavior Expectations

The Meet Behavior Expectations Plan must include the following:

1. The Leadership Plans to be a visible presence.
2. The process for ensuring contractors provide their Zero Tolerance Policy.
3. Details on how the contractor leadership is aligned with Dow expectations, particularly related to Field Supervisor/Foreman accountability for work crew performance.
4. Description of Rest Policy for all during Turnaround that indicates forced days off.
5. Measures that ensure employees are fit for duty.
6. Housekeeping management system/plan.
7. Recognition system that includes safety incentives and penalties, recognition for individuals and contractor companies, on the spot recognition plans, and alignment of Capital and Maintenance Recognition Plans.
8. Contracting strategies that include: knowing the Contractor Company's injury rate, assessing their overall safety performance, and their plan for managing at risk or new employees.
9. Assignment of Dow reps to ensure short term contractors such as truck drivers or vacuum truck operators work safely.
10. Validation of line of fire training.
11. Management System that indicates expectations relating to pre-task analysis, ensures all workers, including operators, use this, having job specific pre-task cards available, and a recognition system for good pre-task cards.
12. Expectations for interventions.
13. Behavior based observation system including data collection, analysis of data, follow-up on findings or trends, and expectations regarding numbers of observations each person must conduct.

Hazard Control

The Hazard Control Plan must include the following:

1. The process for reviewing each contractor's safety plan.
2. Giving contractor access to Dow Safety Standards.
3. Confirming contractor training programs.
4. Checking competency to perform specific jobs.
5. Special emphasis areas identified due to issues from facility's or site's previous experience.
6. Details for confirming tower contractor's ability to manage work and workers.
7. Defining at risk individuals and at risk companies and measures taken to minimize incidents.
8. Preparing Dow employees leveraged from other facilities to work safely in this facility.
9. Inspecting leased and contractor portable equipment for proper set-up and grounding.

10. Inspecting safety equipment and tools brought in by contractors to confirm they are not altered and are safe to use.
11. Measures established to minimize affect of weather on safety of workers.
12. Measures to ensure that grinders are safe.
13. Confirming insulators will not leave exposed sharp edges after portions of insulation cladding is removed.
14. Measures to ensure job plans are in place to avoid use of hammer wrenches.
15. Measures to properly plan jobs that involve manual lifting.
16. Management system to meet Life Critical Guards Policy.
17. Engineering approval for hanging or cantilevered scaffolds.
18. Details to safely manage compressor deck openings.
19. Fall prevention plans in place for all.
20. Awareness of scaffold changes and need to revise full prevention plans.
21. Assigning a Dow Scaffold Focal Point.
22. Barricades Management Policy.
23. Emphasis for forklift safety.
24. Detailed spotters policy.
25. Surveying the process area as part of job planning for piping that can be easily snagged or struck by mobile equipment.
26. Completing job hazard review, SHEL Reviews for the high risk jobs.
27. Process for scope changes to ensure safety.
28. Completing job hazard reviews for operator tasks.
29. Addressing ergonomic issues with hard to open or close valves.
30. Ensuring Safe Work Permit (SWP) issuers ask SWP acceptors to confirm the work crew understands the hazards of job to be performed.
31. Ensuring SWP is part of the job planning and scheduling.
32. Documenting work not requiring SWP.
33. Using a system to positively identify all line break points to ensure workers do not break into wrong lines.
34. Continuous atmospheric monitoring during Confined Space Entries (CSE).
35. Process for creating and approving CSE's and process for re-authorizing multi-day CSE's
36. Confirming chemical cleaning contractors have trained employees with proper procedures to complete work without incident.
37. Measures in place to avoid affecting downwind workers.
38. Communication and verification of hands free hydroblasting policy.

Industrial Hygiene

The Industrial Hygiene Plan must include the following:

1. Identifying all relevant agents (chemical, physical, ergonomic, noise).
2. Identifying all new chemicals such as solvents or other cleaning agents that require additional approvals/MSDS.

3. Establishing exposure assessment procedures and specifying testing/monitoring equipment.
4. Ensuring test/monitoring equipment is available and arranging the related training.
5. Reviewing the IH Plan with the Site IH Focal Point.
6. Verifying job plans contain steps to prevent exceeding exposure limits due to long shift duration.
7. Confirming the CSE procedures contain specific atmospheric testing/ventilation instructions.
8. Confirming contaminated equipment management procedures are established.
9. Verify control procedures to lead paint, asbestos, and ceramic fibers.
10. Specifications for breathing air equipment.
11. Details to verify plant preparation procedure have adequate steps to control odors/exposure, and a management system for temporary hose or tubing use.

Environmental and Waste Management

The Environmental and Waste Management Plan must include the following:

1. Confirming the spill prevention plan includes learning from past events.
2. Confirming the pre-start up/commissioning plans include leak checks, verification of plugs in drains/vents, positions of valves, etc.
3. Detailing measures to minimizing flaring before, during, and after the Turnaround.
4. Details for coordinating communications with the local community.
5. Details for coordinating waste management, including inventory tracking and cost control.

Process Safety

The Process Safety Plan must include the following:

1. Instructions established for shift operations to clear non-essential personnel from process areas during cleaning and preparation, re-inventory, and start-up.
2. Ensure start-up and shutdown procedures are updated and accurate.
3. Details for confirming workers know what is in service and what is not if sections of the plant will be in operation.
4. Temporary building placement and occupancy rules must be approved.
5. Instructions established for shift operations of how and when to notify adjacent blocks of shutdown, start-up, or abnormal conditions.
6. Using a fire protection system special procedure for any fire protection systems that are inoperable during the Turnaround.
7. Measures in place to address interruptions to critical utilities such as nitrogen instrument air or potable water.
8. Preventing cold liquid full hydrocarbon lines isolated without pressure relief.
9. Completing NDT inspection of cracking furnace decoke lines prior to shutdown.

10. Additional temporary gas detection systems arranged.
11. Proper handling of pyrophoric material.
12. Confirming oxygen will be removed from equipment prior to introduction of hydrocarbons.
13. Details confirming operating procedures adequately address low temperature embrittlement prevention where the potential exists.
14. Details ensuring nitrogen hazards are understood by all facility staff.
15. Details for confirming testing of field welds.
16. Detailing measures to track control loops/alarms that have been disabled during the turnaround.
17. Confirming written head torquing and flange assembly procedures will be used.
18. Details of quality assurance program.
19. Ensuring critical instruments, which cannot be tested during normal operations will be tested during the Turnaround.
20. All hot taps challenged then reviewed by all necessary experts.

Emergency Response and Security

The Emergency Response and Security Plan must include the following:

1. Block evacuation procedures considering large numbers of workers.
2. Details for confirming evacuation procedures will be reviewed with contractor foremen.
3. Details for accounting for all personnel.
4. Details for verifying response plans are in place for critical jobs.
5. Details for confirming sufficient safety showers and eye wash facilities in place and functional.
6. Details for determining fire protection systems available with adequate pressure.
7. Details for arranging adequate security coverage.
8. Details for providing First Aid Facilities.

Once the review is completed, actions are documented and follow-up occurs until all actions are complete.

Summary

The Review Process Best Practices, as shown earlier in this paper, spells out the process in more detail. The Turnaround is now ready to be executed and excellent EHS results are expected. The execution will occur with sufficient oversight, details of which are included in the EHS Plan and the Review Process.

() Strategic Chemistry

Troy Schuenemann, Cody Nath and Sean Sears Sr., Refined Technologies, Inc., The Woodlands, TX

Optimizing the clearing and cleaning of process units during turnarounds. Primary discussion involves the three key areas related to an effective, timely cleaning plan. Chemistry, planning and execution. Minimization of effluent is also addressed. Overall timeliness, from oil out to vessel entry, is also optimized by shortening or eliminating certain steps in the*clearing*process.

AIChE Paper Number: 127d

Turnaround Excellence - Key Success Factors

R.P. (Bobby) Vichich

Vice President – Turnaround Consulting Services

AP-Networks

Prepared for Presentation at the 2012 Spring Meeting

Houston, Texas April 4, 2012

Turnaround Excellence - Key Success Factors

R.P. (Bobby) Vichich

Vice President – Turnaround Consulting Services

AP-Networks

Abstract: Turnaround Excellence is more than a state of being… it is a journey. Achieving and sustaining excellence is wholly dependent on a site approach of shared accountability towards Turnaround Excellence and must be a continually improving process. Starting well in advance of "feed out", a long range plan must be established that optimizes business economics with turnaround interval, duration, and upstream/downstream interfaces. This strategic plan provides the framework for the development of a Turnaround Premise, which in essence is a performance contract between the business and the facility (or asset team). The Turnaround Premise defines the key parameters of turnaround success as well as expectations for achieving a state of optimum turnaround readiness.

In this technical paper, a compelling case will be presented for aggressively driving towards an optimum state of readiness by showing industry data relationships between turnaround readiness and outcomes. The paper will describe the preparation practices that are critical for achieving optimal readiness. The paper will also focus particular attention to the key deliverables and interfaces of operations, maintenance and reliability plant personnel in the context of turnaround excellence.

The days of plant turnarounds being considered a natural extension of a facility's maintenance cycle are long gone. Many in the Petrochemical, Refinery and Energy industry today can recall this period, where a turnaround, regardless of complexity or significance was approached merely as a "necessary part" of the assets lifecycle. Minimal time was spent preparing, the concept of optimization was reserved only for business feedstock planning, and the use of tools or technology more sophisticated than a "Big Chief Tablet" schedule and a field supervisors' pocket notebook was unheard of.

During this period, it was commonplace for a turnaround to be considered "successful" so long as there were no lost time safety incidents. Of course, back then success was very loosely defined, schedule and cost targets were not challenging, safety performance was the result of reactive incident management, environmental regulations were drastically less restrictive, and reliability metrics were insignificant and rarely met.

In contrast, today's turnarounds are complex events that require entire plant cooperation and focus and involve work scopes that far exceed the traditional "major maintenance" era. Today's turnaround work scopes are often dominated by safety, environmental and reliability improvements, plant expansions and unit debottlenecks.

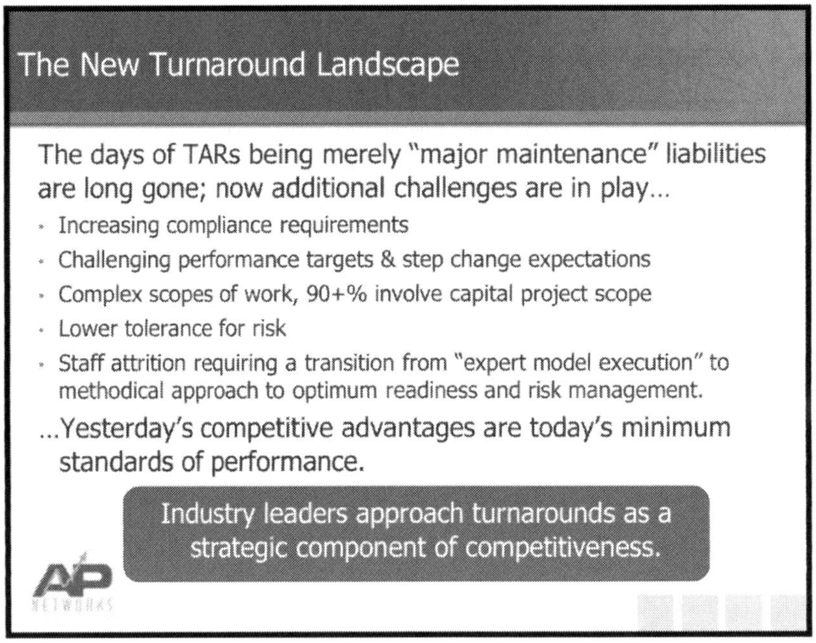

Figure 1

In addition to best-in-class cost and schedule targets, today's turnarounds are exposed to extremely challenging safety, environmental, operability, quality and even community affairs targets. Present day turnarounds are expected to deliver step change performance with each cycle. In order to endure the "*natural selection*" phenomenon, effectively adapting to external influences and continuously improving are the difference between thriving business margins and desperate survival.

Our industry is constantly responding to regulatory pressures, restrictions and constraints. These pressures combined with lower risk tolerance have transformed yesterday's competitive advantages into today's minimum performance standards. In order to satisfy the present day business environment of optimistic targets and single digit margins, turnarounds are a necessary and integral component of the short and long range business planning process. In short... business viability and manufacturing competitiveness is much defined by the ability to predictably deliver superior turnaround performance. As such, industry leaders approach turnarounds as a strategic component of competitiveness.

The industry's turnaround performance statistics show that there is still significant improvement required to achieve predictably competitive turnaround results. In particular:

- 82% of turnarounds do not satisfy all performance expectations; and
- 1 in 4 turnarounds significantly under-perform in more than one success criteria dimension; and are deemed a failure (or "train wreck").

Referring to Figure 2, the average high complexity turnaround exceeds cost and schedule targets by more than 20% with a range of predictability that is +/- 25-40% around the mean.

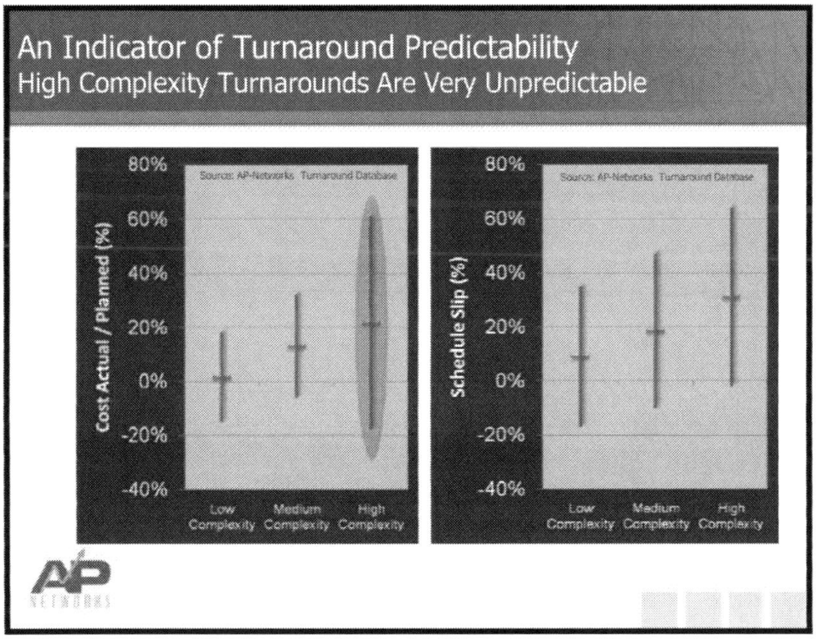

Figure 2

***Note**: AP-Networks' proprietary Complexity Calculator© tool defines turnaround complexity as a function of turnaround man-hours, amount of project capital work and turnaround interval.*

Data shows that the underlying reason for this sub-par performance is that many in the industry simply are not ready to execute the high complexity turnaround. They have not applied the rigor, focus, nor organizational energy required during the preparation phases to achieve an optimal state of readiness.

There is a strong correlation between a turnarounds' state of readiness and its outcomes. Turnarounds that deploy the industry's best turnaround preparation practices are able to achieve high levels of readiness, as measured by AP-Networks' Turnaround Readiness Index (TRI©). As shown in Figure 3, an optimal TRI© increases not only the probability that a turnaround will achieve its targets, but also that it will perform better than others in the industry relative to cost and schedule outcomes.

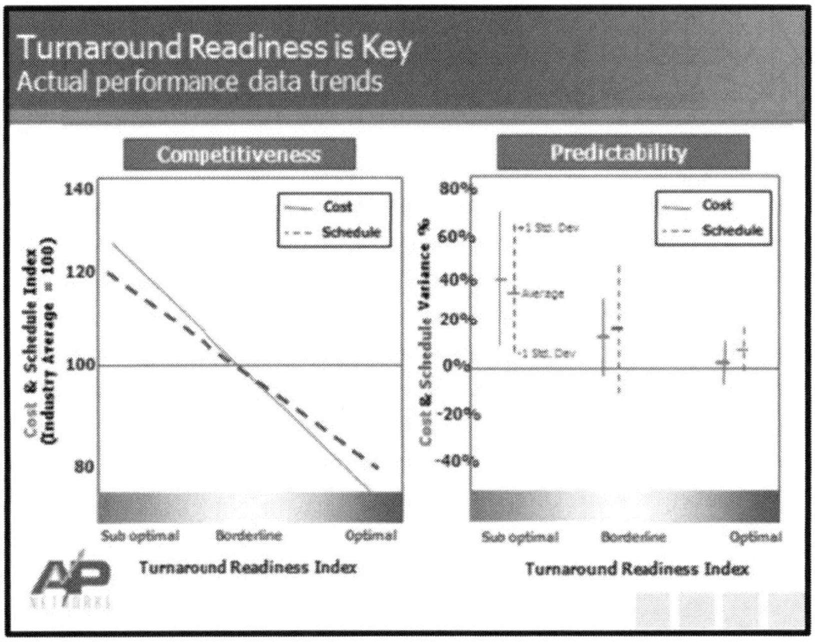

Figure 3

Hundreds of millions of dollars have been expended in recent years to prepare for turnarounds; and noticeable gains have been realized. However, less than adequate performance still riddle the large majority of industry's turnarounds. So if readiness is strongly correlated to outcomes and the industry as a whole has increased the importance of preparations, why are there still so many disappointing turnarounds?

The answer is that while most embrace "the concept" of turnaround preparations, only the industry's leaders have implemented a framework of People, Processes and Tools to ensure high quality preparation deliverables are produced at an intricate level of detail. Lack of such a framework is the "silent killer" that unveils itself during the execution phase as a continual barrage of insurmountable obstacles that overtake even the most capable turnaround execution organizations.

Efficiently preparing with the quality and attention to detail necessary to produce a high level of readiness is the paramount issue facing the turnaround industry today. From our research and industry experience, the solution to this issue involves the following:

1. Well defined turnaround premise;
2. Fully implemented standard turnaround work process;
3. Disciplined capable organization; and
4. Deployment of the industry's best practices.

This paper is written to provide insights into the fourth item – the Industry's Best Practices. However, in order to ensure the context of the full solution is firmly understood, the fundamentals of items one through three must first be summarized.

The turnaround premise is in effect the "contractual terms" for the turnaround steering team that defines the boundaries of the turnaround (Interval, scope criteria, level of acceptable risk, etc); and outlines quantifiable metrics by which success will be measured. In simple terms, the premise document describes how the turnaround will contribute to the strategic objectives of the facility and the business.

Documented in various formats and deployed via different tools, many of industry's leaders have adopted a standard turnaround work process that contains appropriate level of detail and key elements to deliver high degrees of turnaround readiness. Experience shows that the critical success factor for achieving desirable results is much less determined by work process sophistication as it is by the quality of implementation. Effective work process implementation requires far more than a process manual, a training course, and formal rollout session to be successful. It must be fully integrated into an organization's routine business practices and environment. Organization styles vary broadly from matrix, self lead structures to the command and control of yesteryear. As such, effective implementation involves a deliberate strategy designed to gradually weave the process deliverables into the natural work teams' culture; and is often deployed via tools that "bring the process to life" - versus allow it to be yet another dust collecting reference manual on stagnant office bookshelves. Effective implementation is achieved when the turnaround preparation activities become a normal piece of the daily business rhythm at the facility.

Premises and processes are important, but highly ineffective without people. People are the engine that delivers high states of readiness which ultimately enables a turnaround to deliver against its stated premises. An actively engaged management team that heightens turnaround success as a top priority is the difference between a list of plant resources and a disciplined capable organization. Without exception, this fully participative site leadership team is the inspiration that creates and maintains a level of discipline; and is the fuel source that provides the stamina to maintain a "we will win" attitude.

The final key to the solution for sustainably producing predictably competitive turnaround outcomes is to identify and deploy the industry's best practices. Defined as *"a technique or methodology that based upon experience and research has proven to reliably lead to a desired result*", best practices are selected among alternatives and implemented because they maximize business value. A critical variant to understand at this juncture is that a best practice is identified by its' technique or methodology; not its' implementation. Proper implementation must be considerate of the unique culture defined by each site's People, Process and Tools environment to assure an effective deployment.

The number of papers, manuals, text books and periodicals that describe one or many of the industry's best practices is unlimited. As such, this paper focuses on seven best practices that affect operations, maintenance and reliability plant personnel in the context of turnaround excellence. Following the outline provided in Table 1, each of these seven best practices are described in greater detail.

Table 1: Best Practice Outline

Practice/Technique/Methodology	**Applicable to...**		
	Ops	Maint	Rel
1. Equipment strategies and performance data	R	R	R
2. Scope development and optimization	R	R	R
3. Oils/Utilities planning	R	I	C
4. Blind/Isolation management	R	I	I
5. Shutdown and startup and decontamination procedures	R	I	I
6. Inspection test planning	I	I	R
7. Quality management	C	I	R

***Note**: "Applicable to..." letters refer to standard RACI definitions. R=Responsible; C=Consult; I=Inform*

1. Equipment Strategies and Performance Data

   Well documented equipment reliability strategies are the cornerstone of modern day reliability-centered asset management programs. These strategies not only document the reliable operating window of each major piece of equipment within the asset, but also describe the unique combination of tasks required to assure economically reliable operations. In addition to routine and non-routine maintenance tasks, equipment strategies also define time-based overhaul tasks and frequencies that must be programmed into turnaround scope definitions and budgets.

   Equipment performance data is a best practice that involves the collection of operating data about the equipment in relation to real time operating conditions. Basic types of performance data to be collected include:

- Operator logbook information using local instrumentation;
- Time-based condition monitoring by certified craftsmen; and
- Operating conditions by process engineers.

Performance data enables equipment performance trending, degradation tracking, end of life predictions, and bad actor analysis. From these sorts of analyses incremental solutions, design changes and/or technology upgrades are developed. Without detailed, trendable equipment performance data, technically sound turnaround work scope decisions are not possible, and turnaround work scope optimization defaults to emotion and organizational rank.

The responsibility to deliver this information varies, depending on the type and timing. However, in general, the reliability department is responsible for defining equipment strategies, the maintenance department is responsible for documenting and being knowledge of equipment condition, and the operations teams are responsible for logging real-time equipment operating data and providing alerting experts when abnormalities are noted.

2. Scope Development and Optimization

Competitive turnaround performance is not possible without an aggressive scope control and optimization effort. A prerequisite to controlling and optimizing scope is to first ensure that it is developed in accordance to industry best practice. In particular, this includes:

- Comprehensive, site wide scope collection effort;
- Descriptive, specific and plannable scopes of work;
- Aggressive screening vs. the scope criteria (documented in the Premise) by the appropriate site department;
- Order of magnitude estimating of each scope item; and
- Closed work list 12 months before feed out (and formal documented scope control is initiated).

Once the work list is closed, all non-mandatory work list items that passed through the initial screening are challenged according to risk tolerance and economic criteria. The best practice is to perform a quantified risk-based scope challenge. Using proprietary tools, external facilitators challenge each discretionary work list item and (with the turnaround team) assign risk factors, likelihood/impact ratings, and business specific incremental economics. The output of this scope challenge, which requires detailed equipment performance data, is a calculated "Mitigation Index" for each work list item. The work list is then sorted from high to low Mitigation Index and the affordable/risk acceptable work scope can then be defined.

Typically about 200 work list items can be reviewed in one week. Experience shows that externally facilitated scope challenges yield an average scope reduction of 15%.

Developing, challenging and optimizing the turnaround work scope is ultimately a collaborative effort that must be shared plant-wide. So, the responsibility for performing this effort completely and consistent with industry best practice is shared among operations, maintenance and reliability departments (as well as others).

3. Oils/Utilities Planning

An Oils and utilities management plan is a time based schedule of stream dispositions, flow rates, tankage levels, and utility usage for the entire plant - as impacted by the turnaround. Since this plan involves feedstock rates and compositions (in addition to and product rates and mix), its development is typically spearheaded by business planning departments. This plan is typically quite comprehensive and requires information about turnaround duration at the "determining" unit level of detail.

Once developed and challenged, this plan must be kept evergreen as it forms the basis for how the plant will operate and how suppliers and customers will be impacted by the turnaround. As the turnaround operating procedures are finalized, the utilities portion of the plan is updated to include more granular information about plant utility balances, shed lists, external utility purchase requirements and hydraulic limitations.

During the shutdown, commissioning and start up phases of a turnaround, the blowdown, flare and slop systems are often over taxed. Hence, these systems must be analyzed for hydraulic limitations and the need for temporary facilities to ensure the operating procedures and plans are feasible. As such, this is a highly technical plan that takes time to develop and should be started during the early stages of turnaround preparations (18 months before feed out).

Although the input of business planning, process engineering and reliability persons is essential, operations is ultimately responsible for preparing the Oils and Utilities Plan for the turnaround.

4. Blind/Isolation Management

The ability to flawlessly manage energy and equipment isolation is an essential element for ensuring an incident free turnaround execution. Significant positive strides have been made in the past few years relative to managing energy and isolation.

The best practice starts with the expectation and commitment to a single list of isolation devices, and the implementation of a comprehensive system for managing insertion/ removal or closing/opening of each device - most commonly referred to as Blind/Isolation Management. Since numbering and location of blinds and isolation devices occur during detailed planning of each work list item, the process for managing blinds/isolation must be defined early in the turnaround preparation phase.

Industry practice is to design an intelligent blind/isolation device numbering system that ensures each device/blind location is unique. Then, as work scope planning and operating procedures are developed, the location of each required device is noted on a common set of P&IDs. These locations are then entered into a master blind/isolation database that for each individual record identifies all the turnaround work and operating procedures that require the use of that location/device. Then, as the detailed planning phase is completed, individual blind/isolation lists can be printed and inserted as a final document into each job package.

With this level of attention to detail during planning, the actual turnaround execution is set up for success. The best practice for managing blinds/isolation (including LOTO) during execution is to appoint one operator per shift as the isolation coordinator. The function of this role is to manage the status of each device/location within the blind/isolation management database and to officially authorize the physical change of status of each device (open/close) or location (insert/remove). Although the concept of managing blinds seems simple, this additional level of control has proven highly effective; in particular, for complex situations like:

- Using a single device to isolate/blind for multiple job; or
- Ensuring that properly rated blinds is utilized (especially, important for hydro-testing).

The core purpose of Blind/Isolation Management is to ensure that the all forms of energy are controlled so that mechanical work can be performed safely. Since operations is responsible for controlling the interface between energy sources and work efforts, operations is fully responsible for the Blind/Isolation Plan.

5. Shutdown, Start Up and Decontamination Procedures

Having operating procedures is an OSHA mandate that all US facilities satisfy. However, the best practice for turnarounds is to review and update the procedures that will be exercised and ensure that all operators involved in the turnaround are trained on these procedures and fully aware of the overall operations turnaround plan. To this end, the industry's best typically assign a veteran operations representative to the turnaround team at least 12 months prior to feed out; and one key deliverable that this person produces in the first 6 months is a fully reviewed, updated and endorsed set of operating procedures for the turnaround.

Once developed, the "down" and "up" procedures are summarized and input into the execution schedule as physical activities with duration, resources and logic. Typically summarized into a series of activities with a maximum duration of one shift, these activities are logically sequenced with each other to represent the unit (or systems) shutdown and start up sequences.

Without exception, one best practice that has enabled aggressive turnaround duration improvements is the concept of "systemization". In short, systemization is the process of segregating the turnaround unit(s) in a way that relates to the process flow. In this concept, a system is defined a collection of process equipment and piping that is LOTO'd together. Typically, each system is shutdown, decontaminated and commissioned together and logically tied within the execution schedule to the mechanical work that will occur within the confines of the system LOTO boundaries.

The many process systems are then logically related to one another to represent the sequence of operations "down" and "up" activities. This complex array of tied activities that are grouped into systems and logically tied to feed-out and feed-in milestones enables the identification of "critical path systems". This provides greater opportunity from optimize the turnaround duration as compared to just analyzing the mechanical critical path. This practice of schedule optimization through the application of systemization is shown pictorially in Figure 4.

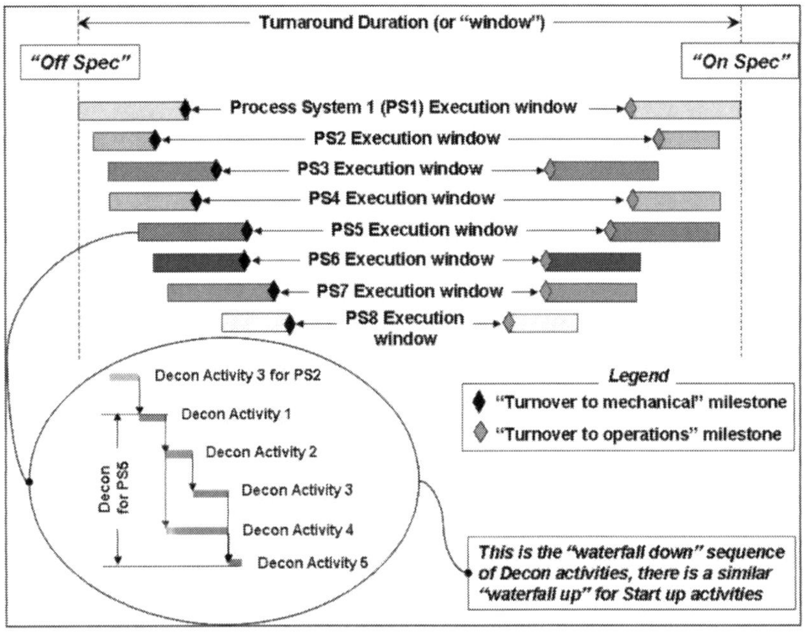

Figure 4

When systemization is fully implemented, the systems as defined by operations are used by the entire turnaround team to manage & control work execution and provide intelligence to the prioritization of Prestart up Safety Reviews (PSSR) and unit recommissioning.

The final component of the procedure best practice is decontamination. This practice has experienced significant enhancements in recent years, and the majority of today's turnarounds are applying some sort of decontamination chemical to assist with the removal of hydrocarbon and/or other hazardous chemicals. Since decontaminating the unit is a dominant component of the shutdown phase of the turnaround, the methods and procedures for providing the safest atmosphere for mechanical work must be defined early (general understanding 12 months prior to feed out).

The decontamination procedure(s) is the final output of several interactive discussions. Development of this complex procedure requires the following minimum participation:

- Operations (and/or decontamination specialist) to provide intimate knowledge of the process configuration and equipment;
- Process Engineering to provide details about the process chemistry, equipment specific contents and to ensure that process specific contaminants (killers) are not introduced/produced during decontamination;
- Reliability Engineering to provide information on the equipment pressure envelope and to ensure that equipment metallurgy will not be adversely impacted during decontamination;
- Decontamination contractor to provide industry experience and information on the decontamination chemicals and decontamination process requirements.

A best practice decontamination procedure prescribes in detail the critical parameters of the decontamination process: minimum/maximum flow rates, temperatures, pressures, injection points and rates, blowdown dispositions, sample points, value positions, process step completion variables (sample results, duration, visual color, etc). Once these are defined in detail, the remaining critical component of a best practice decontamination procedure is a clearly defined responsibility protocol for each step and decision. In particular: What is the decontamination contractor expected/permitted to do, and what is the responsibility of the owner/operator?

As noted above, the development of operating procedures, especially decontamination, requires the input of several parties; however, for turnaround preparation assurance, operations is fully responsible for ensuring that operations procedures are in place, reviewed and representative of turnaround requirements.

6. Inspection Test Planning

Often referred to as inspection scope details, the practice of inspection test planning is one of the key variables that must be well defined in order to have a complete turnaround scope of work. Although not overly complex, the component that differentiates the best from the others is the timely completion of inspection test plans. In particular, the best practice is to provide the full details of any required inspections as part of the work scope prior to scope freeze (refer to Best Practice #2 – Scope Development and Optimization).

Inspection test plans should be prepared for each work scope item that requires any type of non destructive inspection. A well prepared scope-specific inspection test plan contains the following minimum pieces of information:

- Type of inspection method;
- Quantity to be inspected (ft^2, linear footage, size of grid, # components or tubes, # points, etc);
- Location of inspection and access requirements;
- Duration and number of inspection resources;
- Quality control/assurance requirements and/or hold points; and
- Photos (as available/required).

The reliability department, and the inspection team in particular, is responsible for delivering the inspection test plans to the turnaround planning team.

7. Quality Management

Managing quality during a turnaround must start early in the preparation phase and includes two primary components: asset integrity assurance and control/assurance of mechanical work quality ("mechanical" is used to cover all work performed by craft persons during the turnaround).

Relative to asset integrity, inspection and reliability personnel utilize the turnaround to collect data, make repairs and install replacement components in the interest of ensuring the integrity of the operating equipment and compliance to local and federal codes and regulations. For many turnarounds, this is the primary driver for taking the unit(s) out of service in the first place. Best practice is to only include in the scope of work the equipment and scope required to ensure asset integrity that cannot be cost justifiably performed outside of a turnaround. Deployment of mature a Reliability Based Inspection (RBI) program is the key differentiator of a turnaround work scope that is reliably sound and financially responsible from one that is merely regulatory driven and time based.

The second component of turnaround quality management involves the deployment of a comprehensive set of hold points and checks to ensure a high level of craft quality. Best practices for managing quality of mechanical work during the turnaround include:

- Deliberate combination of quality control and quality assurance;
- Collaboration between contractor quality control personnel and owner inspectors;
- Prescription of critical parameters, such as: pneumatic or hydro-test envelopes and pressures, weld procedures, torque values, material specifications, clearances, etc.;
- Appropriate verification steps, such as: hold points, intermediate weld checks, testing checks, torque checks, Positive Material Identification (PMI), measurements, etc.;
- Intentional control requirements, such as: flange tagging, pinging bolts, weld x-ray, visual gasket confirmation, etc.;

- Minimum (random) assurance protocols that escalate if deviations are identified, such as: sampling of gaskets before install, bolt pinging, occasional retest of pressure, etc.;
- Prescribed procedures such as: heat exchanger tube testing, use of temporary materials/gaskets, PSSR, vessel closure, loop checks, turnover of work to operations for commissioning, etc.; and
- Final verifications prior to energizing with hazardous materials, such as: leak testing flanges, low point bleeder checks, testing overspeed trips and other "final element" instrumentation, etc.

The reliability department must fully accept responsibility for implementing a comprehensive Quality Management Plan. Without exception, other departments, such as operations, must contribute to the overall completed plan and it's successful implementation.

Obviously, there are many, many fundamental components of a high quality turnaround execution plan and the sheer number of "basic blocking and tackling" preparation tasks can totally consume an organization. The proven leading indicator of turnaround performance, Turnaround Readiness Index ("TRI"), is a complex measurement of how well the industry's best practices are knitted into the fabric of a disciplined and capable site organization. As described in this paper, optimum readiness does not require perfection in all aspects of turnaround preparation; but does require diligence in four fundamental areas.

In order to maintain pace with (or accelerate ahead of) the evolving minimum standards of performance in today's turnaround industry, owners are urged to articulate their own pathway to turnaround excellence. They must realize that turnaround excellence is so much more than a state of being, and that this "Journey" requires the integration of their fundamental strengths into the continuous improvement program that builds on each turnaround cycle. Owners must demonstrate a relentless commitment to the "Journey" in order to realize predictably competitive turnaround performance.

About the Author:

Robert P. (Bobby) Vichich - Vice President of Turnaround Services

Bobby is the Vice President of Turnaround Services and is located in AP-Networks' Houston office. Prior to joining AP-Networks in 2005, Bobby was a direct hire employee of Exxon, ExxonMobil and Lyondell/Equistar for more than 18 years. In addition to numerous maintenance, reliability and operations roles, Bobby served in various turnaround functions spanning from operations preparedness, mechanical planning, event scheduling, project controls, execution management and even corporate procurement management.

In addition to leading AP-Networks' Turnaround Consulting Practice, Bobby is a Recognized Industry Expert in the petrochemical industry, with particular focus in turnaround preparation methodologies, best planning and scheduling practices, and emerging turnaround management principles. Most recently, Bobby has written white papers and delivered presentations on the following topics: *Turnaround Competitiveness: The Secret Ingredient; Leading Indicators of Turnaround Performance Outcomes; Turnaround Performance Excellence; Breaking the Barriers of Traditional Learning; Turnaround Benchmarking: The Right Dataset; Turnaround Best Practices and Emerging Trends; and Team Alignment: An In Depth Look.*

Bobby holds a Mechanical Engineering degree from Virginia Tech and his many years of experience as a client with top-tier companies provide a unique perspective on being a leader in a world-class consultancy.

This is a panel discussion, and therefore a paper is not available.

A panel of industry experts will discuss topics of interest to Ethylene Producers with respect to turnaround excellence. Topics will be solicited from session attendees during the paper portion of the session. Prepared discussion topics will include: How are late scope changes handled, Is there differentiation between capital projects and maintenance project execution, How is work authorization handled during T/A, Discuss turnover process to maintenance, Discuss mechanically complete turnover to operations, etc.

() Acetylene Converter Safety Systems

Nelson Shen, Technip USA, Claremont, CA

Acetylene is produced as a by-product during the pyrolysis cracking of hydrocarbons in ethylene plant furnaces. Typically the acceptable acetylene contaminant level in the polymer grade ethylene product is in the range of 1 ppmv or lower. Since acetylene will fractionate with ethylene, it needs to be removed before the C2 fractionation step. An acetylene converter (or acetylene hydrogenation reactor) is a widely used unit operation for the conversion of the acetylene to ethylene and/or ethane in the ethylene plant.

Acetylene hydrogenation (in an ethylene plant) is a highly exothermic reaction. If reactor operating conditions are not carefully controlled, the reaction temperature excursions may occur and will cause safety issues in the plant. An integrated reactor safety system is essential for the safe operation of the ethylene plant.

AIChE Paper Number: 150c

EMERGENCY BLOCK VALVES

A GUIDE TO SELECTION

Joseph A. Viviano, P.E.

Supervising Engineer, Process Engineering

Technip

Prepared for Presentation at the 2012 Spring National Meeting

24th Ethylene Producers' Conference

Houston, Texas, April, 1 - 5, 2012

Emergency Block Valves

A Guide to Selection

Joseph A. Viviano, P.E.

Supervising Engineer, Process Engineering

Technip

Abstract: An Emergency Block Valves (EBV) is used as a means of isolating flammable or toxic substances in the event of a leak or fire. The installation, and use, of EBV's in many hydrocarbon services will significantly reduce the potential of fire and explosion damages caused by loss of hydrocarbon containment situations; and a general improvement in overall plant safety for both capital assets, and operating and maintenance personnel. In Hydrocarbon, and / or toxic services; EBV's will serve to mitigate the potential of significant environmental releases.

This paper is to be presented as a general design guide to aid in understanding of *"WHAT"* an EBV is, "*WHO*" is responsible of definition of selection basis, *"WHEN"* an EBV should be considered, "*WHERE*" an EBV should be installed, and "*WHY*" they should be installed. The items to be covered include the code basis of EBV's (API RP-553, NFPA 58 specific requirements for LPG service), the types of valves to be used; the selection of manual or automatic valve actuation; guidelines for determining if EBV's are recommended based on equipment services, and hydrocarbon inventories; general installation requirements; integration with other safety instrumented systems within the processing unit; and justification of installation cost.

Introduction

Safety. One word, one idea, should be foremost in our minds throughout our range of daily activities. We have heard it all before; "safety is number 1", "the person most responsible for safety is you". However, we need to be especially aware of the instances when an individual's decision affects the safety of others. Similar to the selection and design of any safety system; the selection, design, and installation of Emergency Block Valves (EBV's) may have long term impact on the safety of many individuals. The use of EBV's can significantly reduce the impact of an accidental release of hydrocarbons, by limiting the potential of damage and personnel injury caused due to subsequent fire or explosion.

The primary purpose of EBV's is to limit the release of large quantities of dangerous materials, and the potential for damage or injury which may occur from such a release. In addition, limiting the quantities of materials released may also result in an economic benefit; both in loss of valuable process fluids, and reduced costs of subsequent repair / replacement of damaged capital items (equipment, piping, instrumentation, etc.).

The main basis of this paper is the definition of EBV requirements as defined in API RP-553. However, other codes and standards which also describe emergency isolation are also referenced. The main codes and standards used to complete the basis of this paper are as follows:

API RP-553	"Refinery Control Valves"
NFPA 30	"Flammable and Combustible Liquids Code"
NFPA 58	"Liquefied Petroleum Gas Code"
API RP-2510	"Design and Construction of LPG Installations"
NFPA 85	"Boiler and Combustible Systems Hazard Code"
API RP-560	"Fired Heaters for General Refinery Service"
API RP-556	"Instrumentation and Control Systems for Fired Heaters and Steam Generators"
NFPA 497	"Recommended Practice for the Classification of Flammable Liquids, Gases, or Vapors and of Hazardous (Classified) Locations for Electrical Installations In Chemical Process Areas"
API RP-500	"Recommended Practice for Classifications of Locations for Electrical Installations at Petroleum Facilities Classified as Class I, Division 1 and Division 2"
MSDS	"Material Safety Data Sheets"

The intention of this paper is to use the 5 W's to develop a general guide to establish a basis for design and selection of EBV's by the following:

"What" is an EBV? (Definition and Valve Types Considered)
"Who" defines the basis of design and selection? (Codes / Standards)
"When" an EBV should be considered? (Hydrocarbon Volume Requirements)
"Where" should an EBV be installed? (Locations relative to Equipment)

"Why" should they be installed? (Safety / Cost Justification)

Asking these questions, and using the material presented; the reader can develop a design basis for establishing if an EBV is needed, the type selected, and where it should be installed.

WHAT? - Definition of an Emergency Block Valve

This paper has been titled "Emergency Block Valves", or EBV. However it should be understood that many names are often used by designers, operators, owners; and even various codes, standards, and recommended practices. NFPA 58, for LPG facilities, refers to Emergency Isolation Valves (EIV's). NFPA 85, for fired heaters, refers to "safety shutoff valves" in fuel gas systems. Then there are the codes and standards which define various other terms for storage, loading, and other fluid handling facilities. Across the petrochemical industry many other names, like "cutoff valve", "chopper valve", "trip valve", "SIS valve" and various combinations of names including the words, safety, emergency, isolation, trip, and fire are used to identify devices used for the same purpose; reduction of the amount of dangerous materials (combustible, flammable, hazardous, toxic) released to the environment due to an accident.

For simplicity, this paper will use the basic definition as provided in API recommended practice (RP) 553 - Refinery Control Valves (2007 edition).

> *API RP-553 paragraph 7.2.1: "Emergency Block Valves are used to control a hazardous incident. These are valves for emergency isolation and are designed to stop the uncontrolled release of flammable or toxic materials."*

Types of Valves

In addition to a plethora of names; and EBV may also take on many forms. While API RP-553 is titled "Refinery Control Valves", EBV's do not need to be actual control valves, or for that matter, to even have any instrumented functions. The only requirement of RP-553 is the valve used is of a fire-safe design as per API spec 6FA "Fire test for valves"; or an equivalent standard. However, even this may not be a fixed requirement depending on the specific installation, valve location relative to the likely fire area, or the amount of other fire protection (such as water sprays) provided.

Valve Selection (the Good, the Bad, and the Ugly)

Practically any valve can be used to create an EBV; for the most crude to the highly sophisticated. Some refining services still use reverse mounted swing check valves; with a fusible link and an external actuator. LPG service shut-off valves are also available with remote cable pulls and fusible links. Most EBV classifications in RP-553 are simply manual valves (gate / ball / butterfly), and may have a motor or pneumatic actuator. Excess flow check valves (spring type), and regulators (differential pressure based) are also used in some services in various hydrocarbon process industries.

As unit capacity, and the valve diameter required, increases; standard type on / off valves are more likely to be required. Often the design will require an actuator; and may incorporate various instrumented functions. A typical basis for EBV's in today's olefins units would incorporate a fire-safe manual valve with a DCS enabled power actuator, and some additional SIS functionality (open / close limit switches; interlocks; etc.). Additional fire protection of power and control wiring would also be incorporated into the design.

Valve Categories per API RP-553

API RP-553 paragraph 7.1 defines the general valve type as follows:

> *"Valve type is dependent upon the distance from the leak source. Any valve in the fire zone should be fire-safe. A gate valve, metal-seated ball valve, or a high-performance butterfly valve is considered to be fire-safe."*

RP-553 classifies the selection on EBV's into four categories simply defined as types "A", "B", "C" and "D". The definitions of each type are included in Table 1, and summarized as follows:

Type "A": A manually operated block valve.

The RP-553 definition includes the valve be "fire-safe", installed at the equipment, and used when ignition of the leaking fluid is not expected. This implies an operator would need to enter the area where leaking fluid exists in order to operate the valve. A more appropriate design for a manual valve would be to locate the valve a safe distance from both the leak source and the potential fire zone. A typical use of a Type "A" EBV would be a battery limit block valve used to stop flow of hydrocarbons, or other dangerous material, from entering the unit after a breach of containment has occurred.

Type "B": A manually operated block valve with location requirements.

Essentially this is Type "A" with location requirements added due to the expected ignition of the leaking fluid. RP-553 states the valve should be installed a minimum of 25 feet for the ignition source (a typical API fire radius), no larger than 8 inches in diameter, no pressure class rating greater than 300 lb. with access via platform with stairs and elevation no greater than 15 feet. As with the type "A" valve; consideration should be given to having the valve placed a safe distance from the potential leak and fire source; or more than 50 to 100 feet away. Increased distance from the fire source should negate the size, pressure class, and elevation requirements; as these are likely based on the intensity of the fire, and the personnel exposure permissible at a 25 to 50 foot distance. As with a type "A" EBV, a typical use of a Type "B" EBV would be a battery limit block valve used to stop flow of hydrocarbons, or other dangerous material, from entering the unit after a breach of containment has occurred.

Type "C": a power operated type "B" valve.

Essentially this is a manual valve with addition of an actuator. Although power operated is not specifically defined; later sections refer to selection and design of electric motor and pneumatic actuators. The control initiation is still manual (local pushbutton) and located at / near the valve in an accessible location. Depending on the actual valve location relative to the fire zone, consideration needs to be given to fireproofing the actuator and power wiring to allow operation during exposure to a fire. A pneumatic valve / actuator combination should be designed as a fair safe system; or fail closed on loss of air supply.

Type "D": defined as an EBV with remote controls.

This type does not specify either the type of valve or the actuator; only that the control is remote. It removes the valve location requirements and adds the control location should be a minimum of 40 feet from the leak source and outside the fire zone. Further definition includes that the actuator and portion of control wiring / tubing within the fire zone be fireproofed. Fireproofing is later defined as being per API 2218 - *Fireproofing Practices in Petroleum and Petrochemical Processing Plants.*

Table 1 - API RP-553 EBV Valve Types (Category)

Valve Type	Fire-Safe (*)	Operation	Pressure Class	Diameter	Distance from Source	
					(Valve)	(Control)
"A"	Yes	Manual (no ignition)	N.D.	N.D.	@ Equipment	NA
"B"	Yes	Manual (ignition)	≤ 300 #	≤ 8"	≥ 25 Ft	NA
"C"	Yes	Power (**)	> 300 #	> 8"	≥ 25 Ft	At Valve
"D"	Yes	Power (**)	N.D.	N.D.	No Restriction	Remote ≥ 40 Ft
* Fire-Safe valve may not be required if located outside the Fire-Zone						
** Actuator, power, cables fire-proofed						

When positive isolation of the process stream is not likely; and the valve cannot be located a safe distance from the potential leak source and fire zone; a type "D" valve is the most likely selection. In today's modern plant designs with advanced DCS control system, it is likely that remote actuation from the DCS operations panel in the control room, or remote instrument enclosure will be incorporated into the design. This will enable the board operators to keep control of the intended operation, as well as prevent the potential exposure of any personnel to an area near the fire zone.

WHO? - Establishing the Basis for Needing an EBV

The first thing required in the review or development of a unit design is to determine the basis to define the need for EBV's. Even though there are some specific service requirements for EBV's (LPG systems, Fired Heater Fuel Gas, Loading / Unloading operations, etc.); the majority of selection basis is defined by recommended practices, and good engineering principles.

Many companies have their own engineering and safety standards for establishing when an EBV is required. These may appear in a verity of design guides or standards; including fire protection, piping, instrumentation, safety, operations. Establishing the need for an EBV may also be the result of a hazard review (Hazop, LOPA, SIL); or be required by HSE standards. Various combinations of standards may also be required to define the complete design. As an example; the need for an EBV may be defined by a safety standard, with fire protection of the device defined in the fire protection standards, and the actuation and controls defined in an instrumentation standard.

The main point here is to check the standards of the facility owner first. Any design basis or selection criteria contained in the owner's standards are likely to be compiled as the owner's interpretation of applicable codes. Local site requirements, insurance requirements, or even environmental permit requirements can affect the specific need in any particular unit.

What if owner standards are not available?

If owner standards are not available, the designer can default to a number of selection basis documents from various NFPA sections or API Recommended Practices. For general service, RP-553 provides guidance on the hydrocarbon inventory which will suggest an EBV be installed; as well as the general selection basis of valve type. When the fluid processed is considered to be LPG (mostly C3 or C4 hydrocarbons), NFPA 58 and API RP-2510 may be used. Requirements specific to the designs of fired heaters would utilize NFPA 85 and API RP-560.

When a wide range of flammable / combustible materials are encountered; NFPA 30 can be used to establish the threshold value of fluid quantities which will trigger consideration of installing an EBV. NFPA 30 breaks down the maximum permitted storage volume of flammable liquids by combinations of flash points and boiling points.

It is also important to establish if the proposed EBV is located within a fire zone; and fire protection of the valve and instrument components is required. To establish if an EBV is to be installed within a fire affected zone, RP-553 basis can be used (25 foot radius typical of many API based fire applications). A more conservative approach would be to use the classification area as per NFPA 497/ API RP-500. This will establish a fire zone which will account for the fluid properties (flammable / combustible class) and source pressure. Of course all EBV installations can simply be treated as requiring fire-safe valve designs with fire protection of auxiliary components; negating the need for any detailed review or backup documentation.

When establishing the design basis for EBV requirements; the following should normally be considered:

1. Check owner's design specification first
2. Use RP-553 for installations without owner specifications
3. Consider NFPA 30 for more detailed treatment of fluid volume
4. Consider NFPA 497 RP-500 for establishing fire protection requirements
5. Use Additional standards for specific applications (NFPA 58 / API RP-2510 for LPG service; NFPA 85 / API RP-560 /API RP-556 for fired heaters; review codes for special cases like loading / unloading at waterways; etc.)

EBV's for Toxic Materials

Special consideration should be given to the need for EBV's when handling toxic materials. Although olefins units do not typically have streams with large quantities of truly toxic materials; there are occasions where high concentrations of toxic components (such as H2S) may present a significant health hazard in event of a leak.

Toxic materials need to be covered on a case by case basis. The need is likely to be triggered by a HSE review of materials being processed in the unit, and concentrations above safe exposure limits. Material Safety Data Sheets (MSDS) for the component considered will give both the safe and health threatening concentrations. Once a high concentration of a toxic component is identified, and estimate of the leak quantity would need to be determined (such as the escape rate via a 3/4 inch bleed valve). The leak rate would then be used to establish a concentration profile within a reasonable distance from the leak point. The need for an EBV would be determined based on comparison of allowable emergency exposure concentration to the calculated dispersion profile.

WHEN? - Establishing if an EBV is needed?

Determining when an EBV should be considered may be the most difficult question to answer in review of plant design. Since the process streams of most plants, and especially olefins units, are comprised of flammable/ combustible hydrocarbons; how is the need for an EBV determined? Should there be a different basis for various fluids? Should other design factors be addressed? These questions, and others, are likely to require answers from individuals design a unit. In addition, documentation defining the selection criteria will likely be required for input to safety and hazard reviews.

To develop a basis of design and selection, various standards will need to be consulted. Each may yield a different basis; and the designer will need to select the most applicable to the specific application considered. In order to make a selection the basis of standards and codes presented in the previous section should be used. The steps taken would then be as follows:

1. Review owner's standards

This is where the designer needs to do a little legwork, and crack open the books. As previously mentioned, many owners have available standards. When a specific EBV / isolation standard is not available; check Safety, HSE, Instrumentation, and Fire Protection standards. A little time to review them can result in a clear design basis, and prevent having to re-design after HAZOP or other reviews.

2. API RP-553 Design Criteria

When specific owner's standards do not exist; the first source of information would be RP-553 (Section 7 of this RP contains the basis for EBV design). Separate design criteria are included for compressors, pumps, vessels, and fired heaters. A summary of the criteria contained in RP-553, and some interpretation of the requirements, is as follows:

Compressors: EBV's should be considered for installation when the following criteria are met.

a) Install suction and discharge EBV when horsepower is greater than 200.
b) Install EBV's on interstage equipment when liquid inventory is more than 1000 gallons (4 cubic meters).

The installation of EBV's on compressor lines is generally to allow isolation of significant high pressure gas volumes in event of damage or failure of the seal system. The EBV can also act as equipment isolation, and prevent need for removal of process system inventory to perform repairs. EBV's should also be considered on any induction / extraction lines as found in many C2 or C3 refrigeration systems. An EBV would not normally be considered for self contained interstage equipment (no side stream flows); other than liquid removal lines when liquid volume is greater than 1000 gallons.

Pumps: For pumps with seals, EBV's should be considered for installation when the following criteria are met.

a) Install an EBV in the pump suction line when the liquid inventory in the upstream vessel is more than 2000 gallons (8 cubic meters) of either light ends, hydrocarbons with an operating temperature above the auto-ignition point, or hydrocarbons with an operating temperature above 600 degrees F (xx degrees C).
b) Install an EBV in the pump suction line when the liquid inventory in the upstream vessel is more than 4000 gallons (16 cubic meters) of liquid hydrocarbons.

The pump suction EBV is mainly installed to prevent the release of fluids from the upstream vessel. Therefore, the default basis would be to install a single EBV close to the vessel liquid outlet. For services with multiple pumps (either multiple operating or operating / spare configurations); consideration should be given installing an EBV on each individual pump suction line. The use of individual EBV's would be more likely as pump flow increases, pump horsepower increases, the flash point of the material processed decreases, or an unplanned failure (closure) of the EBV can lead to a major unit upset / relief scenario. Individual EBV's may also be desirable if they are part of an SIS system, and require periodic testing.

Vessels: EBV's should be considered when the following criteria are met.

a) Install an EBV on vessels containing light ends or toxic materials.
b) Install an EBV on vessels containing liquids heavier than light ends; with their operating temperature above their flash point.

Vessel EBV's will generally be installed on the liquid outlet lines only. RP-553 does not specify a volume for vessel inventory; but the pump suction vessel criteria can be assumed to apply.

For vapor lines, or vessels in vapor only service, total vessel isolation from the normal vapor path would be considered where an upset condition could lead to a vessel failure (exclusive of normal relief conditions). A prime example of this would be olefins plant acetylene reactor systems, where normal flow path isolation would be used to prevent adding reactive materials. These would normally be used in conjunction with other safety systems, such as a flare dump valve.

The EBV criteria contained in the vessel section of RP-553 is the only mention of toxic material control. When toxic materials are encountered, several things should be checked to determine if emergency isolation is required. The first thing is to know the concentration and volume of material contained within the process equipment. Secondly, establish a potential leak path and location (likely to be a bleed valve). Check with owner HSE for exposure guides, or check applicable Material Safety Data Sheets (MSDS). The MSDS will usually contain information on allowable short term / emergency exposures. Additional information may also be indicated on an evacuation radius. The NFPA 497 / RP-500 classification areas could be used as the initial distances for calculation of exposure concentration resulting from a leak. A simple basis would be to uniformly distribute the leak volume over the volume covered by the classification area. A more refined method would be to use dispersion modeling to determine the concentration gradients.

Systems containing toxic materials will likely require a controlled access area. The boundary of this area will be the location of the maximum allowable concentration without protective equipment. Additional access control may also be required for maximum allowable concentrations when protective equipment (respirator) is used.

Fired Heaters: EBV's should be considered when the following criteria are met.

a) Install an EBV on each fuel gas or oil line
b) Install an EBV on each feed line which contains a flammable fluid.

Isolation of fuel and feed lines for fired heaters are covered in detail in NFPA 85, API RP-560 / 556. The recommendation here is to default to these other documents, in addition to any applicable owner's standards on combustion control and burner management.

Lines Crossing Battery Limits and Other Locations not covered by RP-553

While the criteria included in RP-553 will cover the majority of design cases likely to in any particular process unit; other applications for EBV installations may exist. The most significant of these would be lines crossing the unit battery limits.

The typical battery limit manual block valve would be considered an EBV by RP-553. However the valve size, class, and location may not be strictly within the complete criteria. Typical pipe spec valves used for the battery limit block may also not be of a fire-safe design. In today's modern highly automated plants with remotely located control rooms, and few field operators; consideration should be given to including power actuated EBV's on hydrocarbon feed lines to the unit.

Table 2 - API RP-553 Design Criteria

Equipment Type	Power	Inventory		Notes
	(HP)	(Gallons)	(Cubic Meter)	
Compressor	≥ 200	≥ 1000	≥ 4	
Pump	-	≥ 2000	≥ 8	Light Ends / Auto-Ignition Temperature> 600 °F
	-	≥ 4000	≥ 16	Liquid Hydrocarbons
Vessel	-	YES	YES	Light Ends / Toxics
		YES	YES	Operating Temperature above Flash Point

Fired Heater	-	YES	YES	Each Fuel Gas / Fuel Oil Line
		YES	YES	Each Flammable Material Feed Line

3. NFPA 30 - Storage Volume Basis

When the contents of a facility are mainly large quantities of combustible, or flammable hydrocarbons; additional definition (above that of RP-553) may be in order. NFPA 30 is a good source to use for establishing a consistent design basis. This code can be used to classify the fluids into several categories, as well as establishing both a volume criteria, and a fire affected zone.

The primary basis is to use the data provided in NFPA 30; referencing storage limitations for outside storage. The assumption is the released material is in an area with no special fire protection (sprinklers, deluge, etc), other than hydrant / monitor stream coverage. Therefore, the released quantity would be equivalent to a storage volume. The values included in NFPA for minimum separation distances to a property line (or uncontrolled access), is interpreted as the area affected by the resultant fire, or fire zone. The storage volume values from NFPA 30 are tabulated in Table 3; and NFPA liquid classification basis in Table 4.

Table 3 - NFPA Storage Volume and Separation Distance

(Based on NFPA 30 - Table 15.3)

Liquid Class	Maximum Volume		Separation Distance	
	(Gallons)	(Cubic Meters)	(Feet)	(Meters)
IA	1,100	4.2	50	15.2
IB	2,200	8.4	50	15.2
IC	4,400	16.8	50	15.2
II	8,800	33.6	25	7.6
III	22,000	83.3	15	3

Table 4 - NFPA Flammable / Combustible Class Definition

(Based on NFPA 30)

Liquid Class	Flash Point				Boiling Point	
	(°F)	(°C)	(°F)	(°C)	(°F)	(°C)
IA			< 73°F	< 22.8°C	< 100°F	< 37.8°C
IB			< 73°F	< 22.8°C	≥ 100°F	≥ 37.8°C
IC	≥ 73°F	≥ 22.8°C	< 100°F	< 37.8°C		
II	≥ 100°F	≥ 37.8°C	< 140°F	< 60°C		
III	≥ 140°F	≥ 60°C				

From the above, it is readily seen that for lighter hydrocarbons (class I), the limiting volume is typically similar to the RP-553 values of 2000 or 4000 gallons. However, use of the NFPA basis will allow larger volumes of heavier hydrocarbons (class II and III) before considering installation of an EBV.

4. Determining the Fire Zone with Respect to the EBV Location

Since the primary purpose of most EBV installations is to prevent the spread of flammable / combustible materials, most EBV's are likely to be subject to fire exposure. This would be especially true for EBV's installed in vessel outlet lines (as close to / or on, the vessel outlet flange), or individual pump suction lines. As stated earlier in this document, RP-553 recommends that EBV's be of a fire-safe design (low leakage post fire exposure), and actuators / control cable / power cable be fire proofed as required. However, all EBV's may not be located with an area reasonably affected by a fire resulting from the service they are protecting. In order to determine if the EBV is within the fire affected zone several criteria can be used. These are as follows:

Fire Zone Criteria (NFPA 497 / RP-500)

API General Criteria	25 Ft radius from leak source / fire (40 Ft minimum for remote control)
NFPA 30 Separation Distance	50 Ft for Class I Liquids 25 Ft for Class II Liquids 15 Ft for Class III Liquids
NFPA 497 / API RP-500	Various based on material (typically between 15 Ft and 25Ft - more for shelters, or non-ventilated areas)

The primary purpose of establishing the fire zone is to determine the requirements for fire-safe design of the EBV, and the need for fire proofing of control components. Any site specific fire protection requirements, or owner safety standards, should always be consulted before using general code values. An EBV located within the fire affected area, which is protected by deluge or water spray (potential for large pumps / compressors) a fire-safe valve may not be required. Control and power wiring would need to be reviewed on an individual basis. If the potential for the EBV to be affected by a fire case is questionable, or the boundyies of the fire affected zone are unclear; a full fire-safe valve selection and fire proofed control and power systems should be used.

5. LPG System Design by NFPA 58 / API RP-2510

Specific attention should be given to design situations which may be interpreted as an LPG installation. NFPA 58 defines Liquefied Petroleum Gas (LPG) as:

"Any material having a vapor pressure not exceeding that allowed for commercial propane that is composed predominately of the following hydrocarbons, either by themselves or as mixtures: propane, propylene, butane (normal butane or isobutane), and butylenes)."

This definition can apply to some of the light hydrocarbon systems within ethylene units. As always, the designer should check with owner's standards, including site operating permits and local requirements, to see if any systems could be classified as LPG. If this is the case; the volume limits associated with NFPA 58 / RP-2510 would be used to determine if an EBV is required. The criteria for NFPA 58 compatibility is as follows:

Emergency Valve Criteria per NFPA 58 - Section 6.12

Install an Emergency Shutoff Valves when:

a) Storage system capacity greater than 4000 gallons (15.1 cubic meters)
b) Liquid Transfer Line 1-1/2 inch or larger
c) Vapor equalizing line 1-1/4 inch or larger

The NFPA 58 criteria above (4000 gallons capacity threshold) is somewhat less stringent then either API-553 (2000 gallons for light ends), or NFPA 30 (1,100 gallons of class IA liquids). However, if the system in question is legally classified as LPG, the NFPA 58 criteria would likely meet the minimum code requirements.

If a system is classified as LPG, the designer should note that other requirements for excess flow check valves, and isolation may be needed. Especially for loading / unloading situations (either loading arms or hoses), or system designs interfacing with transportation systems (containers, tank trucks, rail cars, barges, or ships) and beyond the intention of this paper.

6. Additional Code / Standards Requirements

This paper has been created based on using general criteria from API RP-553 as a starting point to guide the designer in determining the selection of EBV installation. It is not intended to cover every design case or any site specific requirements. Every plant site is likely to have some degree of owner specifications, local regulations, and maybe even environmental permit or insurance requirements.

In previous sections, codes and standards other than RP-553 have been mentioned. When determining if an EBV is required for a specific design, some questions to ask would be;

Can the system be classified as LPG?
Can the system be classified as a process fired heater / steam generator?
Can the system be classified as a thermal oxidizer / incinerator?
Does the system involve loading / unloading facilities?
Does the system include transportation facilities (DOT / Coast Guard / etc)?

Does the system process toxic / hazardous / radioactive materials?

These are just some special circumstances which may be involved in determining the correct criteria for establishing EBV design basis. A yes answer to any of the above questions could add additional specific design requirements to the system.

WHERE? - Establishing the location to install an EBV

Once it has been determined an EBV should be included in a system design; the next step is to determine where the valve should be installed. Guidance from RP-553, and other references, is limited. The decision is left more to the designer, and various practices may be utilized depending on the specifics of the individual installation.

Determining the location to install an EBV involves a bit of application of common sense design. Remembering the primary definition and purpose of an EBV (*stop the uncontrolled release of flammable or toxic materials)* the first choice for installing would be directly mounted on the outlet flange of the source equipment. While this may be a reasonable choice for many vessels, it will likely be incompatible for EBV's connected to pumps, compressors, or fired equipment. The location of the valve needs to address the maintenance and operation requirements of the hydrocarbon containing equipment, as well as the same for the EBV itself. If the EBV is to be part of a SIS; consideration will also need to be given to system testing.

In order to minimize the release of materials, the EBV should be located as close as possible to the source volume, and upstream of the potential leak point. On this basis, the equipment nozzle flange would be the most ideal location. However, this is often a poor overall design choice, when considering unit operation, maintainability, and even constructability. The EBV is therefore, likely to be located a reasonable distance from the material source, and the following should be considered in setting the location.

Minimize the volume contained in the piping between the source and the EBV

Prevent installing any piping takeoffs (other than isolation blinds and bleeds) between the source and the EBV

Install the EBV in a position which facilitates construction, operation, and maintenance (usually in a horizontal pipe run - depending on valve and actuator style)

Is the valve in the fire zone? / Can it be within deluge - water spray area of the equipment?

Will an unintended EBV closure cause a major plant upset? (separate lines / multiple EBV's?)

Does the valve installation allow for periodic stroke / control system testing?

Going through this simple set of questions will aid in determining the best location for the specific EBV application. Tradeoffs will always be a part of the design process, and knowing which criteria are the most important for the specific application will help guide the decision making process.

WHY? - The Benefits of Installing EBV's

This is the easiest question to answer. Directly from our definition of an EBV, the purpose is to control the flow of dangerous materials. With regard to flammable / combustible materials this means having the ability to isolate fuel from a fire (or fuel from a potential fire). Almost any hazard review will establish that being able to control, or stop, the release of fuel to a fire will result in a safer design; or safer with respect to potential injury of personnel, potential of damage to capital installations, or even potential environmental impact. Clearly the benefit of installing EBV's can be summarized as follows:

Reduce the potential for injury to personnel due to limiting the magnitude or duration of a plant fire

Reduce the potential for damage to capital equipment. piping, instrumentation, etc., due to limiting the magnitude or duration of a plant fire

Reduce the potential for damage caused by environmental release of hydrocarbons, or products of combustion

Reduce the potential for injury to personnel in the unit, adjacent units, or outside a facility boundary due to limiting the quantity of toxic, hazardous materials

The need for EBV's from a hazard review / safety standpoint is not likely to be debated. However the installation of can by justified solely from a capital standpoint. This evaluation can be made by using either published safety review criteria, or developing project specific guidelines. A review of this type may result in indicating an EBV could reduce the potential damage to the affect area of a facility by as much as 10 percent; and reduce repair time by several days.

Using one published source, a review of a typical C3 splitter system indicated a manually actuated EBV would reduce the expected fire damage by 2 percent, and save 1 day of repair time. If the EBV installation is coupled with SIS control and fire and gas detection; the expected facility damage could be reduced by as much as 10 percent, and downtime reduced by 2 to 4 days. Using this basis, the following tabulation (Table 5) is presented as typical cost analysis.

Table 5 - Cost Analysis of EBV Installation (fire damage basis)

Installed Value	Damage Value (*)	Savings		
	No EBV	With EBV	With EBV/SIS	1 day downtime (**)
$1,000,000	$700,000	$14,000	$70,000	$300,000
$5,000,000	$3,500,000	$70,000	$350,000	$300,000
$10,000,000	$7,000,000	$140,000	$700,000	$300,000
* Damage value is estimated @ 70% of Installed Value (Typical)				
** Daily Production Value of $300,000 assumed				

From this analysis, it is readily seen that selection and installation of EBV's can be a cost effective method of providing increased safety within hydrocarbon processing units.

Concluding Remarks

From the preceding, it can be readily seen that the selection of EBV installation can be a simply process. It only requires the following a few steps, including the answers obtained by asking WHAT, WHO, WHEN, WHERE, and WHY. The answers obtained, with the use of established codes and recommended practices form the basis of a typical design.

While the first goal of installing any EBV is the increased safety of an operating plant; the installation can be viewed as insurance to limit facility damage cost.

AIChE Paper Number: 150D

NOX RISK MITIGATION FOR THE CO-FEEDING OF NATURAL GAS TO THE SASOL CTL COMPLEX

Michael Schnobel

Sasol Technology, Secunda, South Africa

Prepared for Presentation at the 2012 Spring National Meeting

Houston, TX, April 1-5, 2012

NOX RISK MITIGATION FOR THE CO-FEEDING OF NATURAL GAS TO THE SASOL CTL COMPLEX

Michael Schnobel

Sasol Technology, Secunda, South Africa

Abstract: Natural gas has been a co-feedstock to the reformers in the Sasol CTL complex since 2004. A more beneficial way is to re-route natural gas to the cryogenic separation unit and thus only reform the separated methane in the natural gas feed.

However, introducing NOx into the cryogenic separation unit where conjugated diolefins from the tail gas of the CTL complex are present creates the risk of explosive NOx hydrocarbon gum formation. The steps taken on the project to quantify and mitigate the NOx safety risks are described below.

The first step was to establish whether NOx is present in the natural gas feed via dry colorimetry. Although the majority of the results were below the detection limit, the presence of NO_x in the natural gas feed could not be ruled out, so a NOx guard bed was added to the design. An adsorption technology was chosen versus the more conventional catalytic process to act as a NO_2 guard bed. In order to monitor the effectiveness of the guard bed several cold liquid streams downstream of the guard bed will be analysed for NO_2. From theoretical calculations it was established that the separated liquids will contain up to ten times higher concentrations of NO_2 compared to the gaseous inlet stream. Detection of NO_2 in the liquid streams will thus indicate a high level of NO_2 in the feed if the removal system fails. NO will fractionate with methane, hence it will not come into contact with conjugated diolefins.

In addition, a procedure has been proposed to periodically wash the system with methanol to remove NOx hydrocarbon gum if it is formed. The procedure has been developed in such a way that the impact on production will be minimal. The methanol wash sample will be analyzed for any indications of NOx hydrocarbon gum and the frequency of methanol wash will be adjusted based on the analytical results of NOx downstream of the guard bed. As guidance, a calculated maximum amount of 3 lb NOx accumulation will be allowed before a methanol wash must be carried out.

Introduction

Sasol has been operating two cryogenic plants with two trains each in the Secunda Coal-To-Liquid (CTL) facility from as far back as 1980 and 1982, respectively [1]. In the Fischer-Tropsch (FT) synthesis unit, synthesis gas ($CO+H_2$) is reacted over an iron catalyst to produce low molecular mass alkenes and liquid products in the gasoline and diesel fuel range [2].

The light hydrocarbon fraction enters the cryogenic unit after an acid gas removal for further separation. The dangers of explosive gum formation due to reaction of NOx with conjugated diolefins in cryogenic plants in the Cold Box have been widely discussed [3] – [5]. Sasol has been aware of this danger in their CTL facility in Secunda/South Africa. The possibility of NOx and conjugated dienes entering the cryogenic plants and forming explosive NOx gums has been monitored since then, but no indications of NOx entering the cryogenic plants from the FT synthesis units have been observed nor any signs of NOx gums formation.

In 2004 natural gas started to be transferred from Mozambique to Sasol in Secunda, amongst others, via an 865 km long pipeline [6]. The natural gas is co-fed with methane rich gas to auto thermal reformers in order to increase the production output of the Secunda facilities. However, co-feeding natural gas with the existing feedstock of methane rich gas is not without issues. There is a difference in the gas composition between the two feedstock. The existing methane rich gas feedstock consists only of methane, argon, nitrogen, hydrogen and carbon monoxide, with only trace amounts of hydrocarbons heavier than methane. Natural gas on the other hand consists of methane, nitrogen and heavier alkanes. No olefins or diolefins are present in natural gas [7]. Reforming alkanes other than methane has some distinct disadvantages. One disadvantage is higher oxygen consumption, because the reforming reaction becomes more endothermic the longer the alkane chain length [8]. Another disadvantage is the lower H_2:CO ratio in the reformed gas, because of the increasingly lower H:C ratio in longer chain alkanes. Sasol decided therefore to go ahead with a new project, which provides an alternative way of co-feeding natural gas [9]. In this project natural gas is routed to the existing cryogenic plant and after pre-treatment and cooling it is co-fed to all four existing trains each with its own demethanizer column. The advantage of rerouting natural gas to the cryogenic plant is that methane in natural gas is separated from the other alkanes and only methane and nitrogen are routed to the reforming plant. The recovered alkanes in the natural gas can be used in the existing product streams. Ethane and propane are cracked to ethylene, which is converted to polyethylene. Butane is used in LPG and all other alkanes are blended with fuel products.

However, since conjugated diolefins enter the cryogenic plant as a product from the FT synthesis and any NOx in the natural gas will enter the retrofitted cryogenic plant as well, the danger of explosive gum formation had to be investigated prior to the design of the modification of the plant. The 3 lb recommendation [5] was used as a guideline for the maximum amount of NOx that can enter the cryogenic plant before deriming or another method of removing the accumulated NOx is used.

Flushing with methanol has been suggested as an effective way to remove NOx or gum from the cryogenic plant [3]. In this method flushing with methanol is carried out after the plant has been warmed up to ambient temperature. However, flushing with methanol during warm up also has been reported [4]. This method has the advantage that NOx gum is flushed out before it is warmed up and possibly becomes unstable. The disadvantage of this method is the thermal stress as a result of adding a warm solvent into the cold equipment and piping. In order to measure if any NOx or gum is being removed, it has been reported that the effluent is analysed for nitrogen [10].

This paper discusses the analytical techniques to detect NOx, the steps that were taken in the design to prevent explosive gum formation and changes in the operation of the modified cryogenic plant to prevent explosive gum formation.

Analytical Results

The presence of NOx and conjugated diolefins had to be investigated first. NOx analysis of natural gas was carried out using an analyzer system from C.I. Analytics, using a dry colorimetric technique. The natural gas sample was supplied in a 1 gal (US) sampling cylinder with inert lining. The analysis cannot distinguish between NO and NO_2 and is therefore reported as NOx.

Sampling took place over a period of fifteen months and the results are presented in Figure 1. The lower detection limit was given as 5×10^{-9} (vol/vol) or 5 ppbv.

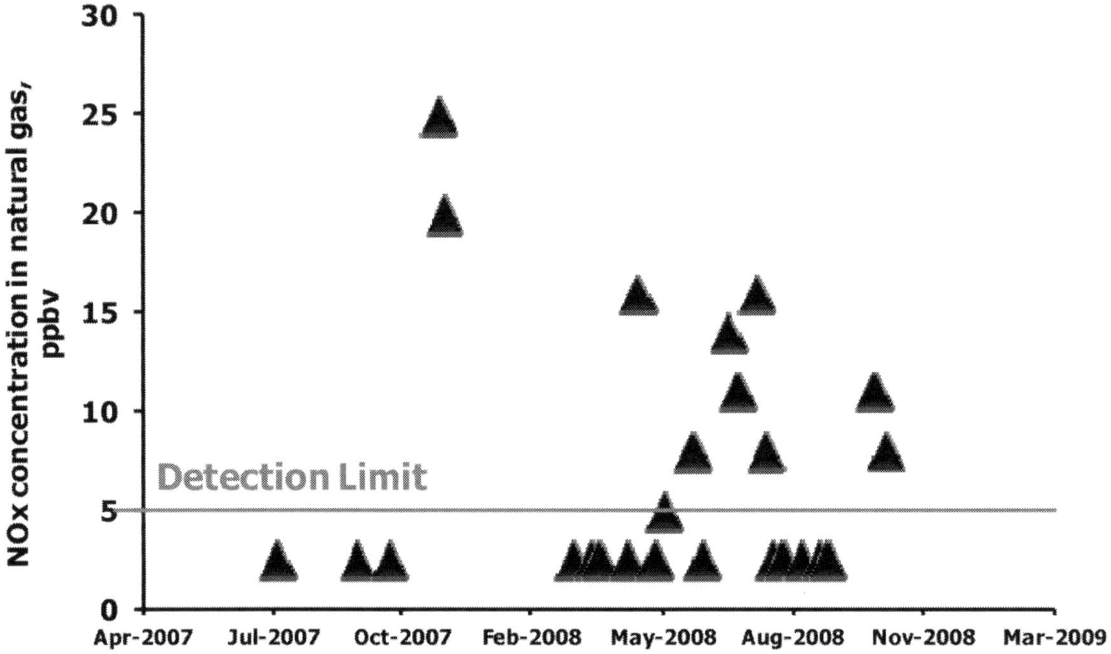

Figure 1: Measured NO_x concentration in natural gas

The results in Figure 1 show that the NOx content in the natural gas that enters Sasol's CTL facility in Secunda/South Africa showed a maximum concentration of 25 ppbv. This is an indication that NOx is present in the natural gas feed stream. However, multiple measurement points were below the detection limit.

Conjugated diolefins were detected in the product stream downstream of each demethanizer column. The relevant results (average of four streams) are shown in Table 1. The analysis results show that conjugated diolefins must have passed through the demethanizer columns.

Table 1. Analysis of dienes downstream of demethanizer columns

Component	Fraction, % (wt)
1,3 Butadiene	0.011
Trans1,3 Pentadiene	0.002
Cis 1,3 Pentadiene	0.011
Methyl-1,3 Cyclopentadiene	0.002

Other impurities besides NO_X that were identified for removal from the natural gas before entering the retrofitted cryogenic plant were water, CO_2, and mercury. Water and CO_2 could freeze and lead to blockages in the cryogenic plant. Mercury could accumulate in the aluminum heat exchangers in the cryogenic plant and lead to corrosion and leakage from the aluminum heat exchangers. Table 2 shows the measured level and the acceptable levels of water and CO_2 entering the cryogenic plant.

Table 2. Analysis of water, CO_2 and mercury in natural gas

Compound	Measured	Maximum allowable
H_2O	100 x 10^{-6} (vol/vol)[1)]	0.1 x 10^{-6} (vol/vol)
CO_2	300 x 10^{-6} (vol/vol)[1)]	30 x 10^{-6} (vol/vol)
Hg	6 x 10^{-9} (mole/mole)	0.001 x 10^{-9} (mole/mole)

Notes: 1) 95% percentile

Plant Design

The existing cryogenic plant uses an internal recycle of methane rich gas as reflux to the demethanizer columns. Figure 2 shows the process configuration around the demethanizer column of the existing plant.

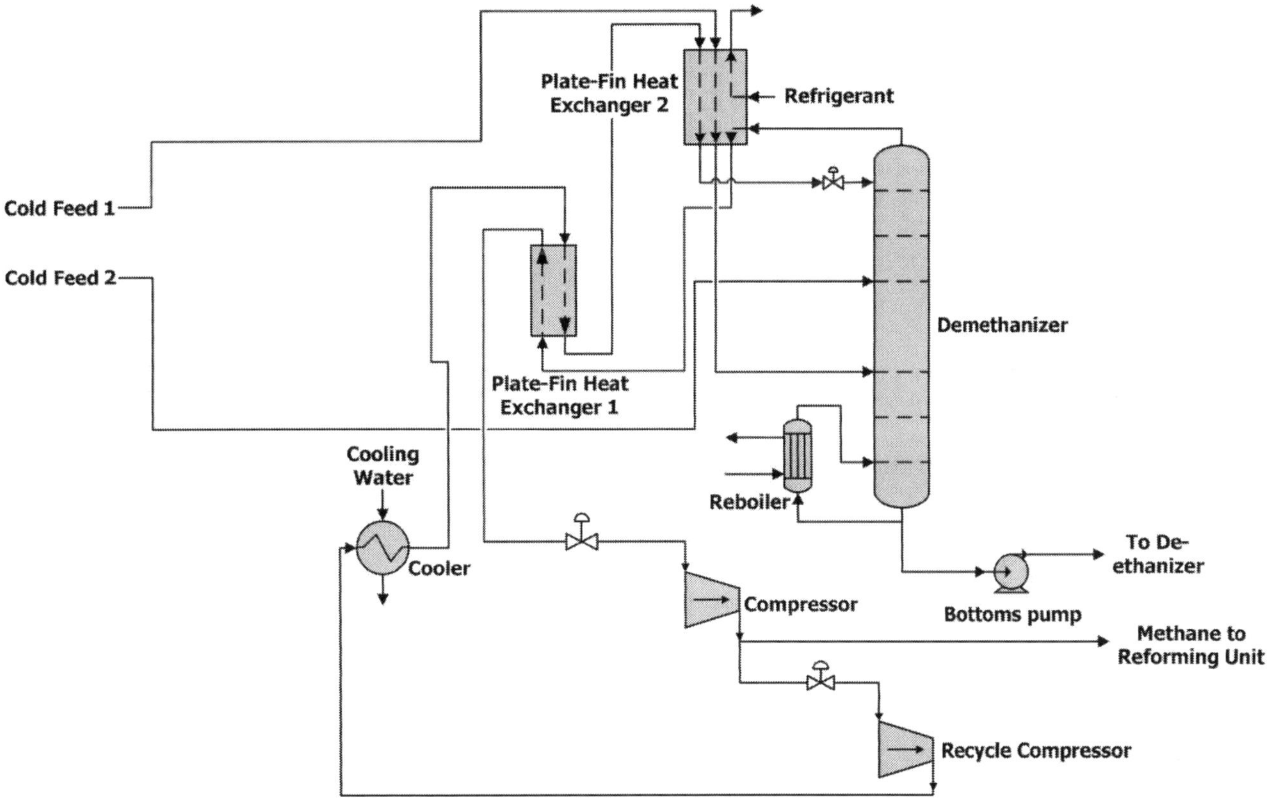

Figure 2. Cryogenic plant around the demethanizer column before the modification

The natural gas is taken from the tie-in point in the Reforming Unit at 42 bar and is routed via an approximately 500 m long new pipeline to the Cryogenic Unit. The natural gas is expanded to 34 bar and passes the natural gas Separator 1. This vessel is equipped with a cyclone separator to prevent liquid droplets passing through to the mercury guard bed.

Mercury is removed to below detection limit in the Mercury Guard Bed in order to protect the cryogenic unit with its aluminum equipment against mercury attack. Downstream of the mercury guard bed the natural gas is dried in one of the two adsorbers (Adsorber 1 or 2), while the second one is regenerated. The molecular sieve will also remove CO_2 in the same vessel. A 4Å molecular sieve was chosen as adsorbent to remove water, CO_2 and NOx to below the maximum allowable levels.

The molecular sieve will be regenerated with hydrogen rich gas (HRG) at 25 bar. The HRG will be heated in the Regeneration Gas Heater to 210°C by using HP steam at 42 bar as heating medium and then routed to the adsorber vessel which is in regeneration mode. Downstream of the adsorber vessel the HRG is cooled in the Regeneration Gas Cooler to 50°C by air cooling and is then routed to Separator 2.

Any dust from the molecular sieve and mercury guard bed is prevented from entering the downstream section by a filter. The treated natural gas is then equally distributed and routed to the two identical trains.

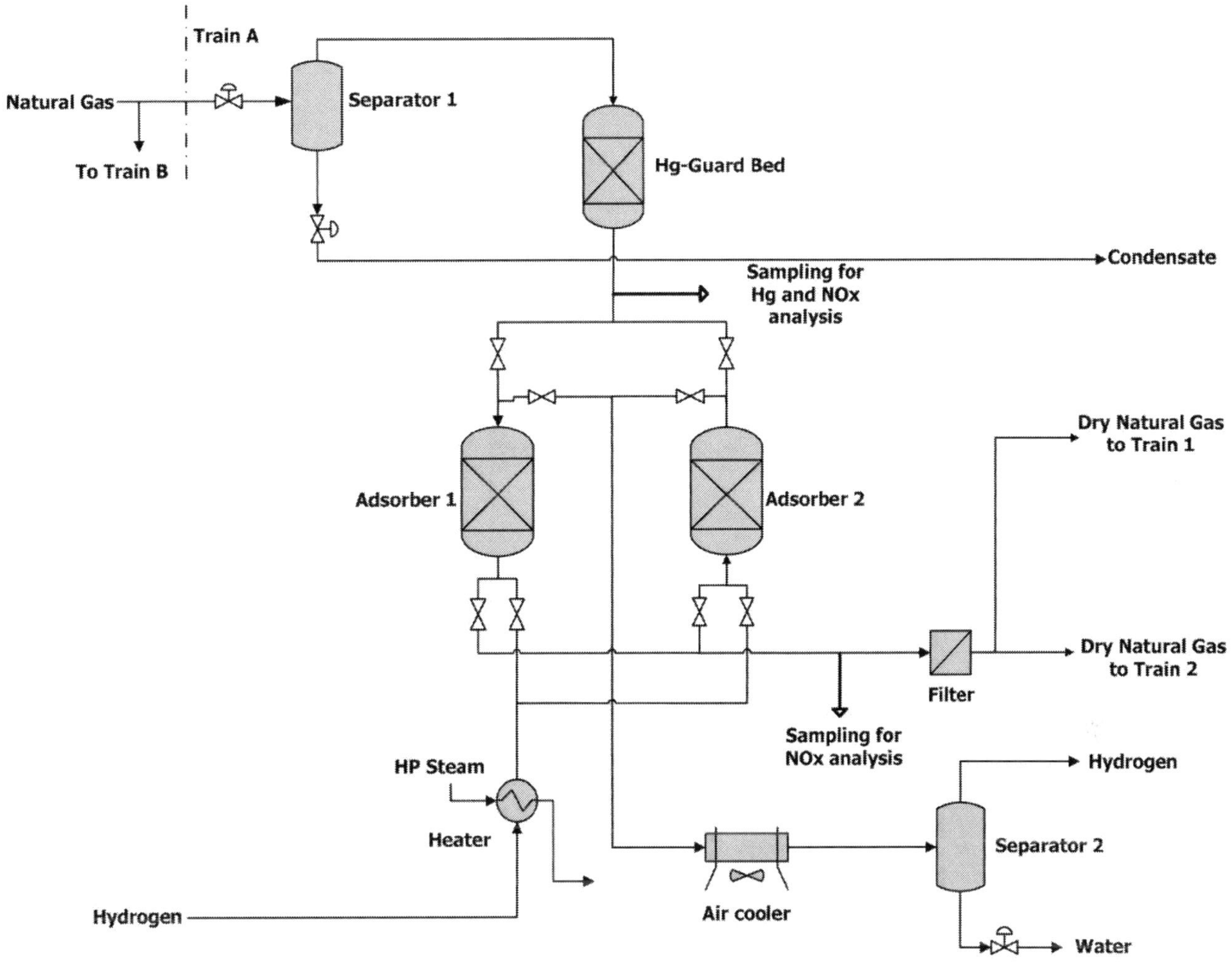

Figure 3. Natural Gas pre-conditioning unit (Train A)

The aim in the plant design for the retrofitted cryogenic unit was to integrate as much of the existing equipment as possible and minimize capital costs. Figure 4 shows the process configuration of the retrofitted cryogenic unit.

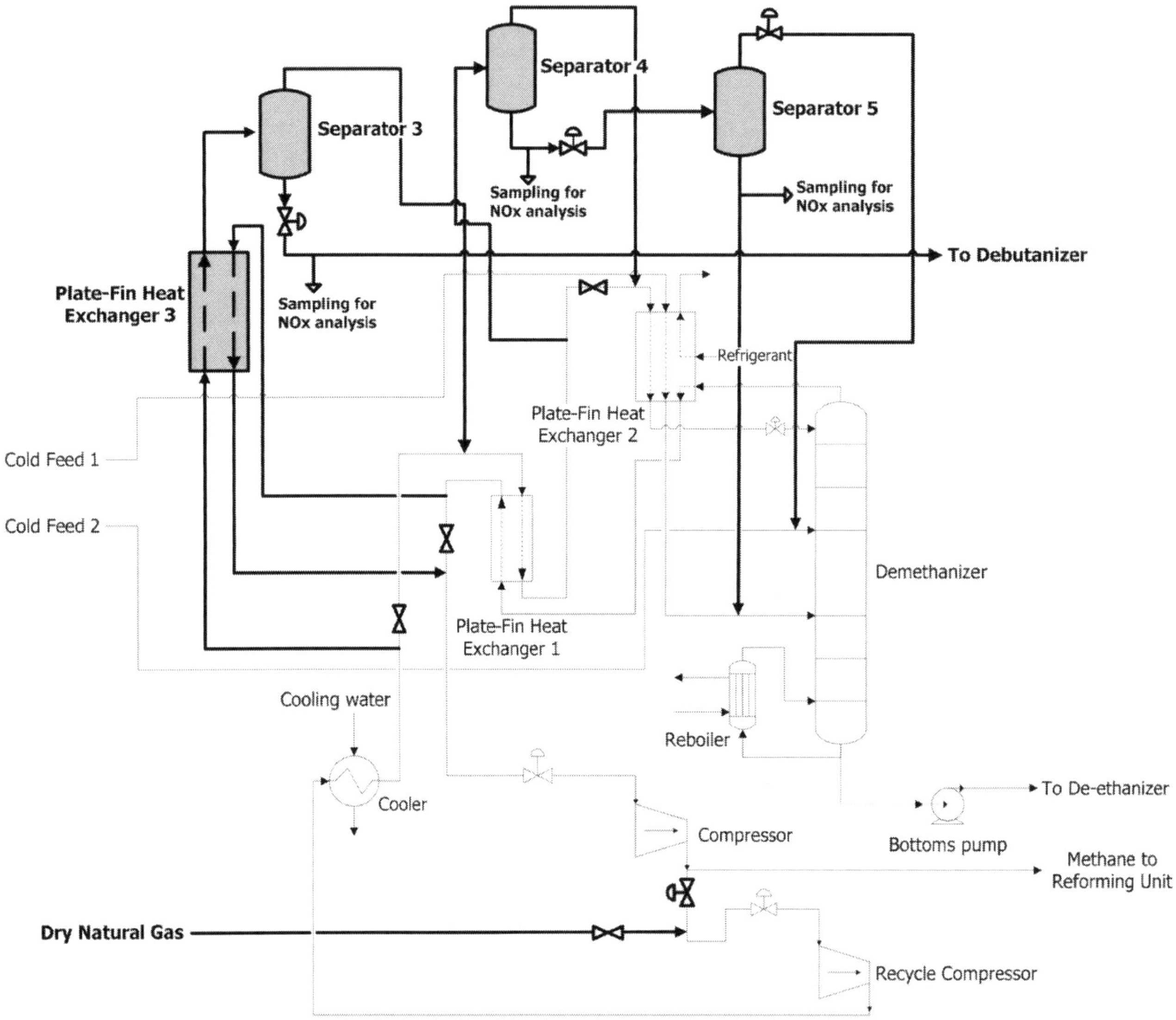

Figure 4. Modified cryogenic plant (new plant addition = bold) around demethanizer column with dry natural gas feed (Train 1)

In order to incorporate the dry natural gas stream into the cryogenic plant certain modifications were necessary. Instead of using methane rich gas from the compressor discharge, dry natural gas is fed to the suction of the recycle compressor. The natural gas is compressed in the existing Recycle Compressor to 50 bar and cooled in the existing cooler to 40°C. In the new Plate-Fin Heat Exchanger 3 the natural gas is further cooled to approximately -27°C using LP MRG coming from the existing Plate-Fin Heat Exchanger 1. A small amount of hydrocarbons condense in the Plate-Fin Heat Exchanger 3 (approximately 2 wt%) and are separated in the new Separator Drum 3. From there the condensate is routed to an existing separator drum and consequently the C_{4+} components end up primarily in the Debutanizer bottom product.

The overhead gas from Separator Drum 3 is routed to the inlet of the existing Plate Fin Heat Exchanger 1, where it is partly condensed and cooled to -73°C. Downstream of the Plate Fin Heat Exchanger 1 the two-phase hydrocarbon stream with a liquid fraction of 23 wt% is separated in the new Separator Drum 4. The overhead gas is routed to the existing Plate-Fin Heat Exchanger 2 where it is liquefied and subcooled to -132°C using LP-C_2H_4 refrigerant and Demethanizer overhead gas as cooling media. Then the liquid hydrocarbon stream is expanded and used as reflux to the existing Demethanizer Column which is operated at 3 bar.

The liquid fraction from Separator 4 is expanded from 48 bar to 3 bar over a valve and as a result partially vaporized and cooled to -133°C. The two-phase stream is then separated in the new Separator 5. The liquid fraction is routed to the existing feed pipeline to tray 11 of the Demethanizer and the overhead gas is routed to the existing feed pipeline to tray 15 of the Demethanizer.

One unique feature of the design of the plant modification is that the operation can revert back to the original operating mode with recycling methane rich gas to the demethanizer column. In such case no liquid is separated in Separators 3, 4, and 5 and the valves in the bottom outlet of Separator 3, and 4 stay closed.

Using the 3 lb recommendation [5], which is generally regarded as an acceptable amount of NOx accumulation between warm-ups, the time to reach this amount of accumulation per train was calculated for two average NOx concentrations, namely 1 ppbv and 7 ppbv. The calculation was carried out for NOx being either NO or NO_2. The results are shown in Table 3.

Table 3. Estimated time to reach 3 lb accumulation for different NOx concentrations in natural gas

Concentration of NOx in natural gas	Time to reach 3 lb accumulation
1 ppbv [1)]	5 years, 10 months
1 ppbv [2)]	4 years, 5 months
7 ppbv [1)]	12 months
7 ppbv [2)]	8 months

Note 1: NOx = NO, Note 2: NOx = NO_2

Since each train in the cryogenic unit is on a 4 year turnaround cycle, the NOx concentration in the natural gas downstream of the adsorbers is only acceptable up to approximately 1 ppbv. Computer simulations using Linde's OPTISIM® program were carried out in order to establish if NOx will be at higher concentrations in other streams in case the NOx content in the natural gas feed is below or close to the detection level. If such streams are identified then the analysis result can be used to calculate the NOx concentration in the natural gas feed. The NOx concentration in the natural gas was an input parameter into the simulation and was fixed at 7 ppbv, which is marginally above the detection limit (see Figure 1). The two extreme cases of NOx being either NO or NO_2 were simulated. The results of the computer simulations are shown in Table 4.

Table 4. Simulation results of NO and NO_2 in different streams of the modified cryogenic unit

Stream	Amount of NO	Amount of NO_2
Feed to Separator 3[1)]	7 ppbv	7 ppbv
Bottoms from Separator 3	0 ppbv	175 ppbv
Vapor from Separator 3	7 ppbv	4.8 ppbv
Feed to Separator 4	7 ppbv	4.8 ppbv
Bottoms from Separator 4	0 ppbv	23.9 ppbv
Vapor from Separator 4	7.8 ppbv	0 ppbv
Feed to Separator 5	0 ppbv	23.9 ppbv
Bottoms from Separator 5	0 ppbv	56.7 ppbv
Vapor from Separator 5	0 ppbv	0 ppbv

Note 1: Input into simulation

The results in Table 4 show that the NO_2 concentration is highest in the liquid streams. Analyzing the liquid streams from Separator 3, 4, and 5 will therefore help in confirming or establishing the NO_2 concentration in the natural gas downstream of the adsorbers. The results further show that NO is not found in any stream in considerably higher concentrations. In addition, NO will enter the demethanizer column only with the reflux stream (vapor from Separator 4) into an area of the column where no conjugated diolefins are found, since conjugated diolefins are only measured in the bottom product of the column. It is known that NO approximately fractionates with methane [11]. This leads to the conclusion that NO in the natural gas will not lead to gum formation in the cryogenic unit.

In order to verify the results in Table 4, Sasol carried out simulations independently using ASPEN PLUS®. It was assumed as mentioned above that NO cannot lead to gum formation, therefore only NO_2 was taken into account. In order to show a different scenario, the NO_2 concentration in the natural gas feed stream to Separator 3 was decreased to 1.7 ppbv, which is well below the detection limit (see Figure 1). The results of the computer simulation are shown in Table 5.

Table 5. Simulation result of NO_2 in different streams of the modified cryogenic unit

Stream	Amount of NO_2
Feed to Separator 3[1)]	1.7 ppbv
Bottoms from Separator 3	58 ppbv
Vapor from Separator 3	1.6 ppbv
Feed to Separator 4	1.6 ppbv
Bottoms from Separator 4	7 ppbv
Vapor from Separator 4	0.3 ppbv
Feed to Separator 5	7 ppbv
Bottoms from Separator 5	17 ppbv
Vapor from Separator 5	0 ppbv

Note 1: Input into simulation

The simulation results in Table 5 confirm the results in Table 4 that NO_2 is found in highest concentrations in the liquid streams from the three separator drums. The highest concentration of NO_2 in the liquid stream from Separator 3 is confirmed also. These results show that in the case of natural gas entering the cryogenic unit with a NO_2 concentration being below the detection limit, it is more likely to detect NO_2 in the liquid streams exiting Separator 3, 4, and 5.

Gas samples for NOx analysis will be taken upstream and downstream of the adsorbers (see Figure 3). In order to confirm the NOx concentration downstream of the adsorbers, sampling points were installed on the bottom outlet of the Separators 3, 4, and 5. The sampling points are indicated in Figure 4. The liquid samples will be vaporised and transported in a sample cylinder to the laboratory for NOx analysis.

Future Plant Operation

The danger of explosive gum formation was not only addressed in the design of the modified plant but also in the planning of the future plant operation. The adherence to the 3 lb NOx accumulation recommendation will be monitored by regular NOx analysis. Sampling will take place upstream and downstream of Adsorber 1 and 2 (see Figure 3) in order to evaluate the extend of NOx removal in the molecular sieve. In addition a total of three samples will be taken from the liquid streams of Separator 3, 4, and 5 (see Figure 5). The sampling frequency will be once a week, alternating between Train A and B. The analysis results will be used by the plant operator together with the flow rate of the natural gas to calculate the amount of NOx that entered the cryogenic unit since the previous sampling. Different scenarios have to be considered in the NOx analysis result. If the NOx analysis of the natural gas is above detection level, the value will be used together with the natural gas average weekly flow rate to calculate the amount of NOx entering the cryogenic unit. The analysis result will be validated by the NOx analysis result of the liquid streams leaving the three separator drums. If the NOx analysis of the natural gas is below the detection level and the liquid streams of the separator drum analysis are above detection level, the NOx level in the natural gas will be calculated from the analysis results of the liquid streams. If the all liquid streams show NOx concentrations below the detection level then the NOx concentration in the natural gas is negligible.

Once the calculation indicates that the NOx accumulation could have reached 3 lb, the natural gas feed to the cryogenic unit will be discontinued and the reflux stream to the demethanizer column reverts back to methane rich gas. At the next opportunity a methanol wash will be carried out. Methanol wash will be applied to areas where NOx and conjugated dienes could be present at temperatures near or below -100°C. Three pathways for methanol flushing have been identified (see Figure 5). The first path is from the bottom of Separator 3 to the Debutanizer column, where methanol will be collected at the bottom of the column. The second path is from the bottom of Separator 4 to the Demethanizer column. Methanol will be collected at the bottom of the Demethanizer column. The third path is from the bottom of Separator 5 to the Demethanizer column. Methanol is again collected at the bottom of the column. Each wash will be carried out separately in order to be able to narrow down the position of NOx gum formation in case it is formed.

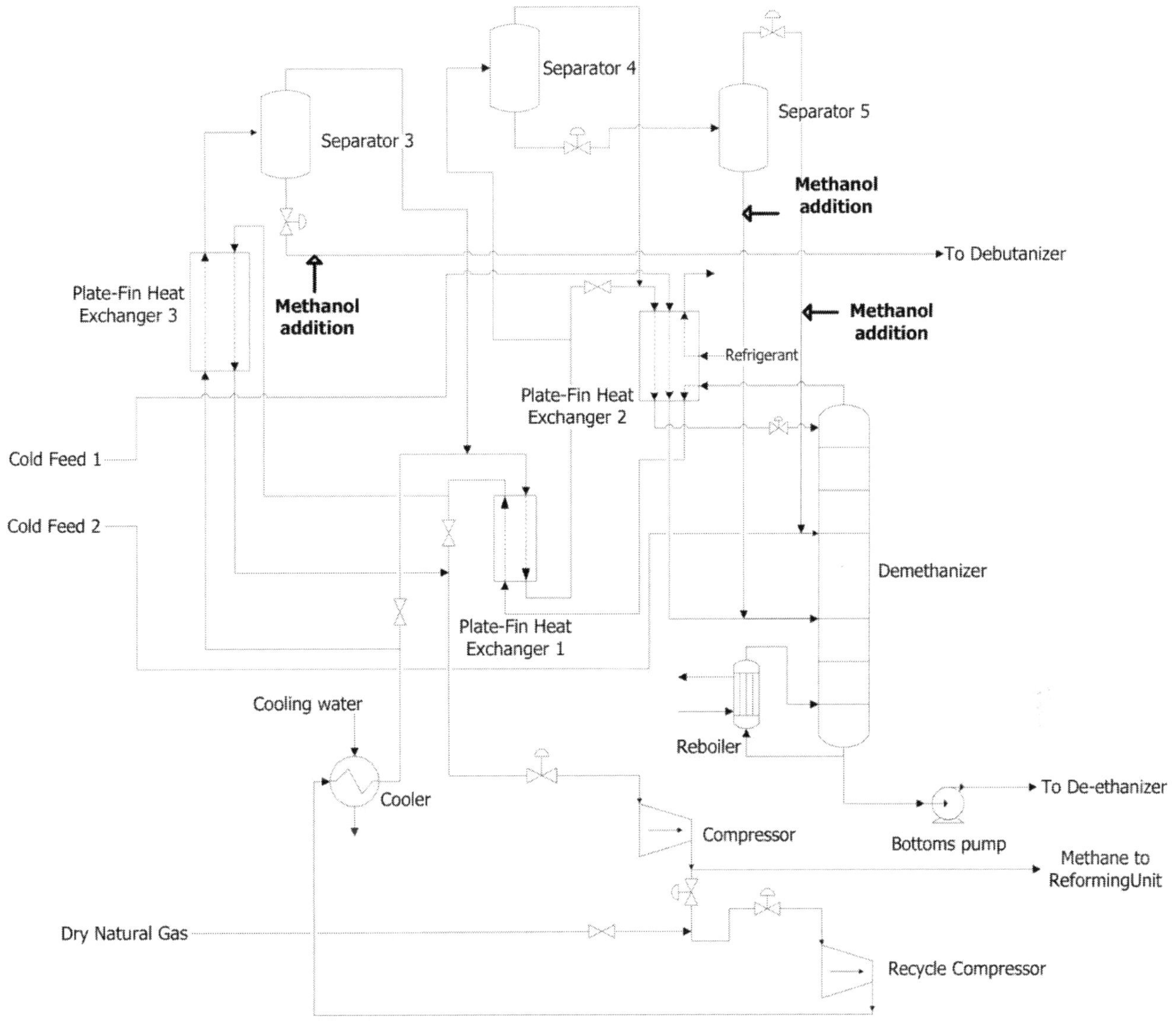

Figure 5. Position of Methanol addition points

In order to establish whether any NOx or gum was present, the methanol will be analyzed for nitrogen according to methods described in literature [10]. The analysis result for nitrogen in the used methanol will determine whether methanol can be re-used or must be disposed of.

Conclusions

The investigation for the re-routing of natural gas to the cryogenic unit at the Sasol Secunda CTL facility showed that there is a potential danger of explosive NOx gum formation. This was shown by natural gas analysis which showed the presence of NOx and demethanizer product analysis which showed the presence of conjugated diolefins. It was decided to incorporate NOx risk mitigation steps into the design and operation of the modified unit.

Simulation of the modified plant operation showed that NO will remain in the vapor phase and enter the demethanizer column in the reflux stream. NO would not come into contact with conjugated diolefins and poses therefore no danger for explosive gum formation. However, in case of NOx being NO_2, there is a danger of NOx gum formation. This is due to NO_2 mainly separating into the liquid phase of the separator drums which means it can come in contact with conjugated diolefins.

In order to detect small concentrations of NO_2 in the natural gas, sampling points in the liquid streams of the separator drums were installed. Since the NO_2 concentration in the liquid streams will be much higher than in the gas phase, it will enable the operator to determine the amount of NO_2 entering the cryogenic unit and monitor whether the 3 lb accumulation recommendation is adhered to. Upon reaching 3 lb accumulation of NO_2 according to the calculation, methanol wash in the identified areas of the plant will be carried out to remove NO_X or gum residues. The design has allowed for reverting back to the current way of operation until an opportunity arises to carry out the methanol wash.

Acknowledgements

The support by Sasol Technology and the simulation work by The Linde Group are gratefully acknowledged.

References

1. J. Collings (2002), "An epic in two parts", in *Mind over Matter, The Sasol Story: A half century of technological innovation*, Corporate Affairs Department of Sasol Limited, Johannesburg.
2. M. Dry (2004), "Chemical concepts used for engineering purposes", in *Fischer-Tropsch Technology*, Vol. 152, Chapter 3, Andre Steynberg and Mark Dry, ed. Amsterdam, Elsevier B. V.
3. W. H. Henstock (1986), "NOx in the Cryogenic Hydrogen Recovery Section", *Plant/Operations Progress*, Volume 5, No. 4, pp. 232 – 237.
4. R. T. Halle (1993), "Potential Hazards of Nitrogen Oxide Compound Accumulations in Cryogenic Ethylene Recovery Facilities", *The Fifth Annual Ethylene Producers Conference*, AIChE Spring National Meeting, Houston, Texas.
5. D. R. Simms (1997), "Update on the Activities of the NOx Task Group", Ethylene Plant Safety, AIChE Spring National Meeting, Houston, Texas.
6. I. Madamombe (2007), "Pipeline benefits Mozambique, South Africa", *Africa Renewal*, Vol. 21, No. 3, October, pp. 18 - 20.
7. G. Hammer (2003), "Natural Gas", in *Ullmann's Encyclopaedia of Industrial Chemistry*, 6th Edition, Weinheim, Wiley-VCH.

8. V. Subramani, P. Sharma, L. Zhang and K. Liu (2010), "Catalytic Steam Reforming Technology for the Production of Hydrogen and Syngas", in *Hydrogen and Syngas Production and Purification Technologies*, Ke Liu, Chunshan Song and Velu Subramani, ed. Hoboken, New Jersey, John Wiley & Sons, Inc.
9. W. S. Ernst and J. F. Van Tonder (2011), "Production of Hydrocarbons from Natural Gas", *U.S. Patent, 7,879,919 B2*.
10. W. H. Henstock, G. L. Baker, P. J. Thornton, D. Hines, D. R. Slim, M. Jeansonne and J. L. Airey , "Recommended Method for Evaluation of Residues from Cryogenic Processing Equipment for the Possible Presence of NOx Gums and NOx Salts", Developed by the NOx Analytical Subcommittee of the NOx Task Force of the Ethylene Producers' Committee of the AIChE.
11. W. H. Henstock and D. R. Simms (2007), "NOx Safety in Ethylene Plants", *EEPC Ethylene Seminar*, Rome.

AIChE Paper Number: 150e

OVERVIEW OF A NOX WASH FOR THE CRYOGENIC SECTION OF AN ETHYLENE PLANT

Mangesh Butle

Saudi Yanbu Petrochemical Company A SABIC Affiliate

Bhuvnesh Laud

Saudi Yanbu Petrochemical Company A SABIC Affiliate

Blair Margot

Saudi Yanbu Petrochemical Company A SABIC Affiliate

Ralph Sare

Saudi Yanbu Petrochemical A SABIC Affiliate

Prepared for Presentation at the 2012 Spring National Meeting

Houston, TX April 1-5, 2012

OVERVIEW OF A NOX WASH FOR THE

CRYOGENIC SECTION OF AN ETHYLENE PLANT

Mangesh Butle

Saudi Yanbu Petrochemical Company A SABIC Affiliate

Bhuvnesh Laud

Saudi Yanbu Petrochemical Company A SABIC Affiliate

Blair Margot

Saudi Yanbu Petrochemical Company A SABIC Affiliate

Ralph Sare

Saudi Yanbu Petrochemical A SABIC Affiliate

Abstract: Recent industry experience has reinforced the potential hazard associated with NOx in the cryogenic section of an ethylene plant. This paper provides an overview of a recent NOx wash performed by Saudi Yanbu Petrochemical Company (Yanpet), a SABIC Affiliate. The paper discusses the justification for doing the wash, the work planning process, the wash procedure, the wash results, and overall learnings.

General Information

Saudi Yanbu Petrochemical Plant (Yanpet) operates two steam crackers in Yanbu, Saudi Arabia. Yanpet is a Joint Venture between Sabic and ExxonMobil. The facility has been in operation since 1985 and has undergone several expansions. The newer cracker referred to in this paper started up in 2000 and cracks ethane, propane and light naphtha feeds.

Purpose of This Paper

At the March 2009 Ethylene Producer's Conference, paper 108e [1] was presented describing a NOx related cold box incident in a plant with no prior history of finding NOx compounds. This paper and the discussions that followed prompted both JV partners and our licensor (Lummus Technology) to strongly recommend solvent washing the cold box during the 2011 turnaround. Up to this point in time, Yanpet considered their risk of NOx hazards to be low and had no plans for a NOx wash. The strong recommendations from our partners and licensor were sufficient justification to conduct a solvent wash.

This paper will provide an overview of our recent NOx wash with the objective of sharing the planning and execution experiences with others who may be considering a NOx wash for the first time. Both JV partners and our licensor provided guidance and assistance in developing the solvent wash plan. The topics to be covered include:

- NOx Risk Factors for Yanpet
- Overview Followed by Discussion of the Wash Activities
- Wash Results
- Learnings

Disclaimer:

The material given here is presented in the interests of sharing data related to safe operations. Neither the AIChE, the North American EPC of the AIChE, the authors nor their employers make any guarantee as to the accuracy or completeness of the written materials.

NOx Risk Factors at Yanpet

Table 1 illustrates Yanpet's perception of NOx risk for the cracker referred to in this paper.

Table 1 – NOx Risk Factors at Yanpet

Higher NOx Risk Factors	Lower NOx Risk Factors
+ Additive with -NO group in the C4/C5 recycle stream.	– No refinery gas processing
+ Demethanizer first sequence with backend converter	– No cryogenic hydrogen system
+ Ammonia due to BFW and additives in dilution steam system	– No history of finding brown gas, blue liquid, or NOx salts
	– No unexplained plugging in Cold Train

NOx Wash Activity Overview

Below is an overview of the recent NOx wash activities.

1. Identify the Locations to Wash
2. Calculate the Volume of Solvent Required
3. Select the Solvent
4. Design a Solvent Collection System
5. Evaluate Mechanical Integrity
6. Develop and Execute the Wash Procedure
7. Send the Spent Solvent to an Incineration Facility
8. Analyze the Spent Solvent Samples to Determine the Quantity of NOx

The following paragraphs explain these activities and discuss factors Yanpet considered in the wash planning. Woven into the discussion are instances where the execution differed markedly from the plan.

1. Identifying Locations to Wash

NOx compounds can accumulate in the chilling train in locations with liquid at temperatures colder than -100C. Vapor areas are unlikely to have enough residence time for NOx to form [2]. NOx salts have been found mostly in regions between -100 and -130C. NOx gums have been found mostly in regions colder than -130C. Typical locations for NOx accumulation at Yanpet are shown in Figure 1 and include:

- Mixed or liquid phase piping colder than -100C.
- The area below the HHLL in drums colder than -100C.
- Piping downstream of control valves colder than -100C due to flashing.
- Condensing exchangers colder than -100C.
- Vaporizing exchangers colder than -100C.
- Low points, drains, dead legs, and instrumentation in any of the above locations.

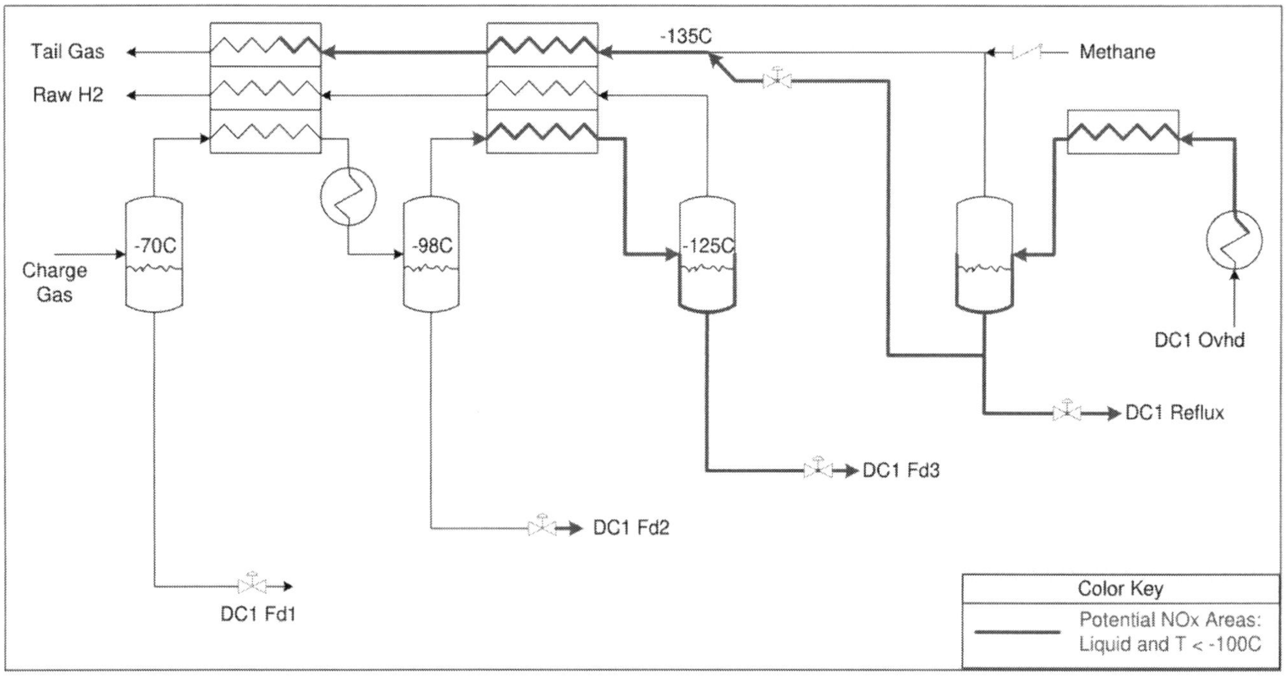

Figure 1: Locations to Solvent Wash

The cold box vendor PIDs were used to identify bleeders, drains and dead legs in the cold box. Plant PID's might not show all the piping. Missing piping was identified and included in the wash plan.

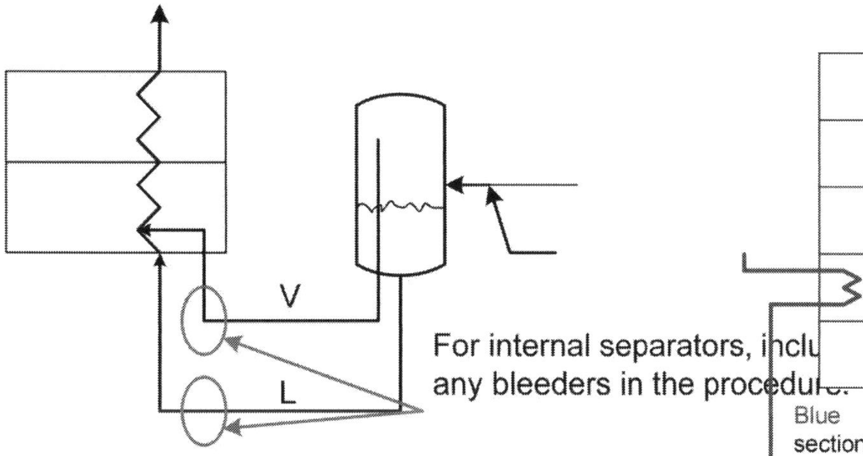

Figure 2: Typical separator drum internal to the cold box

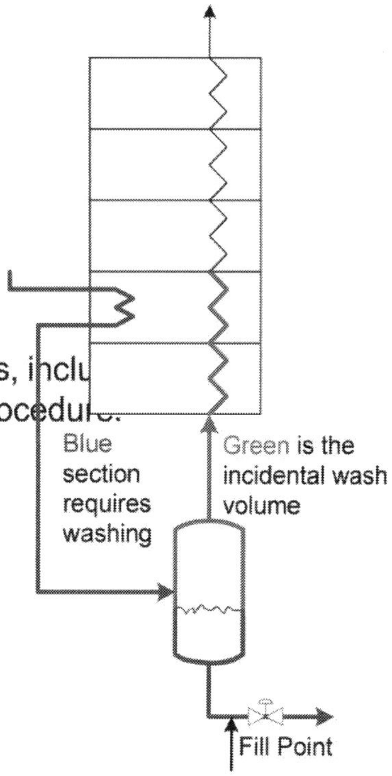

Figure 3: Wash and Incidental Volumes

Liquid and vapor piping on separator drums internal to the cold box were reviewed to ensure all small connections were included in the wash plans, see Figure 2. The BASF FINA incident[1] involved a small connection to a vapor line on a similar separator. The bleeder was shown on the vendor PIDs but not on the plant PIDs.

2. Calculate the volume of solvent required

Due to layout, a soak wash includes equipment that does not require washing. The wash plan divided the system into 3 sections. Using drawings showing vertical orientation, the volumes to be washed and the incidental volumes were identified as shown in Figure 3. For piping downstream of control valves, the plan included the horizontal piping and only 2 meters of vertical to avoid overflowing solvent into the Demethanizer. For each section, the volume of solvent was calculated using piping isometrics, elevation drawings, datasheets, etc.

Thirty one cubic meters of solvent were required for all 3 sections. The plan included two washes making the total solvent requirement 62 cubic meters as shown in Figure 4.

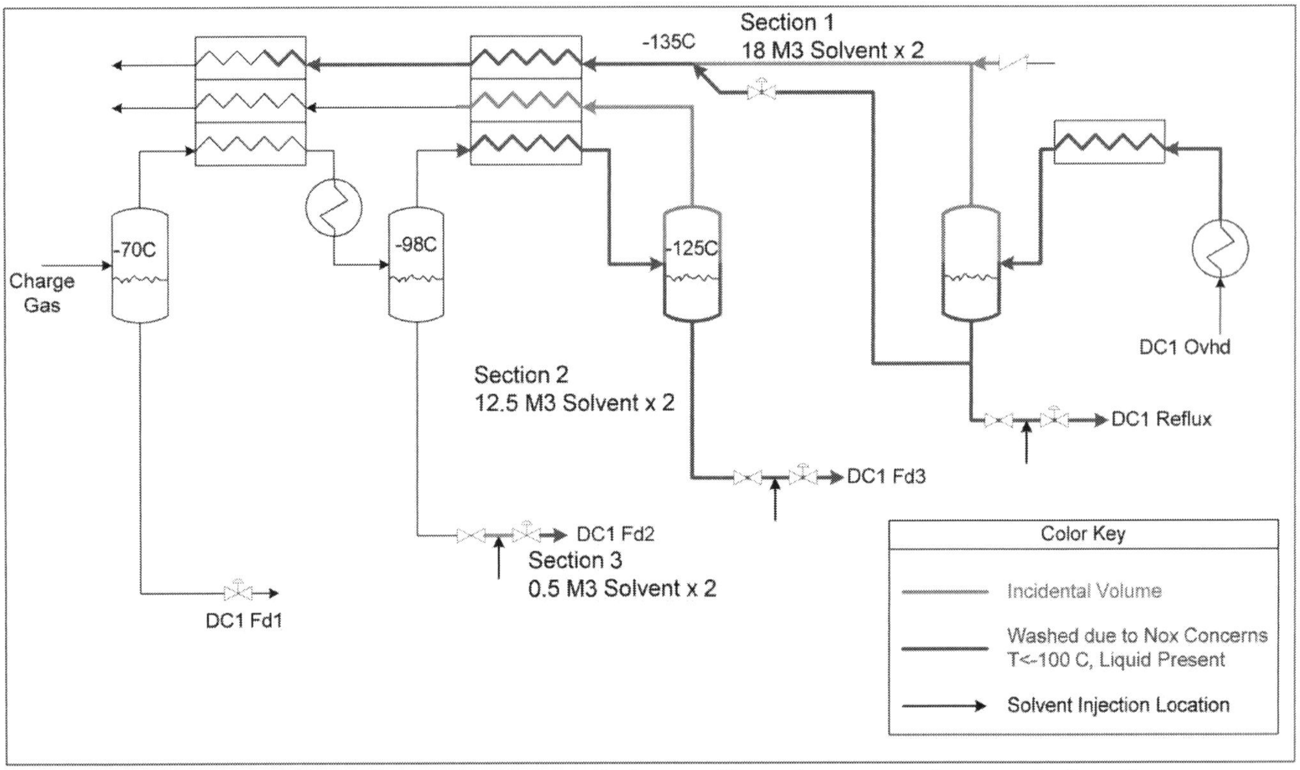

Figure 4: Wash Plan Showing Solvent Requirement for each Section

Two methods were used to measure the amount of solvent injected during the wash. The level drop in the methanol tank was monitored with a DCS based level instrument in the control room. Outside a pressure gauge was used to measure the liquid head of the solvent.

Monitoring the drop in the methanol tank was accurate and worked well for the larger volumes. The head gauge was effective for sections downstream of control valves where the volume was too small to be reliably detected by the tank level instrument. Pulsations from the reciprocating injection pump, and fluctuating flare pressure made the head gauge ineffective for larger volumes.

3. Select the Solvent

The freeze point of methanol is -98C [7]. Although this temperature was adequate for the temperatures during the wash, it is not suitable for the temperatures during normal operation. n-Propanol has a freeze point of -126C [7]. A 50/50 v solution of methanol/n-propanol was selected to help avoid freezing on startup. Draeger style tubes were ordered to sample for residual alcohol during the pre-startup drying out.

Our existing methanol tank, pump, and distribution system, supplemented with stainless steel hose were used during the wash activities. Left over mixed alcohol was expected to be compatible with pump seals that normally use only methanol. The methanol/n-propanol solution was mixed by the supplier and delivered in isotanks which were unloaded before and during the wash.

4. Design and Implement a Solvent Collection System

Three 25M3 isotanks manifolded together at top and bottom to fill as one tank were used for the spent solvent collection system. The vapor spaces were tied together and vented to the flare. The original plan called for a 2" vent to flare. Isolation difficulties necessitated a change to a smaller connection which proved to be a constraint during draining. There was vapor off-gas from the spent solvent during the draining steps. The isotanks have PSV's that lift at 4 bar. During draining the pressure in the isotanks reached 1.3 bar, ~ 0.3 bar above flare pressure.

The wash plan did not include the capability to measure level in the isotanks. The isotanks were large enough to hold the 62M3 of spent solvent we expected. Stainless steel isotanks were used in case the spent solvent was cold, but the coldest spent solvent temperature observed was 0C.

The isotanks came with threaded nozzles using British Standard threading. The plant required National Pipe Threading (NPT). Rethreading was completed during the initial setup. The isotanks had a "hidden" cable-operated valve on the bottom that required opening before putting them in service.

5. Evaluate Mechanical Integrity

- Piping Supports

Once the volumes that require filling were known, the piping and vessel supports were checked for the additional weight of the solvent. The team was concerned extra weight on the spring hanger pipe supports would stress the equipment nozzles. This concern was addressed by locking four spring hangers prior to filling with solvent, and unlocking them after completion of the wash. Locking the spring hangers prevented the springs from compressing under the weight of the solvent.

- Vessel Supports

The vessel supports were determined to be adequate for the solvent weight by plant staff and also confirmed by the cold box supplier.

- Thermal Stresses Within Cold Box Exchangers due to Temperature Differential

Plate-fin heat exchangers are vulnerable to failure due to thermal stress caused by local temperature differences. ALPEMA [4] recommend a maximum 30C difference between local metal temperature and any medium introduced into the cold box. Since the temperature of the solvent is ambient (25C) and the temperature inside the cold box is -50C or colder, potential damage to the exchangers was a concern.

Converting an available ethylene chiller to solvent cooling service was considered to minimize this temperature difference. This would have required keeping the ERC and PRC compressors online until after the wash. Instead the plan injected solvent into the vessels and backflowed into exchangers as shown in Figure 5. It was expected this would provide enough cooling to keep the thermal stress to an acceptable level. This method cooled the solvent to 7C before it contacted the exchanger, making it 42C warmer and outside the ALPEMA guideline.

ALPEMA also recommend the rate of temperature change in the cold box not exceed 2C per minute during startup and shutdown. The rate of change stayed within this guideline except for one 10 minute period during the fill of Section 2 when it exceeded 4C per minute.

6. Develop the step by step procedure

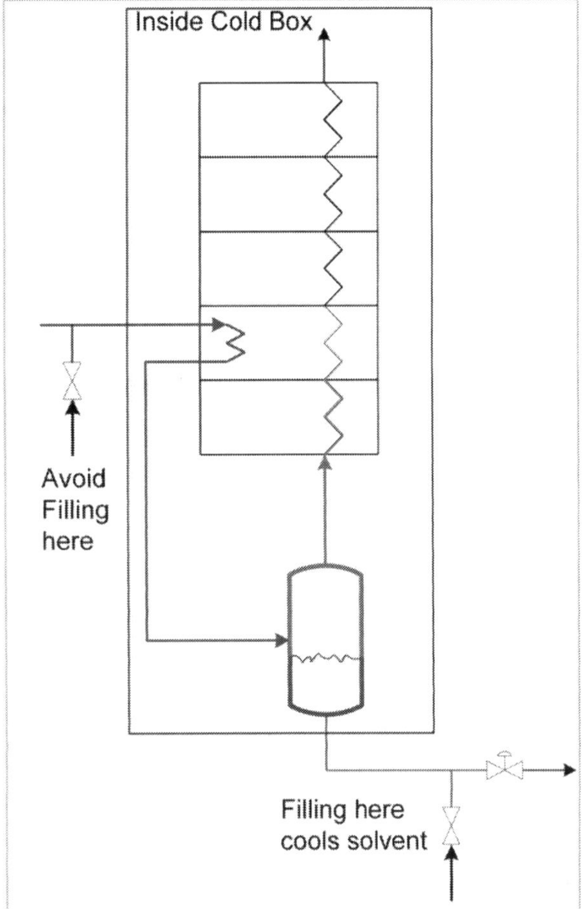

Figure 5: Preferred Filling Location is into Drums

NOx gums are known to become unstable at temperatures well below ambient. To minimize the risk of an incident, the wash should be started as cold as practical. The team was concerned incomplete hydrocarbon freeing could potentially overpressure and exceed the capacity of the spent solvent collection system. The plan set -50C at the 3rd DM Feed Separator as the maximum temperature to allow before starting solvent injection. Based on data from past shutdowns the plan expected to have ~24 hours for HC freeing.

Chart 1 shows the warm-up trend for the cold box during the HC freeing. Before starting the pressure depressure, the cold box was warming at 0.3C per hour. The warming rate increased to 2.5C per hour during the pressure depressure cycles. The plan called for 4 cycles of HC freeing, but due to nitrogen shortage only 3 cycles were completed before the system warmed to -50C.

Chart 1: Cold Box Temp Trend during HC Freeing

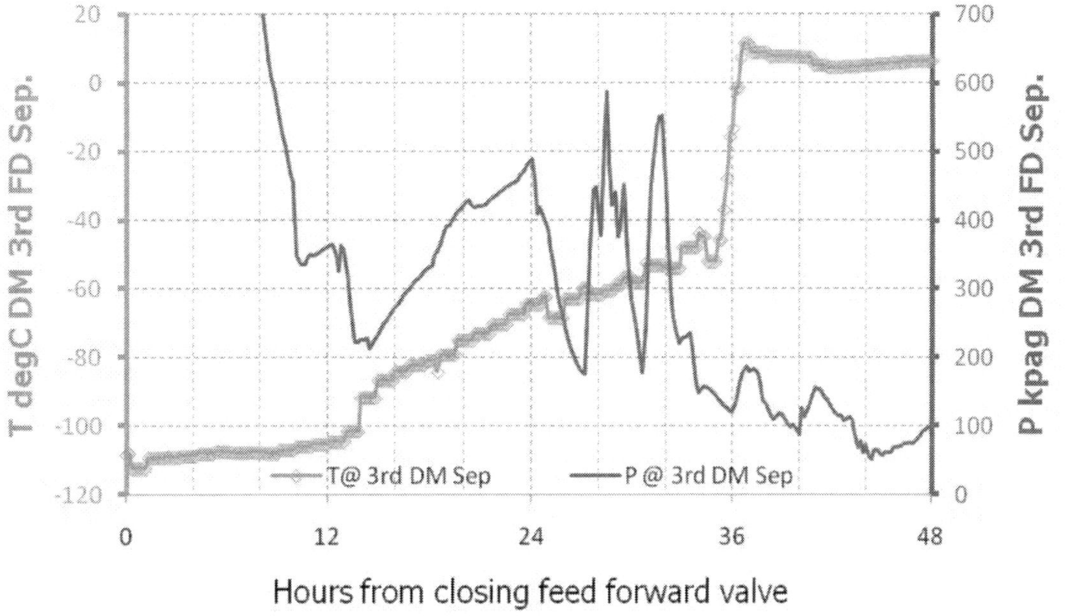

During the hydrocarbon freeing step depressuring was done in the direction reverse to normal flow to avoid sweeping heavies into NOx areas. Depressuring was done mostly from vessel bottoms to remove as much liquid as possible.

Each section was filled sequentially to use the Methanol tank level instrument in quantifying the amount of solvent used. Section 2 the 3rd DM feed separator had the most potential for NOx gums and was washed first. Section 1 the DM Reflux section was washed next as it also had potential for gums. Section 3 the 2nd DM feed separator control valve was washed last due to its low potential for gums. Section 3 could have NOx salts which become unstable at temperatures near ambient for ammonium nitrite[3].

The solvent filling step followed immediately after the HC freeing step. No blinds were installed. The systems to be filled were vented to flare from high points and left open during the filling, soaking and draining steps. Each section was filled in sequence with minimal delay. The inside operator monitored the methanol tank level during filling and communicated when to stop the injection. Outside personnel monitored the progress of the filling with the head gauges. During filling the clean solvent was sampled for later analysis.

After soaking for 12 hours the spent solvent was sampled using glass sample bottles. A color test [5] was used to determine if additional washing was required. All sections required a 2nd wash except for the 3rd Stage DM feed separator which only needed a 2nd wash for the piping downstream of the control valve. The soak time for the 2nd wash was also 12 hours. At the end of the 2nd soak, all samples were clear with no trace of yellow. We did not do a 3rd wash.

During the draining step, outside personnel drained spent solvent from each bleeder, low point, dead leg, and instrumentation location identified on the wash procedure checklist. The locations to be drained were identified with numbered metal tags. The drain locations were connected by hose to the collection system.

7. Send the Spent Solvent to the Incinerator Facility

Alcohol dissolves NOx compounds and is a heat sink to absorb heat from any reaction preventing the temperature from increasing. When disposing of the spent alcohol waste it is important that the spent alcohol be burned as a liquid. The spent alcohol should not be vaporized, heated, or evaporated to dryness. Residual solvent with NOx compounds may become unstable. The incineration facility needs to be aware of the potential hazards of the dissolved NOx compounds. They also need to be aware of the hazardous properties of the solvent from the MSDS sheets. We communicated these issues to the incineration facility via the environmental department.

Within the plant for similar reasons, spent alcohol should not be sent to the flare vaporizer or the flare header since the alcohol may evaporate leaving behind NOx compounds.

8. Analyze the Spent Solvent Samples to Determine the Quantity of NOx

The spent solvent samples were analyzed by an outside laboratory. They used liquid chromatography to determine the amount of nitrate and nitrite in the samples. Additionally they analyzed the samples for ammonia. The samples were not analyzed for total nitrogen or gums.

Wash Results

The amount and type of NOx salt at each sample location was calculated from the volume of solvent used to wash each section, and by assuming sample results were representative for the entire section. The results are summarized in Table 2 below.

Table 2 - Yanpet Wash Results

Section	NH4NO3 (g)	NH4NO2 (g)
1. DM Reflux Section	2.8	3.0
2. DM 3rd Feed Sep	0.5	0
3. DM 2nd Feed Sep CV	0.2	0.1

Learnings

The results for each section were below the 10 grams [6] of NOx considered sufficient to rupture piping, but demonstrated the existence of NOx compounds in the unit. The planning and the wash activities raised the level of NOx hazard awareness in the plant, and highlighted the importance of continued training and education. The plant plans to further define and develop NOx mitigation plans for both crackers.

Future NOx washes can be improved by better understanding the potential for damage to the cold box exchangers.

References

1. Hack, W.R., "Investigation of NOx Related Cold Box Incident". Paper 108e presented at the 21st Ethylene Producers Conference at the 2009 Spring National Meeting of the AIChE.
2. Halle, R.T., "Hazards of NOx Accumulations in Cryogenic Recovery Facilities". Paper 15e presented at the 5th Ethylene Producers Conference at the 1993 Spring National Meeting of the AIChE.
3. Simms, D. R., "Update on the Activities of the NOx Task Group". Paper presented at the 9th Ethylene Producers Conference at the 1997 Spring National Meeting of the AIChE.
4. The Standards of the Brazed Aluminum Plate-Fin Heat Exchanger Manufacturers' Association, ALPEMA, 2nd Edition, 2000
5. Grenoble, D., *et al*, "Safety Tutorial – Management of NOx in Cryogenic Section of Ethylene Plants". Paper 71a presented at the 19th Ethylene Producers Conference at the 2007 Spring National Meeting of the AIChE.
6. Grenoble, D., "Report on Activities of the NOx in Ethylene Recovery Task Group". Paper 55a presented at the 16th Ethylene Producers Conference at the 2004 Spring National Meeting of the AIChE.
7. De Haan, S. "An Overview of Self-Inflicted Contaminants", Paper 30a presented at the 18th Ethylene Producers Conference at the 2006 Spring National Meeting of the AIChE.

AIChE Paper Number: 150f

AN APPROACH TO RISK ASSESS NOX IN THE CRYOGENIC SECTION OF ETHYLENE PLANTS

Michael Moran

ExxonMobil Chemical Company

Thomas Pavia

ExxonMobil Chemical Company

Prepared for Presentation at the 2012 Spring National Meeting

Houston, TX April 1-5, 2012

AN APPROACH TO RISK ASSESS NOX IN THE CRYOGENIC SECTION OF ETHYLENE PLANTS

Michael Moran

ExxonMobil Chemical Company

Thomas Pavia

ExxonMobil Chemical Company

ABSTRACT

Assessing the safety risks of NOx in a cryogenic section of an ethylene plant is difficult because the hazard is a result of complex chemistry, a function of many site specific parameters, and limited available industry data. This paper presents an approach for conducting a semi-quantified risk assessment. The paper discusses key site parameters that influence the risks associated with NOx and how the limited available industry data can be used. In addition, the paper will present an event tree that can be used as an example for the risk assessment. The combination of a semi-quantitative overall risk assessment and a quantitative understanding of the factors that influence that risk can help sites to be better positioned to manage the NOx hazard.

INTRODUCTION

Assessing the safety risks of NOx in a cryogenic section of an ethylene plant is difficult because the hazard is a result of complex chemistry, a function of many site specific parameters, and limited available industry data. This paper presents an approach for conducting a semi-quantified risk assessment. The paper discusses key site parameters that influence the risks associated with NOx and how the limited available industry data can be used. In addition, the paper will present an event tree that can be used as a template for the risk assessment.

RISK ANALYSIS PROCESS OVERVIEW AND TOOLS

A systematic process is followed to effectively assess and manage risk using a scenario based approach. The steps of the process typically involve:

1. Identifying hazards, and "brainstorming" initiating events that could lead to an incident
2. Developing hypothetical risk scenarios and estimating the probability and consequences that may be associated with those scenarios
3. Assessing the significance of risk based on the relative position of the scenario on a Risk Matrix
4. Identifying relevant prevention or mitigation steps to reduce risk
5. Defining the remaining risk after one or more of the identified risk reduction steps are considered

Two main tools used in the risk assessment process were a Risk Matrix and an Event Tree.

The Risk Matrix, shown below in Figure 1, is a tool to quantitatively communicate a potential risk scenario that could result in an incident. The Risk Matrix shown has increasing consequence on the y-axis and decreasing probability on the x-axis. The highest risk scenarios are those in the upper left corner and the lowest risk scenarios are those in the lower right corner. The value of the Risk Matrix lies in helping to consistently convey relative risk significance of various scenarios.

Figure 1: Risk Matrix

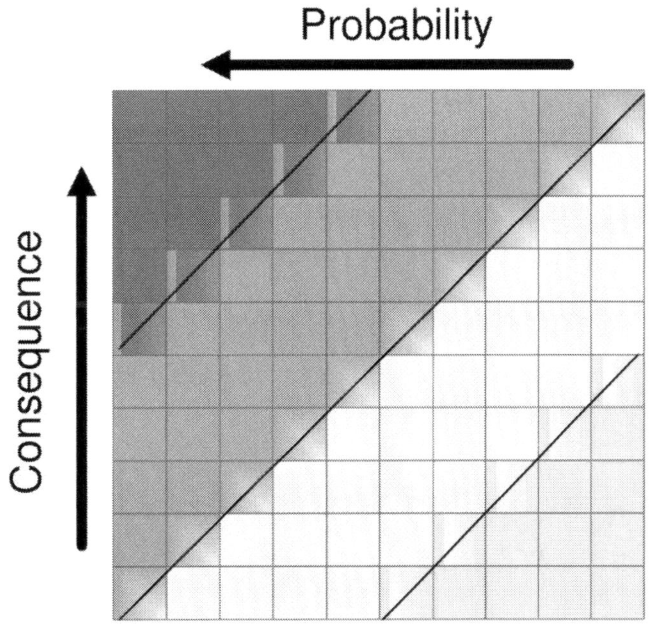

An Event Tree, shown below in Figure 2, is a useful aid in establishing and "quality checking" scenarios. An event tree is a graphical representation of potential incident pathways, and outcomes of scenarios, following a "hypothetical" initiating event. The event tree model depicts the logical interrelationships between potential system successes and failures, or dependent events as they respond to the initiating event. Each end point (right side of the event tree) presents a potential scenario when worked back to the initiating event (left side of the event tree).

Figure 2: Event Tree

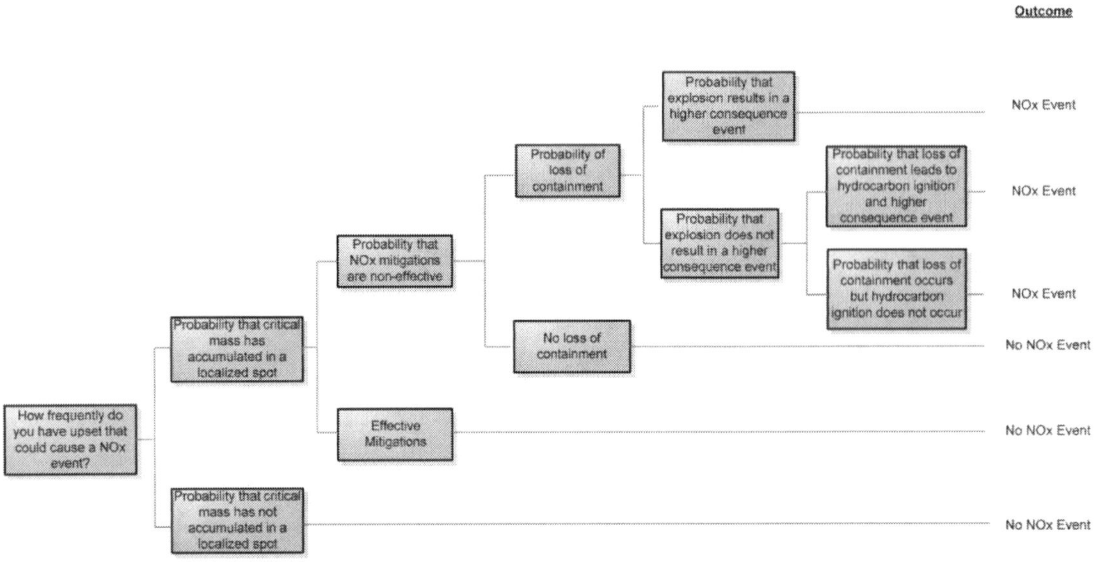

EVENT TREE OVERVIEW

The paper provides an overview of the Generic NOx Hazard Risk Event Tree, showing the structure and key considerations in the scenario. Also, the paper discusses the considerations that go into developing the inputs into the event tree for the frequency and probabilities.

The event tree starts with an initiating event; how often is the cold box exposed to having a NOx related event. Next the event tree consists of a number of conditions that if met could lead to a higher consequence NOx event. During the risk assessment process a probability is assigned to each condition. A list of conditions is provided below:

1. That a critical mass of NOx salts and/or NOx gums exist in one location in the cold box
2. That the plant's NOx mitigation plan is non-effective
3. That a NOx explosion would result in a loss of containment
4. That a NOx explosion would result in a higher consequence event
5. That if there was a loss of containment that the hydrocarbon would ignite and result in a higher consequence event

INITIATION EVENT FREQUENCY

NOx events in industry have occurred during plant upsets, shutdowns, and start-ups. History of plant operating data can be used to determine the frequency that a plant experiences those conditions. Quantitative criteria were developed to assess if those conditions existed. The criteria were comprised of a relative change in temperature of the cold box section of the plant, the cold box section reaching a specific temperature, or a relative change in pressure of the cold box section.

PROBABILITY OF CRITICAL MASS OF NOx

The next judgment is to assign a probability that a critical mass of NOx has accumulated in a localized spot in the cold box. There is little historical data for the basis of this judgment and therefore practical engineering judgment needs to be used. The probability will be based on the amount NOx in the cold box feed, the amount of NOx that accumulates in the cold box, and the time since the previous cold box NOx wash or warming. The amount of NOx in the cold box feed can be determined with an on-line analyzer or spot samples. The amount that accumulated in the cold box can be estimated from analyzing the spent alcohol from previous NOx cold box washes. If data for a specific plant is not available, an estimate can be developed either by benchmarking with other plants that have data or by using data that has been published in literature.

PROBABILITY THAT NOx MITIGATIONS ARE NON-EFFECTIVE

Two common approaches that are used to mitigate the NOx hazard are:

1. Preventing heavy diolefins (e.g. butadiene) from reaching the colder sections of the cold box
2. Washing the cold box to remove the accumulated NOx compounds

Other mitigations could include management of change and cold box feed quality monitoring.

The input into the event tree is a judgment as to how effective the plant's NOx mitigation plan is in preventing NOx events. Both a weighting factor for each component of a plant's NOx mitigation plan and the effectiveness of that component can be assigned. The probability that a plant's NOx mitigation is non-effective can be based on the overall score of the weighting factors and the assessment of the plant's execution effectiveness.

PROBABILITY OF LOSS OF CONTAINMENT

Developing an assessment of the probability that a plant will have a loss of containment if:

+ the cold box was in a condition that a NOx event could occur
+ the cold box had a critical mass of NOx in a localized spot
+ the plant's NOx mitigation plan was non-effective

is a difficult and complex judgment.

A consideration is if the cold box accumulates only NOx salts or if the cold box also accumulates NOx gums. Knowing whether the NOx accumulation is NOx salts only or contains NOx gums can be assessed by either lab analysis of the spent alcohol from previous cold box washes or can be estimated by knowing the operating conditions of the cold box.

Also, assessing the frequency of NOx events in industry under the conditions listed above can be a useful input in developing an estimate of the probability of the loss of containment. To develop an estimate of the frequency of NOx events industry data was used and propriety assumptions that included:

+ Total number of steam cracking operating years
+ Number of plant operating years with NOx
+ Number of plant operating years with effective NOx mitigation plans
+ Number of industry initiating events for a NOx event
+ Number of industry reported NOx events

With an understanding of the type of NOx compounds that accumulate and an estimate of the frequency of NOx events, a probability of loss of containment can be developed.

PROBABILITY OF HIGHER CONSEQUENCE EVENT DUE TO EXPLOSION

An estimate of the probability of a higher consequence event occurring by just the internal explosion of the NOx compounds can be developed by:

+ Understanding the types of NOx compounds accumulated in the cold box
+ Estimating the amount of NOx compounds accumulated in the cold box
+ Assessing the frequency of higher consequence NOx events in industry

Understanding the types of NOx compounds was discussed above.

Estimating the amount of NOx compounds can be based on the amount NOx in the cold box feed and the time since the previous cold box NOx wash or warming. The amount of NOx in the cold box feed can be determined with an on-line analyzer or spot samples. The amount that accumulated in the cold box can be estimated from analyzing the spent alcohol from previous NOx cold box washes. If data for a specific plant does not exist, an estimate can be developed by either benchmarking with other plants that have data or by using data that has been published in literature.

The frequency of higher consequence NOx events in industry can be estimated by reviewing the number of industry reported NOx events and their severity.

With an understanding of the type of NOx compounds that accumulate, an estimate of the amount of NOx compounds in the cold box, and an estimate of the frequency of higher consequence NOx events, a probability of a higher consequence event due to just the explosion can be developed. This approach is straight forward and does provide a probability, but the size of the underlying data set is relatively small.

PROBABILITY THAT HYDROCARBON WILL IGNITE AND RESULT IN A HIGHER CONSEQUENCE FIRE

Determining the probability that the hydrocarbon will ignite subsequent to a loss of containment and result in a higher consequence NOx event is essentially a typical process safety assessment. Potentially the only steam cracking specific consideration is the structure of the cold box, whether the cold box is free standing or is enclosed in a traditional cold box structure.

EVENT TREE VALIDATION AND TESTING MITIGATION EFFECTIVENESS

The structure of the event tree is now complete, and an initial assessment of the risk is available. At this stage of the risk assessment, it would be appropriate to test sensitivities of the frequency and probabilities used in the event tree. Validation of the event tree results after probabilities are assigned based upon the historical information is an important step in a semi-qualitative risk assessment. One needs to test to make sure that event tree scenarios with extremely low probability have not occurred numerous times in the historical data as well as checking that scenarios with higher expected probabilities have actually occurred in the data set. As was highlighted in the discussion above, there is less historical data on the success or failure of some of the branches in the event tree then would exist for failure of an instrument or failure to respond to an alarm that is typically used in risk assessments.

The next step in the risk assessment is to identify relevant prevention or mitigation steps to reduce the risk and to define the remaining risk after one or more of the identified risk reduction steps are considered.

The risk can then be plotted on the risk matrix and communicated to the organization.

CONCLUSIONS AND SUMMARY

Doing a semi-quantitative risk assessment of the process safety hazard associated with NOx in the cryogenic section of ethylene plants is possible. A NOx risk can then be compared to the sites other risks. In addition, a semi-quantitative risk assessment can be used to support the appropriateness of a site's NOx management plan. Specifically, it can be used to determine if there is a need to conduct cold box washes and add facilities to isolate heavy-diolefins from the colder sections of the cold box.

(153a) Integrating Gas Turbines with Cracking Heaters - Impact on Emissions and Energy Efficiency

Erwin Platvoet, Process Burner Engineering, John Zink Company LLC, Tulsa, OK

Turbine Exhaust Gas (TEG) contains high levels of oxygen, typically 15 vol%, due to gas turbine blade material temperature limits. As such it can be used as an oxidant for combustion in cracking furnaces and reformers. Its high temperature and flow rate results in a significant reduction of fuel consumption of the furnaces while increasing the high pressure steam production. The resulting high combined energy efficiency makes this an attractive option when energy costs are high.

This paper will discuss some of the theoretical considerations of NOx formation, and correlate differences between air and TEG firing with adiabatic flame temperature and oxygen concentration.

In order to validate emissions predictions an industrial-scale test was conducted at the John Zink Test Facility with two floor burners operating with TEG. The results were compared with ambient air tests and show good agreement with the predictions. They also give some insight if and when reburn of NO is significant.

AIChE Paper Number 153b

Development of a Differential Pressure Prediction For an Oil Quench Tower

Bernardo M. Corripio

Sr. Global Technology Manager

Quench Technology

The Dow Chemical Company

Prepared for Presentation at the 2012 Spring National Meeting

Houston, TX,

April 5th, 2012

Development of a Differential Pressure Prediction For an Oil Quench Tower

Bernardo M. Corripio

Hydrocarbons Sr. Global Quench Technology Manager

The Dow Chemical Company

Abstract

The fouling of oil quench tower internals is a recurrent problem for ethylene plants and it is important to monitor the rate of fouling so that mitigating action can be taken and tower cleaning can be planned. The most useful instruments for monitoring the fouling of the OQT are differential pressure measurements across some or all of the sections of packed beds or trays. When the measured pressure drop across a section of internals is higher than design and growing, this is usually due to fouling of that section. Unfortunately, normal variations in plant operation, including feed rates and feed slates, reflux flows, temperatures and pressures cause enough variation in the measured differential pressure that it can be very difficult to see an increase due to fouling until it is very large; this reduces the time available to plan a response to the fouling.

A useful tool for improved monitoring of fouling progression is a prediction for the differential pressure which accounts for variation due to process variables. An accurate prediction can be compared to the measured pressure drop across a section of internals to determine if there is a growing deviation between the measured and predicted values. By eliminating the normal process variation, fouling can be detected when it is less advanced and appropriate action taken.

This paper will describe the development of a regression equation for prediction of the differential pressure across a section of trays in an oil quench tower. Topics included will be the approach to selecting variables for the regression, a demonstration of its accuracy and a description of some success using the prediction and finally, some weaknesses of the equation.

Introduction

Dow operates multiple ethylene plants with oil quench towers (or primary fractionators) and has experienced fouling of the internals of these distillation columns, both trays and packing. This fouling has in the past led to operability problems due to high pressure drop and loss of separation in the column. These problems resulted in production and product yield losses, because the clearing of oil quench towers is time-consuming and in some cases plant turnaround intervals were shortened due to these operability issues.

One of Dow's plants experienced a serious fouling issue in the oil quench tower which was not initially recognized as fouling; the primary initial symptom was a high differential pressure (dP) across the trays of the column. However, because the dP measurement had high normal variation due to changes in cracking furnace feedslate, plant load and other process variables, and because fouling had never before been an issue for this distillation column, the dP was not seen as excessive until the fouling had significantly progressed and seriously impaired the operation of the plant.

Following a scheduled plant turnaround allowed for the cleaning of the oil quench tower trays, a regression model was built to predict the differential pressure across the clean trays. The intent was to track the deviation between the measured and predicted dP to provide early warning should the fouling recur. As will be described, the tray fouling did in fact recur; the dP model proved to be of sufficient quality and accuracy to allow early recognition and mitigation of the fouling. An unscheduled plant turnaround was avoided.

The emphasis of this paper is on the successful implementation of a regression model and its utility. While some discussion and tips for successful regression modeling are discussed, the main point is that this type of modeling can be accurate enough to be useful in the operation of a plant dealing with serious fouling of its oil quench tower and is meant as an encouragement to those fighting similar issues and the inherent difficulties in modeling this system.

Challenges in Modeling Oil Quench Tower dP

There are many difficulties to be overcome in the creation of a predictive model for the differential pressure across distillation column internals and there are unique challenges posed by an ethylene plant oil quench tower. These challenges include:

- Wide variation in feed flow to the tower
- Significant variation in tower pressure due to changes in cracked gas compressor suction pressure and feed flow
- Wide variation in composition due to large yield differences and dilution steam usage for different cracking feedstocks; a consequence of the composition changes are large ranges for other process variables such as temperature and reflux flow
- A large range of components in the feed stream, from hydrogen and methane to heavy fuel oil hydrocarbons
- Daily variation due to ambient temperatures where air or cooling tower water exchangers are used
- Significant differences in tray liquid holdup due to changes in process variables

The variation in some key process variables from the baseline period used to construct the plant model can be seen in the following charts. The model is required to give accurate prediction for significant ranges – up to 50% turndown in total feed flow to the column and in typical reflux flow, for example. In addition, the use of multiple feedstocks on the furnaces feeding this system results in dramatic changes in column feed composition; for this system, the furnaces will operate cracking 20-100% of one of the two primary feedstocks.

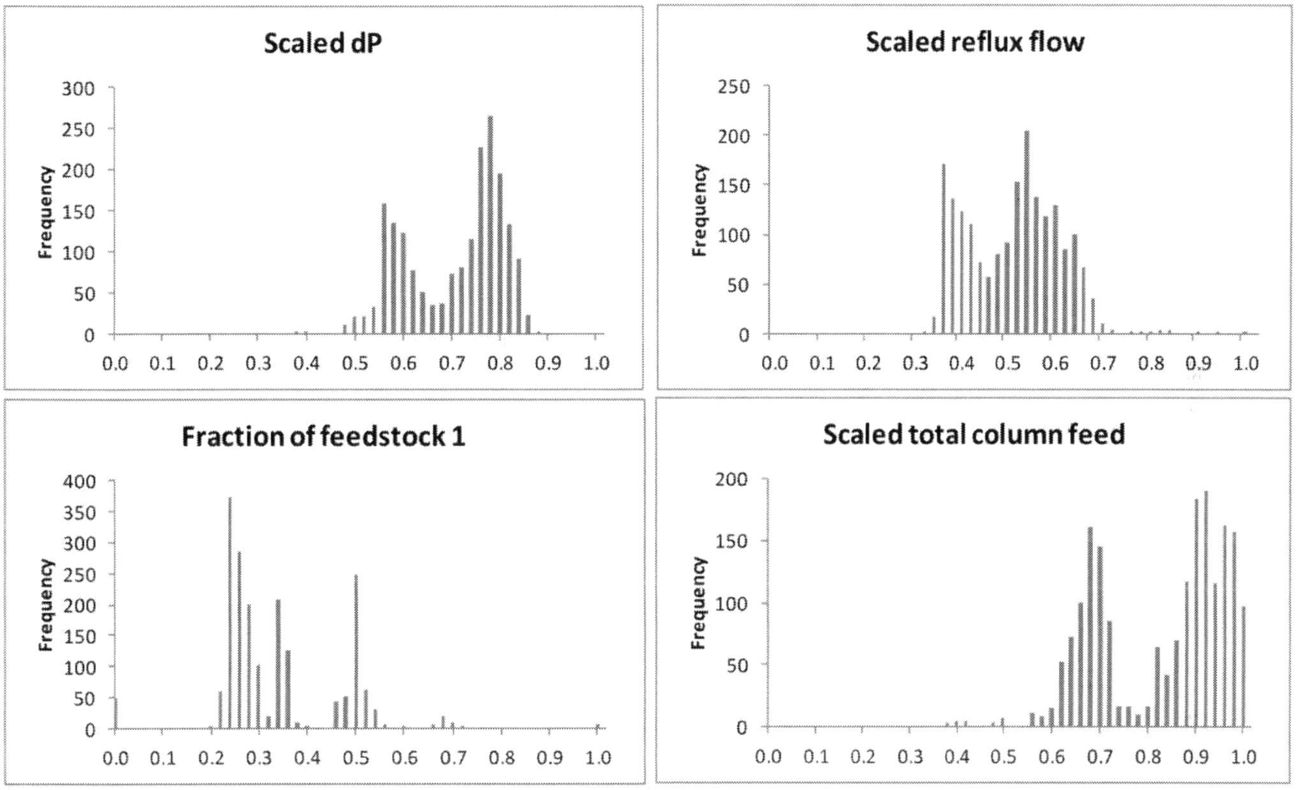

Figure 1: Range of differential pressure and key process variables during baseline period

A regression model was chosen for the differential pressure, primarily for the speed and simplicity of development; it also was not clear that the modeling effort would be successful or useful and so the significant investment in time and manpower required for a first-principles model was not justified. Regression models have weaknesses that must be accounted for when interpreting the model predictions; these will be discussed later in this paper. No attempt has been made to develop a first-principles model for the differential pressure of this column. Given that it was possible to develop a useful model of this quench oil system, however, it is possible that a first-principles model will be attempted in the future, primarily to overcome the maintenance problems of a regression model which will be described below.

Building the Regression Model

Selection and Linearization of Independent Variables

The key factor in the successful construction of an accurate regression model is the proper selection of the variables to be used in the regression. In order for the correlation to be valid and useful as a prediction, it is important that the variables used are independent to the greatest degree possible and that the quantity to be modeled truly depends on the chosen independent variables.

There are few truly independent variable in any plant (Friedman [1]) since the control system introduces dependencies between variables; for example, the differential pressure across the distillation trays of the oil quench tower clearly depends on both the vapor feed rate to the column and on the reflux flow to the top tray – however the reflux flow also has some dependence on the vapor feed rate to the column since the temperature the reflux is used to control will vary with the feed rate. For this work, variables were chosen that were known to have, from operating experience and engineering analysis, significant effects on the differential pressure.

Clearly there can be problems with the inclusion of either too many or too few variables. It is possible that variables which can significantly affect the dP measurement would not be included because they do not vary significantly during the baseline period and so appear not to affect the goodness of fit. In this case, it is likely that the prediction would be initially accurate but later unreliable once the excluded variables change due to a change in plant operation or control scheme.

If too many variables are included there are two potential problems: 1) it is possible that variables with no significant effect on the differential pressure would be included – should one or more instruments for these variables later drift significantly, this would introduce unnecessary error into the prediction; 2) it is possible to include a variable which is correlated with the modeled variable but not because the latter is dependent on the former but rather because both are dependent on the same independent variables; this could result in an artificially high correlation coefficient for the baseline data without accounting for all the important independent variables, resulting in a model with a high correlation but poor predictive power.

The preceding paragraphs emphasize the importance of using experience and engineering analysis to determine which variables are truly independent and have important effects on the modeled variable. A scattergun, brute-force attempt to build a regression model by randomly adding variables to improve the correlation will likely result in a model that predicts poorly, even if the initial match is good. A model with a large number of variables will also be more difficult to maintain and likely fall into disuse.

For this work, the variables selected were based on observation and analysis of the oil quench tower differential pressure over a period of months along with a very elementary engineering analysis that asked how the dP depends on different variables; additionally it was determined whether that dependence ought to be linear or non-linear and whether the variables are directly or inversely correlated. For distillation trays, per Kister [2], , the differential pressure is a sum of the dry tray pressure drop and the pressure drop through the aerated mass, so the key variables are those which determine vapor flow through the column and those which determine the liquid properties and froth height on the trays. Because the model is a regression and not a first principles model, quantities that do not change (e.g. tray configuration) need not be considered and the absolute values of the vapor and liquid flow properties (e.g. velocity and molecular weight) need not be absolutely determined. Instead readily available plant variables which determine the vapor and liquid flows and characteristics can be used. Thus feed mass flow, temperature, pressure, and cracked gas composition can be used as independent variables. In fact, it is not necessarily the case that a complete analysis of the cracked gas is required since what is critical is any change in the composition; thus variables which determine the feedslate might be sufficient to determine the cracked gas composition for the regression.

The model calculations were done using multi-linear regression. Obviously, some of the effects of the independent variables on the differential pressure are likely to be nonlinear. Where the engineering analysis indicated this should be the case, the data was transformed prior to performing the regression. For example, the vapor velocity is expected to have a second-order effect on the dry tray pressure drop (Kister [2]), so the baseline values of the variables which determine this velocity were squared for the calculation; the squared variables will then have a linear effect and be used in a linear regression. It is important to note that if the regression equation is later programmed to calculate the predicted differential pressure continuously, this same transformation must be done on the live data in the program.

Results of the Regression

The regression was executed using the multi-linear regression tool in the data analysis package in Microsoft Excel®. A total of six independent variables were used in the final equation. Some of these variables had nonlinear effects as described above and were transformed prior to executing the regression. Several other variables were regressed to determine if they improved the match and later rejected; also, some of the chosen variables were regressed with slightly different transformations to improve the linearization and the fit before selection of their final form.

A total of 1940 hourly average points formed the baseline data for the differential pressure and the six independent variables, capturing 80 days of operation after startup following tower and tray cleaning. The baseline period began with a plant startup following a complete cleaning of the oil quench tower and its trays. It was fortunate that during the baseline period there were significant periods of operation across almost the entire operating range of most of the variables of interest, some distributions of which can be seen in Figure 1; this led to a wide confidence region for the final equation.

The dP predicted by the regression equation agrees very well for the baseline period as evident from several measures. The correlation coefficient (adjusted R^2) was 0.964. The average absolute error between the predicted and measured dP was less than 2%. Only 28 of the nearly 2000 points used for the calculation differed from the predicted value by more than 10%. A plot of predicted versus measured dP values can be found in Figure 2.

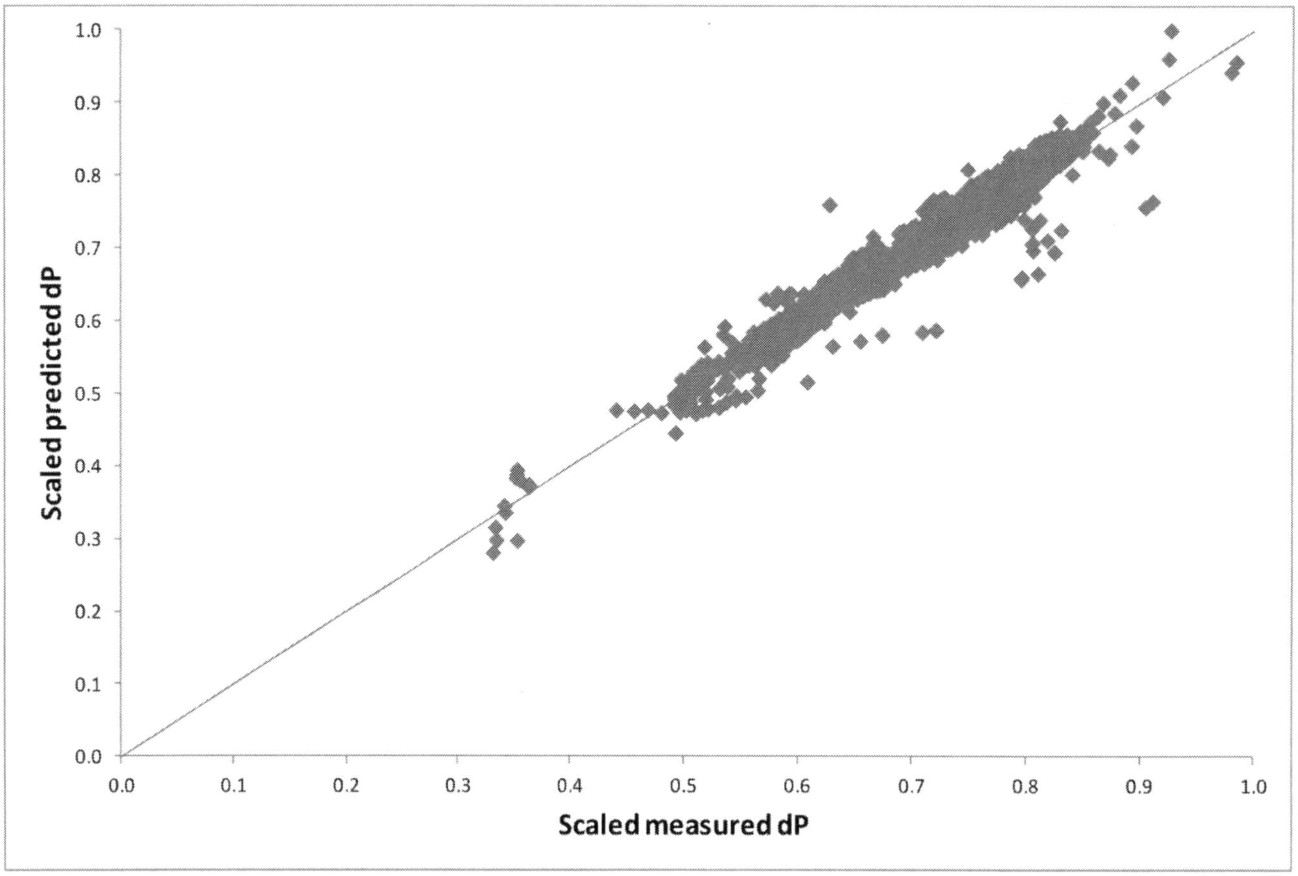

Figure 2: Agreement of predicted with measured dP during baseline period

Model Maintenance Requirements

Any process model requires some attention and maintenance to maintain its accuracy and user confidence in its output. One disadvantage of a regression-based model is that it generally requires more maintenance than a first-principles model (Friedman [1]). The fundamental physical and chemical relationships and the equations describing them used in a first-principles model rarely change, so a change to the physical process usually requires only minor adjustments. However for an empirical model this generally requires completely rebuilding the equation. In the case of the oil quench tower dP, for example, replacing the trays with a different design would require a repeat of the entire data collection and regression. In this example, the engineering analysis of the relevant independent variables would not change, but the existing equation would be useless.

In this work, while only six independent variables were ultimately used in the multi-linear regression, some of these independent variables were sums of multiple plant meters; data from a total of 35 plant instruments was used. In addition, some of the data used came from an online calculation performed by the plant process information system, so the model is also dependent on the ongoing integrity of that program.

In short, development of a model of this type can be exceedingly useful, as will be discussed in the following section, but it does require a commitment to identifying a model owner to ensure the maintenance of the model. Since the equation was programmed into the plant process information system and displayed to operations, it is also possible that it could do great harm in the future if it leads to false conclusions about the tower because of a lack of maintenance. The accuracy of the model led operating personnel to trust it, which is risky if the model should later prove to give inaccurate information.

Use of the Model by Operations to Identify and Manage Tray Fouling

The main intent in developing the predictive dP model was to provide a means of early detection should the tray fouling recur. This was accomplished by programming the regression equation into the plant process information system to display to operators and engineers the predicted differential pressure for any plant operating condition. The difference between the measured and predicted dP (delta-dP) was also calculated and both the predicted dP and the delta-dP were logged in the PI history. This allowed for trend display in the hope that, should fouling recur, it would be identified early enough to take corrective action.

In the event, fouling did in fact recur on the oil quench tower trays before the next turnaround. Figure 3 shows a trend of the measured and predicted dPs taken from the actual operator display. The data displayed show several weeks of data over the course of a fouling event as the two measures diverge – the difference between the two was believed and later confirmed to be caused by fouling of the trays.

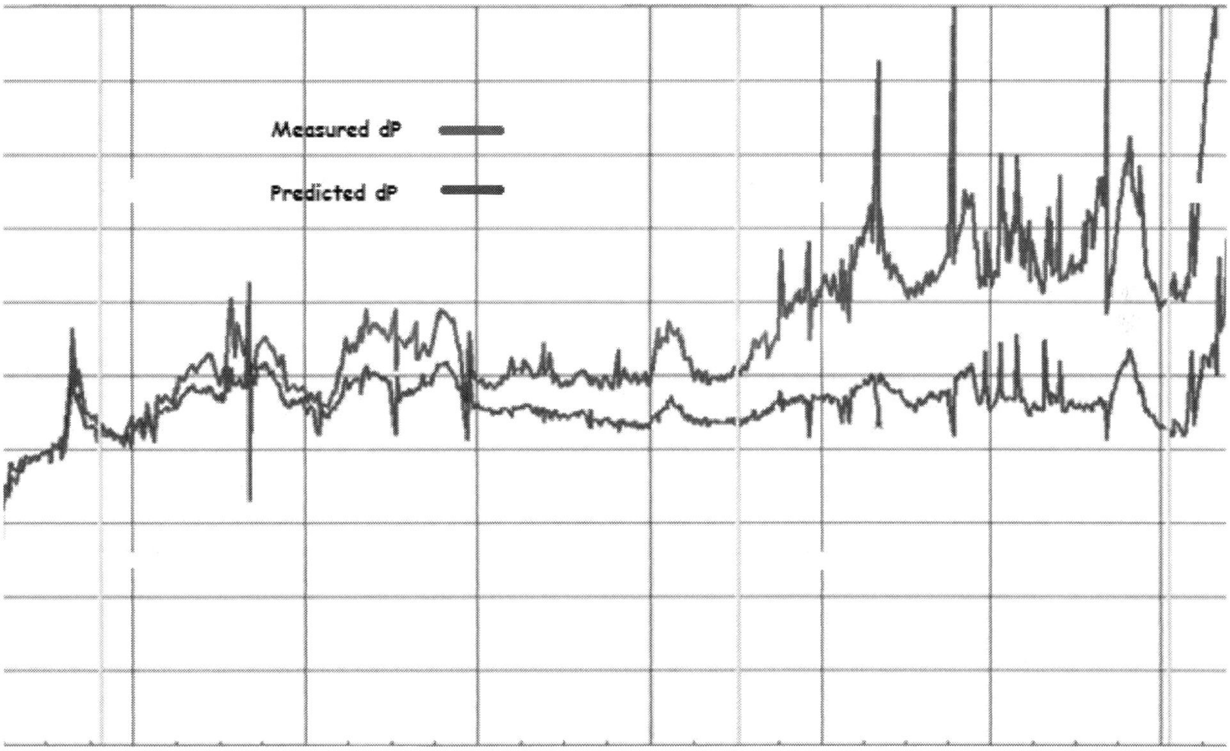

Figure 3: Operator display showing deviation of measured and predicted dP measurements

It was fortunate that, in this application, a method for online mitigation and partial reversal of the fouling had been identified that allowed operating personnel to take action to prevent the fouling from leading to a plant outage. The details of this method are beyond the scope of this paper and the details have not been approved for release. What is relevant to this discussion is that the method would not have been implemented in time to prevent a plant outage without the early warning provided by the regression model of the dP.

The reversal of the fouling by the method mentioned in the preceding paragraph could be seen in the deviation of the measured and predicted dPs. There were multiple occurrences of fouling and mitigation over the course of months preceding the next planned plant outage and quench clearing; each time, the two dP trends would deviate as the column fouled and then close as the mitigating steps were taken. Figure 4 displays these trends during a mitigation over the course of a few days. The measured dP can be seen coming on-scale from a tray flood condition and then actually dropping below the predicted dP, which was unexpected since it indicates that the model is inaccurate. The next spike in the measured dP indicates an instrument calibration, after which the measured dP is slightly above the predicted value, as expected; this was interpreted as the fouling having been partially, but not completely, removed. The dP had clearly drifted over time, which had the effect of masking some of the fouling. After this experience, the instrument was calibrated on a weekly schedule. The key conclusion from these data is that operators and engineers were able to monitor this deviation and plan the mitigation before it resulted in production losses or a plant outage.

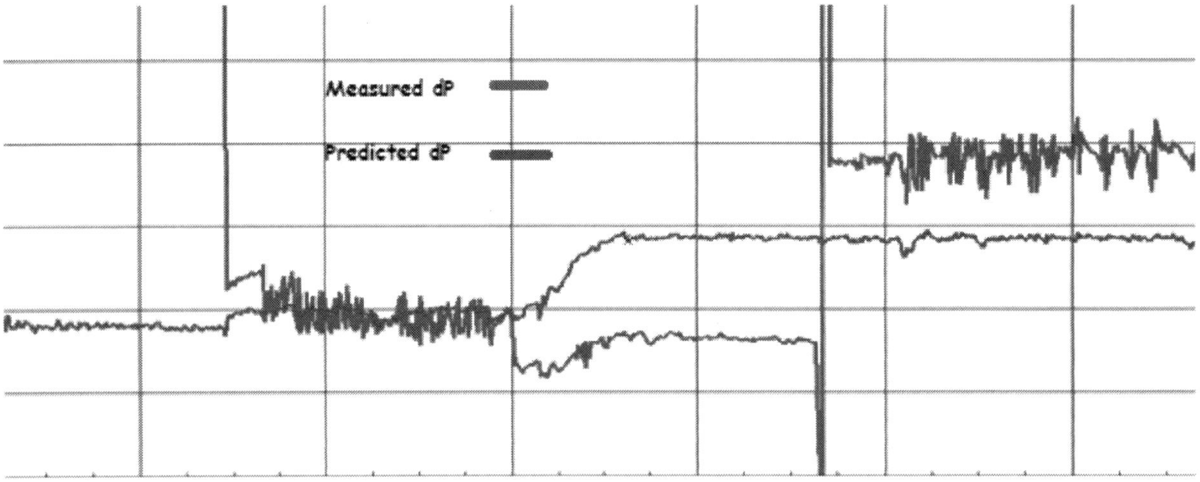

Figure 4: Operator display of fouling mitigation and instrument calibration

During the next outage, modifications were made to the system to reduce or eliminate future fouling; as these modifications included changes to some column internals, they required an update of the regression model as mentioned earlier in the discussion on model maintenance.

Summary

One of Dow's ethylene plants experienced tray fouling in an oil quench tower (primary fractionator), the first symptom of which was an elevated differential pressure across the trays. In order to monitor the tower and provide early detection of any future fouling, a prediction model for the differential pressure was developed using a multi-linear regression of key independent process variables. This model was accurate enough to provide real-time information to operators and engineers that was useful in detecting and managing the fouling problem.

There is nothing in the development of the regression which restricts its use to trays; the principles used in this work, especially the importance of engineering analysis and experience in determining which independent variables have a true influence on the dependent variable, are just as applicable to packed section differential pressure modeling. In fact, the experience gained from this work has been helpful in building or improving differential pressure models in some of Dow's other oil quench towers, including those with packed beds instead of trays.

The success of this effort should serve as encouragement to other ethylene producers experiencing similar problems – modeling such towers, while challenging, can be accomplished with reasonable time and effort and provide a useful tool for plant operation.

References

1. Friedman, Y. Zak (2006) "Choosing Inferential Modeling Tools", Hydrocarbon Processing, January, pp. 15-16.
2. Kister, Henry (1992) Distillation Design, New York, N.Y., McGraw-Hill

(153c) Furnace Stories

Ken Beasley, Consultant, Houston, TX

The cracking furnace is the heart of the ethylene plant, consuming the most energy, adding the most value and with the potential to cause the most damage. Every aspect of the furnace is critical from the specifications through design, construction and including commissioning, start-up, operations, maintenance and shutdown.

Unless the process side and the fire side of the furnace are kept in balance, problems can occur. Three furnace incidents are examined describing situations, causes and follow-ups. The importance of good maintenance, useful checklists and strong operating routines are stressed.

ASSURING ETHYLENE PURITY: RELIABLE SUB-PPM MEASUREMENT OF AMMONIA AND WATER WITH ONLINE TDL ANALYZERS

Xiang (Sherry) Liu
Principal Scientist

HsuHung(Steven) Huang
Senior RD Engineer

Kuan-Ting (Gary) Yeh
Senior RD Engineer

SpectraSensors, Inc.
11027 Arrow Route,
Rancho Cucamonga, CA

Wenhai Ji
Senior Scientist

Alfred Feitisch
VP Analyzers

William Jenko
Director Technical Support

Greg Lankford
Director Business Development

SpectraSensors, Inc.
4333 W Sam Houston Pkwy N
Houston, TX

Patrick M. Killough, Ph.D.
Senior Researcher
Projects and Technology,
Global Solutions Downstream
Shell Global Solutions (US) Inc.
Westhollow Technology Center
PO Box 4327
Houston, TX 77210

Wesley L. Carter
Analyzer Engineering Specialist
Projects and Technology
Process Automation, Control, and Optimisation
Shell Global Solutions (US) Inc.
Woodcreek, 200 N. Dairy Ashford Dr., Houston, TX 77079

KEYWORDS

Ethylene, Process Control, Product Purity, On-line Analyzers, Tunable Diode Laser (TDL) Analyzer, Ammonia (NH_3), Water Vapor (H_2O), SignalSelect^{TM}, SpectraLock^{TM}

ABSTRACT

Accompanying the ever-increasing performance of polymerization catalysts used to produce polyethylene (PE) is a general increase in catalyst sensitivity to impurities in ethylene feed. These can impact the length and branching of polymer chains as well as interchain bonding, which in turn determine important properties such as crystallinity. Furthermore, impurities may poison catalysts irreversibly, necessitating their premature replacement. Thus, the economics of polyethylene production do not merely depend on the cost of ethylene feed. The amount and quality of PE produced by a given amount of catalyst also is critically important, and the purity of ethylene feed directly impacts both.

Methods of measuring ethylene purity include gas chromatography for acetylene, ion mobility for ammonia, and a variety of sensors that indirectly respond to water including vibrating quartz crystals (VQC) and aluminum oxide (AlO_3) capacitance probes. Each of these suffer from limitations ranging from long response times (GC) to response drift, long wet-up or dry-down times, sensitivity to contamination, and loss of responsiveness due to poisoning of or physical/chemical changes in the sensor element (VQC and AlO_3).

Modern TDL (tunable diode laser) analyzer technology overcomes all of these shortcomings by measuring an intrinsic property of water, acetylene, or ammonia molecules: their response to light from a laser at specific wavelengths determined by the chemical bonding between constituent atoms. Being governed by the laws of nature, this effect does not change over time. And because the interaction is instantaneous and highly selective, TDL spectroscopy responds quickly to concentration changes and is not affected by contaminants. Employing a patented differential spectroscopy technique, the same technology employed to measure NH_3 and H_2O in ethylene has been deployed globally to measure H_2O at sub-ppm levels in dryer outlets for natural gas processing and liquefied natural gas operations.

INTRODUCTION

Ethylene is the most produced organic compound in the world. Real-time measurement of certain trace gas species during ethylene production processes is especially important for optimization of production efficiency, production throughput and for end product quality control. Trace pollutants such as ammonia (NH_3) and water (H_2O) need to be accurately monitored and controlled at very low concentrations (usually in the sub-ppm_v or even ppb_v level) in order to produce the highest quality poly-ethylene.

Traditionally, ppb_v level NH_3 measurement had to rely upon ion-mobility spectrometers, using radioactive materials, or upon GCs using an ionization detector. In the recent past, ion-mobility analyzers have been discontinued by the manufacturer. Ammonia GCs remain difficult and costly to use and continue to offer long measurement cycles. Trace H_2O measurement used to use vibrating quartz crystals and aluminum oxide capacitance probes. They suffer from response drift, long wet-up or dry-down times, sensitivity to contamination, and loss of responsiveness due to poisoning of or physical/chemical changes in the sensor elements.

Based on tens of years experience in tunable diode laser (TDL) based gas absorption spectroscopy [1-3], SpectraSensors started to develop the first TDL based sub-ppm_v H_2O analyzer in 2007 and ppb_v NH_3 analyzer in 2008 for ethylene purity monitoring and control applications [5]. These TDL analyzers overcome the shortcomings of the traditional techniques and provide superior sensitivity and repeatability, fast response times, operational robustness, long-term reliability under harsh conditions, no retention or wet-up/dry-down delays, no sensitivity or drift due to contamination from vapor impurities, no field calibration and low maintenance. These significant advantages have positioned TDL analyzers as the most desirable analytic technique for ethylene purity and other challenging petrochemical applications. [4-8]

This paper describes the fundamental design of TDL based NH_3 and H_2O analyzers, and the main improvements that have been implemented since 2007, including spectral line selection, patented SignalSelect™, SpectraLock™ and pressure-variation-compensation strategies, as well as analyzer configurations and validation options. Laboratory tests and field data are also presented to demonstrate the analyzer's excellent sensitivity, repeatability, measurement robustness, and fast response time.

SPECTROSCOPY DESIGN - LINE SELECTION

The absorption lines selected for the NH_3 and H_2O TDL analyzers support ppb_v level detection sensitivity, at atmospheric sample pressure and an absorption pathlength of 28m, contained in a Herriott type sample cell [9-10]. The two respective lines have been chosen because they are relatively isolated from background ethylene absorption interference. Selection of the laser lines for specific TDL applications is typically based on spectroscopy databases such as PNNL [11] and HITRAN [12] using selection routines previously developed by one of the authors [3]. Figure 1 shows the absorption spectra of ammonia and ethylene. In the 1.5-3.0 micron spectral region, the 2.9 micron band of NH_3 offers the strongest absorption, but diode lasers in this region are usually cost prohibitive and lack operational reliability. The 1.9 & 2.3 micron bands of NH_3 are both stronger than the 1.5 micron band, but the ethylene absorption in these two bands is too strong to allow sufficient absorption path length for ppb_v level detection. Therefore, the NH_3 absorption line was selected from the strongest NH_3 lines in the 1.5 micron band which offered minimal absorption from ethylene and the least amount of cross interference from ppm_v H_2O. The H_2O absorption line was selected from the 1850nm to 1900nm spectral region where ethylene has relatively low absorption.

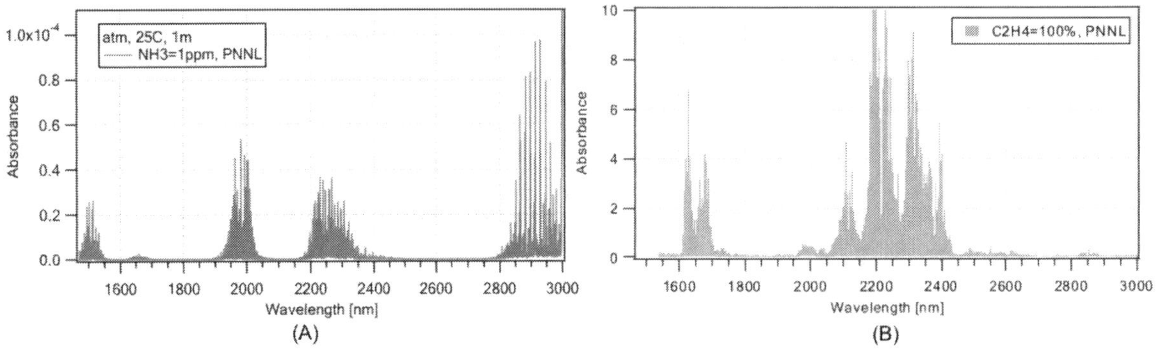

FIGURE 1. ABSOPTION SPECTRA: (A) AMMONIA; (B) ETHYLENE.

SPECTROSCOPY DESIGN – SIGNAL SELECT

However, even selecting those NH_3 and H_2O lines with the least amount of ethylene background interference still prevents any direct measurement of the target 2f signals from sub-ppm NH_3 and H_2O. As can be seen from the example in Figure 2(A), which compares the ethylene TDL 2f spectra with and without 1 ppm_v NH_3 contaminant, the ethylene background absorption dominates the 2f spectrum, making it impossible to detect the NH_3 contamination directly in the TDL spectra. The corresponding 2f spectrum for H_2O in an ethylene background depicts a basically similar situation but is not being shown in this publication.

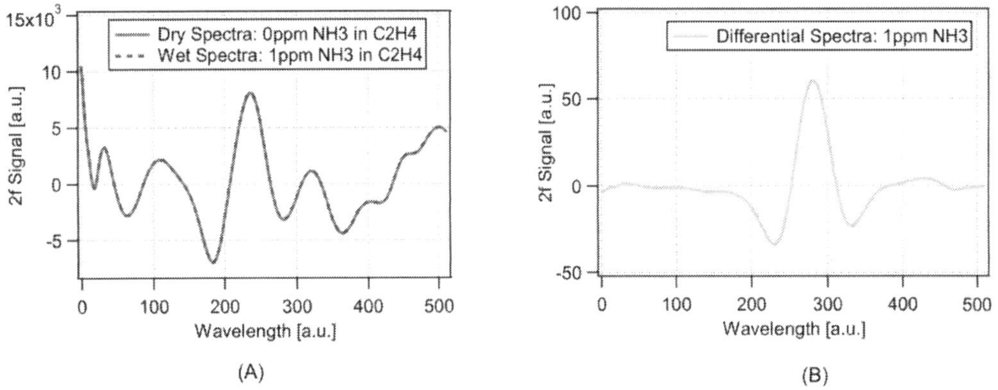

FIGURE 2. 2F SPECTRA OF TRACE NH_3 IN ETHYLENE STREAM: (A) DRY AND WET SPECTRA; (B) DIFFERENTIAL SPECTRA.

A patented spectral subtraction technique, "SignalSelect™" [13-14], has been developed to enable trace level measurement of reactive gases such as NH_3 and H_2O in applications with spectrally overshadowing background streams. As depicted in Figure 2(B), applying "SignalSelect™", enables the extraction and measurement of the unobstructed trace contaminant spectrum, in this example 1-ppm_v NH_3, from the background ethylene spectrum. The "SignalSelect™" spectroscopy technique works similarly for the detection of H_2O in ethylene. A schematic of the stream switching and stream conditioning which underpins SignalSelect™ is shown in Figure 3. Each measurement cycle starts with a so called scrubbing cycle as shown in Fig. 3(A). In this cycle, the sample stream flows through a high-efficiency solid-state scrubber which selectively removes either the trace NH_3 or H_2O, without altering the background ethylene stream itself. The 2f spectrum of the background ethylene stream, called "dry" spectrum, is recorded during the scrubbing cycle. Following recording of the "dry" spectrum, the analyzer automatically switches back to measuring the unmodified "wet" sample stream which contains ethylene and the trace NH_3 or H_2O, bypassing the scrubber as shown in Fig. 3(B). Subtraction of the recorded dry spectrum from the live wet spectrum generates a live, unobstructed differential spectrum of the trace NH_3 or H_2O, which is free of ethylene interferences, as illustrated in Fig. 2 (A) & (B). The concentration of the NH_3 or H_2O in the ethylene stream is then easily calculated from the live differential spectrum.

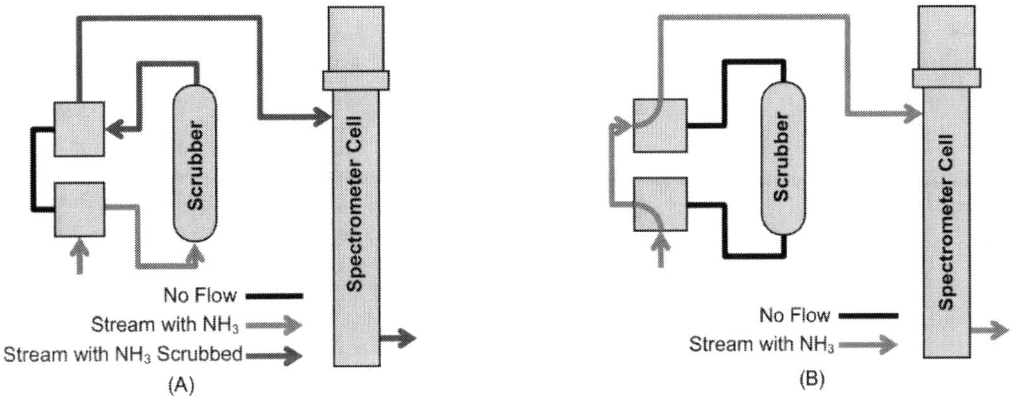

FIGURE 3. ILLUSTRATION OF THE STREAM SWITCHING FOR SIGNAL SELECT: (A) SCRUBBING CYCLE; (B) LIVE MEASUREMENT CYCLE.

SignalSelect™ is not only an efficient tool for TDL applications where trace gas spectra become obscured by components of the background sample stream but it also cancels common mode electronic and optical noise in the wet and dry spectra, which increases the measurement signal-to-noise ratio and improves detection sensitivity. The dry spectra update interval can be user programmed from minutes to hours, as required by the desired long term measurement repeatability. In addition, "SignalSelect™" can automatically compensate for any alteration of the actual 2f spectra due to changes in temperature, pressure or stream composition. The analyzer data algorithm monitors the 2f spectra real time and can automatically initiate dry spectra updates when the spectral distortion exceeds a user preset level. Applying the updated dry spectra to "SignalSelect™" then automatically compensates for any impact of ambient temperature or pressure changes, and/or any background stream composition changes. Although not applicable to ethylene purity monitoring, compensating for background composition changes is important for many TDL gas processing and petrochemical applications [7-8].

The scrubbers for NH_3 [15] and H_2O both employ solid state media which selectively remove either NH_3 or H_2O from the sample stream, by forming non-volatile, non-hazardous solid salts or binding the moisture, without altering the background ethylene stream. Both types of scrubbers are designed to last 2 to 10+ years under standard analyzer operating conditions. Scrubber cycles can be optionally deactivated, for example, when high concentrations of NH_3 or H_2O occur during process upsets, or when high levels of NH_3 or H_2O need to be measured. The analyzer data algorithm automatically calculates the scrubber consumption based on the measured NH_3 or H_2O concentration, the sample flow rate and the time usage of the scrubber. The analyzer will give an alarm when the scrubber approaches the pre-programmed percentage of its lifetime. A visual breakthrough indicator installed at the NH_3 scrubber exit will change its color upon contact with NH_3. The scrubbers can also be easily validated by feeding the analyzer with a known level of premixed NH_3 or H_2O in N_2 or ethylene backgrounds.

SPECTROSCOPY DESIGN – SPECTRA LOCK

SpectraLock™ is a patent pending enhancement of the basic "SignalSelect™" technology, which improves the analyzer operational robustness and its measurement fidelity with respect to its original factory calibration. Under certain circumstances, a spectral shift of the laser, due to electronic noise or temperature changes, can potentially lead to a distortion and shift of the differential spectra, during a single cycle of measuring sub-ppm_v NH_3 or H_2O levels in a spectrally dominating ethylene background. Applying SignalSelect™ directly to distorted and shifted differential spectra, without any correction, could lead to a concentration measurement error. Figure 4(A) compares the wet 2f spectra of trace NH_3 in ethylene background without any spectral shift with the spectra resulting from a laser shift in the order of $5X10^{-4}$ nm (0.5 count). Such a spectral shift could already create differential spectra with false peaks, as illustrated by Fig. 4(B). Using the distorted differential spectra directly for concentration calculations, the NH_3 analyzer would read 8.7 ppm_v, which is 1.3 ppm_v lower than the correct 10ppm_v NH_3 concentration! The proprietary SpectraLock™ algorithm (patent pending) detects spectral shift in the order of $1X10^{-5}$ nm (<0.01 count) and reconstructs the corresponding zero-shift spectra to ensure an accurate measurement. Engaging the SpectraLock™ algorithm, the NH_3 analyzer reads accurately at 10.0 ppm_v.

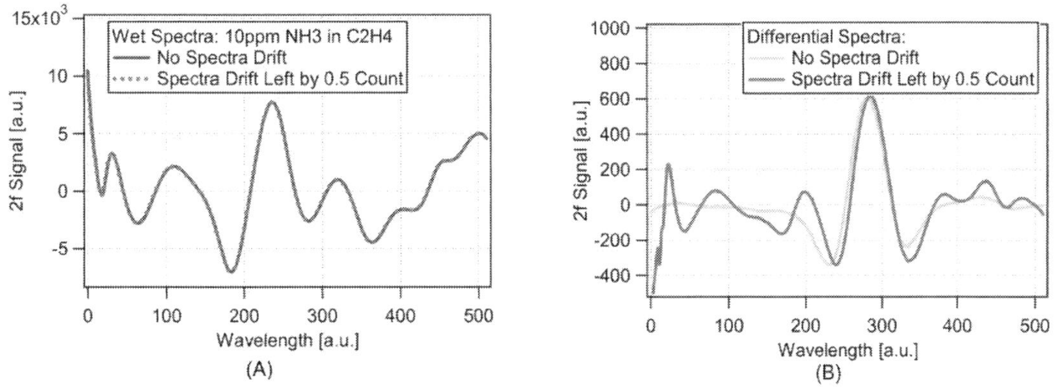

FIGURE 4. 2F SPECTRA OF TRACE NH_3 IN ETHYLENE STREAM WITH AND WITHOUT SPECTRA SHIFT: (A) WET SPECTRA; (B) DIFFERENTIAL SPECTRA.

SPECTROSCOPY DESIGN – P VARIATION COMPENSATION

A proprietary "pressure compensation" algorithm further enhances operational robustness and accuracy of the "SignalSelect^{TM}" analyzers by improving their tolerance to sample flow pressure variation during the measurement cycle. Figure 5(A) compares the wet 2f spectra of trace NH_3 in ethylene with only 10mbar pressure change. As can be seen, such a small pressure variation barely causes any visual change in the raw wet spectra, but it slightly changes the differential spectra amplitude and introduces erroneous peaks into the differential spectra as illustrated by Fig. 5(B). When using the distorted differential spectra for concentration calculation, the NH_3 analyzer would read 10.5 ppm_v, which is 0.5 ppm_v higher than the real NH_3 concentration. A proprietary algorithm ([16], patent pending) has been developed to reconstruct the no-pressure-change spectra using the measured raw 2f spectra, the resultant distorted differential spectra and the measured pressure values. Engaging the pressure compensation algorithm, the NH_3 analyzer reads accurately at 10.0 ppm_v.

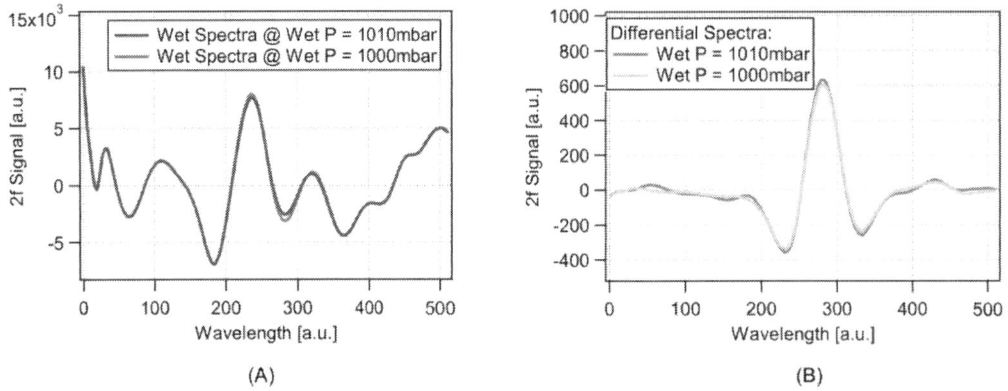

FIGURE 5. 2F SPECTRA OF TRACE NH_3 IN ETHYLENE STREAM WITH MINUTE P VARIATION: (A) WET SPECTRA; (B) DIFFERENTIAL SPECTRA.

SYSTEM CONFIGURATION & ANALYZER VALIDATION

Both, the NH_3 and the H_2O analyzer, have the sample handling system (SHS) and the spectrometer cell mounted in an insulated and electrically heated stainless steel enclosure. The temperature inside the enclosure is controlled to 50±0.1°C. Maintaining the sample stream at this elevated temperature improves measurement repeatability by minimizing micro-condensation of ethylene at flow orifices, by minimizing tubing wall adsorption and desorption of NH_3 and H_2O in the SHS, by stabilizing the optical alignment of the Herriott cell, and by eliminating any variation of the 2f signal with temperature. The layout of the SHS components inside the enclosure has been optimized to provide the best analyzer performance.

Both analyzers are designed to operate near atmospheric sample pressure so that the sample gas can be easily vented to flare. Operating the sample stream near atmospheric pressure also avoids the drawbacks arising from vacuum operation, such as susceptibility to actual and/or virtual leaks and pressure and modulation current variation.

Each NH_3 or H_2O analyzer is equipped with an integrated validation system which allows verification of the analyzer accuracy and reliability at the time of manufacture, at commissioning and throughout the life of the analyzer. The H_2O validation system is typically configured to use an on-board permeation tube, delivering about 1 ppm_v H_2O to the dried ethylene stream generated by the dryer, which is being used in the "SignalSelect™" stream switching system. Gas temperature and gas flow rate through the permeation tube are both tightly controlled to guarantee consistency of the generated H_2O level. As an option, the validation system can also be configured for use of premixed cylinders with trace NH_3 or H_2O in ethylene or nitrogen background. The analyzer can store several different calibration profiles generated at the factory to match premix cylinder balance gases. The analyzer has the ability to recognize the presence or absence of ethylene and will apply the appropriate calibration profile generated at the factory. Permeation tube based validation for NH_3 is presently under development, being site tested. To date, all site testing for NH_3 in ethylene has been accomplished by using NH_3 in nitrogen premixed cylinders. Testing for NH_3 in ethylene was done at the SpectraSensors laboratories where the mixtures were generated using mass flow controllers.

LAB TEST RESULTS

Figure 6 shows the concentration step test results of a trace NH_3 analyzer for ethylene purity monitoring. The superior linearity of the concentration measurement is being illustrated by the inset plot of the averaged analyzer reading on each step versus the real NH_3 concentration. The reading repeatability on each step ranges from ±14 to ±27 ppb_v over the entire 0 – 5 ppm_v range. Here, repeatability is defined as twice the standard deviation of the analyzer readings (2σ), which corresponds to a 95% confidence interval. The repeatability of the concentration measurement can be improved by applying moving averages to the raw readings shown in Fig. 6. The response time of the raw readings shown in Fig. 6 ranges from 51s – 68s, most of which is due to the sample transport time through the sampling lines and the spectrometer cell. Since the analyzer always updates the concentration measurement every 17s, irrespective of any data averaging, a data post processing algorithm, rapid change monitoring (RCM), can apply moving averages to the raw readings when no rapid changes are being detected, while automatically returning to non-averaged data display when reading change rates exceed a preset value. With the RCM algorithm enabled, the analyzer can deliver single digit ppb_v repeatability readings during stable process conditions, while still supporting the fast 51s – 68s response time for any NH_3 concentration change faster than a preset value.

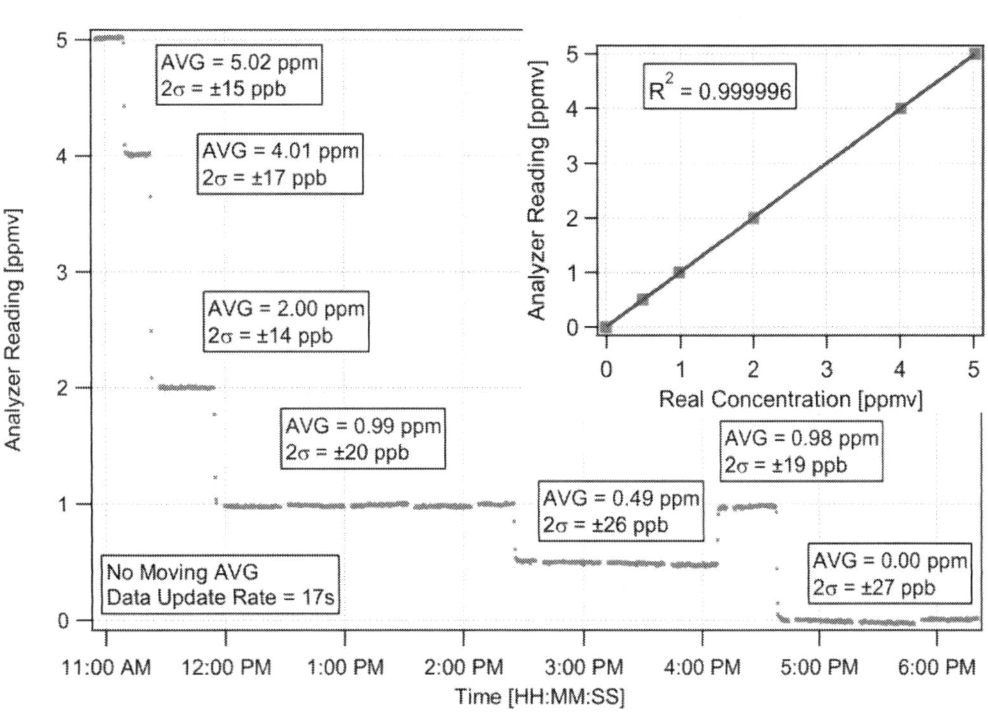

FIGURE 6. STEP TEST RESULTS OF THE TRACE NH_3 ANALYZER.

Figure 7 zooms in the 2.5 hour long 1 ppm_v test result shown in Fig. 6. The repeatability will be improved by 20% by applying a moving data average of 8. Even without moving average, the analyzer performs well within its reading repeatability spec of ±50 ppb_v, which is highlighted by the green bar. It should be noted that the measurement cycle for these lab tests was set to about half an hour in order to examine the cycle to cycle reading repeatability, which is dominated by the SHS design. During field operation, the measurement cycle can be extended to support less blind time and even better measurement repeatability.

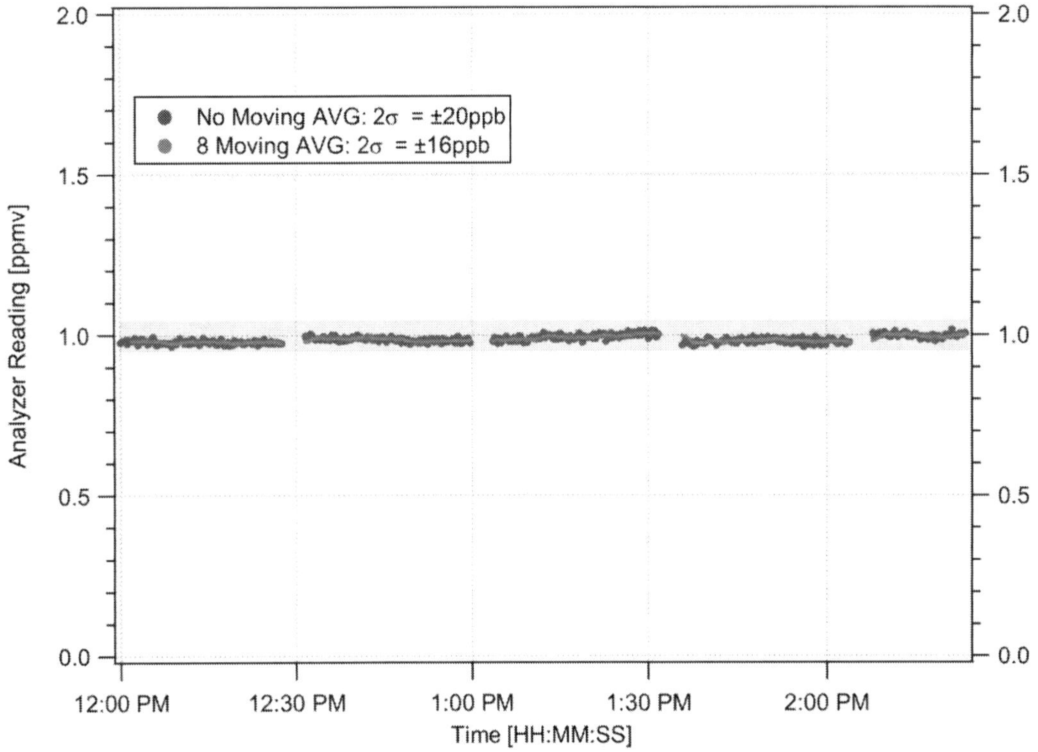

FIGURE 7. LONG TERM 1 PPM_V TEST RESULT OF THE TRACE NH_3 ANALYZER.

The trace H_2O analyzers for ethylene purity monitoring perform similarly to the trace NH_3 analyzers. Figure 8 shows the long term 0.5 ppm_v test result of a H_2O analyzer. The analyzer performs well within its reading repeatability spec of ±200 ppb_v, as highlighted by the green bar.

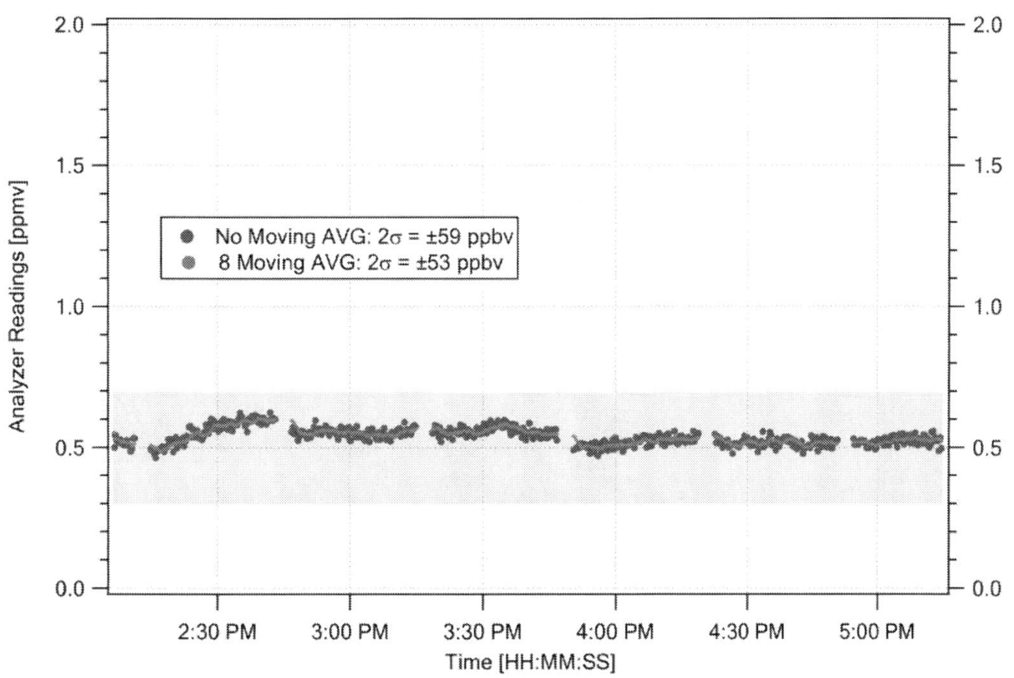

FIGURE 8. LONG TERM TEST RESULT OF THE TRACE H_2O ANALYZER.

FIELD OPERATION RESULTS

Site test data for NH_3 was done at Shell Global Solutions (US) Inc.'s Westhollow Technology Center with a Kin-Tek® permeation tube oven configured with an additional dilution module to generate NH_3 in nitrogen at sub-ppm$_v$ levels. Figure 9 shows the 2f differential spectra generated for 3 ppm$_v$ to 117 ppb$_v$ NH_3.

Figure 10 isolates the 117 ppb$_v$ spectrum and shows that there is sufficient signal to noise to reach an approximate 25 ppb$_v$ limit of detection.

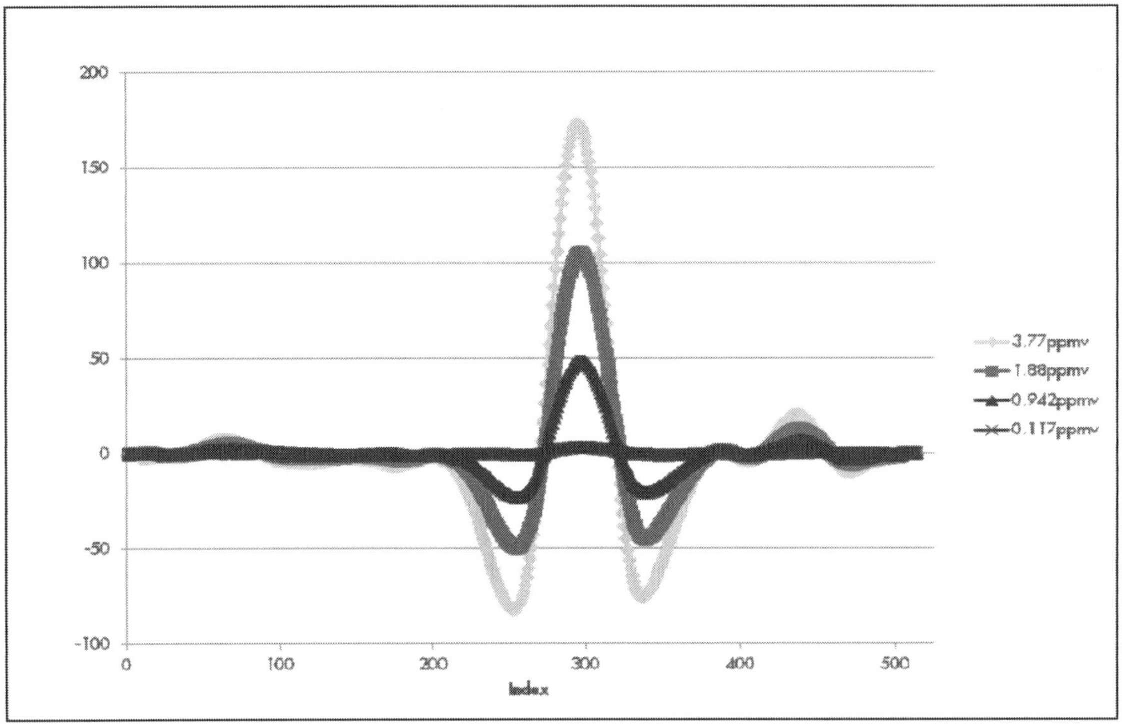

FIGURE 9. 2F DIFFERENTIAL SPECTRAL RESULTS FOR NH_3 IN NITROGEN

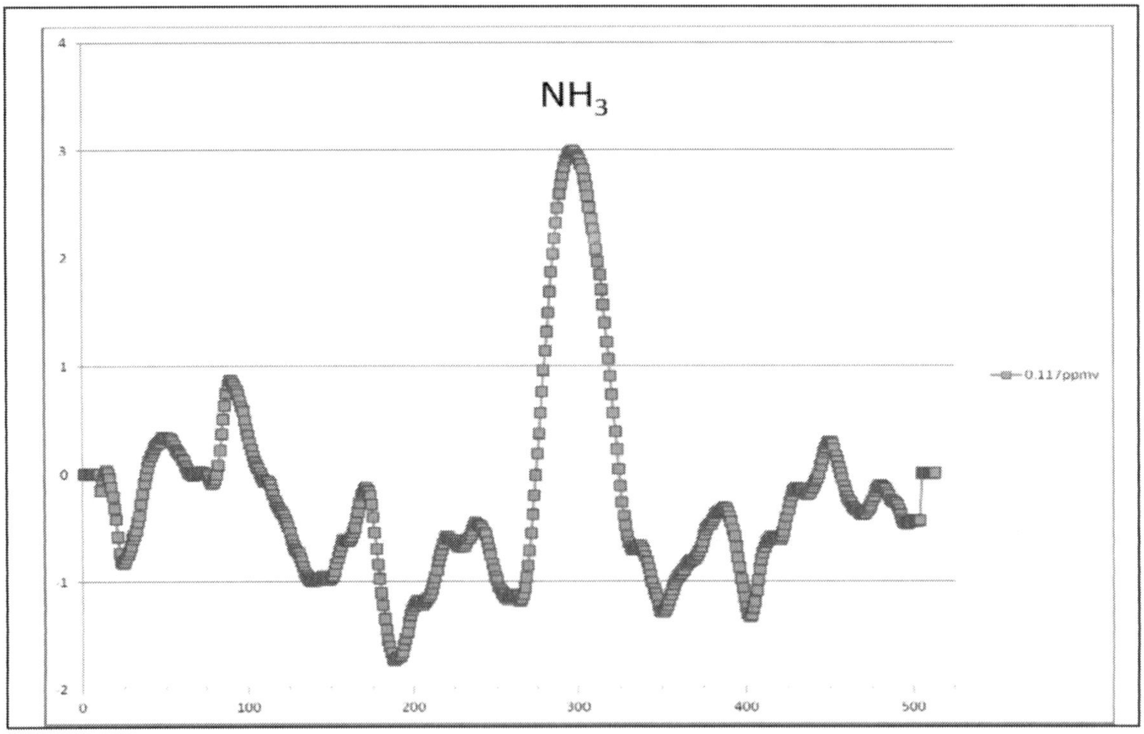

FIGURE 10. 2F DIFFERENTIAL RESULTS FOR 117 PPB_V NH_3 IN NITROGEN

CONCLUSIONS

Real-time measurement of trace pollutants such as NH_3 and H_2O in ethylene streams is a critical and challenging task for the production of high quality ethylene. TDL based analyzers developed by the authors in recent years have demonstrated superior detection sensitivity, response times and long-term reliability in measuring sub-ppm_v levels of NH_3 and H_2O in ethylene sample streams. This paper describes the fundamental design of TDL NH_3 and H_2O analyzers for ethylene streams and advances that have been implemented over recent years to improve operational robustness and measurement accuracy. These advances include spectral line selection, patented SignalSelect™ spectroscopy, SpectraLock™ and pressure-variation compensation, and permeation and pre-mix based validation capability. The discussed laboratory test results demonstrate that the TDL NH_3 analyzers can deliver measurement repeatability of much better than ±50ppb_v over the full range of 0-10ppm_v. The TDL H_2O analyzers can deliver measurement repeatability of much better than ±200 ppb_v over the full range of 0-10ppm_v. The field operation results demonstrate that the TDL NH_3 analyzer does in fact perform lower than ±50ppb_v. The results presented in this paper underscore that the described TDL gas analyzers can provide reliable sub-ppm_v measurements suitable for monitoring and control of ethylene purity. By far the biggest challenge in the deployment of the NH_3 and H_2O TDL analyzer systems into an operating facility has been the development of credible field validation techniques. This study demonstrates that these field validations are possible using NIST traceable standards.

ACKNOWLEDGEMENTS

The authors thank Adam Chaimowitz, John Lewison, Lutz Keller, Nathan St John, Keith Helbley and other colleagues for their contributions to the development and testing of all these analyzers. Shell Global Solutions gratefully acknowledges William Jenko's contributions to the successful field testing of these instruments.

REFERENCES

1. May, R., "System and method for water vapor detection in natural gas", US patent 6657198B1, Dec. 2, 2003.
2. Liu, X., Jeffries, J.B., Hanson, R.K., and etc., "Development of a tunable diode laser sensor for measurements of gas turbine exhaust temperature," Applied Physics B, 82, pp. 469-472, 2006.
3. Zhou, X., Liu, X., Jeffries, J.B., and Hanson, R.K., "Development of a sensor for temperature and water vapor concentration in combustion gases using a single tunable diode laser", Meas. Sci. Technol. **14**, pp. 1459-1468, 2003.
4. Liu, X., Zhou, X., Feitisch, A. and Sanger, G., "Tunable diode laser absorption spectroscopy based trace moisture detection in natural gases", 52nd Analysis Division Symposium of the Instrumentation, Systems and Automation Society, Houston, TX, April 15-19, 2007.
5. Liu, X., Zhou, X. and Feitisch, A., "Tunable diode laser analyzers for ethylene production and quality control", 53rd Analysis Division Symposium of the Instrumentation, Systems and Automation Society, Calgary, Alberta, CA, April 20-24, 2008.

6. Liu, X., Zhou, X., Ji, W. and Feitisch, A., "Advanced NH_3 and CO_2 TDL gas analyzers for petrochemical process control and product qualification", 54th Analysis Division Symposium of the Instrumentation, Systems and Automation Society, Houston, TX, April 20-23, 2009.
7. Liu, X., Ji, W., and Feitisch, A., "Development of H_2S, H_2O, NH_3 and C_2H_2 TDL analyzers for petrochemical applications in optically interfering hydrocarbon streams", 55th Analysis Division Symposium of the Instrumentation, Systems and Automation Society, New Orleans, LA, April 25-29, 2010.
8. Liu, X., Ji, W., Feitisch, A., Jenko, B. And Trygstad M., "Advancing TDL Technology: Applied spectroscopy in service of process control objectives", 56th Analysis Division Symposium of the Instrumentation, Systems and Automation Society, Houston, Texas, April, 2011.
9. Herriott, D., Kogelnik, H. and Kompfner, R., "Off Axis Paths in Spherical Mirror Interferometers", Appl. Opt. 3, pp. 523-526, 1964.
10. Herriottt, D. and Schulte, Harry J., "Folded Optical Delay Lines", Appl. Opt. 4, pp.883-887, 1965.
11. Sharpe, S.W., Johnson, T.J., Sams, R.L., Chu, P.M., Rhoderick, G.C., and Johnson, P.A., "Gas-Phase Databases for Quantitative Infrared Spectroscopy", Applied Spectrosc. 58, pp. 1452-1461, 2004. https://secure2.pnl.gov/nsd/nsd.nsf/Welcome.
12. Rothman, L.S., Gordon, I.E., Barbe, A., and etc., "The HITRAN 2008 molecular spectroscopic database", J. Quant. Spectrosc. Radiat. Transfer. 110, pp. 533-572, 2009. Available at http: //cfa-www.harvard.edu/hitran/.
13. Zhou, X., Liu, X., Feitisch, A. and Sanger, G., "Reactive gas detection in complex backgrounds," US Patent # 7,704,301, Apr. 27, 2010.
14. Zhou, X., Liu, X., Feitisch, A. and Sanger, G., "Reactive gas detection in complex backgrounds," US Patent # 7,819,946, Oct. 26, 2010.
15. Baum, M.M., Moss, J., Zhou, X., Kwan, A., Feitisch, A., Liu, X., "Scrubber for reactive gases," US Patent # 7,829,046, Nov. 9, 2010.
16. Liu, X., Zhou, X., Feitisch, A. and Sanger, G., "Pressure invariant trace gas detection," US Patent # 7,508,521, March 24, 2009.

153e

Emissivity of Cracking Furnace Coils, Novel Temperature Measurement Solutions for Cracking Furnaces

Ben Brown

Sr. Applications Development Engineer

BASF Corporation

Dane DeRouen

Furnace Engineer

BASF Corporation

Dave Sankey

Furnace Specialist

BASF Corporation

Prepared for Presentation at the 2012 Spring National Meeting

Houston, TX, April 1-5, 2012

Emissivity of Cracking Furnace Coils, Novel Temperature Measurement Solutions for Cracking Furnaces

Ben Brown

Sr. Applications Development Engineer

BASF Corporation

Dane DeRouen

Furnace Engineer

BASF Corporation

Dave Sankey

Furnace Specialist

BASF Corporation

Abstract: Accurate measurements of skin temperature on cracking furnace coils have become necessary for improved process control. Mechanical and thermal constraints of furnaces have made infrared measurements necessary, but concerns about inaccuracies of these instruments have surfaced. The perceived inaccuracy has been driven in large part by an uncharacterized and misunderstood emissivity value. BASF Temperature Sensing, in partnership with BASF FINA Petrochemicals LP, has researched the emissivity of typical furnace coils. The surfaces of sample coils were treated to simulate three stages of coil aging. The study was performed using three pyrometers at different near infrared wavelengths.

In addition to the test results, a brief description of emissivity, its dependence on wavelength, and its impact on infrared temperature measurements will be presented. Important considerations for measuring skin temperatures of furnace coils will also be discussed.

Introduction

Optical Temperature Measurements

Pyrometry has been in use since the 18th century in some form or another. A famous potter, Josiah Wedgewood, is credited with inventing the pyrometer in the 1780's to determine the temperature inside his kilns. This technique uses an element placed inside the kiln, which glows a known color at given temperatures.

The first modern optical pyrometer was developed in 1901 by L. Holborn and F. Kurlbaum. This invention is a disappearing filament pyrometer. In this device, a thin filament is viewed by an operator with a high-temperature process in the background. The electrical current through the filament is adjusted until it glows with the same intensity and color as the background, and thus "disappears". The process temperature is derived from the electrical current value. This is a subjective technique. It proves repeatable with a given operator, but not reproducible between operators.

Figure 1: Historic Disappearing Filament Pyrometer

With advancements in electronics technology, pyrometers have become less subjective by using a thermopile or photodiode as the sensing component. The current across a photodiode, for instance, changes relative to the intensity of the light to which it is exposed. The current is proportional to the intensity of light at a certain color range detectable by the photodiode.

The purpose of a pyrometer is to provide a non-contact method to collect the energy from thermal radiation, or irradiance, of a target surface and calculate its temperature. A pyrometer has three primary components: a detector, an amplifier circuit, and an optical system. The optical system collects the thermal radiation from the target and images the radiation onto a detector.

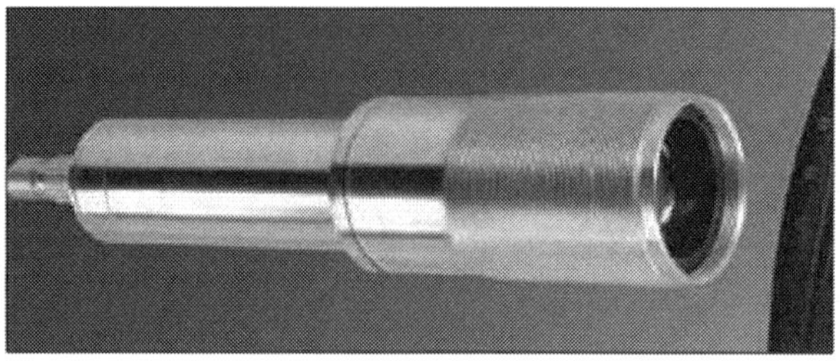

Figure 2: *EXACTUS®* Digital Temperature Sensor by BASF

The amplifier circuit is used to translate the irradiance (j) of the object into a temperature value (T). In its simplest form, this is achieved by using the Stefan-Boltzmann law, described by **Equation 1**.

$$j = \varepsilon \sigma T^4$$

Equation 1: Stefan-Boltzmann Law

In this equation, **j** is the irradiance; **ε** is the emissivity of the object (discussed later in this document); **σ** is the Stefan-Boltzmann constant, or proportionality constant; and **T** is the temperature.

The Stefan-Boltzmann law, however, is a generalization, and it includes all of the emitted energy from an object. In reality, however, all of the energy from an object can not be detected. Only specific colors, or wavelengths, of light can be detected. The broad range of usable wavelengths of light is described in the **Figure 3**.

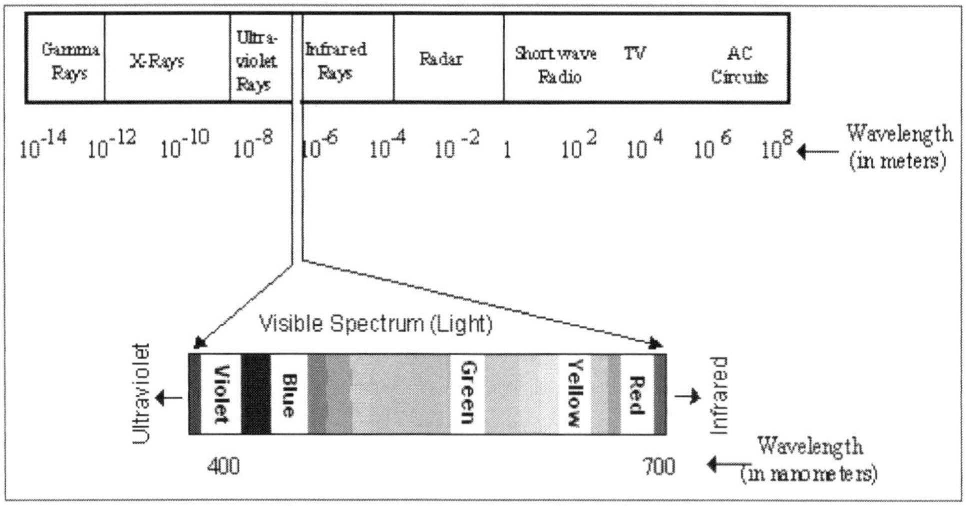

Figure 3: Wavelengths of Light

Max Planck took this a step further, and he segmented this relationship into individual colors, or wavelengths, of light. Planck's law can be written in the form denoted by **Equation 2**.

$$I'(\lambda, T) = \frac{2hc^2}{\lambda^5} \frac{1}{e^{\frac{hc}{\lambda kT}} - 1}$$

Equation 2: Planck's Law

In this equation, **I'** is the irradiance; **h** is the Planck constant; **c** is the speed of light; **λ** is the wavelength of light; **k** is the Boltzmann constant; and **T** is the temperature. The chart in **Figure 4** illustrates this relationship at given temperatures. The x-axis represents wavelength, or color, of light, and the y-axis represents the irradiance. The units for irradiance are kept arbitrary for simplicity. Irradiance is the energy per unit time radiated per unit area of emitting surface in the normal direction per unit solid angle per unit frequency by a blackbody.

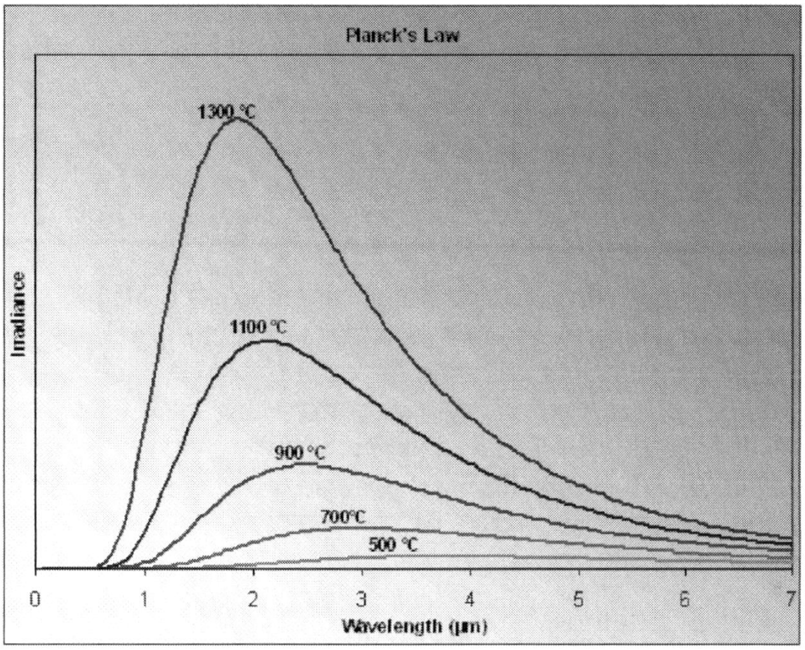

Figure 4: Planck's Law Charted

As illustrated in the chart, light emitted from a heated body varies with wavelength. By using a detector that is only sensitive to a distinct wavelength or range of wavelengths, the temperature can be approximated more easily.

Planck's law describes the light energy emitted from a blackbody at absolute temperatures. A blackbody is an object that is perfectly black. That is, it absorbs all light energy that falls on it, and it reflects nothing. Although closely approximated, a perfect blackbody does not exist in nature.

Emissivity Fundamentals

While a blackbody approximation of optical temperature measurement works in some cases, the limiting factor is emissivity. Emissivity, simply put, is a body's ability to emit energy by radiation. The value of emissivity, denoted by **ε**, is simply a ratio of the energy emitted by an actual surface at a given temperature divided by the energy emitted by a blackbody at the same temperature.

$$\varepsilon = I_{actual}/I_{blackbody}$$

Equation 3: Emissivity

The emissivity value for a perfect blackbody would be 1, but all surfaces in nature have a value less than 1. For instance, charcoal has an emissivity of approximately 0.96 for common wavelengths. If charcoal is heated, it glows noticeably bright. Polished gold, however, has an emissivity of approximately 0.02 for the same wavelengths. If gold is heated, it glows dimly, and its appearance does not change noticeably until it is exposed to a significant amount of heat.

Emissivity is also related to reflectivity. If a surface is reflective, or "shiny", the emissivity value is low. The relationship for an opaque material can be described simply as **Equation 4**, where **ρ** is reflectivity.

$$\varepsilon + \rho = 1$$

Equation 4: Emissivity and Reflectivity Relationship

This relationship is important for Tube Metal Temperature (TMT) measurements. When measuring the temperature of a coil inside a hot furnace using an optical device, reflectivity must be taken into consideration. For example, if a coil has an emissivity of 0.8, according to **Equation 4**, 80% of the light measured is emitted from the surface of the coil, but 20% of the light measured is reflected from its surroundings. In a cracking furnace, the problem is compounded by the fact that the surroundings are usually at a higher temperature than the coils. In this scenario, the background must be subtracted. To perform this calculation correctly, the emissivity value must be very well known.

To complicate matters further, emissivity changes with detection wavelength and, often, temperature. The exact emissivity value of a target is difficult to know, which can be a limiting factor for traditional pyrometry. Several techniques are used to overcome emissivity uncertainties, but using an accurate emissivity value is the best-case scenario for optical temperature measurements.

Emissivity of Cracking Coils Experiment

Background

Temperature measurements of cracking coils are critical to process control and safety for ethylene furnaces. Measurement points such as Tube Metal Temperature (TMT) and Coil Outlet Temperature (COT) are among the most critical for this process. Given the thermal and mechanical constraints inside the furnace, these measurements benefit from non-contact methods such as infrared pyrometry. Pyrometers also allow for fast response to temperature changes and alarm conditions. To ensure accurate measurements, the emissivity value of the coil surface must be known as precisely as possible.

The cracking coils are normally constructed of an alloy formulated to perform well in high-temperature, oxidizing environments. The coil surface is initially shiny, with a relatively low emissivity, when manufactured. Once in the furnace environment, the coils quickly become oxidized and the emissivity starts to rise. Over the lifetime of the coil, the surface of the coils changes noticeably from a lighter to a darker color. This perceived change of emissivity over time is a large concern for optical measurements of cracking coils.

Several types of hand-held and fixed-spot pyrometers are used to measure coils at regular intervals during the cracking process. One of the most popular wavelengths for measuring coils is 3.9 µm. This wavelength is selected to enable the pyrometer to view objects through flames, since this wavelength avoids the emission and absorption wavelengths in common burner gasses. However, it often proves unreliable in practice to view coils through flames when present. Another concern with using a 3.9 µm wavelength is the emissivity error dependence. **Figure 5** shows the temperature error at a given coil temperature when a 10% error in emissivity value is entered into the calculation for different common wavelengths.

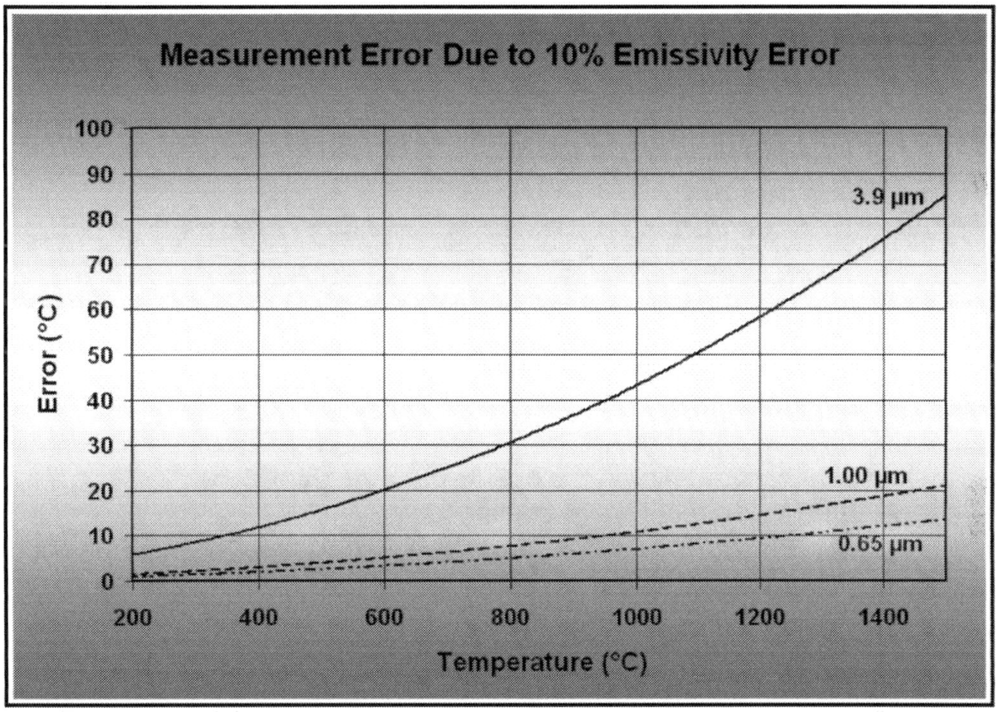

Figure 5: 10% Emissivity Error – Temperature Error by Wavelength

As an example of the 10% error illustrated in **Figure 5**, a coil has an actual emissivity of 0.9. The temperature of the coil is measured by a pyrometer using an emissivity value of 1 to calculate the temperature. In this scenario, a coil at an actual temperature of 800 °C would measure approximately 769 °C using a pyrometer operating at a wavelength of 3.9 µm, yielding an error of roughly 31 °C. However, the same conditions would measure approximately 792 °C using a pyrometer operating at wavelength of 1.0 µm, yielding an error of roughly 8 °C.

Design

The purpose of this experiment is to investigate thoroughly the emissivity value of cracking coils. Several measurements are made of coils at varying oxidation levels, wavelengths, and temperatures. The study includes measurements of coil surfaces at two different aging conditions to study the effect of aging on the emissivity value. Three different wavelengths; 0.65 μm, 0.9 μm, and 1.55 μm; are utilized to analyze the effect of wavelength on emissivity. The study also includes three different temperatures, 600 °C, 800 °C, and 1200 °C. These temperatures were recommended by BASF FINA Petrochemicals as typical operating temperatures of cracking coils.

Sample Description

The samples for the experiment were provided by BASF FINA Petrochemicals LP in Port Arthur, Texas. The samples were sections of a used coil, recently extracted from a furnace for replacement. The coil is constructed of 35/45 CrNi, a centrifugally cast Chromium Nickel alloy, typical for ethylene cracking processes.

Six samples are extracted from the coil sections. The width of each sample is 0.5 inch square and the thickness remains the wall thickness of the coil. The top surface of three samples is ground to simulate a new coil surface. **Figure 6** shows three samples, labeled 1, 2, and 3. Sample 1 is a ground sample, Sample 2 is a ground sample after soaking at 500 °C for 30 minutes, and Sample 3 is untreated, maintaining the original surface of the used coil.

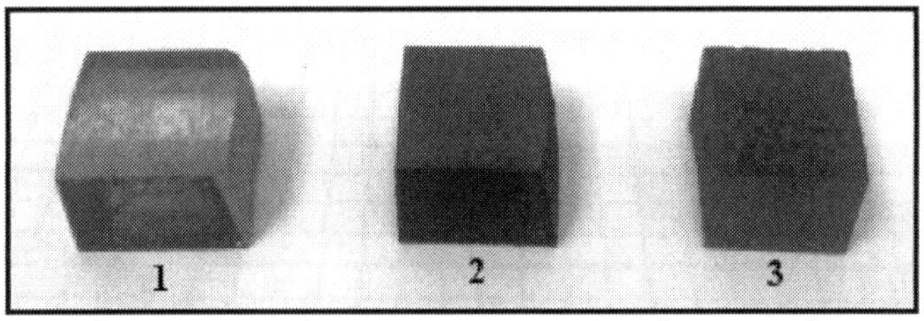

Figure 6: Three Emissivity Study Samples

A small hole was drilled ¾ through the side of each sample to create a blackbody for the insertion of a carefully calibrated Type R thermocouple.

When introduced to a 500 °C environment, the sample immediately oxidized. Therefore, in use, the coils would oxidize as soon as the furnace is started up. Because the surface of the machined sample oxidized so quickly, only the oxidized condition was used for the study.

Experimental Setup

Figure 7 gives a labeled illustration of the experimental setup.

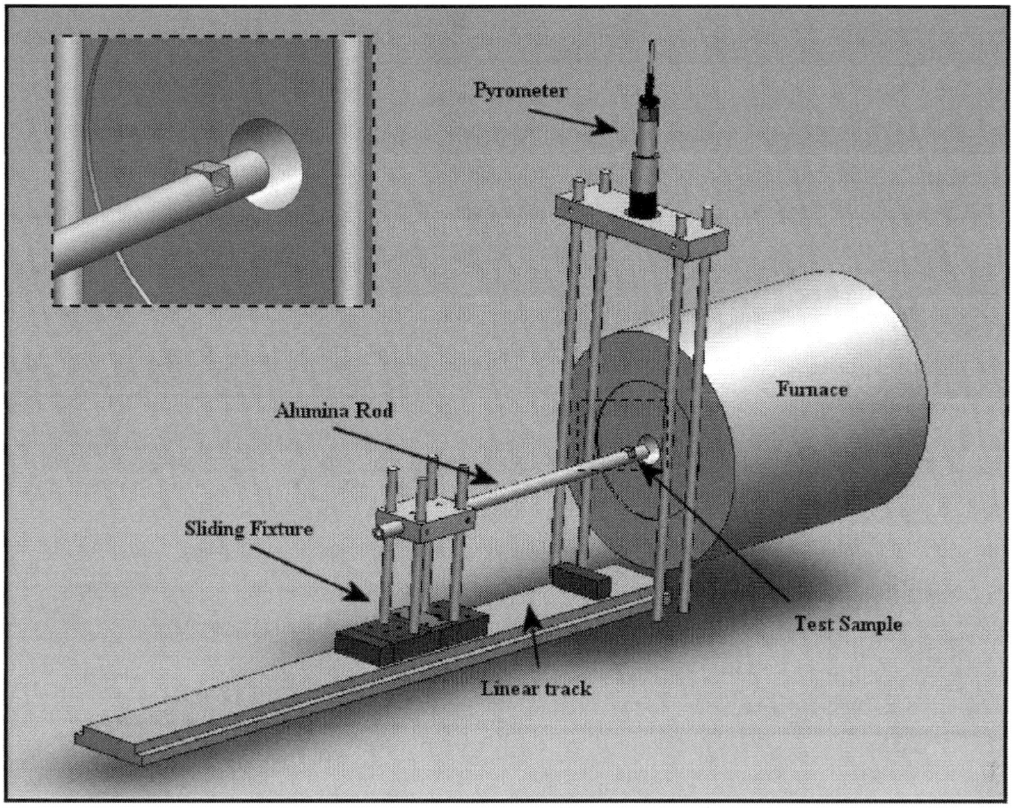

Figure 7: Experimental Setup, Labeled

A tightly controlled cylindrical lab furnace with a 2" cavity is used to heat the samples to the desired temperatures for testing. The samples are placed in a cut-out cavity of a hollow alumina rod for insertion into the furnace. A sliding fixture on a linear track holds the alumina rod, so the sample can be quickly, repeatedly, and safely inserted into and extracted from the cylindrical furnace cavity. A fixture is placed over the sample so that a pyrometer can view the top surface for measurement. The fixture is designed so that pyrometers using different wavelengths can be quickly and precisely replaced

For the pyrometer in this experiment, the BASF *EXACTUS®* Optical Temperature Sensor was selected. This instrument was selected due to its ultra-fast detection speed of 1000 readings per second. It also performs with high accuracy, repeatability, and drift characteristics, which provides trustworthy data.

Experiment Procedure

The following procedure is followed to collect the experimental data.

1. A sample is placed into the machined notch in the alumina rod.
2. The Type R thermocouple is fed through the alumina rod and inserted into the drilled hole in the sample for a reference temperature.
3. The furnace is heated to 600 °C.
4. The pyrometer using a 1.55 μm wavelength is placed on the fixture.
5. When the furnace is thermally stable, sample is inserted into the furnace cavity by moving the sliding fixture.

6. When the sample is thermally stable, using the Type R thermocouple measurement as criteria, the sample is quickly extracted from the furnace, and the pyrometer measurements and the thermocouple measurements are recorded.
7. Steps 4 through 6 are repeated for a 0.9 µm wavelength instrument then a 0.65 µm wavelength instrument.
8. Steps 3 through 7 are repeated at 800 °C and 1200 °C.
9. Steps 1 through 8 are repeated for the next sample.

Data Analysis

The emissivity of the samples can not be measured inside the cavity of the furnace, because the cavity acts as a blackbody radiator. Therefore, the temperature measurements are taken as the sample is cooling immediately after being extracted from the furnace. The measured temperature decay is analyzed to estimate the temperature of the sample when it is in the furnace cavity.

Using this method, the sample extraction of the sample must be both quick and repeatable. For this reason, the selected pyrometer must have a very high measurement rate with a high accuracy. The *EXACTUS®* optical temperature sensor meets or exceeds both these requirements.

Figure 8 shows a chart of the data from the 1550 nm pyrometer (dotted line) as compared to the data from the inserted thermocouple (solid line). The data to the left of the graph is collected when the sample is inside the furnace. The data in the center is collected as the sample is extracted from the furnace, and the sample is moved under the pyrometer. The sample is then inserted back into the furnace resulting in the data to the right.

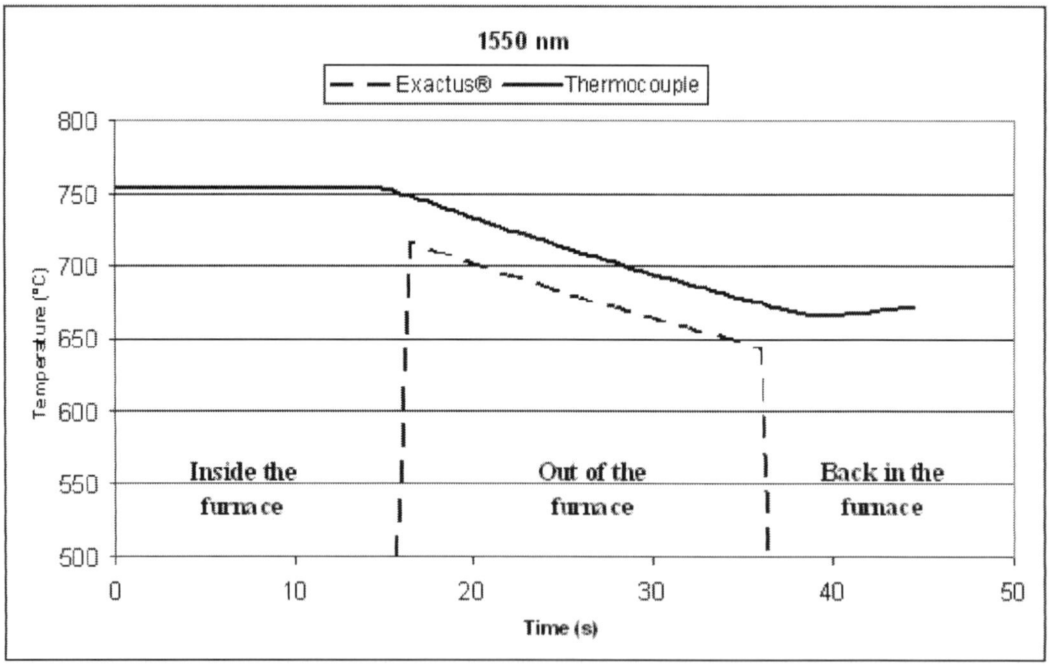

Figure 8: Chart of Experimental Data – 1 Measurement Point

The pyrometer data in the region of validity, between approximately 20 and 35 seconds according to the chart in **Figure 8**, is fit with a second order polynomial regression. The point in time that the sample was extracted from the furnace cavity is then determined using the thermocouple data. This value is applied to the regression equation to closely estimate the temperature of the sample as it would be measured by the pyrometer at the point in time that it was extracted from the furnace cavity.

This temperature is used in the emissivity calculation along with the temperature measured by the thermocouple at the same time. By using Planck's Law (**Equation 2**), the irradiance, or theoretical blackbody energy, is calculated for the actual temperature of the sample measured by the thermocouple. The same calculation is performed on the temperature measured by the pyrometer. The emissivity is given by the ratio of these two energy values. This can be explained simply by **Equation 5**.

$$\varepsilon = I_{measured}/I_{actual}$$

Equation 5: Emissivity Calculation

Results

The emissivity values in this wavelength region for the two samples are given in **Table 1**. These values are calculated using the emissivity values determined by all three wavelengths across all three temperatures. Therefore, the value is reported as "spectral emissivity" or, in other words, a general emissivity over all utilized wavelengths.

Sample	Emissivity
Machined surface	0.83
Older coil surface	0.86

Table 1: Spectral Emissivity Results

The values in **Table 1** closely match historical documentation on the emissivity of cracking coils for pyrometers using a 1 µm wavelength. The two values estimate the change in emissivity, as measured at approximately 1 µm, over the lifetime of a coil. At 1000 °C, the change over time of a 0.03 emissivity value would equate to a change in measured temperature of 4.1 °C using a 1 µm wavelength. This same change using a 3.9 µm wavelength would equate to a measured temperature change of 16.3 °C.

As expected, however, the calculated emissivity value does vary with the wavelength of light detected. **Figure 9** shows a chart of the calculated emissivity using the three different wavelengths, 1550 µm, 900 µm, and 650 µm, respectively, for the two different surfaces.

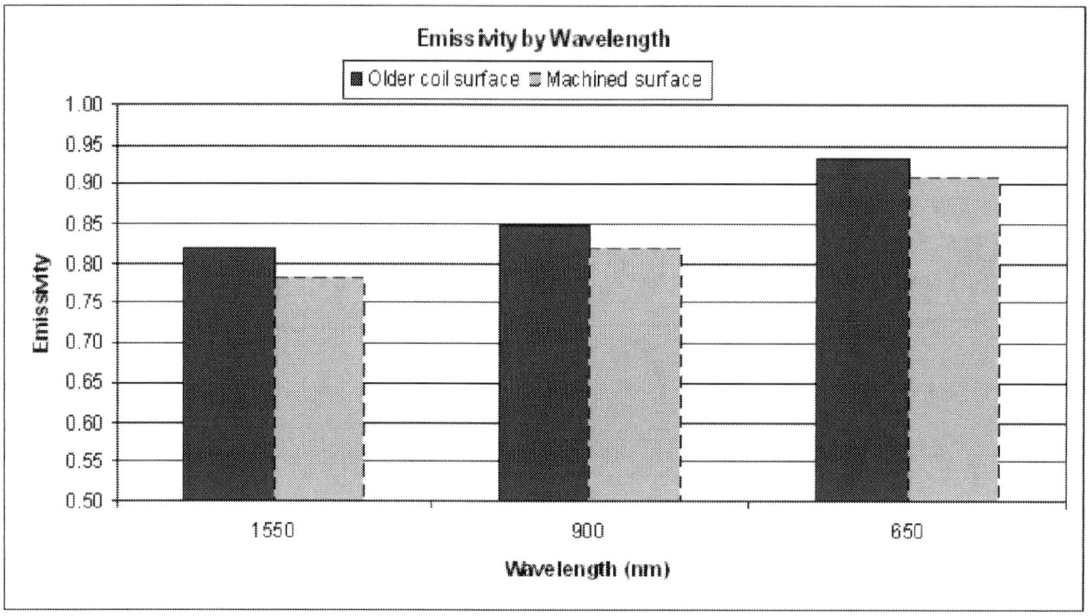

Figure 9: Calculated Emissivity by Wavelength

For these samples, the calculated emissivity value did not vary noticeably over the temperature range of 600 °C to 1200 °C. For each temperature value, the calculated emissivity was within +/- 0.02 of the average.

Conclusions

The importance of accurate temperature measurements of cracking coils has increased as higher demand for olefins has pushed the limits of furnace yields. Optical measurements are necessary for avoiding mechanical and thermal constraints in the furnace.

Emissivity is a critical variable for accurate optical measurements. The experiment has shown that the emissivity of the sampled cracking coils is dependent upon two variables: the age of the coil and the wavelength used by the optical device to measure the coil temperature.

When choosing an optical device to measure the temperature of cracking coils, measurement performance factors, such as repeatability and drift, must be taken into consideration. However, it is also essential to understand the wavelength or wavelength range used by an optical device. This will affect the accuracy of the measurement due to emissivity as well as measurement error due to changing or unknown emissivity.

References

1. L. Michalski et al (2001), *Temperature Measurement, Second Edition*. Wiley, pp. 162-208.
2. Boltzmann, L. (1884) "Ableitung des Stefan'schen Gesetzes, betreffend die Abhängigkeit der Wärmestrahlung von der Temperatur aus der electromagnetischen Lichttheorie". *Annalen der Physik und Chemie, Vol. 22*, pp. 291-294.
3. Planck, M. (1914). *The Theory of Heat Radiation*. Masius, M. (transl.) (2nd ed.). P. Blakiston's Son & Co., pp. 6, 168.

AIChE Paper Number

Galvanic Corrosion Between Dissimilar Metals

Richard Rolke

Hydrocarbons Global Improvement Leader

Dow Chemical Company

Prepared for Presentation at the 2012 Spring National Meeting

Houston, TX,

April 1st - 5th, 2012

Galvanic Corrosion Between Dissimilar Metals

Richard Rolke
Hydrocarbons Global Improvement Leader
Dow Chemical Company

Abstract:
A Dow USGC cracker suffered a tube failure on a cracked gas compressor inner stage cooler after ten plus years of service. To further extend the equipment life, the decision was made to replace the tube bundle and upgrade the metallurgy of the tubes from Admiralty brass to Duplex stainless steel. The new tube bundle was placed in service, but failed after only eight months of operation. The premature failure was found to be caused by deep pitting in the carbon steel tubesheet. The dissimilar metals of the Duplex tubes and carbon steel tubesheet, with river water acting as the electrolyte, formed a galvanic cell which preferentially corroded the carbon steel tubesheet. This paper will describe the details of the failure and failure mechanism.

Introduction

The event described in this paper occurred at the Light Hydrocarbons #2 cracker located within the Dow Chemical complex in Plaquemine, Louisiana. This light feed cracker was built in 1969 and has a nominal ethylene capacity of 500 kta. During the spring of 2011, a cracker gas compressor inner stage cooler suffered a tube failure after only 11 months of service. The exchanger was taken out of service to prevent any possible effect upon the environment.

Equipment Background

This heat exchanger was a shell and tube design with cracked gas on the shell side and river water cooling on the tube side. This heat exchanger tube bundle had been recently replaced during a cracker turnaround. The previous bundle had been in service since 1991. In 2006, this bundle had an unexpected tube leak after 15 years of service. Repairs were made to the exchanger and it was placed back in service, but it was realized that the exchanger was nearing the end of life. The decision was made to fabricate a new replacement tube bundle for installation during the next available cracker turnaround. The design was expected to be like for like with a few minor improvements that were expected to further extend the life of the exchanger.

Original tube bundle design specifications (see figure #1):

Diameter:	3'6"
Tubes:	3096 1" diameter rolled tube to tubesheet joints
Tube metallurgy:	Admiralty brass
Tubesheet Material:	carbon steel

The design improvement from the original heat exchanger included upgrading the tube metallurgy from Admiralty brass to Duplex 2205 stainless steel to improve the resistance to river water erosion. Due to the increased hardness of the Duplex steel, the tubes could no longer be rolled into the carbon steel tubesheet, but needed to the strength welded.

Figure #1. Heat exchanger details

Heat Exchanger Failure Analysis

Inspection of the failed heat exchanger bundle found that the source of the process leakage was through numerous pin holes originated in the carbon steel area of the tubesheet exposed to the river water (see figure #2). The carbon steel tubesheet had a very small area of exposure to the river water due to the weld overlay around the tube the tubesheet joint. Each area of exposed carbon steel tubesheet showed aggressive corrosive attack and this caused several leak paths to develop between the river water and the process (see figure #3).

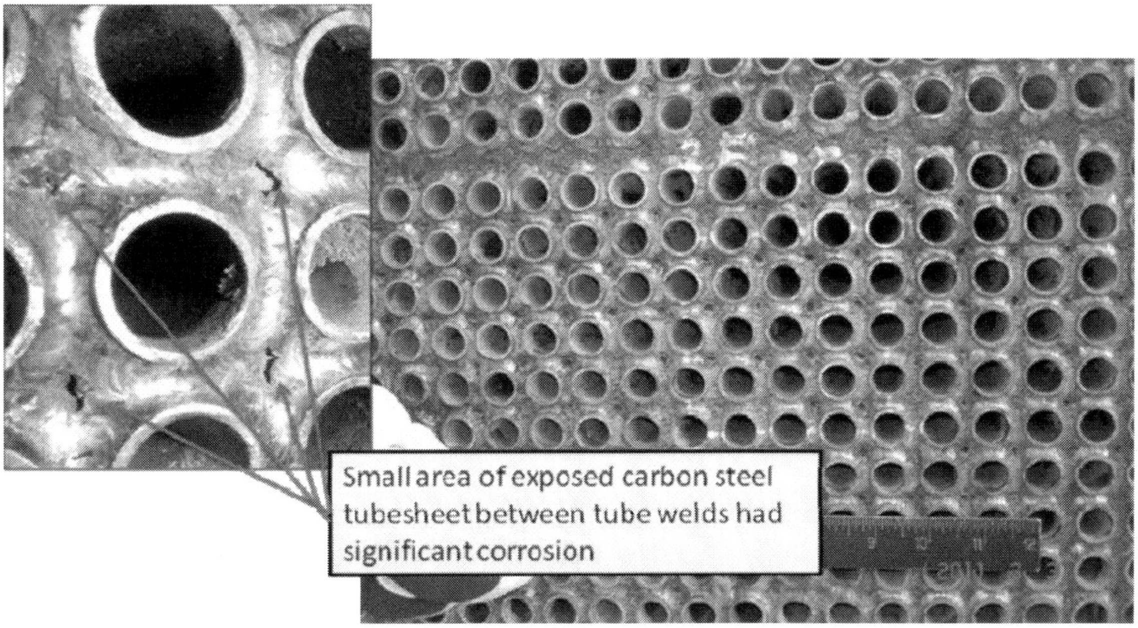

Figure #2. Exchanger failure analysis

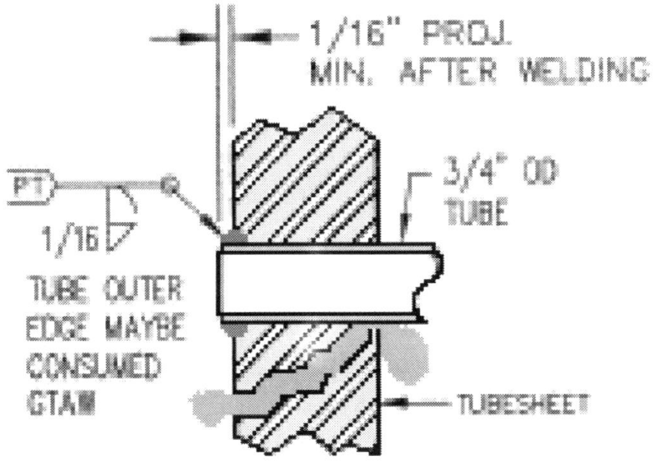

Figure #3. Exchanger failure analysis

What is galvanic corrosion

Galvanic corrosion occurs when two dissimilar conducting materials, metallic or nonmetallic, are in electrical contact through an aqueous or moist environment. A galvanic current is formed when two materials are in electrical contact and connected by an electrolyte. The driving force of this galvanic current is the inherent potential difference between the dissimilar materials. The resulting reaction forms a galvanic cell similar to a battery. All metals can be affected with the exception of most noble metals.

There are three essential components of a galvanic cell, which can be viewed as very similar to a fire triangle (see figure #4).

Materials possessing different potentials

1. the anode which liberates electrons and corrodes typically by metal dissolution and/or metal oxide formation
2. the cathode which gains electrons and typically liberates hydrogen to the electrolyte and/or reduces oxides the presence of an electrolyte
3. a fluid that can conduct current, which forms the electrical connection between the anode and the cathode

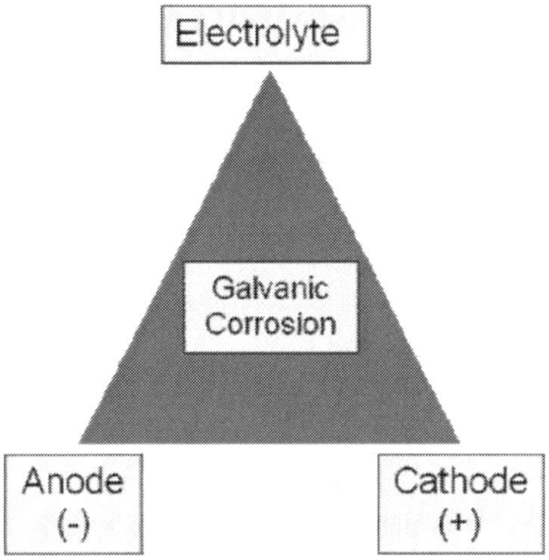

Figure #4. Galvanic corrosion triangle

Key factors affecting galvanic corrosion

There are several factors that have significant impact upon galvanic corrosion rates: the potential difference between the dissimilar materials, the nature of the electrolyte, and the geometric relationship between the dissimilar materials.

Galvanic corrosion rates are directly related to the electrical current caused by a natural potential difference or electromotive source (emf) between different metals. The following chart (figures #5) ranks the relative potential differences in salt water. The higher up the chart, the more active the material; anodic, while lower down the chart are a less active material; cathodic. The larger the potential difference between the materials, the higher the potential difference and corresponding galvanic corrosion rate.

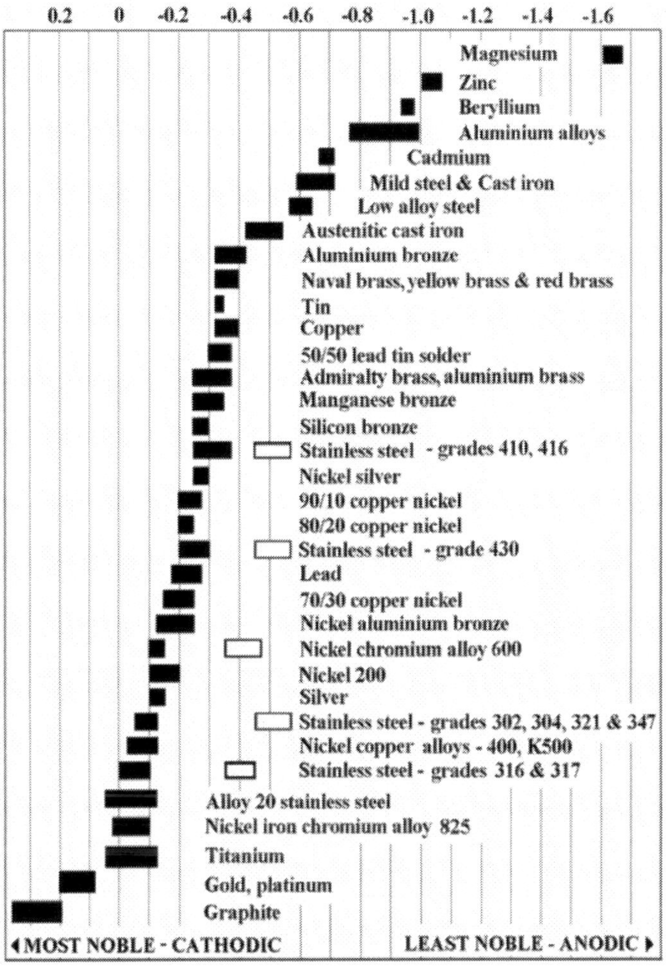

Figure #5. Material potential chart in salt water

Galvanic corrosion rates are also very dependent upon the nature of the environment (electrolyte) connecting the dissimilar metals. Different electrolytes and conditions will affect the position of metals in the galvanic series. Conditions affecting electrolytes include Acid vs. Base, Flowing vs. Static, and presence of impurities, inhibitors or oxygen scavengers.

Other factors affecting galvanic corrosion are the geometric relationship of the dissimilar component metals. The intensity of galvanic corrosion is affected by the relative size of the anodic to cathodic surface area exposed to the electrolyte. Large cathodic areas coupled to smaller anodic areas will cause severe dissolution of the more active or anodic metal. The reverse is also true; large anodic areas coupled to smaller cathodic areas produces little galvanic current. A design rule of thumb is a target 40 – 1 area ratio of anode to cathode.

Galvanic corrosion rates of the anode will normally be significantly higher immediately adjacent to the connection to the cathode. This is illustrated in Figure #6.

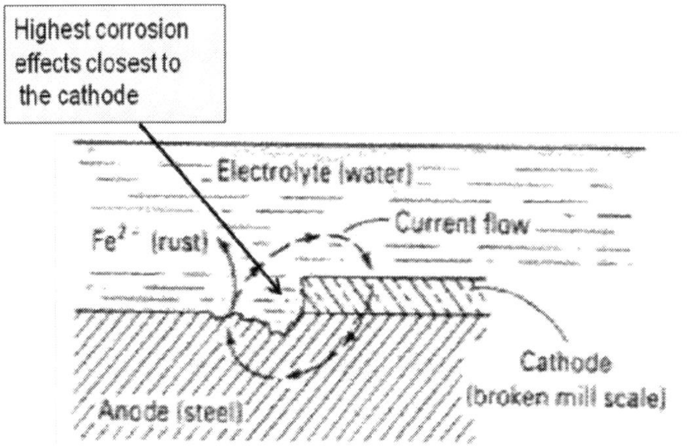

Figure #6. Galvanic corrosion detail

Heat Exchanger Failure Cause

The failure of the cracker gas compressor inner stage cooler was caused by galvanic corrosion of the carbon steel tubesheet. The river water was the electrolyte completing the galvanic cell between the dissimilar metals: duplex steel tubes (cathode) and carbon steel tubesheet (anode). See figure #8.

A few design changes to this exchanger led to the failure in less than 11 months of operation versus the previous exchanger's life span of 20 years. The high galvanic corrosion rate was driven by several factors;

1. The change in the tube metallurgy from Admiralty brass to Duplex stainless steel increased the potential difference between the dissimilar metals thereby accelerating the corrosion rate; Duplex stainless steel tubes (cathode) and the carbon steel tubesheet. See figure #7.
2. Strength welding the tube to tubesheet joint versus rolling reduced the amount of exposed anode to the electrolyte thereby concentrating the galvanic corrosion rate to a very small area. This ratio was greater than 100 to 1 cathode to anode, which is well above the design guidelines.

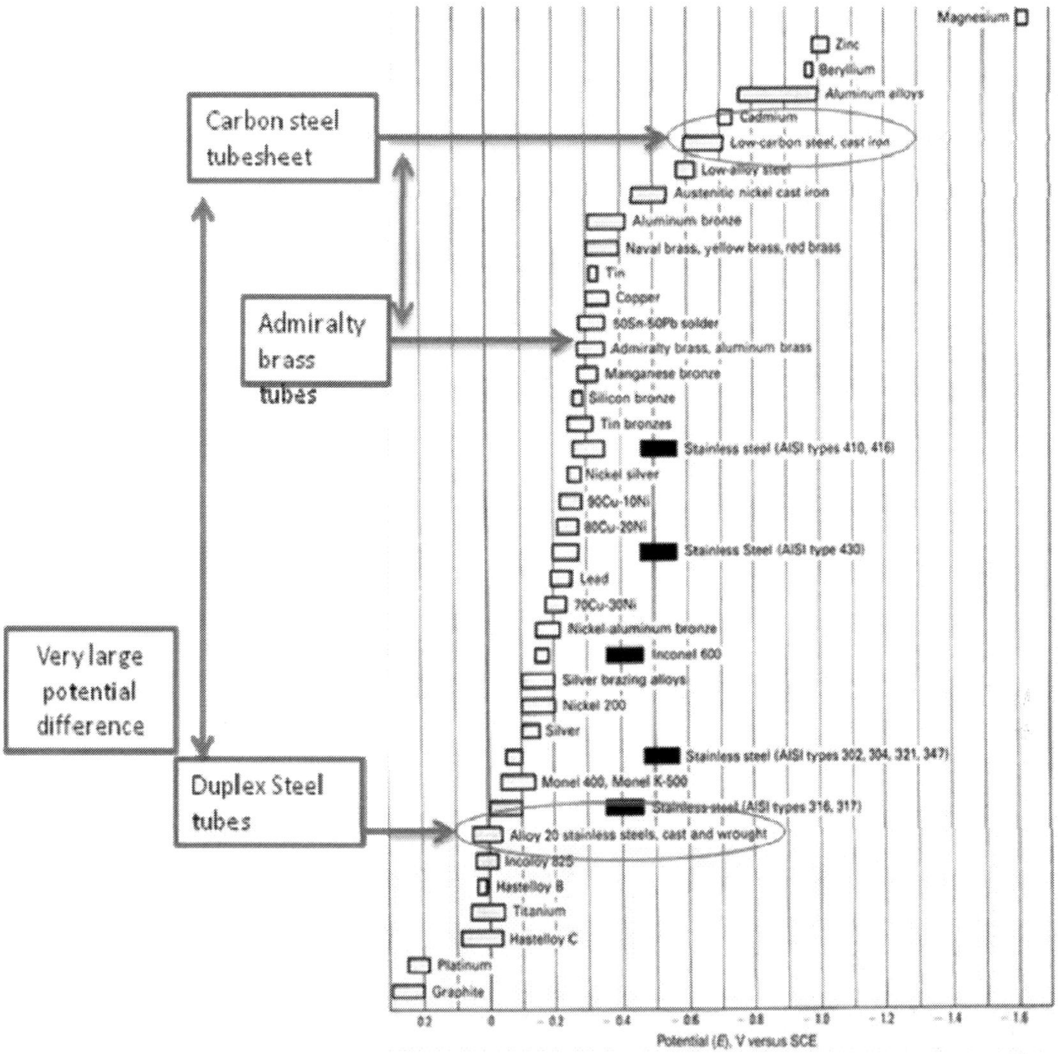

Figure #7. Metal potential chart for exchanger failure

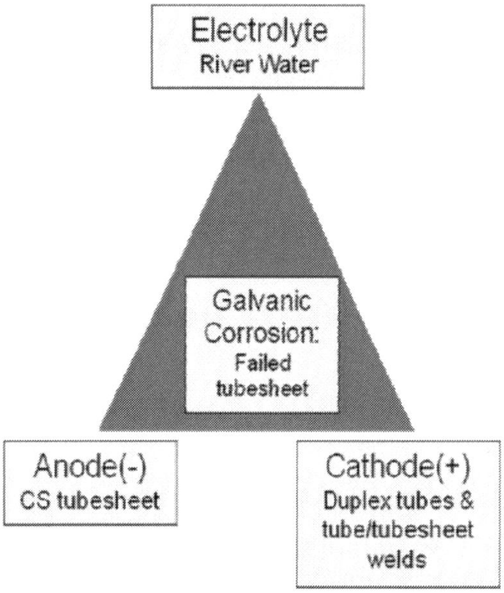

Figure #8. Galvanic corrosion triangle for exchanger failure

Conclusions

In conclusion, design changes made to the new exchanger meant to increase the equipment life actually led to the premature failure. Upgrading the metallurgy of the tubes in this case caused the heat exchanger to fail prematurely. The Duplex stainless steel (cathode) had greater potential difference with carbon steel than previous tube metallurgy; Admiralty brass. Combine this change with welding the tube to tubesheet joints, which significantly reduced the exposed area of the carbon steel (anode) tubesheet, significantly increased the corrosion rate and led directly to the rapid exchanger failure. In this case, cladding of the tube sheet to eliminate the dissimilar metals would have eliminated the galvanic cell from forming and eliminated this exchanger failure mechanism.

References

1. Failure Analysis and Prevention, Vol. II, ASM Handbook, ASM International
2. Damage Mechanisms Affecting Fixed Equipment in the Refining Industry API recommended practice 571; 2nd Edt. April 2011

AUTHOR INDEX

AUTHOR INDEX